Name Reactions
in Heterocyclic Chemistry

Name Reactions in Heterocyclic Chemistry

Edited by

Jie-Jack Li

Pfizer Global Research & Development

Scientific Editor:

E. J. Corey

Harvard University

WILEY-INTERSCIENCE

A JOHN WILEY & SONS, INC., PUBLICATION

Published by John Wiley & Sons, Inc., Hoboken, New Jersey.
Published simultaneously in Canada.

For general information on our other products and services please contact our Customer Care
Department within the U.S. at 877-762-2974, outside the U.S. at 317-572-3993 or fax 317-572-4002.

Wiley also publishes its books in a variety of electronic formats. Some content that appears in print,
however, may not be available in electronic format.

Library of Congress Cataloging-in-Publication Data is available.

ISBN 0-471-30215-5

Printed in the United States of America.

10 9 8 7 6 5 4 3 2 1

To Alexandra

Table of Contents

Foreword

Part of the charm of synthetic organic chemistry derives from the vastness of the intellectual landscape along several dimensions. First, there is the almost infinite variety and number of possible target structures that lurk in the darkness, waiting to be made. Then, there is the vast body of organic reactions that serve to transform one substance into another, now so large in number as to be beyond credibility to a non-chemist. Further, there is the staggering range of reagents, reaction conditions, catalysts, elements and techniques that must be mobilized in order to tame these reactions for synthetic purposes. Finally, it seems that new information is being added to the science at a rate that outstripped our ability to keep up with it. In such a troubled setting any author, or group of authors, must be regarded as heroic if, through their efforts, the task of the synthetic chemist is eased.

The field of heterocylic chemistry has long presented a special problem for chemists. Because of its enormous information content and variety, it is not well taught to chemistry undergraduate or graduate students, even in simplified form. There is simply too much material for the time available. And yet, the chemistry of heterocyclic compounds and methods for their synthesis form the bedrock of modern medicinal chemical and pharmaceutical research. It is important for medicinal chemists to be broadly knowledgeable across a wide swath of heterocyclic chemistry. Those who specialize narrowly do so at their own peril. If you grant me the accuracy of all of the above, you likely will share my conviction that there is a need for high-quality, up-to-date, and authoritative books on heterocyclic synthesis that are helpful for the professional research chemist and also the advanced student. This volume, *Name Reactions in Heterocyclic Chemistry* is a model of what such books should be. Written concisely and with great skill and care by Dr. Jie Jack Li and a distinguished group of experts in the field of heterocyclic chemistry, this is a book that will be tremendously useful and helpful to synthetic and medicinal chemists, on whose shelves it will surely find a place. On behalf of these users, myself included, I send thanks and congratulations.

E. J. Corey

E. J. Corey
May 1, 2004

Preface

Since the infancy of organic chemistry, the practitioners in the field have often associated reactions with the chemists who discovered it. Even with the advent of IUPAC nomenclature, name reactions are still intimately intertwined with our profession, becoming a part of our daily language. Therefore, getting acclimated with this jargon is an integral part of the training to earn proficiency in organic chemistry.

On the other hand, heterocycles are of paramount importance to medicinal and agricultural chemists. This comprehensive and authoritative treatise provides a one-stop repository for name reactions in heterocyclic chemistry. Each name reaction is summarized in seven sections:

1. Description;
2. Historical Perspective;
3. Mechanism;
4. Variations and Improvements;
5. Synthetic Utility;
6. Experimental; and
7. References.

I also have introduced a symbol [R] to highlight review articles, book chapters and books dedicated to the respective name reactions.

I have incurred many debts of gratitude to Prof. E. J. Corey of Harvard University, who envisioned this project in the summer of 2002. What he once told me:— *"The desire to learn is the greatest gift from God."*—has been a true inspiration. Furthermore, it has been my greatest privilege as well as a pleasure to work with a stellar collection of contributing authors from both academia and industry. Some of them are world-renowned scholars in the field; some of them have worked intimately with the name reactions that they have written; some of them even took part in the discovery of the name reactions that they authored in this manuscript. As a consequence, this book truly represents the state-of-the-art for *Name Reactions in Heterocyclic Chemistry*. We will follow up with the second volume to complete the series on heterocyclic chemistry.

Jack Li
April 24, 2004

Contributing authors:

Nadia M. Ahmad
School of Chemistry
University of Nottingham
University Park
Nottingham
NG7 2RD, UK

Dr. Dawn A. Brooks
Lilly Research Laboratories
Eli Lilly and Company
Lilly Corporate Center
Indianapolis, IN 46285

Prof. James M. Cook
Department of Chemistry
University of Wisconsin—Milwaukee
3210 North Cramer Street
Milwaukee, WI 53211-3029

Dr. Timothy T. Curran
Department of Chemical R&D
Pfizer Global Research & Development
2800 Plymouth Road
Ann Arbor, MI 48105

Dr. Paul Galatsis
Department of Chemistry
Pfizer Global Research & Development
2800 Plymouth Road
Ann Arbor, MI 48105

Prof. Gordon W. Gribble
Department of Chemistry
6128 Burke Laboratory
Dartmouth College
Hanover, NH 03755

Dr. Daniel D. Holsworth
Department of Chemistry
Pfizer Global Research & Development
2800 Plymouth Road
Ann Arbor, MI 48105

Dr. Andrew Hudson
Ligand Pharmaceuticals
10275 Science Center Road
San Diego, CA 92121

Prof. Jeffrey N. Johnston
Department of Chemistry
Indiana University
800 East Kirkwood Avenue
Bloomington, IN 47405-7102

Dr. Jie Jack Li
Department of Chemistry
Pfizer Global Research & Development
2800 Plymouth Road
Ann Arbor, MI 48105

Dr. Jin Li
Research Technology Center
Pfizer Global Research & Development
Eastern Point Road
Groton, CT 06340

Dr. Chris Limberakis
Department of Chemistry
Pfizer Global Research & Development
2800 Plymouth Road
Ann Arbor, MI 48105

Christopher M. Liu
Department of Chemistry
University of Michigan
930 North University Avenue
Ann Arbor, MI 48109-1055

Dr. Adrian J. Moore
School of Sciences
Fleming Building
University of Sunderland
UK SR1 3SD

Prof. Richard J. Mullins
Department of Chemistry
Xavier University
3800 Victory Parkway
Cincinnati, OH 45207-4221

Prof. Brian J. Myers
Department of Chemistry
and Biochemistry
Ohio Northern University
525 South Main Street
Ada, OH 45810

Peter A. Orahovats
Department of Chemistry
University of Michigan
930 N. University Avenue
Ann Arbor, MI 48109-1055

Dr. Michael Palucki,
Department of Process Research
Merck & Co., Inc.
Rahway, NJ 07065-0900

Dr. Derek A. Pflum
Department of Chemical R&D
Pfizer Global Research & Development
2800 Plymouth Road
Ann Arbor, MI 48105

Prof. Christian M. Rojas
Department of Chemistry
Barnard College
3009 Broadway
New York, NY 10027

Dr. Subas Sakya
CNS Chemistry
Pfizer Global Research & Development
Eastern Point Road
Groton, CT 06340

Prof. Kevin M. Shea
Department of Chemistry
Clark Science Center
Smith College
Northampton, MA 01063

Jennifer M. Tinsley
Department of Chemistry
University of Michigan
930 North University Avenue
Ann Arbor, MI 48109-1055

Prof. David R. Williams
Department of Chemistry
Indiana University
800 East Kirkwood Avenue
Bloomington, IN 47405-71020

Prof. John P. Wolfe
Department of Chemistry
University of Michigan
930 N. University Avenue
Ann Arbor, MI 48109-1055

Acronyms and Abbreviations

)))))..ultrasound

◯–.. polymer support

Ac..acetyl

AcOH...acetic acid

ADP.. adenosine diphosphate

AE .... asymmetric epoxidation reaction

AFO.. Algar–Flynn–Oyamada

AIBN..2,2'-azobisisobutyronitrile

Alpine-borane® ....*B*-isopinocamphenyl-9-borabicyclo[3.3.1]-nonane

AME...acetyl malonic ester

AMNT .... aminomalononitrile *p*-toluenesulfonate

Ar .... aryl

ATP...adenosine triphosphate

AUC ....area under curve

B:...generic base

9-BBN .... 9-borabicyclo[3.3.1]nonane

BFO... benzofurazan oxide

TBHP .... *tert*-butyl hydrogen peroxide

BINAP..2,2'-bis(diphenylphosphino)-1,1'-binaphthyl

Bn...benzyl

Boc .... *tert*-butyloxycarbonyl

BOP........... benzotriazol-1-yloxy-tris(dimethylamino)-phosphonium hexafluorophosphate

BPO...benzoyl peroxide

Bu .... butyl

BZ reaction... Barton–Zard reaction

CAN .... ceric ammonium nitrate (ammonium cerium(IV) nitrate)

CTAB..cetyl trimethylammonium bromide

CB-1 ....cannabinoid receptor-1

Cbz...benzyloxycarbonyl

CNS...central nervous system

COX-2 .... cyclooxygenase II

CSA..camphorsulfonic acid

CuTC...copper thiophene-2-carboxylate

DABCO..1,4-diazabicyclo[2.2.2]octane

dba...dibenzylideneacetone

DBU ....1,8-diazabicyclo[5.4.0]undec-7-ene

DCB .... dichlorobenzene

DCC 1,3-dicyclohexylcarbodiimide

DCM..dichloromethane

DDQ 2,3-dichloro-5,6-dicyano-1,4-benzoquinone

DEAD .... diethyl azodicarboxylate

DEPC .... diethyl phosphorocyanidate

DET.. diethyl tartrate

Δ ....solvent heated under reflux

DIC.. diisopropylcarbodiimide
DHPM...3,4-dihydropyrimidin-2(1*H*)-one
(DHQ)$_2$-PHAL .... 1,4-bis(9-*O*-dihydroquinine)-phthalazine
(DHQD)$_2$-PHAL ....1,4-bis(9-*O*-dihydroquinidine)-phthalazine
DHT 5α-dihydrotestosterone
DIBAL .... diisobutylaluminum hydride
DMA .... *N*,*N*-dimethylacetamide
DMA .... *N*,*N*-dimethylaniline
DMAP... *N*,*N*-dimethylaminopyridine
DME..1,2-dimethoxyethane
DMF..dimethylformamide
DMFDMA...dimethylaminoformaldehyde dimethyl acetal
DMS.. dimethylsulfide
DMSO.. dimethylsulfoxide
DMSY.. dimethylsulfoxonium methylide
DMT..dimethoxytrityl
DNA.. deoxyribonucleic acid
DNP..2,4-dinitrophenyl
L-DOPA 3,4-dihydroxyphenylalanine
dppb..1,4-bis(diphenylphosphino)butane
dppe..1,2-bis(diphenylphosphino)ethane
dppf ....1,1′-bis(diphenylphosphino)ferrocene
dppp.. 1,3-bis(diphenylphosphino)propane
E1 .... unimolecular elimination
E2 .... bimolecular elimination
E1cb .... 2-step, base-induced β-elimination *via* carbanion
EDG .... electron donating group
ee .... enantiomeric excess
EMME.. ethoxymethylenemalonate
ent.. *enantiomer*
EPP.. ethyl polyphosphate
Eq ....equivalent
Et ....ethyl
EtOAc .... ethyl acetate
EPR (= ESR).. electron paramagnetic resonance spectroscopy
ESR (= EPR).. electronic spin resonance
EWG..electron withdrawing group
FMO.. frontier molecular orbital
FVP .... flash vacuum pyrolysis
GABA ....γ-aminobutyric acid
GC.. gas chromatography
GC reaction ....Gabriel–Colman reaction
H..hours
His..histidine
HIV ....human immunodeficiency virus
HMDS ....hexamethyldisilazine

HMPA .... hexamethylphosphoric triamide
HOMO .... highest occupied molecular orbital
HPLC .... high performance liquid chromatography
IBCF .... isobutylchloroformate
Imd .... imidazole
IPA .... isopropanol
i-Pr .... isopropyl
KCO .... potassium channel opener
KHMDS .... potassium hexamethyldisilazide
KR .... Kostanecki–Robinson
LAH .... lithium aluminum hydride
LDA .... lithium diisopropylamide
LHMDS .... lithium hexamethyldisilazide
LiHMDS .... lithium hexamethyldisilazide
LTMP .... lithium 2,2,6,6-tetramethylpiperidine
LUMO .... lowest unoccupied molecular orbital
M .... metal
M .... moles per liter (molar)
MCR .... multi-component reaction
m-CPBA .... *m*-chloroperoxybenzoic acid
Me .... methyl
Mes .... mesityl
mL .... milliliters
MMPP .... magnesium monoperoxyphthalate hexahydrate
mmol .... millimoles
MO .... molecular orbital
MOA .... mechanism of action
MOM .... methoxymethyl
MRSA .... methicillin-resistant *Staphylococcus aureus*
MVK .... methyl vinyl ketone
MWI (μν) .... microwave irradiation
NAD$^+$ .... nicotinamide adenine dinucleotide (oxidized form)
NADH .... nicotinamide adenine dinucleotide
NBS .... *N*-bromosuccinimide
NCS .... *N*-chlorosuccinimide
NIS .... *N*-iodosuccinimide
NMDA .... *N*-methyl-D-aspartate
NMO .... *N*-methylmorpholine-*N*-oxide
NMP 1-methyl-2-pyrrolidinone
NMR .... nuclear magnetic resonance
Nu .... nucleophile
NPY .... neuropeptide Y
NSAIDs .... non-steroidal anti-inflammatory drugs
OA .... osteoarthritis
PCC .... pyridinium chlorochromate
PDC .... pyridinium dichromate

PDE .... phosphodiesterase
PG .... prostaglandin
pGlu .... pyroglutamic acid
Ph .... phenyl
PK .... pharmacokinetics
pKa .... −Log acidity constant
PKC .... protein kinase C
PPA .... polyphosphoric acid
PPE .... polyphosphate ester
PPI .... proton pump inhibitor
4-PPNO 4-phenylpyridine-*N*-oxide
PPP .... 3-(3-hydroxyphenyl)-1-*n*-propylpiperidine
PPSE .... polyphosphoric acid trimethylsilyl ester
PPTS .... pyridinium *p*-toluenesulfonate
Pro .... proline
PSI .... pounds per square inch
PTC .... phase transfer catalyst
PTSA .... paratoluenesulfonic acid
Py .... pyridine
Pyr .... pyridine
RA .... rheumatoid arthritis
RNA .... ribonucleic acid
rt .... room temperature
Salen .... *N,N′*-disalicylidene-ethylenediamine
SET .... single electron transfer
S_NAr .... nucleophilic substitution on an aromatic ring
S_N1 .... unimolecular nucleophilic substitution
S_N2 .... bimolecular nucleophilic substitution
t-Bu .... *tert*-butyl
TBAF .... tetrabutylammonium fluoride
TBD .... 1,5,7-triazabicyclo[4.4.0]dec-5-ene
TBDMS .... *tert*-butyldimethylsilyl
TBDPS .... *tert*-butyldiphenylsilyl
TBHP .... *tert*-butylhydroperoxide
TBS .... *tert*-butyldimethylsilyl
TEA .... triethylamine
Tf .... trifluoromethanesulfonyl (triflic)
TFA .... trifluoroacetic acid
TFAA .... trifluoroacetic anhydride
TfOH .... triflic acid
TFP .... tri-*o*-furylphosphine
TFSA .... trifluorosulfonic acid
THF .... tetrahydrofuran
THIP .... 4,5,6,7-tetrahydroisoxazolo[5,4-c]pyridin-3-ol
TIPS .... triisopropylsilyl
TLC .... thin layer chromatography

TMEDA .... *N,N,N′,N′*-tetramethylethylenediamine
TMG...tetramethylguanidine
TMP .... tetramethylpiperidine
TMS .... trimethylsilyl
TMSCl...trimethylsilyl chloride
TMSCN...trimethylsilyl cyanide
TMSI .... trimethylsilyl iodide
TMSOTf.. trimethylsilyl triflate
Tol.. toluene or tolyl
Tol-BINAP ....2,2′-bis(di-*p*-tolylphosphino)-1,1′-binaphthyl
TosMIC ....(*p*-tolylsulfonyl)methyl isocyanide
TPAP.. tetra-*n*-propylammonium perruthenate
TRH...thyrotropin releasing hormone
Ts...*p*-toluenesulfonyl (tosyl)
TSA ....*p*-toluenesulfonic acid
TsO.. tosylate

Part 1 Three- and Four-Membered Heterocycles 1

Chapter 1 Epoxides and Aziridines 1

1.1   Corey–Chaykovsky Reaction

1.1.1   Description

The Corey–Chaykovsky reaction entails the reaction of a sulfur ylide, either dimethylsulfoxonium methylide (1, Corey's ylide, sometimes known as DMSY) or dimethylsulfonium methylide (2), with electrophile 3 such as carbonyl, olefin, imine, or thiocarbonyl, to offer 4 as the corresponding epoxide, cyclopropane, aziridine, or thiirane.[1–7]

$X = O$, CH_2, NR^2, S, $CHCOR^3$, $CHCO_2R^3$, $CHCONR_2$, $CHCN$

For an α,β-unsaturated carbonyl compound, 1 adds preferentially to the olefin to furnish the cyclopropane derivative, whereas the more reactive 2 generally undergoes the methylene transfer to the carbonyl, leading to the corresponding epoxide. Also due to the difference of reactivities, reactions using 1 require slightly elevated temperature, normally around 50–60°C, whereas reactions using the more reactive 2 can be carried out at colder temperature ranging from –15°C to room temperature. Moreover, while it is preferable to freshly prepare both ylides *in situ*, 2 is not as stable as 1, which can be stored at room temperature for several days.

1.1.2   Historical Perspective

In 1962, Corey and Chaykovsky described the generation and synthetic utility of dimethylsulfoxonium methylide (1) and dimethylsulfonium methylide (2).[8–12] Upon treatment of DMSO with NaH, the resulting methylsulfinyl carbanion reacted with trimethylsulfoxonium iodide (5) to produce dimethylsulfoxonium methylide (1). The subsequent reaction between 1 and cycloheptanone rendered epoxide 6. Similar results were observed for other ketones and aldehydes as well, with a limitation where treatment of certain ketones (e.g. desoxybenzoin and Δ^4-cholestenone) with 1 failed to deliver the epoxides possibly due to their ease to form the enolate ions by proton transfer to 1. Interestingly, Michael receptor 7 reacted with 1 to provide access to the "methylene insertion" product, cyclopropane 8. Meanwhile, thiiranes were isolated in good yields from the reaction of thiocarbonyls and 1, and methylene transfer from 1 to imines took place to afford aziridines.

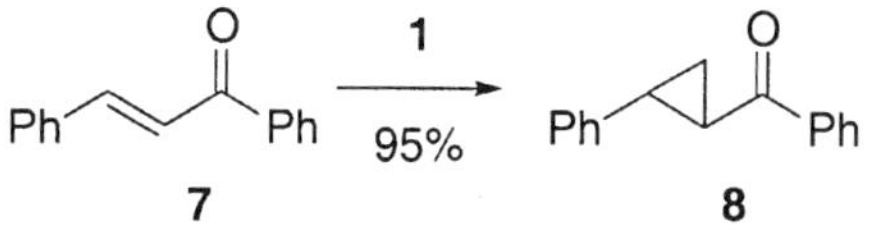

1.1.3 Mechanism

Similar to phosphur ylides, sulfur ylides **1** and **2** possess the nucleophilic site at the carbon atom and the pendant leaving group at the heteroatom (sulfur). Different from the Wittig reaction, the Corey–Chaykovsky reaction does not lead to olefins.

The mechanism of epoxide formation using sulfur ylides[13] is analogous to that of the Darzens condensation. In the Darzens condensation, enolate **9** adds to ketone **10**, forming alkoxide **11**, which undergoes an internal S_N2 to give epoxide **12**. In a parallel fashion, addition of dimethylsulfoxonium methylide (**1**) to ketone **13**, led to betaine **14**, which also undergoes an internal S_N2 to secure epoxide **15**. On the other hand, Michael addition of **1 to** enone **16** gives betaine **17**, which subsequently undergoes an internal S_N2 to deliver cyclopropyl ketone **18**.[14]

Darzens condensation:

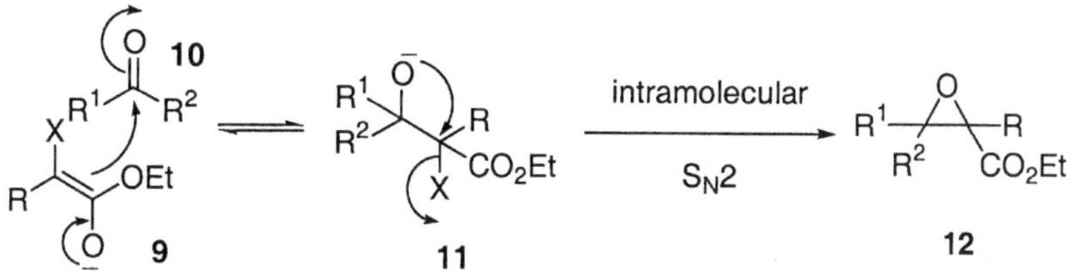

Corey–Chaykovsky reaction:

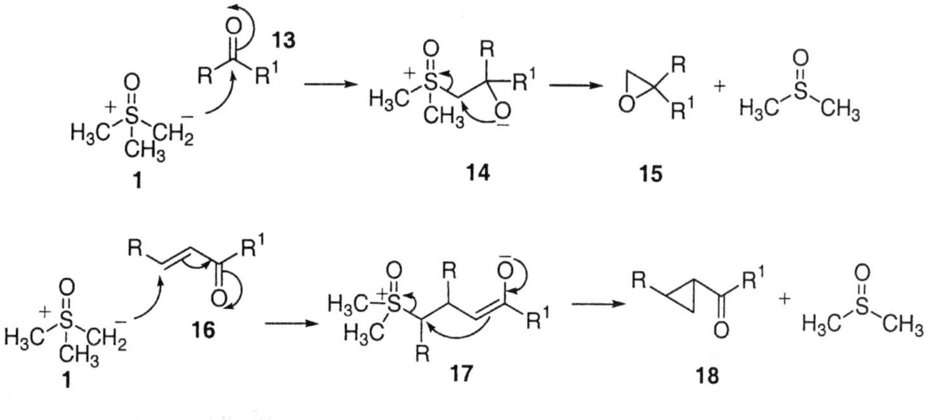

1.1.4 Variations and Improvements

Sulfur ylides **1** and **2** are usually prepared by treatment of either trimethylsulfoxonium iodide (**5**) or trimethylsulfonium iodide, respectively, with NaH or *n*-BuLi.[12] An improvement using KO*t*Bu[13,15] is safer than NaH and *n*-BuLi for large-scale operations.

In addition, NaOMe, and NaNH$_2$, have also been employed. Application of phase-transfer conditions with tetra-*n*-butylammonium iodide showed marked improvement for the epoxide formation.[16] Furthermore, many complex substituted sulfur ylides have been synthesized and utilized. For instance, stabilized ylide **20** was prepared and treated with α-D-*allo*-pyranoside **19** to furnish α-D-cyclopropanyl-pyranoside **21**.[17] Other examples of substituted sulfur ylides include **22–25**, among which aminosulfoxonium ylide **25**, sometimes known as Johnson's ylide, belongs to another category.[18] The aminosulfoxonium ylides possess the configurational stability and thermal stability not enjoyed by the sulfonium and sulfoxonium ylides, thereby are more suitable for asymmetric synthesis.

1.1.5   *Synthetic Utility*
1.1.5.1 *Epoxidation*

Epoxidation of aldehydes and ketones is the most profound utility of the Corey–Chaykovsky reaction. As noted in section 1.1.1, for an α,β-unsaturated carbonyl compound, **1** adds preferentially to the olefin to provide the cyclopropane derivative. On the other hand, the more reactive **2** generally undergoes the methylene transfer to the carbonyl, giving rise to the corresponding epoxide. For instance, treatment of β-ionone (**26**) with **2**, derived from trimethylsulfonium chloride and NaOH in the presence of a phase-transfer catalyst Et$_4$BnNCl, gave rise to vinyl epoxide **27** exclusively.[19]

Isolated carbonyls always give epoxides from the Corey–Chaykovsky reaction. Take the aldehyde substrate as an example. Spiro epoxide **30** was produced from the reaction of trisnorsqualene aldehyde **28** (R$_{20}$ represents the polyene side-chain with 20 carbons) with substituted sulfur ylide **29**, prepared *in situ* from cyclopropyldiphenylsulfonium tetrafluoroborate and KOH.[20] For the epoxidation of ketones, the Corey–Chaykovsky reaction works well for diaryl- (**31**),[21] arylalkyl- (**32**),[22] as well as dialkyl (**33**)[23] ketones. When steric bias exists on the substrate, stereoselective epoxidation may be achieved. For example, treatment of dihydrotestosterone (DHT, **35**) with the Corey ylide **1** followed by TPAP oxidation resulted in only one diastereomeric keto-epoxide **36**.[23]

Stereoselective epoxidation can be realized through either substrate-controlled (e.g. **35** → **36**) or reagent-controlled approaches. A classic example is the epoxidation of 4-*t*-butylcyclohexanone.[12] When sulfonium ylide **2** was utilized, the more reactive ylide irreversibly attacked the carbonyl from the axial direction to offer predominantly epoxide **37**. When the less reactive sulfoxonium ylide **1** was used, the nucleophilic addition to the carbonyl was reversible, giving rise to the thermodynamically more stable, equatorially coupled betaine, which subsequently eliminated to deliver epoxide **38**. Thus, stereoselective epoxidation was achieved from different mechanistic pathways taken by different sulfur ylides. In another case, reaction of aldehyde **38** with sulfonium ylide **2** only gave moderate stereoselectivity (**41:40** = 1.5/1), whereas employment of sulfoxonium ylide **1** led to a ratio of **41:40** = 13/1.[24] The best stereoselectivity was accomplished using aminosulfoxonium ylide **25**, leading to a ratio of **41:40** = 30/1. For ketone **42**, a complete reversal of stereochemistry was observed when it was treated with sulfoxonium ylide **1** and sulfonium ylide **2**, respectively.[25]

In transforming bis-ketone **45** to keto-epoxide **46**, the elevated stereoselectivity was believed to be a consequence of the molecular shape — the sulfur ylide attacked preferentially from the convex face of the strongly puckered molecule of **45**. Moreover, the pronounced chemoselectivity was attributed to the increased electrophilicity of the furanone versus the pyranone carbonyl, as a result of an inductive effect generated by the pair of spiroacetal oxygen substituents at the furanone α-position.[26]

Since chiral sulfur ylides racemize rapidly, they are generally prepared *in situ* from chiral sulfides and halides. The first example of asymmetric epoxidation was reported in 1989, using camphor-derived chiral sulfonium ylides with moderate yields and *ee* (< 47%).[27] Since then, much effort has been made in the asymmetric epoxidation using such a strategy without a significant breakthrough. In one example, the reaction between benzaldehyde and benzyl bromide in the presence of one equivalent of camphor-derived sulfide **47** furnished epoxide **48** in high diastereoselectivity (*trans:cis* = 96:4) with moderate enantioselectivity in the case of the *trans* isomer (56% *ee*).[28]

The Corey–Chaykovsky reaction incited some applications in medicinal chemistry. During the synthesis of analogs of fluconazole, an azole antifungal agent, treatment of **49** with **1** led to the corresponding epoxide, which was subsequently converted to **50** as a pair of diastereomers.[29] Analogously, the Corey–Chaykovsky reaction of ketone **51** gave the expected epoxide, which then underwent an S_N2 reaction with $1H$-1,2,4-triazole in the presence of NaH to deliver **52**, another azole antifungal agent.[30]

1.1.5.2 Cyclopropanation

Due to the high reactivity of sulfonium ylide **2** for α,β-unsaturated ketone substrates, it normally undergoes methylene transfer to the carbonyl to give the corresponding epoxides. However, cyclopropanation did take place when 1,1-diphenylethylene[12] and ethyl cinnamate[13] were treated with **2** to furnish cyclopropanes **53** and **54**, respectively.

$$\text{Ph}\diagdown\diagup\text{CO}_2\text{Et} \xrightarrow[\text{rt, 32\%}]{\textbf{2}, \text{DMSO}} \text{Ph}\triangleright\text{CO}_2\text{Et}$$

54

Dimethylsulfoxonium methylide (**1**) is the reagent of choice for the cyclopropanation of α,β-unsaturated carbonyl substrates. The reaction is generally carried out at more elevated temperatures in comparison to that of **2**, although exceptions exist. The method works for α,β-unsaturated ketones, esters and amides. Representative examples are found in transformations of 2(5*H*)-furanone **55** to cyclopropane **56**[31] and α,β-unsaturated Weinreb amide **57** to cyclopropane **58**.[32]

55   **56**

57   **58**

As in the case of epoxidation, asymmetric cyclopropanation can be accomplished through either substrate-controlled or reagent-controlled approaches. The former approach requires an inherent steric bias in the substrates that often exist in the form of chiral auxiliaries. Substrate **59**, derived from 1-hydroxy pinan-3-one, gave only diastereomer **60** when treated with **1**.[33] Ylide **1** attacked the less shielded face opposite to the *gem*-dimethyl group, and DMSO release with formation of the spirocyclic adduct occurred prior to bond rotation. With regard to chiral α,β-unsaturated bicyclic γ-lactam **61**, the cyclopropanation took place in a highly diasteroselective fashion using anion **22** (dimethylsulfuranylidene acetate), resulting in the *anti*-adduct **62** as the predominant product (**62** : **63** = 99:1).[34]

59   **60**

Reagent-controlled asymmetric cyclopropanation is relatively more difficult using sulfur ylides, although it has been done.[35] It is more often accomplished using chiral aminosulfoxonium ylides. Finally, more complex sulfur ylides (e.g. **64**) may result in more elaborate cyclopropane synthesis, as exemplified by the transformation **65** → **66**.[36]

1.1.5.3 Aziridination

In the initial report by Corey and Chaykovsky, dimethylsulfonium methylide (**2**) reacted smoothly with benzalaniline to provide an entry to 1,2-diphenylaziridine **67**.[12] Franzen and Driesen reported the same reaction with 81% yield for **67**.[13] In another example, benzylidene-phenylamine reacted with **2** to produce 1-(p-methoxyphenyl)-2-phenylaziridine in 71% yield. The same reaction was also carried out using phase-transfer catalysis conditions.[37] Thus aziridine **68** could be generated consistently in good yield (80–94%). Recently, more complex sulfur ylides have been employed to make more functionalized aziridines, as depicted by the reaction between N-sulfonylimine **69** with diphenylsulfonium 3-(trimethylsilyl)propargylide (**70**) to afford aziridine **71**, along with desilylated aziridine **72**.[38]

Asymmetric aziridination from imines using the Corey–Chaykovsky reaction is not well studied. The modest asymmetric induction is possibly due to the weak steric bias a chiral auxiliary exerted on the nucleophilic addition. Another possibility is that the bond rotation of the betaine intermediate may be so fast that it is difficult to achieve high stereoselectivity. Nowadays, asymmetric synthesis from imines is most frequently accomplished by addition of transition metal-catalyzed diazo reagents to the imines in the presence of chiral ligands. At any rate, examples of substrate-controlled aziridine formation using the Corey–Chaykovsky reaction can be found in the transformation 73 → 74 and 74′ where de was only 20%.[39] However, when the p-tolyl group was replaced by a t-butyl group, the de was as high as 90%.

Reagent-controlled aziridination using camphor-derived chiral sulfide 47 has been reported with ee values of 84–98% for the trans isomer although the trans : cis ratio was mediocre.[40]

1.1.5.4 Methylation

C-Methylation products, o-nitrotoluene and p-nitrotoluene, were obtained when nitrobenzene was treated with dimethylsulfoxonium methylide (1).[41] The ratio for the ortho and para-methylation products was about 10–15 : 1 for the aromatic nucleophilic substitution reaction. The reaction appeared to proceed via the single-electron transfer (SET) mechanism according to ESR studies.

N-Methylation of the NH of heterocycles using **1** is also known as exemplified by the methylation of indole.[42] The interesting mechanism is delineated below. *O*-methylation of weak acids such as phenols, carboxylic acids and oximes as well as *S*-methylation such as *N*-phenylisorhodanine, certain thioketones, and dithiocarboxylic acids have also been reported.[43]

1.1.5.5 Heterocycle and carbocycle formation

Corey's ylide (**1**), as the methylene transfer reagent, has been utilized in ring expansion of epoxide **75** and arizidine **77** to provide the corresponding oxetane **76**[15] and azetidine **78**,[44] respectively.

In Corey and Chaykovsky's initial investigation, a cyclic ylide **79** was observed from the reaction of ethyl cinnamate with ylide **1** in addition to 32% of cyclopropane **53**.[10] In a similar fashion, an intermolecular cycloaddition between 2-acyl-3,3-bis(methylthio)acrylnitrile **80** and **1** furnished 1-methylthiabenzene 1-oxide **81**.[45] Similar cases are found in transformations of ynone **82** to 1-arylthiabenzene 1-oxide **83**[46] and *N*-cyanoimidate **84** to adduct ylide **85**, which was subsequently transformed to 1-methyl-$1\lambda^4$-4-thiazin-1-oxide **86**.[47]

In a unique approach to the synthesis of isoxazole derivatives, α-isonitroso ketone **87** was treated with dimethylsulfonium methylide (**2**) to give 5-hydroxyisoxazoline **88**.[48] It was demonstrated that the reaction proceeded through an epoxyoxime intermediate.

In addition to the synthesis of heterocycles, the Corey–Chaykovsky reaction bestows an entry to carbocycles as well. The reaction of (trialkylsilyl)vinylketene **89** with substituted ylide **90**[49] led exclusively to *trans*-4,5-dimethyl cyclopentenone **91**.[50] The substituted ylide **90** here serves as a nucleophilic carbenoid reagent in the formal [4 +1] annulation reaction.

1.1.5.6 Polyhomologation

An ingenious application of Corey's ylide (**1**) was discovered by the Shea group in 1997.[51,52] Using trialkylboranes as initiator/catalyst and **1** as the monomer, a living polymerization led to linear poly*methylene* polymers (as opposed to the common poly*ethylene* polymers). Controlling the initial ratio of ylide **1** and triethylborane leveraged control over molecular weight. Oxidative cleavage of the C–B bond under basic oxidation conditions produced perfectly linear polymethylene **92**. Furthermore, extension of this novel chemistry provided means to build many new polymethylene architectures such as star-shaped polymethylenes, ring expansion of cyclic and polycyclic organoboranes, as well as macrocyclic oligmers and polymers.

$$Et_3B \xrightarrow[\text{2) } H_2O_2/NaOH]{\text{1) 3 eqivs. } \textbf{1}} 3\ Et\text{--}[CH_2]_n\text{--}CH_2OH$$
$$\textbf{92}$$

1.1.6 Experimental

$$93 \xrightarrow[\text{60°C, 3.5 h, 79\%}]{(CH_3)_3S(O)^+I^-\ \textbf{(5)},\ NaH,\ DMSO} 94$$

N-(3-Chloro-4-fluorobenzoyl)-oxa-6-azaspiro[2,5]-octane (94):[53]
A solution of dimethylsulfoxonium methylide (**1**) was prepared, under nitrogen, from sodium hydride (1.52 g of 60% despersion in mineral oil, 37.8 mmol) and trimethylsulfoxonium iodide (**5**, 8.32 g, 37.8 mmol) in anhydrous DMSO (20 mL). A solution of *N*-(3-chloro-4-fluorobenzoyl)-piperidine-4-one (**93**, 9.21 g, 36 mmol) in DMSO (20 mL) was added in 30 min and stirring was maintained at 60°C for 3.5 h. The cooled reaction mixture was poured into ice water and extracted with ethyl acetate. The combined organic layers were washed with water and brine and then dried and concentrated. The residue was purified by a short flash chromatography on silica gel, eluting with $CHCl_3$–EtOAc (9:1), to give 7.68 g of **94** (79%) as an oil which crystallized on standing: mp 75–77°C; ^{1}H NMR (CDCl$_3$) δ 1.50 (m, 2H), 1.92 (m, 2H), 2.74 (s, 2H), 3.87 (m, 1H), 4.19 (m, 1H), 7.18 (t, 1H), 7.32 (m, 1H), 7.51 (dd, 1H); IR (KBr, cm^{-1}) 1620.

1.1.7 References

1   [R] Okazaki, R.; Tokitoh, N. In *Encyclopedia of Reagents in Organic Synthesis;* Paquette, L. A., Ed.; Wiley: New York, **1995**, pp 2139–41.

2   [R] Ng, J. S.; Lin, C. In *Encyclopedia of Reagents in Organic Synthesis;* Paquette, L. A., Ed.; Wiley: New York, **1995**, pp 2159–65.

3   [R] Trost, B. M.; Melvin, L. S., Jr. *Sulfur Ylides;* Academic Press: New York, **1975**.

4   [R] Block, E. *Reactions of Organosulfur Compounds* Academic Press: New York, **1978**.

5   [R] Gololobov, Y. G.; Nesmeyanov, A. N. *Tetrahedron* **1987**, *43*, 2609.

6   [R] Aubé, J. In *Comprehensive Organic Synthesis;* Trost, B. M.; Fleming, I., Ed.; Pergamon: Oxford, **1991**, vol. 1, pp 820–825.
7   [R] Li, A.-H.; Dai, L.-X.; Aggarwal, V. K. *Chem. Rev.* **1997**, *97*, 2341.
8   Corey, E. J.; Chaykovsky, M. *J. Am. Chem. Soc.* **1962**, *84*, 867.
9   Corey, E. J.; Chaykovsky, M. *J. Am. Chem. Soc.* **1962**, *84*, 3782.
10   Corey, E. J.; Chaykovsky, M. *Tetrahedron Lett.* **1963**, 169.
11   Corey, E. J.; Chaykovsky, M. *J. Am. Chem. Soc.* **1964**, *86*, 1640.
12   Corey, E. J.; Chaykovsky, M. *J. Am. Chem. Soc.* **1965**, *87*, 1353.
13   Franzen, V.; Driesen, H. E. *Chem. Ber.* **1963**, *96*, 1881.
14   Mash, E. A.; Gregg, T. M.; Baron, J. A. *J. Org. Chem.* **1997**, *62*, 8513.
15   Wicks, D. A.; Tirrell, D. A. *J. Polym. Sci., Part A: Polym. Chem.* **1990**, *28*, 573.
16   Merz, A.; Märkl, G. *Angew. Chem., Int. Ed. Engl.* **1973**, *12*, 845.
17   Fitzsimmons, B. J.; Fraser-Reid, B. *Tetrahedron* **1982**, *40*, 1279.
18   (a) [R] Johnson, C. R. *Aldrichimica Acta* **1985**, *18*, 3. (b) Johnson, C. R.; Haake, M.; Schroeck, C. *J. Am. Chem. Soc.* **1970**, *92*, 6594. (c) Johnson, C. R.; Janiga, E. R. *J. Am. Chem. Soc.* **1973**, *95*, 7692.
19   Rosenberger, M.; Jackson, W.; Saucy, G. *Helv. Chim. Acta* **1980**, *63*, 1665.
20   Corey, E. J.; Cheng, H.; Baker, C. H.; Matsuda, P. T.; Li, D.; Song, X. *J. Am. Chem. Soc.* **1997**, *119*, 1277.
21   Kulasegaram, S.; Kulawiec, R. J. *J. Org. Chem.* **1997**, *62*, 6547.
22   Cleij, M.; Archelas, A.; Furstoss, R. *J. Org. Chem.* **1999**, *64*, 5029.
23   Maltais, R.; Poirier, D. *Tetrahedron Lett.* **1998**, *39*, 4151.
24   Saito, T.; Suzuki, T.; Takeuchi, K.; Matsumoto, T.; Suzuki, K. *Tetrahedron Lett.* **1997**, *38*, 3755.
25   Mal, J.; Venkateswaran, R. V. *J. Org. Chem.* **1998**, *63*, 3855.
26   Ciufolini, M. A.; Zhu, S.; Deaton, M. V. *J. Org. Chem.* **1997**, *62*, 7806.
27   Furukawa, N.; Sugihara, Y.; Fujihara, H. *J. Org. Chem.* **1989**, *54*, 4222.
28   Saito, T.; Akiba, D.; Sakairi, M.; Kanazawa, S. *Tetrahedron Lett.* **2001**, *42*, 57.
29   Dickinson, R. P.; Bell, A. S.; Hitchcock, C. A.; Narayanawami, S.; Ray, S. J.; Richardson, K.; Troke, P. F. *Bioorg. Med. Chem. Lett.* **1996**, *6*, 2031.
30   Fringuelli, R.; Schiaffella, F.; Bistoni, F.; Pitzurra, L.; Vecchiarelli, A. *Bioorg. Med. Chem.* **1998**, *6*, 103.
31   Janini, T. E.; Sampson, P. *J. Org. Chem.* **1997**, *62*, 5069.
32   Rodriques, K. E. *Tetrahedron Lett.* **1991**, *32*, 1275.
33   Calmes, M.; Daunis, J.; Escale, F. *Tetrahedron: Asymmetry* **1996**, *7*, 395.
34   (a) Groaning, M. D.; Meyers, A. I. *Tetrahedron Lett.* **1999**, *40*, 4639. (b) Romo, D.; Meyers, A. I. *J. Org. Chem.* **1992**, *57*, 6265.
35   Trost, B. M.; Hammen, R. F. *J. Am. Chem. Soc.* **1973**, *95*, 962.
36   (a) Akiyama, H.; Fujimoto, T.; Ohshima, K.; Hoshino, K.; Saito, Y.; Okamoto, A.; Yamamoto, I.; Kakehi, A.; Iriye, R. *Eur. J. Org. Chem.* **2001**, 2265. (b) Chandrasekhar, S.; Narasihmulu, Ch.; Jagadeshwar, V.; Venkatram, Reddy, K. *Tetrahedron Lett.* **2003**, *44*, 3629.
37   Tewari, R. S.; Awatsthi, A. K.; Awasthi, A. *Synthesis* **1983**, 330.
38   Li, A.-H.; Zhou, Y. G.; Dai, L.-X.; Hou, X.-L.; Xia, L.-J.; Lin, L. *J. Org. Chem.* **1998**, *63*. 4338.
39   García Ruano, J. L.; Fernádez, I.; Catalina, M.; Cruz, A. A. *Tetrahedron: Asymmetry* **1996**, *7*, 3407.
40   Saito, T.; Akiba, D.; Sakairi, M. *Tetrahedron Lett.* **2001**, *42*, 5451.
41   Traynelis, V, J.; McSweeney, J. V. *J. Org. Chem.* **1966**, *31*, 243.
42   Kunieda, T.; Witkop, B. *J. Org. Chem.* **1970**, *35*, 3981.
43   [R] Reference 5, pp 2615–2617.
44   Nadir, U. K.; Sharma, R. L.; Koul, V. K. *Tetrahedron* **1989**, *45*, 1851.
45   Rudorf, W.-D. *Synthesis* **1984**, 852.
46   Matsuyama, H.; Takeuchi, T.; Okutsu, Y. *Heterocycles* **1984**, *22*, 1523.
47   Ried, W.; Kuhn, D. *Liebigs Ann. Chem.* **1986**, 1648.
48   Bravo, P.; Ticozzi, C. *Gazz. Chim. Ital.* **1972**, *102*, 395.
49   Corey, E. J.; Jautelat, M. Oppolzer, W. *Tetrahedron Lett.* **1967**, 2325.
50   Loebach, J. L.; Bennett, D. M.; Danheiser, R. L. *J. Am. Chem. Soc.* **1998**, *120*, 9690.
51   Shea, K. J.; Walker, J. W,; Zhu, H.; Paz, M.; Greaves, J. *J. Am. Chem. Soc.* **1997**, *110*, 9049.
52   [R] Shea, K. *J. Chem. Eur. J.* **2000**, *6*, 1113.
53   Vacher, B.; Bonnaud, B. Funes, P.; *et al. J. Med. Chem.* **1999**, *42*, 1648.

Jie Jack Li

1.2   Darzens Glycidic Ester Condensation

1.2.1   Description
Darzens glycidic ester condensation[1] generally involves the condensation of an aldehyde or ketone **2** with the enolate of an α-halo ester **1** which leads to an α,β-epoxy ester (a glycidic ester) (**3**). Thus the reaction adds two carbons to the electrophile; however, the reaction has been primarily developed as a one-carbon homologation method. That is, subsequent to the condensation, the ester is saponified and decarboxylation ensues to give the corresponding aldehyde or ketone **5**.[2]

Various stabilized α-halo anions (diazo ketones, imines, nitriles, phosphonates, silicon, sulfones, etc.) have been employed in the reaction. Methods for the preparation of aziridines using the process have been examined, and asymmetric variants have been reported. Although hydroxide can often be used for generating the anion, a non-nucleophilic base (t-BuOK, LiHMDS, LDA) is generally used in the reaction to avoid S_N2 displacement of the electrophile. The halide of the nucleophilic component of the reaction is typically chlorine — stronger leaving groups (bromine and especially iodine) lead toward γ-keto esters (after saponification/decarboxylation is carried out), a result of intermolecular S_N2 displacement.[3] The diverse nature of the substrates and conditions that can be employed in the reaction precludes further discussion to the general nature of the reaction.[2]

1.2.2   Historical Perspective
Although glycidic esters were first prepared by Erlenmeyer in 1892, Darzens subsequently studied the reaction and demonstrated its usefulness as a synthetic method.[4] In a significant achievement in synthesis during the 1940s, the titled reaction process was used in the industrial reaction pathway to prepare vitamin A (**9**).[5] Thus methyl chloroacetate (**7**) and β-ionone (**6**) were treated with sodium ethylate to give the corresponding glycidic ester.   Upon saponification and decarboxylation, thermodynamically favored trienal **8** is provided, which can be further elaborated to vitamin A.[2,5]

vitamin A (**9**)

1.2.3 Mechanism

Several years ago, there was much debate concerning the mechanism of the Darzens condensation.[2,3] The debate concerned whether the reaction employed an enolate or a carbene intermediate. In recent years, significant evidence that supports the enolate mechanism has been obtained, wherein the stabilized carbanion (**11**) of the halide (**10**) is condensed with the electrophile (**12**) to give diastereomeric aldolate products (**13,14**), which subsequently cyclize via an internal S_N2 reaction to give the corresponding oxirane (**15** or **16**). The intermediate aldolates have been isolated for both α-fluoro- and α-chloroesters **10**.[2,3]

Furthermore, in analogy to the aldol reaction, α-chloro-α,β-unsaturated esters have been observed—likely the result of β-elimination of water from the intermediate halohydrin. For example, when benzaldehyde is condensed with the enolate of **17**, chloride **19** was obtained.[6]

The ratio of products **15** and **16** is dependent on the structures, base, and the solvent. The kinetics of the reaction is likewise dependant on the structures and conditions of the reaction. Thus addition or cyclization can be the rate-determining step. In a particularly noteworthy study by Zimmerman and Ahramjian,[7] it was reported that when both diastereomers of **20** were treated individually with potassium *t*-butoxide only *cis*-epoxy propionate **21** was isolated. It is postulated that the cyclization is the rate-limiting step. Thus, for these substrates, the retro-aldolization/aldolization step is reversible.[7]

An explanation for the stereoselectivity of the reaction involves optimal overlap of the π-orbital of the carbonyl with the developing electron rich p-orbital on C2 during the S_N2 displacement of the chloride by the alkoxide (**24**). Thus, orbital overlap imposes conformational constraints in the transition state that leads to nonbonding interactions disfavoring transition state **25**.[7]

1.2.4 Variations and Improvements

In recent years, several modifications of the Darzens condensation have been reported. Similar to the aldol reaction, the majority of the work reported has been directed toward diastereo and enantioselective processes. In fact, when the aldol reaction is highly stereoselective, or when the aldol product can be isolated, useful quantities of the required glycidic ester can be obtained. Recent reports have demonstrated that diastereomeric enolate components can provide stereoselectivity in the reaction: examples include the camphor-derived substrate **26**,[8] *in situ* generated α-bromo-*N*-

acetyloxazolidinethione **27**,[9] menthol and 8-phenylmenthol esters **28** and **29**.[10] It is noteworthy that Aggarwal recently showed that the camphor derived sulfonium salt **30** could be condensed with various aldehydes in good yields (79–93%), and up to 99% *ee*.[11]

| **26** | **27** | **28** R = H
29 R = Ph | **30** |

Interestingly, phase-transfer catalysts including crown ethers have been used to promote enantioselective variations of Darzens condensation. Tõke and coworkers showed that the novel 15-crown-5 catalyst derived from D-glucose **33** could promote the condensation between acetyl chloride **31** and benzaldehyde to give the epoxide in 49% yield and 71% *ee*.[12] A modified cinchoninium bromide was shown to act as an effective phase transfer catalyst for the transformation as well.[13]

31

33 (5%), PhCHO,
NaOH(aq, 30%),
toluene, –20°C

32
49% yield, 71% ee

33

In a separate report, preparation of the lithium enolate of **31** in the presence of indium trichloride and benzaldehyde provided a 77% yield of **32** with complete *trans* selectivity; however, sequential addition of indium trichloride and benzaldehyde provided Barbier-type products.[14] Organotin enolates have also been used in a Darzens-type reaction.[15]

1.2.5　Synthetic Utility and Applications

The Darzens condensation reaction has been used with a wide variety of enolate equivalents that have been covered elsewhere.[2] A recent application of this important reaction was applied toward the asymmetric synthesis of aziridine phosphonates by Davis and coworkers.[16] In this application, a THF solution of sulfinimine **34** (0.37 mmol, >98% *ee*) and iodophosphonate **35** (0.74 mmol) was treated with LiHMDS (0.74 mmol) at –78 °C to give aziridine **36** in 75% yield. Treatment of **36** with MeMgBr removed the sulfinyl group to provide aziridine **37** in 72% yield.[16a]

34   **36** R = SOMes
                                                          37 R = H

Darzens reaction can be used to efficiently complete the stereoselective synthesis of α′-substituted epoxy ketones. As an example, Enders and Hett reported a technique for the asymmetric synthesis of α′-silylated α,β-epoxy ketones. Thus, optically active α′-silyl α-bromoketone **38** was treated with LDA followed by the addition of benzaldehyde to give α′-silyl epoxyketone **40** in 66% yield with good *de*.[17]

In a separate report, the Darzens reaction was recently used by Barluenga, Concellón, and coworkers for the preparation of enantiopure α′-amino α,β-epoxy ketones. Accordingly, the Z enolate of α′-amino α-bromo ketone **41** was generated with KHMDS at –100°C. Benzaldehyde was added, and *trans* epoxyketone **42** was isolated in 87% yield and >95% de.[18]

Recently, Darzens reaction was investigated for its synthetic applicability to the condensation of substituted cyclohexanes and optically active α-chloroesters (derived from (–)-phenylmenthol). In this report, it was found that reaction between chloroester **44** and cyclohexanone **43** provided an 84% yield with 78:22 selectivity for the axial glycidic ester **45** over equatorial glycidic ester **46** both having the *R* configuration at the epoxide stereocenter.[19]

Of interest is a recent report of a rapid synthesis of efaroxin (**51**), a potent, selective α₂-adrenoceptor antagonist, using Darzens Reaction. Accordingly, α-bromoester **48** was condensed with aldehyde **47**. The glycidic ester (**49**) was then hydrogenated to reduce the more labile epoxide bond to give alcohol **50**. Subsequent standard transformations subsequently lead to a completed 4-step synthesis of efaroxin.[20]

1.2.6 Experimental[21]

A dry 500-ml round-bottomed three-necked flask fitted with a stirrer, internal thermometer, and a pressure-equalized dropping funnel is placed under nitrogen and the flask is charged with 0.148 mole of freshly distilled cyclohexanone and 0.148 mole of freshly distilled ethyl chloroacetate. A solution of 6.0 g of potassium and 125 mL of dry *tert*-butyl alcohol is introduced into the dropping funnel, and the system is exhausted and filled with nitrogen. The flask is cooled with an ice bath, stirring is commenced, and the solution of potassium *tert*-butoxide is added from the dropping funnel over a period of about 1.5 hours, the temperature of the reaction mixture being maintained at 10–15°C. After the addition is complete, the mixture is stirred for an additional 1–1.5 h at about 10°C. Most of the *tert*-butyl alcohol is removed by distillation from the reaction flask at reduced pressure (water aspirator). The oily residue is taken up in ether. The ether solution is washed with water, then with saturated aqueous sodium chloride solution, and is finally dried over anhydrous sodium sulfate. The residue obtained on evaporation of the ether is distilled through a 6-in. Vigreux column to give 83–95% yield of colorless glycidic ester.[21a]

1.2.7   *References*

1.   (a) Darzens, G. *Compt. Rend.* **1904**, *139*, 1214. (b) Darzens, G. *Compt. Rend.* **1906**, *141*, 214.
2.   [R] Rosen, T. In *Comprehensive Organic Synthesis;* Trost, B. M.; Fleming, I., Ed.; Pergamon: Oxford, **1991**, vol. 2, pp 409–439.
3.   [R] Ballester, M. *Chem. Rev.* **1955**, *55*, 283–300.
4.   [R] Newman, M. S.; Magerlein, B. J. *Org. React.* **1949**, *5*, 413–441.
5.   Isler, O.; Huber, W.; Ronco, A; Kofler, M. *Helv. Chim. Acta,* **1947**, *30*, 1911–1927.
6.   Jörlander, H. *Ber.* **1917**, *50*, 1457–65.
7.   Zimmerman, H. E.; Ahramjian, L. *J. Amer. Chem. Soc.* **1960**, *82*, 5459–5466.
8.   Palomo, C.; oirbide, M; Sharma, A. K; González,-Rego, M. C.; Linden, A; Garcia, J. M.; González, A. *J. Org. Chem.* **2000**, *65*, 9007–12.
9.   Wang, Y.-C.; Li, C.-L.; Tseng, H.-L.; Chuang, S.-C.; Yan, T.-H. *Tetrahedron: Asym.* **1999**, *10*, 3249–51.
10.   Takagi, R.; Kimura, J.; Shinohara, Y.; Ohba, Y.; Takezono, K.; Hiraga, Y.; Kojimo, S.; Ohkata, K. *J. Chem. Soc., Perkin Trans. 1* **1998**, 689–98.
11.   Aggarwal, V. K.; Hynd, G.; Picoul, W.; Vasse, J.-L. *J. Amer. Chem. Soc.* **2002**, *124*, 9964–9965.
12.   (a) Bakó, P.; Czinege, E.; Bakó, T.; Czugler, M.; Toke, L. *Tetrahedron: Asym.* **1999**, *10*, 4539–51. (b) Bakó, P.; Vízvárdi, K.; Toppet, S.; Van der Eycken, E.; Hoornaert, G. J.; Tőke, L. *Tetrahedron* **1998**, *54*, 14975–88.
13.   Arai, S.; Shirai, Y.; Ishida, T.; Shioiri, T. *Tetrahedron* **1999**, *55*, 6375–86. (b) Arai, S.; Shioiri, T. *Tett. Lett.* **1998**, *39*, 2145–8. (c) Tanaka, K.; Shiraishi, R. *Green Chemistry* **2001**, *3*, 135–6. (d) Arai, S.; Shoiri, T. *Tetrahedron* **2002**, *58*, 1407–13. (e) Arai, S.; Suzuki, Y.; Tokumaru, K.; Shioiri, T.; *Tetrahedron Lett.* **2002**, *43*, 833–6.
14.   Hirashita, T.; Kinoshita, K.; Yamamura, H.; Kawal, M.; Arai, S. *J. Chem. Soc., Perkin Trans. 1,* **2000**, 825–828. The authors note that the lithium enolate decomposes in the absence of InCl₃.
15.   Shimbata, I.; Yamasaki, H.; Baba, A.; Matsuda, H. *J. Org. Chem.* **1992**, *57*, 6909–6914.
16.   (a) Davis, F. A.; Ramachandar, T.; Wu, Y. *J. Org. Chem.* **2003**, *68*, 6894–8. (b) Davis, F. A.; Liu, H.; Reddy, G.V. *Tetrahedron Lett.* **1996**, *37*, 5473–76. (c) Davis, F. A.; Liu, H.; Zhou P.; Fang, T.; Reddy, G. V.; Zhang Y. *J. Org. Chem.* **1999**, *64*, 7559–67.
17.   Enders, D.; Hett; R. *Synlett* **1998**, 961–2.
18.   Barluenga, J.; Baragatña, B.; Concellón, J.M.; Piñera-Nicolás, A.; Diaz, M. R.; García-Granda, S. *J. Org. Chem.* **1999**, *64*, 5048–52.
19.   Shinohara, Y.; Ohba, Y.; Takagi, R.; Kojima, S.; Ohkata, K. *Heterocycles* **2001**, *55*, 9–12.
20.   Mayer, P.; Brunel, P.; Imbert, T. *Biog. & Med. Chem. Lett,* **1999**, *9*, 3021.
21.   Several excellent synthetic procedures have been published: (a) Hunt, R. H.; Chinn, L. J.; Johnson, W. S. *Org. Syn. Coll.* **1963**, *4*, 459. (b) Burness, D. M. *Org. Syn. Coll.* **1963**, *4*, 649–652 (c) (See reference 4) (d) Allen, C. F. H.; VanAllan, J. *Org. Syn. Coll.* **1955**, *3*, 459.

Brian J. Myers

1.3   Hoch–Campbell Aziridine Synthesis

1.3.1   Description

The Hoch–Campbell aziridine synthesis entails treatment of ketoximes with excess Grignard reagents and subsequent hydrolysis of the organometallic complex.[1-11]

1.3.2   Historical Perspective

In 1934, French chemist Hoch reported that the action of phenylmagnesium bromide on the oxime of propiophenone (**3**) at elevated temperature gave two products.[5-7]   One was aziridine **4** and the other was erroneously assigned as hydroxylamine **5**.   In the subsequent years (1939 onward), Campbell at the University of Notre Dame determined that the purported hydroxylamine **5** was actually β-hydroxylamine **6**.[8-11]   The scope of the Grignard reagents was extended to both aryl and aliphatic Grignard reagents.

1.3.3   Mechanism

The mechanism of the Hoch–Campbell aziridine synthesis was a contentious issue for quite some time.   The incorporation of a double bond into an already highly strained three-membered ring was initially mistakenly thought to preclude the very existence of the 1-azirine system.   However, it was established that when ketoximes were treated with Grignard reagents, a rearrangement took place that involved the migration of the nitrogen atom from one carbon to another.[8]   At the early stage of the Hoch–Campbell aziridine synthesis, the general mode may be reminiscent of the Neber reaction mechanism.[12]   In 1963, Eguchi and Ishii carried out a series of experiments that supported a plausible mechanism that involving an azirine intermediate.[13]   In the 1970s, the Laurent group made great strides in deciphering the mechanism of the Hoch–Campbell reaction.[14-17]   Their results are summarized herein.

When ketoxime **1** is treated with the Grignard reagent, the first action is abstraction of the oxime proton to give **7**, which complexes with another equivalent of Grignard reagent to form **8**.   Intramolecular deprotonation of **8** then gives rise to di-anion **9**, which extrudes OMgX anion to establish vinyl nitrene **10**.   Cyclization of vinyl nitrene **10** delivers the key intermediate azirine **11**.   Addition of another equivalent of Grignard reagent to azirine **11** subsequently affords aziridine magnesium halide **9**, which upon acidic workup gives rise to aziridine **2**.

The aforementioned mechanism is supported by the following experimental data. When oxime **13** was treated with Grignard reagent, 3% of the indole **15** was isolated, indicating the possible existence of nitrene intermediate **14**.[16]   A 2-phenylazirine intermediate, on the other hand, has been isolated and characterized from the reaction under carefully controlled conditions (adding Grignard reagent to the oxime in toluene).[17]

1.3.4   Variations and Improvements

An "intramolecular trapping" variant of the Hoch–Campbell aziridine synthesis was reported by Taguchi et al.[18–21]   When 2-(1-*p*-chlorophenylcyclohexyl)cyclohexanone (**16**) was treated with phenylmagnesium bromide, the resulting azirine **17** did not undergo the usual intermolecular nucleophilic attack of the second equivalent of phenylmagnesium bromide. Instead, due to the proximity of the benzene ring as a potential nucleophile, the intramolecular ring closure of the intermediate was overwhelmingly favored as compared to the normal Hoch-Campbell reaction, resulting in predominantly the 1-benzobicyclononenone **19**. Hydrolysis of iminylmagnesium bromide **19** then gave imine **20**, which was subsequently hydrolyzed to ketone **21** under acidic conditions.

The second variation and improvement involves the use of $N-NR_3^+$ moiety, which in some cases gave better yields than the corresponding ketoxime analogs.[22-24]

The third variation of the Hoch–Campbell reaction is the replacement of the Grignard reagents with $LiAlH_4$. Evidently, the R^3 would be hydride in this case.[25] The mechanism is strikingly similar to that of the Hoch–Campbell reaction except the azirine is attacked by hydride rather than the Grignard reagent.

1.3.5   Synthetic Utility
1.3.5.1   Simple ketoximes

Oxime **26** was prepared from 5,11-dihydro-dibenzo[a,d]cyclohepten-10-one. The Hoch–Campbell reaction of **26** with 3-dimethylaminopropylmagnesium bromide produced aziridine **27** in 46% yield after acidic workup.[26]   Extension of the Hoch–Campbell reaction to steroids has also been reported.[27] Thus, treatment of 3β-hydroxy-5-pregnen-20-one oxime (**28**) with methylmagnesium iodide furnished a mixture of diastereomers, 20α/20β,21-imino-20-methyl-5-pregnen-3β-ol (**29**) in a 50% combined yield and a 3:1 ratio. On the other hand, homo-adamantan-4-one oxime (**30**) was transformed to homo-adamantano[4,5-b]-2′-ethylaziridine (**31**) in 76% yield upon the action of ethylmagnesium bromide.[28] However, the reactions of methylmagnesium iodide and phenylmagnesium bromide gave the corresponding aziridines in only 21% and 22% yield, respectively.

Secondary aziridines bearing a trifluoromethyl group were prepared *via* the Hoch–Campbell reaction of Grignard reagents and oximes bearing a trifluoromethyl substituent.[29] One exception was found for the use of allylmagnesium bromide, which gave homoallylic hydroxylamines.[30] Another exception was found for ethylmagnesium bromide where it served exclusively as a reducing agent.[29] Initial reaction of oxime **32** with ethylmagnesium bromide gave azirine **33**. However, the course of reaction was deviated from the normal Hoch–Campbell reaction since single electron transfer (SET) took place and led to radical anion **34**, which was converted to the reduced aziridine **35** upon workup.

When both α-positions of the oxime possess active hydrogen, the regiochemistry of the Hoch–Campbell reaction prefers the side with more available hydrogens—indicating the process is kinetically controlled.[31] In case of oxime **36**, azirine **37** was not formed.[32] Instead, azirine **38** was obtained exclusively. Addition of the third equivalent of the Grignard reagent delivered aziridine **39** as a mixture of two diastereomers.

1.3.5.2   α-Hydroxy and α-keto ketoximes

The Hoch–Campbell reaction of α-hydroxy ketoximes do not alter the course of the reaction although deprotonation probably took place concurrently for both the alcohol and the oxime. Treatment of oxime **40** afforded aziridine **42** in 30%, presumably *via* the intermediacy of azirine **41**.[33] α-Keto ketoximes would behave similarly to the α-hydroxy ketoximes in the Hoch–Campbell reaction after addition of the first equivalent of the Grignard reagent to the ketone.[34, 35] Therefore, the reaction between α-keto ketoxime **43** and phenylmagnesium bromide gave aziridine **45** in 41% yield, presumably *via* the intermediacy of azirine **44**.

1.3.6 Experimental

$$\begin{array}{ccc} \mathbf{30} & & \mathbf{31} \end{array}$$

Homo-adamantano[4,5-b]-2´-ethylaziridine (31)[28]

To a stirred solution of ethylmagnesium bromide (6.0 mmol) in ether (5 mL) and toluene (5 mL) was added homo-adamantan-4-one oxime (**30**, 2.0 mmol) in toluene (10 mL) at 100–105°C. The mixture was kept at the same temperature for 3 h. The cooled mixture was poured onto an ice-ammonium chloride mixture and the organic layer was separated and the water layer was extracted with ether (2 × 10 mL). The combined organic layer and extracts were dried (Na_2SO_4). Removal of the solvent under reduced pressure at 40°C gave crude product which was purified on a silica gel column eluting with CH_2Cl_2–MeOH to afford homo-adamantano[4,5-b]-2´-ethylaziridine (**31**) in 76% yield. The aziridine had a foul odor peculiar to aziridines. mp = 74–75°C.

1.3.7 References

1.   [R] Freeman, J. P. *Chem. Rev.* **1973**, *73*, 283.
2.   [R] Dermer, O. C.; Ham, G. E. *Ethyleneimine and Other Aziridines,* Academic Press: New York, **1969**, pp65–68.
3.   [R] Deyrup, J. A. In *Small Ring Heterocycles, Part 1, Aziridines, Azirines, Thiiranes, Thiirenes* Hassner, A., ed.; John Wiley & Sons: New York, **1983**, Chapter 1, pp1–214.
4.   (a) [R] Kotera, K.; Kitahonoki, K. *Org. Prep. Proced. Int.* **1952**, *1*, 305. (b) [R] Tanner, D. *Angew. Chem., Int. Ed. Engl.* **1994**, *33*, 599.
5.   Hoch, J. *Compt. Rend. Acad. Sci.* **1934**, *198*, 1865.
6.   Hoch, J. *Compt. Rend. Acad. Sci.* **1936**, *203*, 799.
7.   Hoch, J. *Compt. Rend. Acad. Sci.* **1937**, *204*, 358.
8.   Campbell, K. N.; McKenna, J. F. *J. Org. Chem.* **1939**, *4*, 198.
9.   Campbell, K. N.; Campbell, B. K.; Chaput, E. P. *J. Org. Chem.* **1943**, *8*, 99.
10.   Campbell, K. N.; Campbell, B. K.; McKenna, J. F.; Chaput, E. P. *J. Org. Chem.* **1943**, *8*, 103.
11.   Campbell, K. N.; Campbell, B. K.; Hess, L. G.; Schaffner, I. J. *J. Org. Chem.* **1944**, *9*, 184.
12.   [R] O'Brien, C. *Chem. Rev.* **1964**, *64*, 81.
13.   (a) Eguchi, S.; Ishii, Y. *Bull. Chem. Soc. Jpn.* **1963**, *36*, 1434. (b) Ishii, Y.; Eguchi, S.; Hironaka, T. *Kogyo Kagaku Zasshi* **1965**, *68*, 293.
14.   Alvernhe, G.; Laurent, A. *Bull. Soc. Chim. Fr.* **1970**, 3003.
15.   Chaabouni, R.; Laurent, A. *Bull. Soc. Chim. Fr.* **1973**, 2680.
16.   Bartnik, R.; Laurent, A. *Bull. Soc. Chim. Fr.* **1975**, 173.
17.   Alvernhe, G.; Laurent, A. *J. Chem. Res. (S)* **1978**, 28.
18.   Miyano, K.; Taguchi, T. *Chem. Pharm. Bull.* **1970**, *18*, 1806.
19.   Imai, K.; Kawazoe, Y. *Yakugaku Zasshi* **1974**, *94*, 452.
20.   Imai, K.; Kawazoe, Y.; Taguchi, T. *Chem. Pharm. Bull.* **1976**, *24*, 1083.
21.   Takehisa, E. Y.; Kawazoe, Y; Taguchi, T. *Chem. Pharm. Bull.* **1976**, *24*, 1691.
22.   Chaabouni, R.; Laurent, A. *Synthesis* **1975**, 464.
23.   (a) Diab, Y.; Laurent, A.; Mison, P. *Bull. Soc. Chim. Fr.* **1974**, 2202. (b) Alvernhe, G.; Arsényiadis, S., Chaabouni, R.; Laurent, A. *Tetrahedron Lett.* **1975**, *18*, 355.
24.   Arsényiadis, S.; Laurent, A.; Mison, P. *Bull. Soc. Chim. Fr.* **1980**, II-246.

25.   (a) Kotera, K.; Miyazaki, S.; Takahashi, H.; Okada, T.; Kitahonoki, K. *Tetrahedron* **1968**, *24*, 3681. (b) Kotera, K.; Miyazaki, S. *Tetrahedron* **1968**, *24*, 5677.

26.   Seidlová, V.; Protiva, M. *Collect. Czech. Chem. Commun.* **1967**, *32*, 1747.

27.   Tzikas, A.; Tamm, C.; Boller, A.; Fürst, A. *Helv. Chim. Acta* **1976**, *59*, 1850.

28.   Sasaki, T.; Eguchi, S.; Hattori, S. *Heterocycles* **1978**, *11*, 235.

29.   Quinze, K.; Laurent, A.; Mison, P. *J. Fluorine Chem.* **1989**, *44*, 211.

30.   Felix, C.; Laurent, A.; Lesniak, S.; Mison, P. *J. Chem. Res. (S)* **1991**, 32.

31.   Alvernhe, G.; Laurent, A. *Tetrahedron Lett.* **1971**, *22*, 1913.

32.   Chaabouni, R.; Laurent, A.; Mison, P. *Tetrahedron Lett.* **1975**, *16*, 1343.

33.   Laurent, A.; Marsura, A.; Pierre, J.-L. *J. Heterocycl. Chem.* **1980**, *17*, 1009.

34.   Bartnik, R.; Laurent, A. *Tetrahedron Lett.* **1974**, *17*, 3869.

35.   Bartnik, R.; Diab, Y.; Laurent, A. *Tetrahedron* **1977**, *32*, 1279.

Jie Jack Li

1.4 Jacobsen–Katsuki Epoxidation

1.4.1 Description

The Jacobsen–Katsuki epoxidation reaction is an efficient and highly selective method for the preparation of a wide variety of structurally and electronically diverse chiral epoxides from olefins.[1] The reaction involves the use of a catalytic amount of a chiral Mn(III)salen complex **1** (salen refers to ligands composed of the *N,N'*-ethylenebis(salicylideneaminato) core), a stoichiometric amount of a terminal oxidant, and the substrate olefin **2** in the appropriate solvent (Scheme 1.4.1). The reaction protocol is straightforward and does not require any special handling techniques.

Scheme 1.4.1

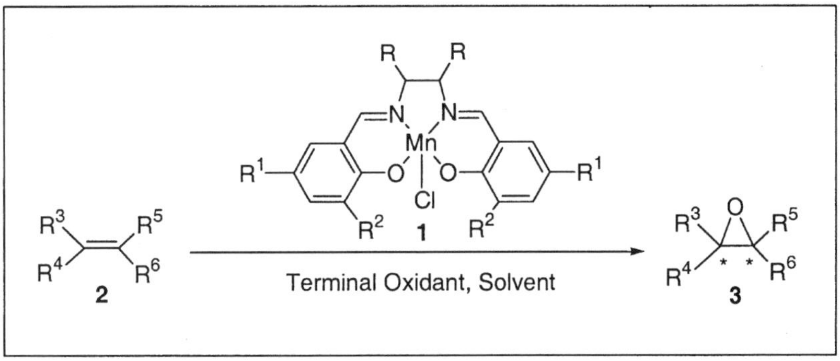

To date, a wide variety of structurally different chiral Mn(III)salen complexes have been prepared, of which only a handful have emerged as synthetically useful catalysts. By far the most widely used Mn(III)salen catalyst is the commercially available Jacobsen catalyst wherein R= $-C_4H_8-$ and $R^1 = R^2 = t$-Bu (Scheme 1.4.1).[2] In contrast, a wide variety of terminal oxidants have been successfully employed. Depending on the terminal oxidant, epoxidation reactions can be performed from –78°C to room temperature. A broad range of olefins undergo epoxidation with good enantioselectivity including terminal, *cis*-disubstituted, *trans*-disubstituted, tri-substituted, and tetra-substituted olefins. However, except for a few isolated examples, the olefin must be conjugated to an aromatic group, an alkyne, or an alkene in order to obtain good stereoinduction. Additives such as pyridine *N*-oxides have been shown to exhibit a profound and often beneficial effect on enantioselectivities, *cis/trans* selectivities in the epoxidation of disubstituted olefins, and catalyst lifetime.

1.4.2 Historical Perspective

In 1990, Jacobsen and subsequently Katsuki independently communicated that chiral Mn(III)salen complexes are effective catalysts for the enantioselective epoxidation of unfunctionalized olefins.[3] For the first time, high enantioselectivities were attainable for the epoxidation of unfunctionalized olefins using a readily available and inexpensive chiral catalyst. In addition, the reaction was one of the first transition metal-catalyzed tranformations that possessed a broad substrate scope and provided epoxidation products in high *ee* without requiring pre-coordination of the substrate to the active catalyst.

Jacobsen and Katsuki continued independently to develop and improve the epoxidation reaction by expanding the substrate scope, increasing catalyst efficiency, and developing a better understanding of the mechanism of reaction. Catalyst design was an integral part of the development of the epoxidation reaction, and was possible only because the Mn(III)salen catalyst can be readily prepared.[4,5] The synthesis of the catalyst can be accomplished in two steps from a salicylaldehyde **4** and a chiral diamine **5** (Scheme 1.4.2). The short modular synthesis of the catalyst provided the opportunity for systematically studying the effect of steric and electronic properties of the various substituents on the salen ligand **6** on enantioselectivity. This attribute rendered the identification of a highly selective and general catalyst practicable. The most commonly used catalyst, the Jacobsen catalyst, is produced on kilogram scale and has been employed on multi-kilogram scale in the pharmaceutical industry.[6]

Scheme 1.4.2

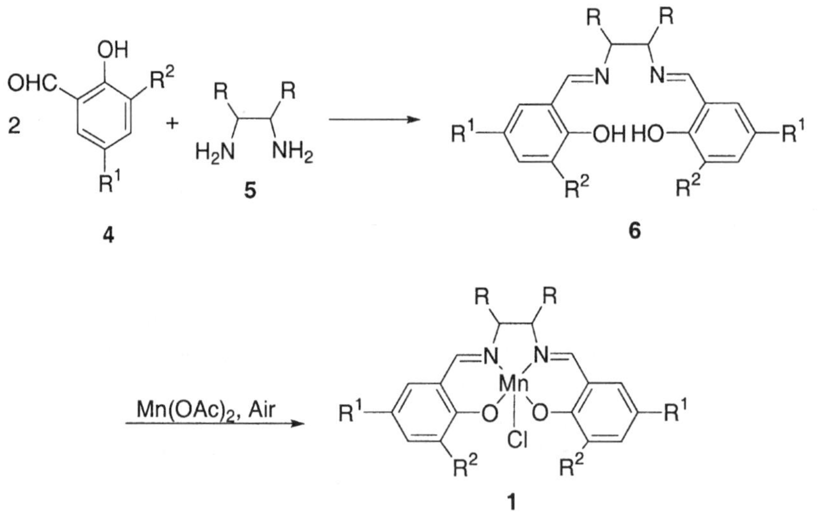

1.4.3 Mechanism

The most widely accepted mechanism of reaction is shown in the catalytic cycle (Scheme 1.4.3). The overall reaction can be broken down into three elementary steps: the oxidation step (Step A), the first C–O bond forming step (Step B), and the second C–O bond forming step (Step C). Step A is the rate-determining step; kinetic studies show that the reaction is first order in both catalyst and oxidant, and zero order in olefin.[7] The rate of reaction is directly affected by choice of oxidant, catalyst loadings, and the presence of additives such as *N*-oxides. Under certain conditions, *N*-oxides have been shown to increase the rate of reaction by acting as phase transfer catalysts.[8]

Scheme 1.4.3

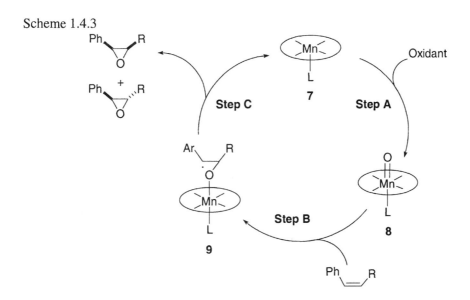

Step C

Step A

7

Oxidant

Step B

8

9

A variety of oxidants have been shown to be effective in generating the putative Mn(V)salen oxo species **8** including sodium hypochlorite, periodates, peracids, peroxides, persulfates, elemental oxygen, and iodosylarenes to name a few. A comparison study of various oxidants in the epoxidation of *cis*-β-methyl styrene in the presence of added *N*-oxide routinely gave the product epoxide with the same sense and degree of enantioselectivity. In contrast, variable enantioselectivities were observed with different oxidants in the absence of *N*-oxides. These results suggest that a common and discrete Mn(V)oxo salen **8** intermediate is preferentially generated in the presence of added *N*-oxide regardless of terminal oxidant.[9] In the absence of added *N*-oxide, a low *ee* pathway perhaps involving either a catalytically active Mn(IV)peroxo-type salen complex **10** or a discrete Mn(IV)oxo salen complex **11** may be operative (Scheme 1.4.4).[10]

Scheme 1.4.4

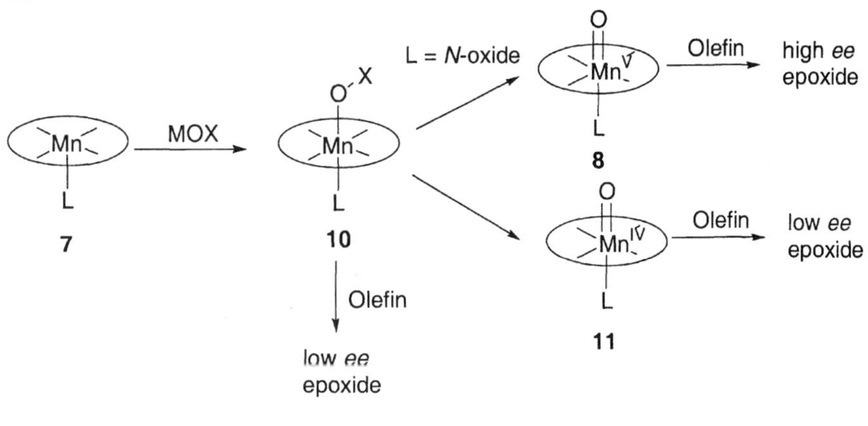

L = *N*-oxide

Olefin → high *ee* epoxide

Olefin → low *ee* epoxide

MOX

7

10

8

11

Olefin

low *ee* epoxide

While generation of a Mn(V)oxo salen intermediate **8** as the active chiral oxidant is widely accepted, how the subsequent C–C bond forming events occur is the subject of some debate. The observation of *trans*-epoxide products from *cis*-olefins, as well as the observation that conjugated olefins work best support a stepwise intermediate in which a conjugated radical or cation intermediate is generated. The radical intermediate **9** is most favored based on better Hammett correlations obtained with σ vs. σ$^+$.[11] In addition, it was recently demonstrated that ring opening of vinyl cyclopropane substrates produced products that can only be derived from radical intermediates and not cationic intermediates.[12]

A concerted [2 + 2] cycloaddition pathway in which an oxametallocycle intermediate is generated upon reaction of the substrate olefin with the Mn(V)oxo salen complex **8** has also been proposed (Scheme 1.4.5).[13] Indeed, early computational calculations coupled with initial results from radical clock experiments supported the notion.[14] More recently, however, experimental and computational evidence dismissing the oxametallocycle as a viable intermediate have emerged.[15] In addition, epoxidation of highly substituted olefins in the presence of an axial ligand would require a seven-coordinate Mn(salen) intermediate, which, in turn, would incur severe steric interactions.[16] The presence of an oxametallocycle intermediate would also require an extra bond breaking and bond making step to rationalize the observation of *trans*-epoxides from *cis*-olefins (Scheme 1.4.5).

Scheme 1.4.5

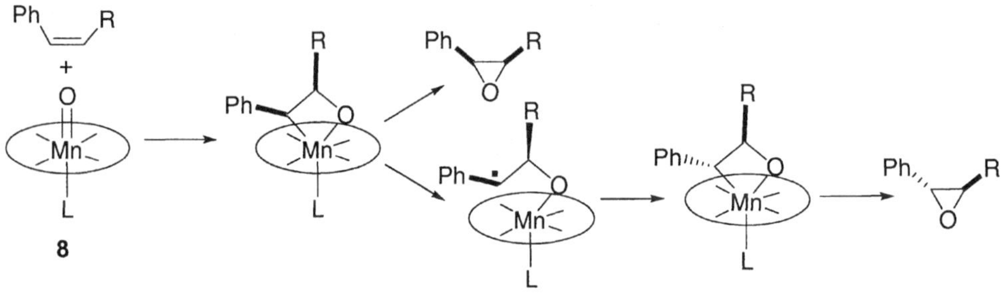

The second C–C bond forming step (step C), while occurring after the first irreversible *ee* determining step (step B), can affect the observed enantioselective outcome of the reaction.[17] If the radical intermediate collapses without rotation ($k_3 \gg k_4$, $k_5 \gg k_6$), then the observed *ee* would be determined by the first C–C bond forming step (k_1 vs. k_2), that is the facial selectivity (Scheme 1.4.6). However, if rotation is allowed followed by collapse, then the rate of both *trans* pathways (k_4 and k_6) will proportionally effect the observed *ee* of the *cis* epoxide (k_3 vs. k_5). Should bond rotation be permissible, the diastereomeric nature of the radical intermediates **9a** and **9b** renders the distinct possibility of different observed *ee*'s for *trans*-epoxides and *cis*-epoxides.

Scheme 1.4.6

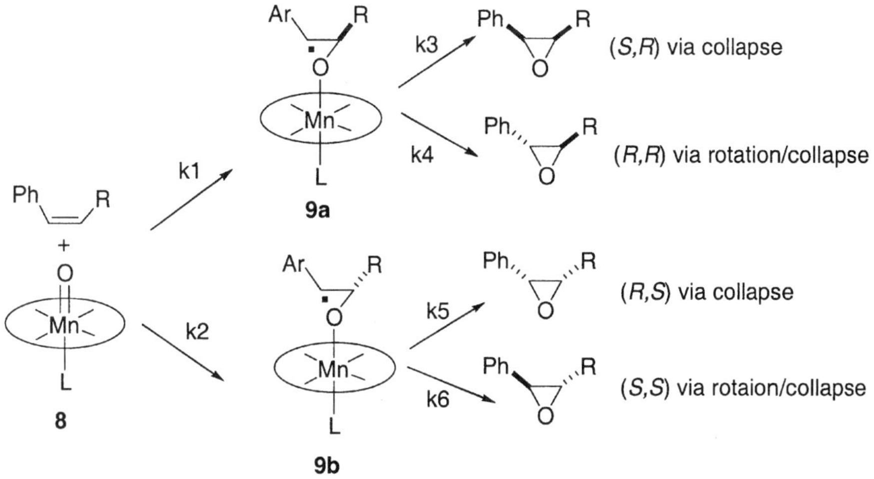

1.4.4   Variations and Improvements
1.4.4.1 Catalyst structure and design

While the Jacobsen catalyst is the most widely used catalyst, a number of other catalysts have been developed through the years that exhibit superior catalyst performance for a given substrate and/or substrate class.[18] The five most prominent catalysts developed to date are shown below, **12–16**. Note that these catalysts all are composed of salen ligands containing electron donating groups. Electron donating groups are thought to stabilize the putative Mn(V)oxo salen complex **8**, thus allowing for a less reactive, more selective catalyst.[19] In addition, the most successful catalysts contain bulky groups at the 3,3′ and 5,5′ position of the salen ring, thus forcing the approach of the olefin near the dissymmetric bridge.

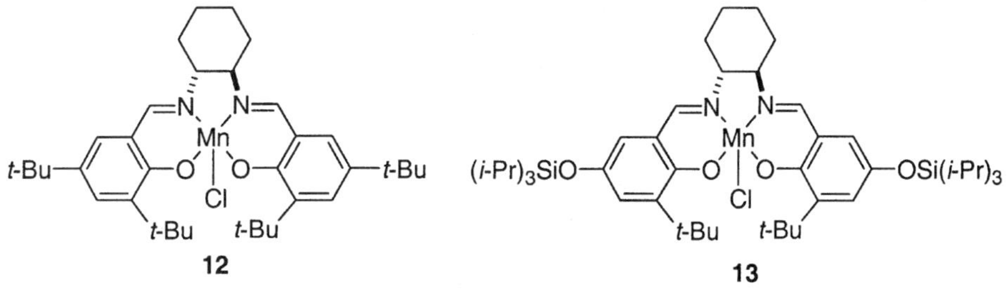

14　　　　　　　　　　**15**

16

1.4.4.2 Additives

One of the most significant developmental advances in the Jacobsen–Katsuki epoxidation reaction was the discovery that certain additives can have a profound and often beneficial effect on the reaction. Katsuki first discovered that *N*-oxides were particularly beneficial additives.[20] Since then it has become clear that the addition of *N*-oxides such as 4-phenylpyridine-*N*-oxide (4-PPNO) often increases catalyst turnovers, improves enantioselectivity, diastereoselectivity, and epoxides yields.[21] Other additives that have been found to be especially beneficial under certain conditions are imidazole and cinchona alkaloid derived salts (*vide infra*).

1.4.4.3 Terminal oxidants

Given that the rate determining step is oxidation of the Mn(III)salen complex by the terminal oxidant, it is not surprising that choice of terminal oxidant can have a profound effect on reaction rates and even enantioselectivity. The choice of oxidant can dictate the temperature at which the reaction can be run, which, in turn, can greatly affect selectivity. In addition, the choice of oxidant will determine if the reaction is homogeneous or heterogenous, and whether the reaction is anhydrous or not. A wide variety of terminal oxidants are now available, including sodium hypochlorite, periodates, peracids, peroxides, persulfates, periodates, elemental oxygen, and iodosylbenzenes, to name a few.

1.4.5 Synthetic Utility

1.4.5.1 Epoxidation of terminal olefins

Based on the mechanism of the Jacobsen–Katsuki epoxidation reaction, the *trans*-pathway would reduce the observed *ee*'s in the epoxidation of terminal olefins. Thus, in order to obtain epoxides with good enantioselectivity, high facial selectivity in the first C–O bond-forming event and high *cis/trans* selectivity in the second C–O bond-forming event must be obtained. In 1994, a low-temperature (–78°C) epoxidation protocol was developed by Jacobsen which resulted in increased enantioselectivity for a variety of olefins.[22] The reaction protocol involved using the combination of *m*-CPBA/NMO as terminal oxidant and additive. Most notably, terminal olefins were epoxidized with good enantioselectivities.[23] Deuterium labeling studies showed that the higher enantioselectivities observed in the epoxidation of terminal olefins were a result of a combination of better facial selectivity and higher *cis/trans* selectivity.

17	18	19	20
86% ee	85% ee	82% ee	80% ee

1.4.5.2 Epoxidation of cis-disubstituted olefins

Table 1.4.1. Epoxidation of *cis*-Olefins to give *cis*-Epoxides.

Entry	Substrate	Catalyst	Conditions[a]	ee	cis/trans	Reference
1	Ph Me	13	a	98	23	15
2	Ph CO₂*i*-Pr	13	b	96	10	11
3		14	c	98	NA	24b
4		13	a	98	NA	19b
5		12	c	98	NA	18
6		13	b[d]	64	NA	24a
7		12	b	94	NA	18

[a]Conditions: (a) 1–5 mol% catalyst, *m*-CPBA/NMO, –78°C, CH₂Cl₂. (b) 1–5 mol% catalyst, NaOCl, 4-phenylpyridine *N*-oxide, 0°C, CH₂Cl₂. (c) 1–5 mol% catalyst, NaOCl, pyridine *N*-oxide, 0°C, CH₂Cl₂. (d) Solvent = diethyl ether

Initial studies on the Jacobsen–Katsuki epoxidation reaction identified conjugated cyclic and acyclic *cis*-disubstituted olefins as the class of olefins best suited for the epoxidation reaction.[24] Indeed a large variety of *cis*-disubstituted olefins have been found to undergo epoxidation with a high degree of enantioselectivity.   2,2′-Dimethylchromene derivatives are especially good substrates for the epoxidation reaction. Table 1.4.1 lists a variety of examples with their corresponding reference.

Historically, the epoxidation of *trans*-olefins typically affords *trans*-epoxides with low enantioselectivity. In 1991, Jacobsen reported that *trans*-epoxides can be obtained from epoxidation of conjugated dienes and enynes containing a *cis*-olefin (Table 1.4.2).[25] Further studies by Jacobsen led to the discovery that addition of cinchona alkaloid-derived salts such as **21** coupled with the use of chlorobenzene as solvent resulted in the formation of the *trans*-diastereomer in high *ee* as the major epoxidation product from the epoxidation of *cis*-olefins.[26] Although it is still unclear as to the exact mechanism by which the cinchona alkaloid-derived salts favor the formation of *trans*-epoxides, clearly, these salts somehow extend the lifetime of the radical intermediate thus allowing for free rotation to occur.

Table 1.4.2. Epoxidation of *cis*-olefins to give *trans*-epoxides

Entry	Substrate	Catalyst	Additive	Solvent	*trans:cis*	*ee* (%)	Reference
1	n-C₆N₁₃ ... Me	12	None	CH_2Cl_2	62:28	90	25a
2	TMS ... (cyclohexyl)	12	None	CH_2Cl_2	84:16	98	25a
3	t-Bu ... Et	13	21	C_6H_5Cl	69:31	84	25b
4	Ph ... Ph	13	21	C_6H_5Cl	> 96:4	90	25b
5	Ph ... Me	13	12	C_6H_5Cl	95:5	81	25b

1.4.5.3 Epoxidation of trans-disubstituted olefins

Although *trans* epoxides can be obtained *via* epoxidation of acyclic *cis*-conjugated olefins under specified conditions, a direct method based on the epoxidation of *trans*-olefins would be valuable. The Katsuki group recently identified catalyst **15** as an efficient catalyst for the direct epoxidation of *trans*-olefins. Crucial to the success of the catalyst is the inherent adoption of a deeply folded conformation coupled with the use of chlorobenzene as solvent. While only a limited number of substrates have been examined to date using catalyst **15**, the results are very promising. For example, *trans*-β-methyl styrene is epoxidized in 91% *ee*, *trans*-β-*n*-butyl styrene in 95% *ee*, and *trans*-stilbene in 87% *ee*.

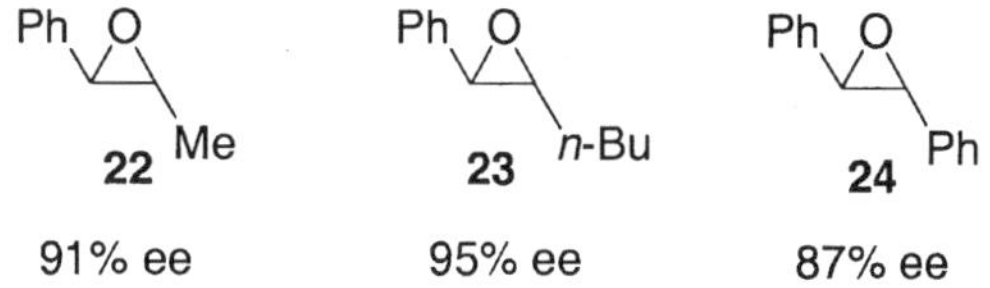

Ph O	Ph O	Ph O
22 Me	**23** *n*-Bu	**24** Ph
91% ee	95% ee	87% ee

1.4.5.4 Epoxidation of 1,1-disubstituted olefins

To date, no efficient and general Mn(III)salen-catalyst exists that effects epoxidation of 1,1-disubstituted olefins with good enantioselectivity.

1.4.5.5 Epoxidation of tri-substituted and tetra-substituted olefins

During the early development of the Jacobsen–Katsuki epoxidation reaction, it was clear that *trans*-disubstituted olefins were very poor substrates (slow reaction rates, low enantioselectivity) compared to *cis*-disubstituted olefins. The side-on approach model originally proposed by Groves for porphyrin epoxidation systems was used to rationalize the differences observed in the epoxidation of the *cis* and *trans*-disubstituted classes (Scheme 1.4.7).[27]

Scheme 1.4.7

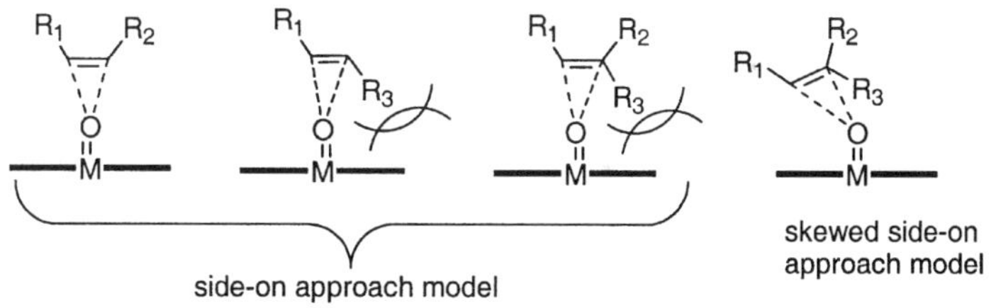

side-on approach model

skewed side-on approach model

This model predicts that tri-substituted and tetra-substituted olefins would also be poor substrates. Thus it was not until 1994 that a study in the epoxidation of higher substituted olefins appeared. Indeed Jacobsen revealed that tri-substituted olefins,[28] and even tetra-substituted olefins can be excellent substrates.[29,30] A new model was put forth that encompasses a skewed side-on approach of tri-substituted olefins to the Mn-oxo complex. The observation that certain tetrasubstituted olefins undergo epoxidation with good enantioselectivity suggests that further studies are needed in order to fully understand the transition state geometry of the catalyst and substrate.

The Jacobsen–Katsuki epoxidation reaction has found wide synthetic utility in both academia and industrial settings. As described previously, the majority of olefin classes, when conjugated, undergo Mn(salen)-catalyzed epoxidation in good enantioselectivity. In this section, more specific synthetic utilities are presented.

Table 1.4.3. Epoxidation of tri- and tetra-substituted olefins.

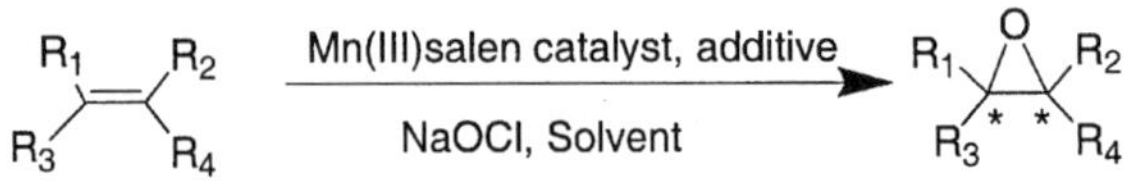

Entry	Substrate	Catalyst	Solvent	Yield (%)	Ee (%)	Reference
1	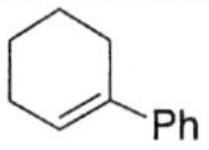	**12**	MTBE	69	93	28
2	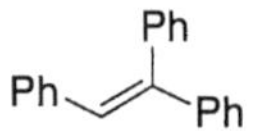	**12**	CH$_2$Cl$_2$	97	92	28
3	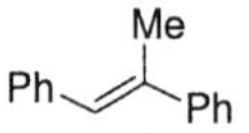	**12**	CH$_2$Cl$_2$	91	95	28
4	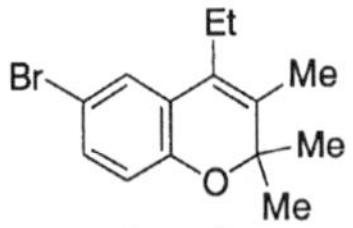	**16**	CH$_2$Cl$_2$	81	97	29
5	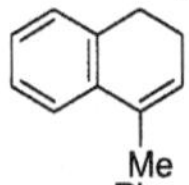	**14**	CH$_3$CN	41	96	30
6	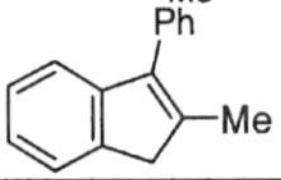	**12**	CH$_2$Cl$_2$	90	90	29

1.4.5.6 Preparation of α-hydroxy ketones and esters
Scheme 1.4.8

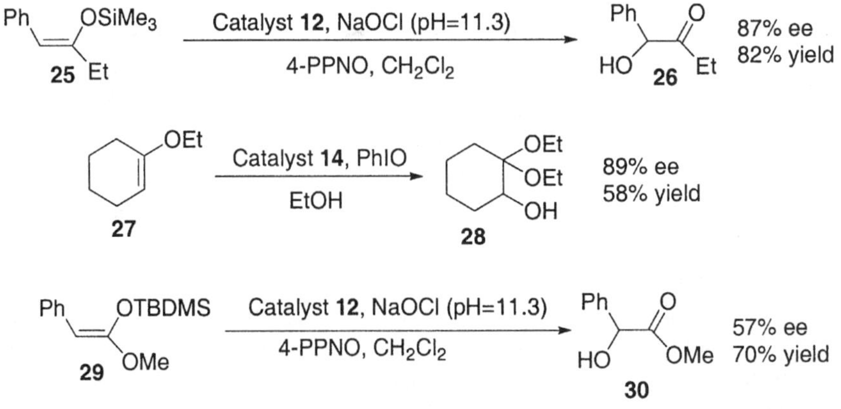

The first asymmetric Mn(salen)-catalyzed epoxidation of silyl enol ethers was carried out by Reddy and Thornton in 1992. Results from the epoxidation of various silyl enol ethers gave the corresponding keto-alcohols in up to 62% *ee*.[31] Subsequently, Adam[32,33] and Katsuki[34] independently optimized the protocol for these substrates yielding products in excellent enantioselectivity.

1.4.5.7 Kinetic resolution

1,2-Dihydronaphthalene is often used as a model olefin in the study of epoxidation catalysts, and very often gives product epoxides in unusually high *ee*'s. In 1994, Jacobsen discovered in his study on the epoxidation of 1,2-dihydronaphthalene that the *ee* of the epoxide increases at the expense of the minor enantiomeric epoxide.[35] Further investigation led to the finding that certain epoxides, especially cyclic aromatically conjugated epoxides, undergo kinetic resolution *via* benzylic hydroxylation up to a k_{rel} of **28** (Scheme 1.4.9).

Scheme 1.4.9

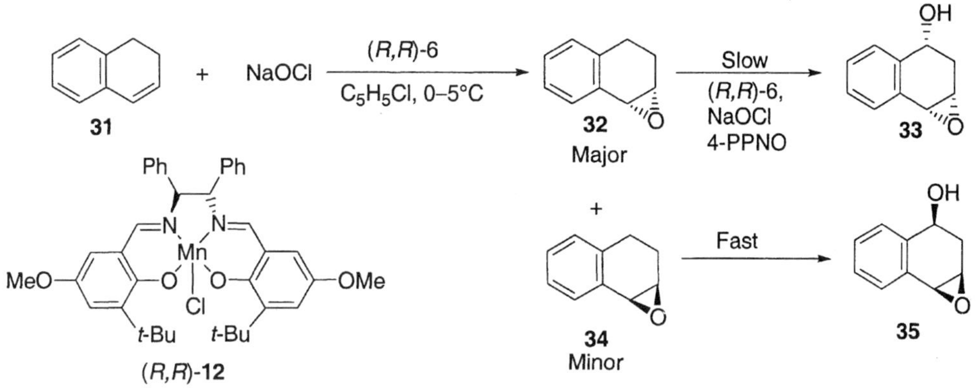

The first application of the Jacobsen–Katsuki epoxidation reaction to kinetic resolution of prochiral olefins was nicely displayed in the total synthesis of (+)-teretifolione B by Jacobsen in 1995.[36]

Scheme 1.4.10

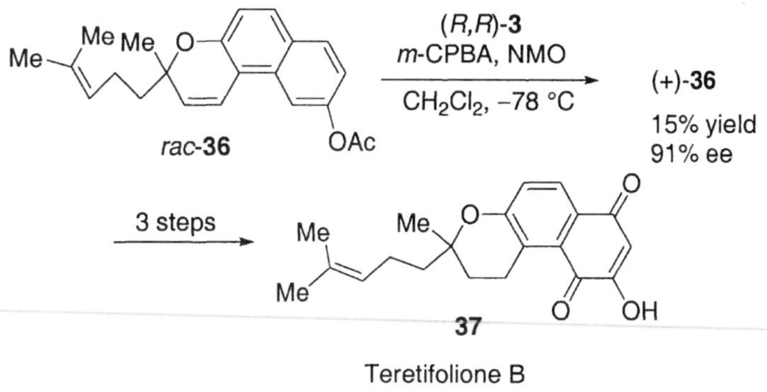

Teretifolione B

1.4.5.8 Jacobsen–Katsuki epoxidation reaction in total synthesis

Scheme 1.4.11

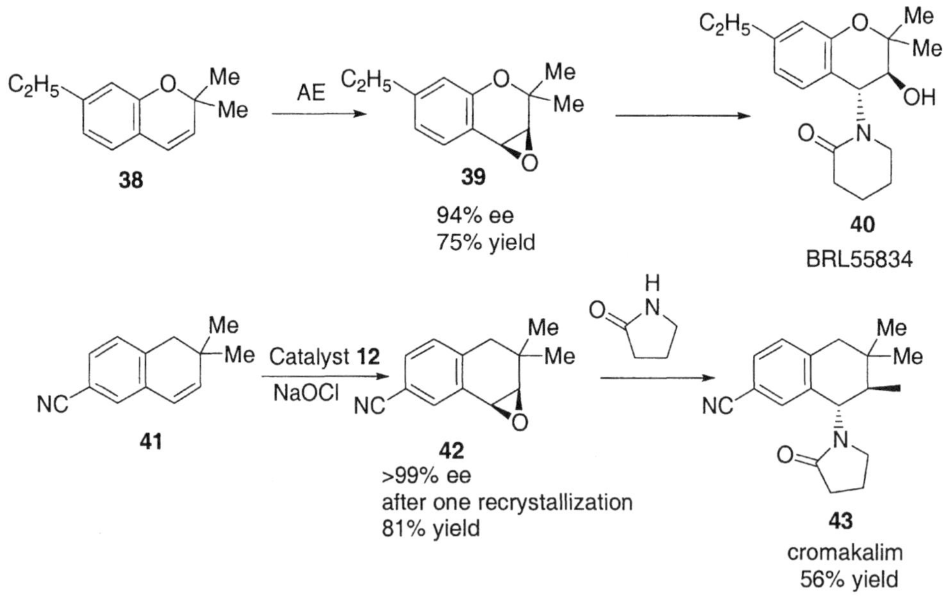

38

39
94% ee
75% yield

40
BRL55834

41

Catalyst **12**
NaOCl

42
>99% ee
after one recrystallization
81% yield

43
cromakalim
56% yield

The Jacobsen–Katsuki epoxidation reaction has been widely used for the preparation of a variety of structurally diverse complex molecules by both academia and the pharmaceutical industry. Summarized below are a few examples.

2,2-Dimethylchromene derivatives typically undergo epoxidation with excellent enantioselectivity (Scheme 1.4.11). The ring-opened product epoxides are potent pharmaceutical pharmacophores. BRL55834 (**40**) is a selective potassium channel activator and is comprised of a 2,2-dimethylchromanol-type structure.[37] Epoxidation of the 2,2-dimethylchromene starting material with catalyst **12** in the presence of catalytic amounts of an N-oxide additive gave the product epoxide in 94% ee and 75% yield. Regioselective ring opening of the chiral epoxide gave BRL55834 (**40**) in 81% yield. The anti-hypertensive agent cromoakalim has also been prepared in a similar fashion.[38]

The asymmetric epoxidation of electron-poor cinnamate ester derivatives was highlighted by Jacobsen in the synthesis of the Taxol side-chain. Asymmetric epoxidation of ethyl cinnamate provided the desired epoxide in 96% ee and in 56% yield. Epoxide ring opening with ammonia followed by saponification and protection provided the Taxol side-chain **46** (Scheme 1.4.12).

cis-2-Aminoindan-1-ol is a structural motif incorporated in various types of ligands for metal catalysis[6] and also comprises the right-hand portion of the HIV protease inhibitor Crixivan® (**50**).[39] Indene was epoxidized by catalyst **12** in 84–86% ee. The product epoxide was then ring-opened via Ritter reaction to provide cis-2-aminoindan-1-ol, which was then used in the eventual preparation of Crixivan® (**50**).

Scheme 1.4.12

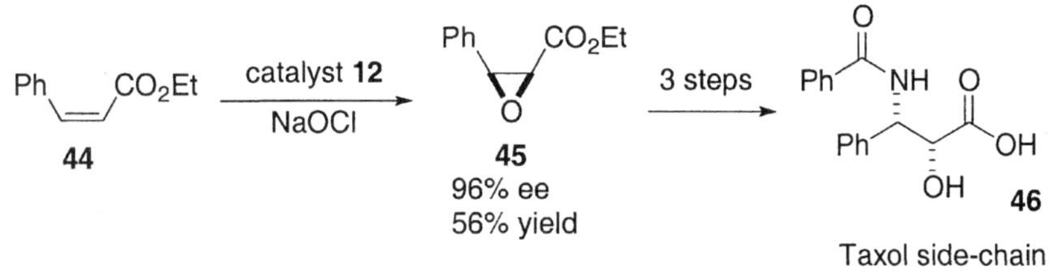

Scheme 1.4.13

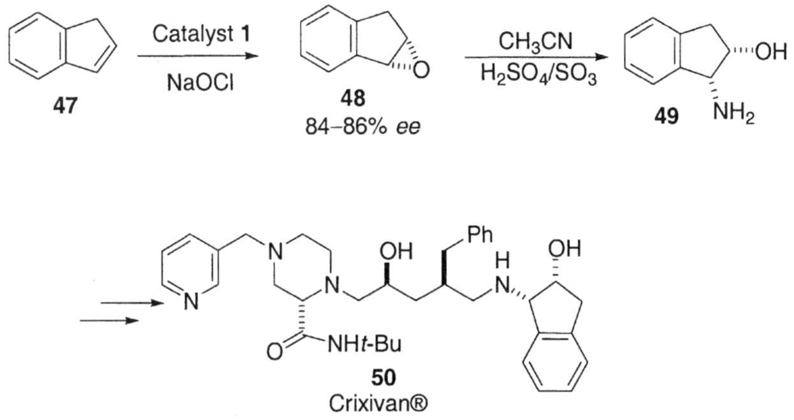

50
Crixivan®

CDP840 is a selective inhibitor of the PDE-IV isoenzyme and interest in the compound arises from its potential application as an antiasthmatic agent. Chemists at Merck & Co. used the asymmetric epoxidation reaction to set the stereochemistry of the carbon framework and subsequently removed the newly established C–O bonds.[40] Epoxidation of the trisubstituted olefin **51** provided the desired epoxide in 89% ee and in 58% yield. Reduction of both C–O bonds was then accomplished to provide CDP840.

Scheme 1.4.14

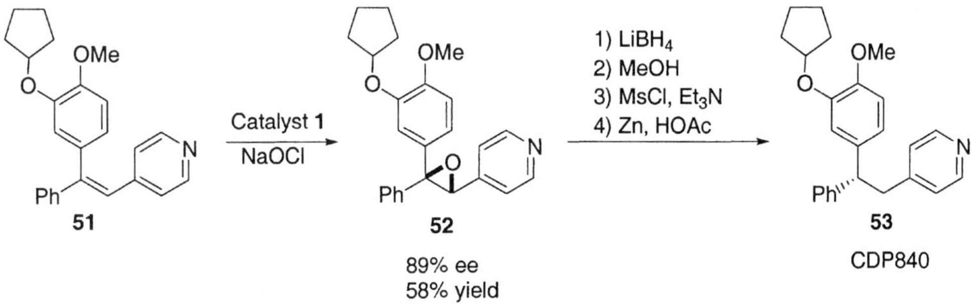

Boger *et al.* prepared Duocarmycin SA *via* asymmetric epoxidation of a cyclic olefin **54**.[41] The stereochemistry set by the epoxidation step was used for subsequent C–C bond forming reactions. Epoxidation of olefin **54** was carried at –78°C to provide the desired product in 92% *ee* and in 70% yield. Reduction of the benzylic C–O bond followed by partial reduction of the middle phenyl ring and a transannular spirocylization provided the activated cyclopropane which was carried on to Duocarmycin SA.

Scheme 1.4.15

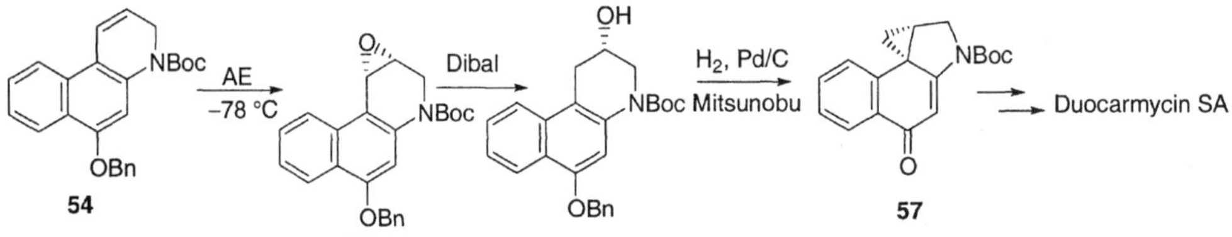

1.4.5   Experimental

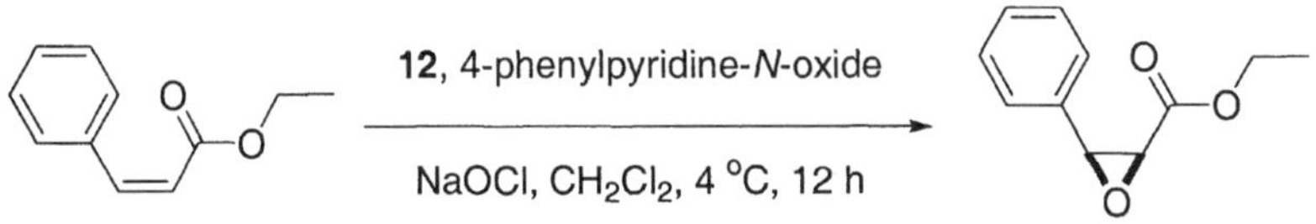

(2*R*, 3*R*)-Ethyl-3-phenylglycidate (45).
To a solution of *cis*-ethyl cinnamate (**44**, 352 mg, 85% pure, 1.70 mmol) and 4-phenylpyridine-*N*-oxide (85.5 mg, 29 mol%) in 1,2-dichloromethane (4.0 mL) was added catalyst **12** (38.0 mg, 3.5 mol%). The resulting brown solution was cooled to 4°C and then combined with 4.0 mL (8.9 mmol) of pre-cooled bleach solution. The two-phase mixture was stirred for 12 h at 4°C. The reaction mixture was diluted with methyl-*t*-butyl ether (40 mL) and the organic phase separated, washed with water (2 × 40 mL), brine (40 mL), and then dried over Na_2SO_4. The drying agent was removed by filtration the mother liquors concentrated under reduce pressure. The resulting residue was purified by flash chromatography (silica gel, pet ether/ether = 87:13 v/v) to afford a fraction enriched in *cis*-epoxide (**45**, *cis/trans*: 96:4, 215 mg) and a fraction enriched in *trans*-epoxide (*cis/trans*: 13:87, 54 mg). The combined yield of pure epoxides was 83%. *ee* of the *cis*-epoxide was determined to be 92% and the *trans*-epoxide to be 65%.

1.4.7   References

1.   For reviews see: [R] (a) Jacobsen, E. N.; Wu, M. H. in *Comprehensive Asymmetric Catalysis*, Jacobsen, E. N.; Pfaltz, A.; Yamamoto, H. Eds.; Springer: New York; 1999, Chapter 18.2. [R] (b) Katsuki, T. *Coord. Chem. Rev.* **1995**, *140*, 189–214. [R] (c) E. N. Jacobsen, in *Comprehensive Organometallic Chemistry II*, Eds. G. W. Wilkinsin, G. W.; Stone, F. G. A.; Abel, E. W.; Hegedus, L. S., Pergamon, New York, 1995, vol 12, Chapter 11.1.
2.   Jacobsen, E.N.; Zhang, W.; Muci, A. R.; Ecker, J. R.; Deng, L. *J. Am. Chem. Soc.* **1991**, *113*, 7063.
3.   (a) Zhang, W.; Loebach, J. L.; Wilson, S. R.; Jacobsen, E. N. *J. Am. Chem. Soc.* **1990**, *112*, 2801. (b) Irie, R.; Noda, K.; Ito, Y.; Matsumoto, N.; Katsuki, T. *Tetrahedron Lett.* **1990**, *31*, 7345.
4.   (a) Larrow, J. F.; Jacobsen, E. N.; Gao, Y.; Hong, Y.; Nie, X.; Zepp, C. M. *J. Org. Chem.* **1994**, *59*, 1939. (b) Larrow, J.; Jacobsen, E. N. *Org. Syn.* **1997**, *75*, 1.

5. Sasaki, H.; Irie, R.; Hamada, T.; Suzuki, K.; Katsuki, T. *Tetrahedron* **1994**, *50*, 11827.
6. [R] Senananyake, C. H. *Aldrichimica Acta* **1998**, *31*, 3.
7. Evidence supporting the Mn(V)oxo salen intermediate: a) Feichtinger, D.; Plattner, D. A. *Angew. Chem. Int. Ed. Engl.* **1997**, *36*, 1718. B) Feichtinger, D.; Plattner, D. A. *Chem. Eur. J.* **2001**, *7*, 591.
8. Hughes, D. L.; Smith, G. B.; Liu, J.; Dezeny, G. C.; Senanayake, C. H.; Larsen, R. D.; Verhoeven, T. R.; Reider, P. J. *J. Org. Chem.* **1997**, *62*, 2222.
9. Michael Palucki, Thesis, 1995 Harvard Univeristy.
10. (a) Adam, W.; Mock-Knoblauch, C.; Saha-Möller, C. R.; Herderich, M. *J. Am. Chem. Soc.* **2000**, *122*, 9685. (b) Bryliakov, K. P.; Babushkin, D. E.; Talsi, E. *J. Mol. Catal. A.* **2000**, *158*, 19. c) Paul J. Pospisil, Thesis, 1995 Harvard Univsersity.
11. Jacobsen, E. N.; Deng, L.; Furukawa, Y.; Martinez, L. *Tetrahedron* **1994**, *50*, 4323.
12. Adam, W.; Roschmann, K. R.; Saha-Möller, C. R.; Seebach, D. *J. Am. Chem. Soc.* **2002**, *124*, 5068.
13. (a) Norrby, P.-O.; Linde. C.; Åkermark, B. *J. Am. Chem. Soc.* **1995**, *117*, 11035. (b) Nuguchi, Y.; Irie, R.; Fukuda, T.; Katsuki, T. *Tetrahedron Lett.* **1996**, *37*, 4533.
14. Linde, C.; Arnold, M.; Norrby, P.-O.; Åkermark, B. *Angew. Chem. Int. Ed. Engl.* **1997**, *36*, 1723.
15. (a) Cavallo, L.; Jacobsen, H. *Angew. Chem. Int. Ed. Engl.* **2000**, *39*, 589. (b) Adam, W.; Roschmann, K. J.; Saha-Möller, C. R.; Seebach, D. *J. Am. Chem. Soc.* **2002**, *124*, 5068.
16. Finney, N. S.; Pospisil, P. J.; Chang, S.; Palucki, M.; Konsler, R. G.; Hansen, K. B.; Jacobsen, E. N. *Angew. Chem. Int.Ed. Engl.* **1997**, *36*, 1720.
17. Zhang, W.; Lee, N. H.; Jacobsen, E. N. *J. Am. Chem. Soc.* **1994**, *116*, 425.
18. Jacobsen, E. N.; Zhang, W.; Muci, A. R.; Ecker, J. R.; Deng, L. *J. Am. Chem. Soc.* **1991**, *113*, 7063.
19. (a) Jacobsen, E. N.; Guler, M. L.; Zhang, W. *J. Am. Chem. Soc.* **1991**, *113*, 6703. (b) Palucki, M.; Finney, N. S.; Pospisil, P. J.; Guler, M. L.; Ishida, T.; Jacobsen, E. N. *J. Am. Chem. Soc.* **1988**, *120*, 948.
20. Irie, R.; Itoh, Y.; Katsuki, T. *Synlett*, **1991**, 265.
21. Senanayake, C. H.; Smith, G. B.; Ryan, K. M.; Fredenburgh, L. E.; Liu, J.; Roberts, F. E.; Hughes, D. L.; Larsen, R. D.; Verhoeven, T. R.; Reider, P. J *Tetrahedron Lett.* **1996**, *37*, 3271.
22. Palucki, M., Pospisil, P. J.; Zhang, W.; Jacobsen, E. N. *J. Am. Chem. Soc.* **1994**, *116*, 9333.
23. Palucki, M.; McCormick, G. J.; Jacobsen, E. N. *Tetrahedron Lett.* **1995**, *36*, 5457.
24. (a) Chang, S.; Heid, R. M.; Jacobsen, E. N. *Tetrahedron Lett.* **1994**, *35*, 669. (b) Sasaki, H.; Irie, R.; Hamada, T.; Suzuki, K.; Katsuki, T. *Tetrahedron* **1994**, *50*, 11827.
25. (a) Lee, N, H.; Jacobsen, E. N. *Tetrahedron Lett.* **1991**, *32*, 6533. (b) Chang, S.; Lee, N. H.; Jacobsen, E. N. *J. Org. Chem.* **1993**, *58*, 6939.
26. Chang, S.; Galvin, J. M.; Jacobsen, E. N. *J. Am. Chem. Soc.* **1994**, *116*, 6937.
27. Groves, J. T.; Myers, R. S. *J. Am Chem. Soc.* **1985**, *108*, 2309.
28. Brandes, B. D.; Jacobsen, E. N. *J. Org. Chem.* **1994**, *59*, 4378.
29. Brandes, B. D.; Jacobsen, E. N. *Tetrahedron Lett.* **1995**, *36*, 5123.
30. Fukuda, T.; Irie, R.; Katsuki, T. *Synlett* **1995**, 197.
31. Reddy, D. R.; Thornton, E. R. *J. Chem. Soc., Chem. Commun.* **1992**, 172.
32. Adam, W.; Fell, R. T.; Mock-Knoblauch, C.; Saha-Möller, C. R. *Tetrahedron Letters* **1996**, *37*, 6531
33. Adam. W.; Fell, R. T.; Stegmann, V. R. *J. Am. Chem. Soc.* **1998**, *120*, 708.
34. Fukuda, T.; Katsuki, T. *Tetrahedron Letters* **1996**, *37*, 4389.
35. Larrow, J. F.; Jacobsen, E. N. *J. Am. Chem. Soc.* **1994**, *116*, 12129.
36. Vander Velde, S. L.; Jacobsen, E. N. *J. Org. Chem.* **1995**, *60*, 5380.
37. Buckle, D. R.; Eggleston, D. S.; Pinto, I. L.; Smith, D. G.; Tedder, J. M. *Bio. Med. Chem. Lett.* **1992**, *2*, 1161.
38. Lee, N. H.; Muci, A. R.; Jacobsen, E. N. *Tetrahedron Lett.* **1991**, *32*, 5055.
39. Vacca, J.; Dorsey, B.; Schlief, W. A.; Levin, R.; Mcdaniel, S.; Darke, P.; Zugay, J.; Quintero, J.; Blahy, O.; Roth, E.; Sardana, V.; Schlabach, A.; Graham, P.; Condra, J.; Gotlib, L.; Holloway, M.; Lin, J.; Chen, I.; Vastag, K.; Ostovic, D.; Anderson, P. S.; Emini, E. A.; Huff, J. R. *Proc. Natl. Acad. Sci. USA* **1994**, *91*, 4096.
40. Lynch, J. E.; Choi, W.-B.; Churchill, H. R. O.; Volante, R. P.; Reamer, R. A.; Ball, R. G. *J. Org. Chem.* **1997**, *62*, 9223.
41. Boger, D. L.; McKie, J. A.; Boyce, C. W. *Synlett* **1997**, 515.

Michael Palucki

1.5   Paterno–Büchi Reaction

1.5.1   Description
The Paterno–Büchi reaction[1-4] is the photo-catalyzed electrocyclization of a carbonyl **1** with an alkene **2** to form polysubstituted oxetane ring systems **3**.

1.5.2   Historical Perspective
In 1909, Paterno and Chieffi[5] noted that mixtures of tri- or tetra-substituted olefins and aldehydes formed trimethylene oxides when exposed to sunlight. Büchi[6] later repeated Paterno's experiments by irradiating 2-methyl-2-butene in the presence of benzaldehyde, butyraldehyde, or acetophenone and rigorously purifying and identifying the resulting products. The reaction thus bears the name of its two primary pioneers and has come to represent any photo-catalyzed [2 + 2] electrocyclization of a carbonyl and an alkene.

1.5.3   Mechanism
The mechanism of the Paterno–Büchi reaction is not well understood, and while a general pathway has been proposed and widely accepted, it is apparent that it does not represent the full scope of reactions. Büchi originally proposed that the reaction occurred by light catalyzed stimulation of the carbonyl moiety **1** into an excited singlet state **4**. Inter-system crossing then led to a triplet state diradical **5** which could be quenched by olefinic radical acceptors. Intermediate diradical **6** has been quenched or trapped by other radical acceptors and is generally felt to be on the reaction path of the large majority of Paterno–Büchi reactions. Diradical **6** then recombines to form product oxetane **3**.

The stability of diradical **6** is often cited as the rationalization for the regiochemistry and relative stereochemistry of the Paterno–Büchi reaction. This "most stable diradical" theory explains the formation of the sterically disfavored 2,3-substituted oxetanes. Quenching of the triplet diradical occurs in such a fashion as to generate the more stable alkyl radical pair and can be predicted using typical rules for radical stabilization. Consideration of the configuration of diradical **6** is the most powerful tool in predicting product distribution and can give a qualitative estimation of products ratios. Generation of 2,3-substituted oxetanes is the major utility of the Paterno–Büchi reaction and it should be noted that this substitution pattern is opposite that of thermal or acid activated [2 + 2] electrocyclizations.

Relative stereochemistry is also dictated by diradical stability but is highly subject to substituent effects. In the case of open chain alkenes, stereochemical information regarding the E-/Z-substitution is lost. In his work with vinyl sulfides, Smith[7] isolated solely *trans* oxetane **9** regardless of the *E/Z* constitution of sulfide **8**. The conformation of diradical **6** dominates the selectivity and the result is the more stable 3,4-*trans* substituted oxetanes. Five- and six-membered cyclic alkenes react exclusively to generate the *cis* products while larger rings react as the acyclic alkenes do.[8] There are some examples of stereospecific Paterno–Büchi reactions in which the stereochemistry of the acyclic alkenes is retained,[3] particularly if the alkene concentration is very high. These reactions are generally believed to spring from singlet state carbonyls with the high alkene concentration serving to accelerate the coupling rate above that of intersystem crossing. The relative stereochemistry of the 2,3-substituents is controlled by the electronic requirements of radical recombination. Griesbeck[9] has demonstrated and rationalized the preference for endo-selectivity for Paterno–Büchi couplings with 2,3-dihydrofuran **11** with aryl and alkyl aldehydes resulting in the substitution pattern in **12**. He noted that selectivity ranged from 1:1 (non-existent) to greater than 98:2 depending on the cycloalkene used. Other studies[10,11] with acyclic alkenes have demonstrated a preference for the corresponding *cis* geometry. The selectivity is reversed for conjugated alkenes such as 1,3-dienes, styrene, and furan **14** which prefer the thermodynamically more stable 2,3-*trans* (*exo*) oxetanes such as **15**.

It is evident from the exceptions noted that the mechanism proposed above does not fully capture the pathways open to the Paterno–Büchi reaction. A great deal of effort has been devoted to deconvoluting all of the possible variants of the reaction. Reactions via singlet state carbonyls, charge-transfer paths, pre-singlet exciplexes, and full electron transfer paths have all been proposed.[3] Unfortunately, their influence on product distribution is not understood and as such represent mechanistic oddities rather than the rule.

1.5.4   Variations and Improvements

Improvements of the Paterno–Büchi reaction have primarily focused on expanding the scope of substrates and developing asymmetric variants to control stereochemistry. The Paterno–Büchi reaction is extremely permissive of substitution on the carbonyl coupling partner, although radical stabilizing groups are preferred for obvious reasons. Alkyl-, alkynyl-, and aryl-ketones as well as glyoxylates have been used successfully. The majority of Paterno–Büchi reactions use aldehydes or ketones as coupling partners, but the reaction is not limited to just these carbonyls; Cantrell[12] and Miyamoto[13] demonstrated that arene carboxylic acid esters **20** are suitable substrates.

A variety of alkenes can serve as radical quenching partners. Because of the wide scope of acceptable alkene partners, relatively complex alkenes can be used successfully. Mono-, di-, tri-, and even tetra-substituted alkenes will undergo Paterno–Büchi reactions. Heteroatom substituted alkenes include enol ethers, silyl enol ethers, enol acetates, acyl and alkyl enamines, and alkenyl sulfides.[7,14,15] The alkenyl sulfides are particularly interesting as the reactions proceed considerably faster than equivalent enol ethers. Cyclic alkenes are often used, particularly furan because of its synthetically useful *exo* face selectivity (*vide supra*) and the convenient functional handles it provides for subsequent reactions.[16] Araki[17,18] has even used glucals as coupling partners. Allenes such as **16**[19] and dienes[20] will also undergo Paterno–Büchi reactions, albeit with low chemoselectivity. There are also reports of alkynes as radical coupling partners to form oxetenes **21**, although these species are usually unstable and degrade quickly.[21]

The most valuable characteristic of the Paterno–Büchi reaction is the ability to set multiple stereocenters in one reaction and the development of diastereocontrolled reactions has been a major theme of research concerning this reaction. Stereocontrol can be envisioned to spring from either the carbonyl or the alkene and be controlled by either the substrate directly or by a chiral auxiliary. Little success has been achieved in substrate-induced selection by the carbonyl; the most successful results were produced by

Zamojski and Jarosz[22] using cyclic ketone **23** with furan that generated a 75:25 ratio of diastereomers **24** and **25**.

23

1) hv (furan)

2) TsOH (Et$_2$O)

X = OH, Y= 3-furyl **24**
X = 3-furyl, Y = OH **25**

24:25 = 75:25

Despite failures in carbonyl substrate-induced selection, auxiliary-induced selection showed good results. Scharf and coworkers[23] performed a systematic study to show that phenylglyoxylates coupled to 8-phenylmenthol **26** gave excellent control of subsequent Paterno–Büchi reactions. With the 8-phenylmenthol moiety blocking the *Si*-face of the glyoxylate and the heterocyclic alkene **27** strongly preferring *endo*-addition, only diastereomer **28** was isolated. The chiral auxiliary could then be recovered in good yield by reduction of the oxetane-substituted ester.

27

hv

C$_6$H$_6$

99%

*ROOC

Ph

26   **28**

Ironically, auxiliary-induced control via the alkene failed to generate synthetically useful selectivities, but direct substrate-induced control did. In particular, chiral silyl enol ethers with stereocenters in the γ-position allowed the synthesis of enantiomerically pure oxetanes. Bach[24] noted that 1,3-allylic strain in molecules such as **29** locked the conformation as shown and caused approaching diradicals to prefer attack opposite R^L with varying selectivities. Depending on the relative steric bulk of R^S and R^L, diastereoselectivity ranging from 61:39 to 95:5 was achieved. It should be noted that chiral groups appended to the α-carbon failed to impart selectivity.

29

1.5.5 *Synthetic Utility*

The oxetane functional unit is a rare but occurring group in natural products and appears both as end products as well as synthetic intermediates.[3,4] Paterno–Büchi reactions can be used to insert oxetanes directly into biologically active compounds, as in the example below by Just.[25] The novel nucleotide oxetanocin **35** is synthesized using a

Paterno–Büchi reaction between 2-methylfuran **31** and an α-hydroxyaldehyde **30** to form the core oxetane. Two of the three stereocenters set in the electrocyclization appear in the final product while the third is selectively reversed with anomeric assistance.

Given the relatively rare appearance of oxetanes in natural products, the more powerful functionality of the Paterno–Büchi reaction is the ability to set the relative stereochemistry of multiple centers by cracking or otherwise derivitizing the oxetane ring. Schreiber noted that Paterno–Büchi reactions of furans with aldehydes followed by acidic hydrolysis generated product **37**, tantamount to a *threo* selective Aldol reaction.[26] This process is referred to as "photochemical Aldolization". Schreiber uses this selectivity to establish the absolute stereochemistry of the fused tetrahydrofuran core **44** of the natural product asteltoxin.[27]

1.5.6 Experimental

(1(R,S),5(R,S))-6(R,S)-n-Octyl-2,7-dioxabicyclo[3.2.0]hept-3-ene (47).[26]
Nonyl aldehyde (32.66 g, 0.23 mol) and furan (200 mL, 187.2 g, 2.75 mol) were mixed in a 250-mL photolysis flask equipped with a quartz immersion well containing a Vycor filter and a 450-W Hanovia Lamp. The system was kept at –20° C with an isopropyl alcohol bath cooled by a Cryocool Immersion Cooler (CC100). Nitrogen was bubbled throughout the duration of the reaction, and the solution was stirred vigorously. Additional furan (150 mL, 140.4 g, 2.06 mol) was added during the course of the reaction. TLC analysis indicated completion of the reaction after 20 h. After evaporation of excess furan ^{13}C and ^{1}H NMR analysis of the resultant oil (48.70 g, ca. 100%) indicated the desired photoadduct had been formed, without contamination from unreacted nonyl aldehyde.

1.5.7 References

1.   [R] Carless, H. A. J. In *Synthetic Organic Photochemistry*; Horspool, W. M., Ed.; Plenum Press: New York, 1984, pp 425.
2.   [R] Jones, G. In *Organic Photochemistry*; Padwa, A., Ed.; Dekker: New York, 1981, pp 1.
3.   [R] Porco, J. A.; Schreiber, S. L. In *Comprehensive Organic Synthesis*; Trost, B., Ed.; Pergamon Press: Oxford, 1991; Vol. 5, pp 151.
4.   Bach, T. *Synthesis* **1998**, *5*, 683.
5.   Paterno, E.; Chieffi, G. *Gazz. Chim. Ital.* **1909**, *39*, 341.
6.   Buchi, G.; Inman, C. G.; Lipinsky, E. S. *J. Am. Chem. Soc.* **1954**, *76*, 4327.
7.   Morris, T. H.; Smith, E. H.; Walsh, R. *J. Chem. Soc., Chem. Commun.* **1987**, 964.
8.   Shima, K.; Sakai, Y.; Sakurai, H. *Bull. Chem. Soc. Jpn.* **1971**, *44*, 215.
9.   Griesbeck, A. G.; Mauder, H.; Stadtmuller, S. *Acc. Chem. Res.* **1994**, *27*, 70.
10.   Bach, T. *Angew. Chem., Int. Ed.* **1996**, *35*, 884.
11.   Schroeter, S. H.; Orlando, C. M., Jr. *J. Org. Chem.* **1969**, *34*, 1181.
12.   Cantrell, T. S.; Allen, A. C. *J. Org. Chem.* **1989**, *54*, 135.
13.   Miyamoto, T.; Shigemitsu, Y.; Odaira, Y. *Chem. Commun.* **1969**, 1410.
14.   Bach, T. *Synlett* **2000**, *12*, 1699.
15.   Bach, T. *Liebigs Ann./Recueil* **1997**, 1627.
16.   D'Auria, M.; Emanuele, L.; Racioppi, R.; Romaniello, G. *Current Organic Chemistry* **2003**, *4*, 1443.
17.   Matsuura, K.; Araki, Y.; Ishido, Y.; Kushida, K. *Carbohydr. Res.* **1973**, *29*, 459.
18.   Matsuura, K.; Araki, Y.; Ishido, Y. *Bull. Chem. Soc. Jpn.* **1972**, *45*, 3496.
19.   Arnold, D. R.; Glick, A. H. *Chem. Commun.* **1966**, 813.
20.   Kubota, T.; Shima, K.; Toki, S.; Sakurai, H. *Chem. Commun.* **1969**, 1462.
21.   Friedrich, L. E.; Bower, J. D. *J. Am. Chem. Soc.* **1973**, *95*, 6869.
22.   Jarosz, S.; Zamojski, A. *Tetrahedron* **1982**, *38*, 1453.
23.   Koch, H.; Runsink, J.; Scharf, H.-D. *Tetrahedron Lett.* **1983**, *24*, 3217.
24.   Bach, T.; Jodicke, K.; Kather, K.; Frohlich, R. *J. Am. Chem. Soc.* **1997**, *119*, 2437.
25.   Hambalek, R.; Just, G. *Tetrahedron Lett.* **1990**, *31*, 5445.
26.   Schreiber, S. L.; Hoveyda, A. H.; Wu, H.-J. *J. Am. Chem. Soc.* **1983**, *105*, 660.
27.   Schreiber, S. L.; Satake, K. *Tetrahedron Lett.* **1986**, *27*, 2575.

Christopher M. Liu

1.6   Sharpless–Katsuki Epoxidation

1.6.1   Description

The Sharpless–Katsuki asymmetric epoxidation reaction (most commonly referred by the discovering scientists as the AE reaction) is an efficient and highly selective method for the preparation of a wide variety of chiral epoxy alcohols.[1] The AE reaction is comprised of four key components: the substrate allylic alcohol, the titanium isopropoxide pre-catalyst, the chiral ligand diethyl tartrate, and the terminal oxidant *tert*-butyl hydroperoxide. The reaction protocol is straightforward and does not require any special handling techniques. The only requirement is that the reacting olefin contains an allylic alcohol.

Scheme 1.6.1

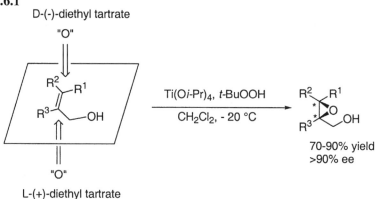

The AE reaction has emerged as one of the most utilized asymmetric transformations in organic synthesis because of its many appealing attributes. First, the starting allylic alcohols are either commercially available or are easily prepared. Second, the generality of the reaction is substantial given the fact that almost any substitution pattern on the allylic alcohol is permissible, and that a wide variety of functional groups are tolerated. Third, the catalyst, ligand, and oxidant are cheap and readily available. Fourth, the product epoxides are valuable chiral building blocks that can be easily elaborated into more complex molecules. And lastly, the stereochemical outcome of the product epoxide can be predicted with almost complete certainty based on the pneumonic diagram shown in Scheme 1.6.1. These attributes have made the AE reaction one of the most widely utilized asymmetric catalytic transformations to date.

1.6.2 Historical Perspective

In 1980, Katsuki and Sharpless communicated that the epoxidation of a variety of allylic alcohols was achieved in exceptionally high enantioselectivity with a catalyst derived from titanium(IV) isopropoxide and chiral diethyl tartrate.[2] This seminal contribution described an asymmetric catalytic system that not only provided the product epoxide in remarkable enantioselectivity, but showed the immediate generality of the reaction by examining 5 of the 8 possible substitution patterns of allylic alcohols; all of which were epoxidized in >90% ee. Shortly thereafter, Sharpless and others began to illustrate the broad scope of the reaction and subsequently demonstrated its wide utility in organic synthesis. Indeed, the AE reaction has emerged as one of the most widely utilized asymmetric transformations. In addition to the asymmetric epoxidation of prochiral allylic alcohols, the reaction protocol has been successfully applied to the kinetic resolution of secondary allylic alcohols[3] and the desymmetrization of meso-bis allylic alcohols.[4] The importance of the discovery and development of this remarkable reaction, along with other asymmetric catalytic reactions, was recognized in part by the presentation of the 2002 Nobel prize in chemistry to K. B. Sharpless, R. Noyori, and J. Knowles.

1.6.3 Mechanism

Elucidating the mechanism of the AE reaction has been the focus of much effort given the importance of the AE reaction in synthetic organic chemistry.[5] Fundamental questions that need to be addressed include the following: what is the rate law/rate determining step, what is the structure of the active catalyst complex, and what are the underlying factors that impart such high enantioselectivity? Answers to all of these questions have been largely addressed through careful experimental[6] and computational studies.[7]

Scheme 1.6.2

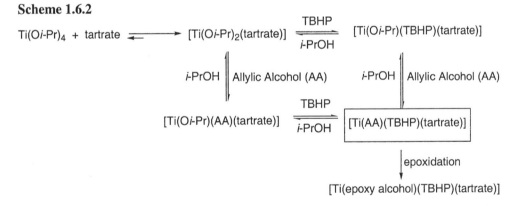

Sharpless *et al.* have shown that a dynamic process exists between various relevant titanium species during the AE reaction via rapid alkoxide exchange.[8] First, complexation of the bidentate tartrate ligand occurs rapidly with Ti(i-OPr)4 with concomitant release of 2 equivalents of iPrOH. Subsequent ligand substitution of another i-OPr alkoxide with TBHP produces the Ti(O-iPr)(TBHP)(tartrate) complex. Likewise, ligand substitution of an O-*i*-Pr alkoxide with the allylic alcohol (AA) produces the Ti(O-*i*-Pr)(AA)(tartrate). Final substitution of either complex with AA or TBHP respectively produces the "loaded" complex Ti(TBHP)(AA)(tartrate). Rate determining intramolecular transfer of the peroxide oxygen to the coordinated allylic alcohol provides Ti(O-*t*-Bu)(EA)(tartrate) where EA is the product epoxy alcohol. Ligand displacement of the product epoxy alcohol and tBuOH with another allylic alcohol and TBHP regenerates the loaded complex. The ability of alkoxides to rapidly ligand exchange on the titanium center allows for the titanium catalyst to effectively catalyze the desired reaction with good efficiency.

$$\text{rate} = k \frac{[\text{allylic alcohol}][\text{Ti(O-iPr)}_2\text{tartrate}][\text{TBHP}]}{[\text{inhibitor alcohol}]^2}$$

A great deal of kinetic information on the AE reaction has been obtained. The rate of reaction is first order in allylic alcohol, Ti(O-iPr)$_2$(tartrate), and TBHP. In addition, the rate is inversely-square dependent on isopropoxide. This reflects the required replacement of two isopropoxide ligands on Ti(O-iPr)$_2$(tartrate) with TBHP and the allylic alcohol. The rate-determining step is oxygen transfer from the peroxide to the olefin.

Studies in varying the amount of tartrate per titanium have shown that exactly 1 equivalent of tartrate per titanium is required for good catalytic activity and selectivity.[6a] The active catalyst is proposed to be the dimeric complex Ti$_2$(tartrate)$_2$(O-i-Pr)$_4$. Sharpless has shown that a protocol using a ratio 1:1.2 equivalents of Ti(O-i-Pr)$_4$ to tartrate ligand is optimal for the AE reaction. This ratio maximizes the concentration of the more highly active and enantioselective catalyst Ti$_2$(tartrate)$_2$(O-i-Pr)$_4$, while suppressing the concentration of the less selective and less catalytically active Ti$_2$(tartrate)(O-i-Pr)$_6$/Ti(O-i-Pr)$_4$ complexes as well as the catalytically inactive Ti$_2$(tartrate)$_2$ complex.

Scheme 1.6.3

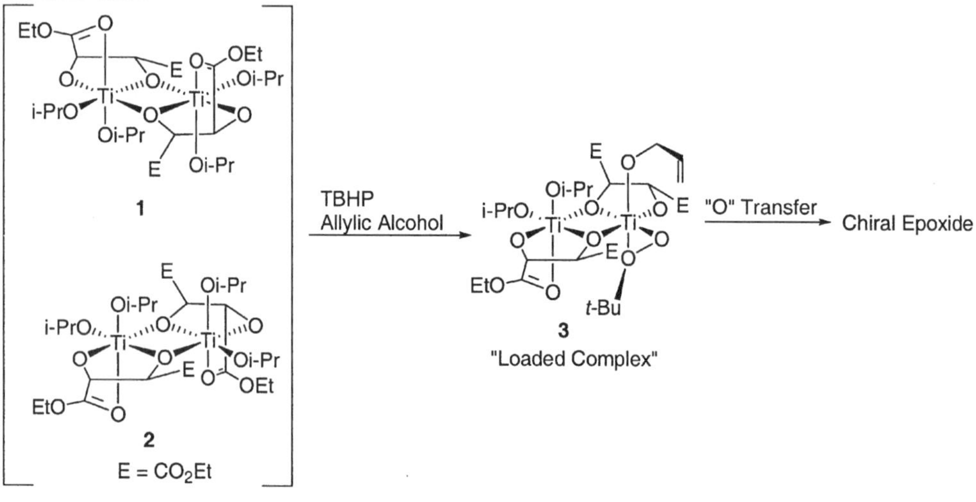

Determining the placement and orientation of tartrate ligand, allylic alcohol, and TBHP about the Ti-catalyst during the critical oxygen transfer step is crucial to understanding why such high enantioselectivities are achieved with a large variety of allylic alcohols. Any proposed model must take into effect the absolute stereochemistry of the product as predicted by the pneumonic diagram in Scheme 1.6.1, since virtually all AE reactions abide by this diagram. Structural evidence supports a dimeric titanium tartrate species as the predominant species in solution.[9] NMR shows the tartrate carbonyls are equivalent thus suggesting fast equilibrium of the two structurally degenerate complexes **1** and **2** (Scheme 1.6.4). Loading of the catalyst with TBHP and allylic alcohol gives the "loaded complex" **3**. The structure of the proposed loaded complex is based on minimization of stereoelectronic interaction of the TBHP and allylic alcohol with the adjacent tartrate ligand and the trajectory required for oxygen transfer of the distal oxygen to the olefin.[10] Relative rate studies of electronically diverse olefins reveals that the olefin acts as the nucleophile. This model also agrees with the predictive outcome as shown by pneumonic diagram in Scheme 1.6.10. It can not be unequivocally ruled out that both titanium species are participating in the epoxidation step. A mechanism involving zwitterionic titanium species has also been proposed.[11]

1.6.4 Variations and Improvements

The AE reaction is general and efficient, thus few substantial improvements have been reported. The most significant improvement to the original conditions is the addition of activated molecular sieves.[12] The addition of activated molecular sieves allows for almost all AE reaction to be performed under catalytic conditions (5–10 mol%). The role of the molecular sieves is thought to sequester any adventitious water or water that may be generated during the course of the reaction via side reactions.

The ability to perform the AE reaction under catalytic conditions via the addition of molecular sieves has greatly enhanced the synthetic utility of the reaction. For water-soluble epoxy alcohols, the catalytic conditions are beneficial for both enantioselectivity and isolated yield. In addition, epoxy alcohols that are susceptible to ring opening via nucleophilic substitution at the C-3 position also greatly benefit from catalytic conditions, since the substitution reaction is known to be promoted by Ti(IV) species.[13]

Table 1.6.1. Epoxidation of 2-methyl-2-propene-1-ol; beneficial effects of catalytic reactions and derivatization.[14]

Entry	Catalyst (%)	Ligand (%)	R =[a]	% yield	% ee[b]
1	Ti(O-t-Bu)$_4$ (100)	DET (100)	H	-	85
2	Ti(O-i-Pr)$_4$ (27)	DET (27)	H	32	94
3	Ti(O-t-Bu)$_4$ (7.6)	DET (10)	H	47	>95
4	Ti(O-i-Pr)$_4$ (5)	DIPT (6)	PNB	78	92(98)
5	Ti(O-i-Pr)$_4$ (5)	DIPT (6)	Tos	69	95
6	Ti(O-i-Pr)$_4$ (5)	DIPT (6)	Nps	60	92

[a] PNB is p-nitrobenzoate, Tos is tosylate, and Nps is 2-naphthylenesulfonyl.
[b] number in parenthesis is the ee after recrystallization.

In conjunction with the addition of molecular sieves, Sharpless *et al.* as also developed an *in situ* derivatization of product epoxy alcohols that were previously difficult to isolate.[15] The derivatization of the product has been accomplished via esterification or sulfonylation of the alcohol functionality. The derivatization is possible only under catalytic conditions given the overwhelming presence of isopropoxide from stoichiometric amounts of Ti(O-i-Pr)$_4$ and the presence of the diol ligand diethyl tartrate.

Advantages of the *in situ* generation include ease of isolation and ee upgrades of crystalline products.[16] Table 1.6.1 shows the beneficial effect of performing the AE reaction under catalytic conditions as well as *in situ* derivatization.

A number of reaction variables or parameters have been examined. Catalyst solutions should not be prepared and stored since the resting catalyst is not stable to long term storage. However, the catalyst solution must be aged prior to the addition of allylic alcohol or TBHP. Diethyl tartrate and diisopropyl tartrate are the ligands of choice for most allylic alcohols. TBHP and cumene hydroperoxide are the most commonly used terminal oxidant and are both extremely effective. Methylene chloride is the solvent of choice and Ti(i-OPr)$_4$ is the titanium precatalyst of choice. Titanium (IV) *t*-butoxide is recommended for those reactions in which the product epoxide is particularly sensitive to ring opening from alkoxide nucleophiles.[13]

1.6.5   *Synthetic Utility*
The AE reaction has been applied to a large number of diverse allylic alcohols. Illustration of the synthetic utility of substrates with a primary alcohol is presented by substitution pattern on the olefin and will follow the format used in previous reviews by Sharpless but with more current examples. Epoxidation of substrates bearing a chiral secondary alcohol is presented in the context of a kinetic resolution or a match versus mismatch with the chiral ligand. Epoxidation of substrates bearing a tertiary alcohol is not presented, as this class of substrate reacts extremely slowly.

1.6.5.1   *Allyl alcohol*
Epoxidation of the simplest allylic alcohol, allyl alcohol **7**, is achieved in 88–92% ee with yields of 50–60% using diisopropyl tartrate as ligand.[12] *In situ* derivatization of the product glycidol **8** via esterification, sulfonylation, or ring opening with nucleophile is an attractive alternative to isolating glycidol.

Scheme 1.6.4

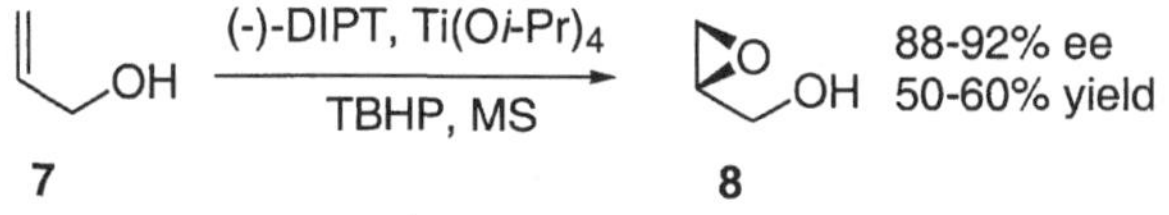

1.6.5.2   *2-Substituted allyl alcohols*
In general, 2-substituted allylic alcohols are epoxidized in good enantioselectivity. Like glycidol, however, the product epoxides are susceptible to ring opening via nucleophilic attack at the C-3 position. Results of the AE reaction on 2-methyl-2-propene-1-ol followed by derivatization of the resulting epoxy alcohol are shown in Table 1.6.1. Other examples are shown below.

Scheme 1.6.5

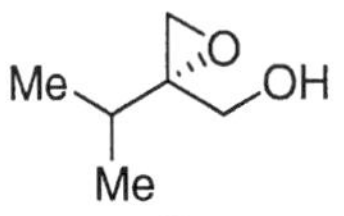

9[17]	10[18]	11[19]
95% ee	>95% ee	88% ee
85% yield	74% yield	77% yield

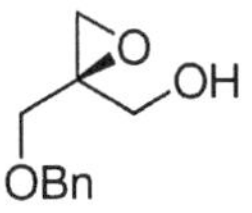

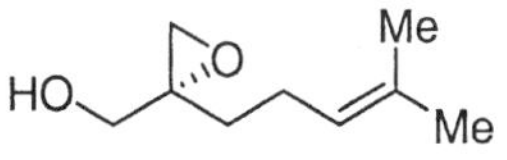

1.6.5.3   3E-Substituted allyl alcohols

This class of substrate is one of the most widely used class of substrates for the AE reaction. The starting allylic alcohols are readily prepared and are generally epoxidized in high enantioselectivity. Epoxide **12** was obtained in high enantioselectivity and utilized in the rapid preparation of (*S*)-Fenfluramine. Epoxide **13** was prepared under the standard AE conditions and then oxidized with DDQ to provide the corresponding 4-hydroxy-2,3-unsaturated carbonyl. These type of structures are found in numerous polyketides, including macrosphelide A. Epoxide **14** was prepared via the AE reaction and utilized in the synthesis of (−)-swainsonine. Epoxides **15** and **16** were utilized in the synthesis of baikiain and the acyl side chain segment of polyoxypeptin A respectively. Finally, epoxide **17** was prepared and subsequently incorporated in the formal synthesis of the proteasome inhibitor (+)-lactacystin.

Scheme 1.6.6

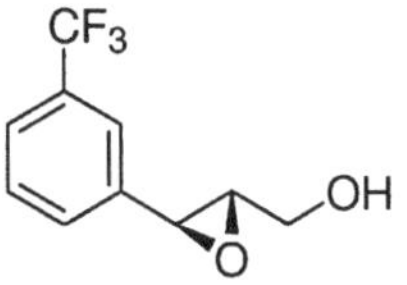

12[20]	13[21]	14[22]
96% ee	96% ee	92% ee
82% yield	97% yield	52% yield

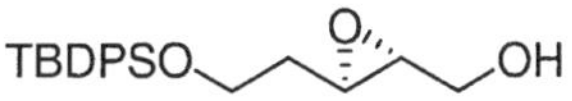

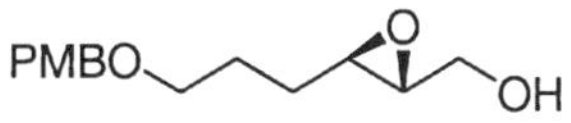

15[23]	16[24]	17[25]
93% ee	95% ee	>95% ee
84% yield	80% yield	94% yield

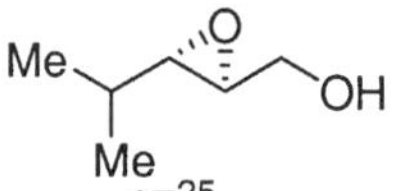

1.6.5.4   3Z-Substituted allyl alcohols

This class of substrate is the only real problematic substrate for the AE reaction. The enantioselectivity of the AE reaction with this class of substrate is often variable. In addition, rates of the catalytic reactions are often sluggish, thus requiring stoichiometric loadings of Ti/tartrate. Some representative product epoxides from AE reaction of 3Z-substituted allyl alcohols are shown below.

Scheme 1.6.7

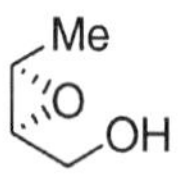

18[17]	**19**[26]	**20**[27]	**21**[28]	**22**[29]
92% ee	66% ee	25% ee	92% ee	78% ee
68% yield	54% yield	75% yield	84% yield	61% yield

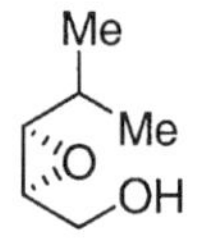

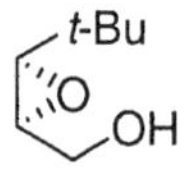

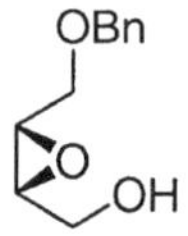

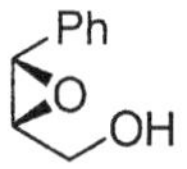

1.6.5.5   2,3E-Disubstituted allyl alcohols

As with *3E*-substituted allyl alcohols, *2,3E*-substituted allyl alcohols are epoxidized in excellent enantioselectivity. Examples of AE reactions of this class of substrate are shown below. Epoxide **23** was utilized to prepare chiral allene oxides, which were ring opened with TBAF to provide chiral α-fluoroketones. Epoxide **24** was used to prepare 5,8-disubstituted indolizidines and epoxide **25** was utilized in the formal synthesis of macrosphelide A. Epoxide **26** represents an AE reaction on the very electron deficient 2-cyanoallylic alcohols and epoxide **27** was an intermediate in the total synthesis of (+)-varantmycin.

Scheme 1.6.8

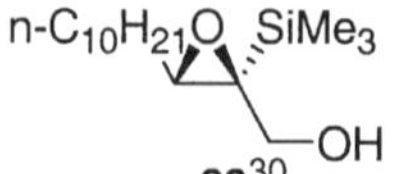

23[30]	**24**[31]	**25**[21]
97% ee	98% ee	96% ee
94% yield	84% yield	95% yield

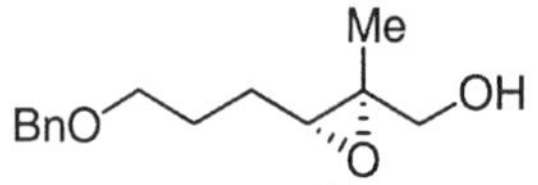

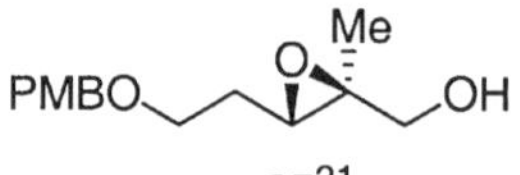

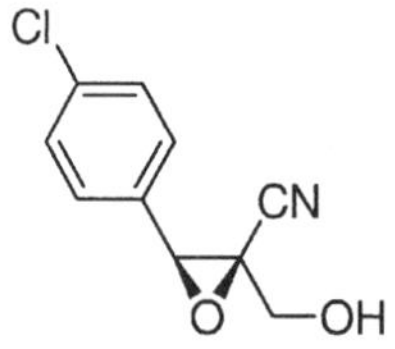

26[32]	**27**[33]
98% ee	90% ee
64% yield	98% yield

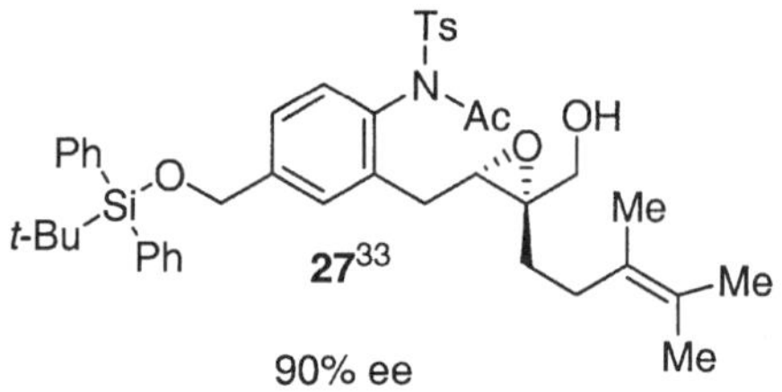

1.6.5.6   2,3Z-Disubstituted allyl alcohols

There are only limited examples of AE reactions on 2,3Z-substituted allyl alcohols. This may be due in part to the difficulty involved in selectively preparing the starting allylic alcohol.

Scheme 1.6.9

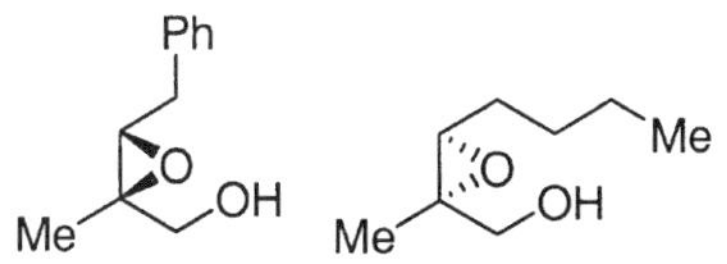

28[34]
91% ee
90% yield

29[2]
89% ee
80% yield

Although the limited examples of AE reactions on 2,3Z-substituted allyl alcohols appear to give product epoxides in good enantioselectivity, the highly substituted nature of these olefins can have a deleterious effect on the reactivity. For example, Aiai has shown that the 2,3E-substituted allyl alcohol **30** can be epoxidized with either (−)-DET or (+)-DET in good yields and enantioselectivity. However, the configurational isomer **32** is completely unreactive using (−)-DET, even after a 34 h reaction time.

Scheme 1.6.10

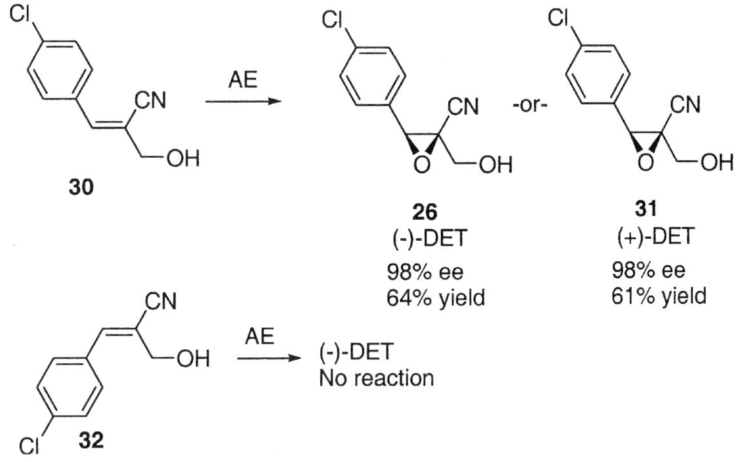

30

AE

26
(-)-DET
98% ee
64% yield

-or-

31
(+)-DET
98% ee
61% yield

32

AE   (-)-DET
No reaction

1.6.5.7   3,3-Disubstituted allyl alcohols

The 3,3-disubstituted allyl alcohols generally undergo the AE reaction with good enantioselectivity. Epoxide **33** was prepared by the AE reaction on an enyne and subsequently elaborated to the core unit of the non-nucleoside reverse transcriptase inhibitor taurospongin A. The highly functionalized epoxide **34** was prepared in good enantioselectivity and yield using (+)-DET as ligand. Interestingly, epoxidation of the isomeric substrate in which the methyl group is E to the alcohol with (−)-DET gave the diastereomeric product **35** in only 31% ee and in 11% yield. Not surprisingly, the reaction was relatively sluggish. Epoxide **36** was obtained in a 9:1 diastereomeric ratio under standard AE conditions and was utilized in the synthesis of *ent*-nakamural A. Epoxide **37** was prepared and trapped *in situ* as the trityl ether. Crystallization of the trityl ether increased the ee to >99%. Epoxide **38** was prepared and utilized in the synthesis of the tricyclic core of Phomactin A.

Scheme 1.6.11

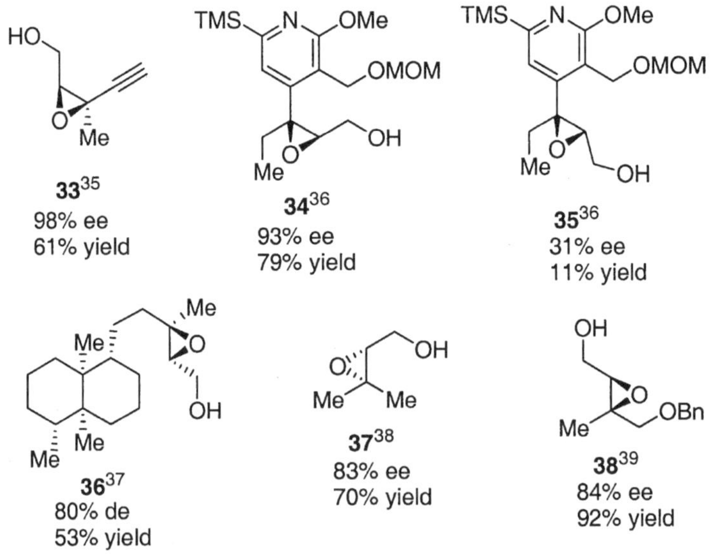

33[35]
98% ee
61% yield

34[36]
93% ee
79% yield

35[36]
31% ee
11% yield

36[37]
80% de
53% yield

37[38]
83% ee
70% yield

38[39]
84% ee
92% yield

1.6.5.8   2,3,3-Trisubstituted allyl alcohols

The AE reactions on 2,3,3-trisubstituted allyl alcohols have received little attention, due in part the limited utility of the product epoxides.   Selective ring opening of tetrasubstituted epoxides are difficult to achieve. Epoxide **39** was prepared using stoichiometric AE conditions and were subsequently elaborated to Darvon alcohol. Epoxides **40** and **41** were both prepared in good selectivity and subsequently utilized in the preparation of (-)-cuparene and the polyfunctoinal carotenoid peridinin, respectively.

Scheme 1.6.12

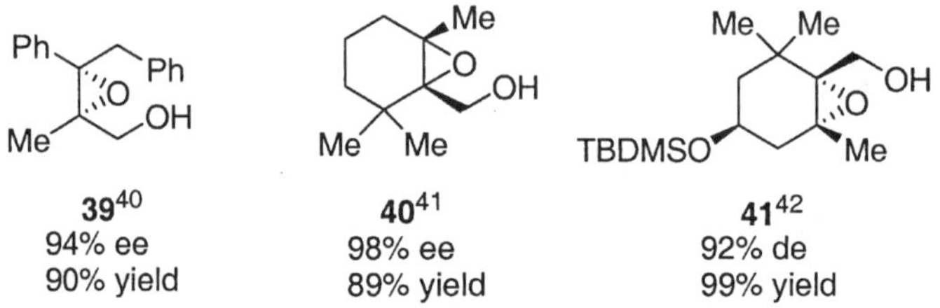

39[40]
94% ee
90% yield

40[41]
98% ee
89% yield

41[42]
92% de
99% yield

1.6.5.9   Kinetic resolution of chiral allylic alcohols

Given the universal predictive value of the pneumonic diagram in Scheme 1.6.1 coupled with the high enantioselectivity observed with a wide variety of allylic alcohols, it is not surprising that the AE reaction is sensitive to preexisting chirality on the allylic alcohol substrate. Indeed, allylic secondary alcohols can undergo effective kinetic resolution to provide enantiomerically enriched allylic secondary alcohols and diastereomerically enriched product epoxides.[43] The relative rates (k_{rel}) of each enantiomer of the starting allylic alcohol with a given enantiomer of tartrate ligand can be significantly different. Thus using 0.6 equiv of TBHP, and carrying out the reaction to 55 ± 5% conversion, chiral secondary allylic alcohols can be obtained in excellent enantioselectivity and in moderate yield.[44] If the epoxy alcohol is desired, then 0.45 equivalents of TBHP is often used. Shown below are a number of racemic allylic alcohols with their relative rates of each enantiomer.[45]

Scheme 1.6.13

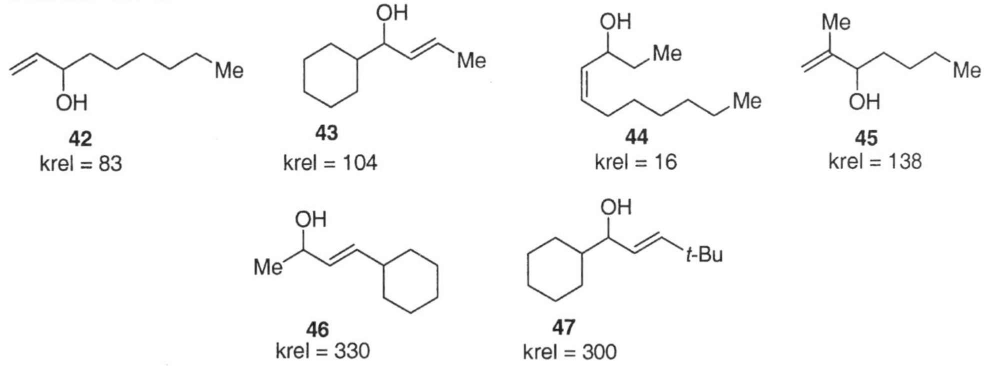

42
krel = 83

43
krel = 104

44
krel = 16

45
krel = 138

46
krel = 330

47
krel = 300

The application of the AE reaction to kinetic resolution of racemic allylic alcohols has been extensively used for the preparation of enantiomerically enriched alcohols and allyl epoxides.   Allylic alcohol **48** was obtained via kinetic resolution of the racemic secondary alcohol and utilized in the synthesis of rhozoxin D. Epoxy alcohol **49** was obtained via kinetic resolution of the enantioenriched secondary allylic alcohol (93% ee). The product epoxy alcohol was a key intermediate in the synthesis of (−)-mitralactonine. Allylic alcohol **50** was prepared via kinetic resolution of the secondary alcohol and the product utilized in the synthesis of (+)-manoalide. The mono-tosylated 3-butene-1,2-diol is a useful C$_4$ building block and was obtained in 45% yield and in 95% ee via kinetic resolution of the racemic starting material.

Scheme 1.6.14

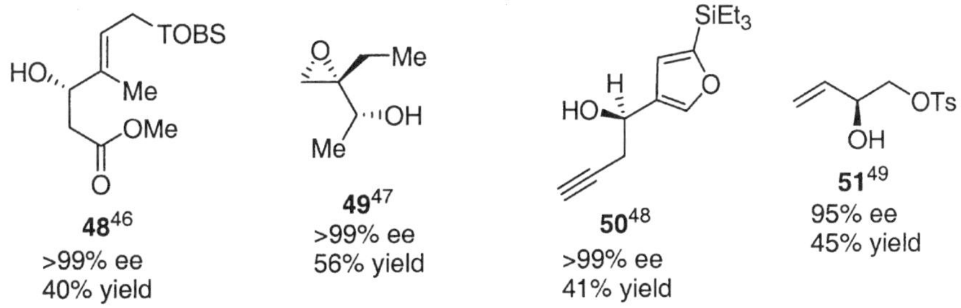

48[46]
>99% ee
40% yield

49[47]
>99% ee
56% yield

50[48]
>99% ee
41% yield

51[49]
95% ee
45% yield

1.6.5.10:   *Unique synthetic applications*
Sharpless and Masumune have applied the AE reaction on chiral allylic alcohols to prepare all 8 of the L-hexoses.[50] AE reaction on allylic alcohol **52** provides the epoxy alcohol **53** in 92% yield and in >95% ee. Base catalyze Payne rearrangement followed by ring opening with phenyl thiolate provides diol **54**. Protection of the diol is followed by oxidation of the sulfide to the sulfoxide via *m*-CPBA, Pummerer rearrangement to give the *gem*-acetoxy sulfide intermediate and finally reduction using Dibal to yield the desired aldehyde **56**. Horner-Emmons olefination followed by reduction sets up the second substrate for the AE reaction. The AE reaction on optically active **57** is reagent

controlled. This four-step reiterative two-carbon extension cycle was used to prepare all 8 L-hexoses.

Scheme 1.6.15

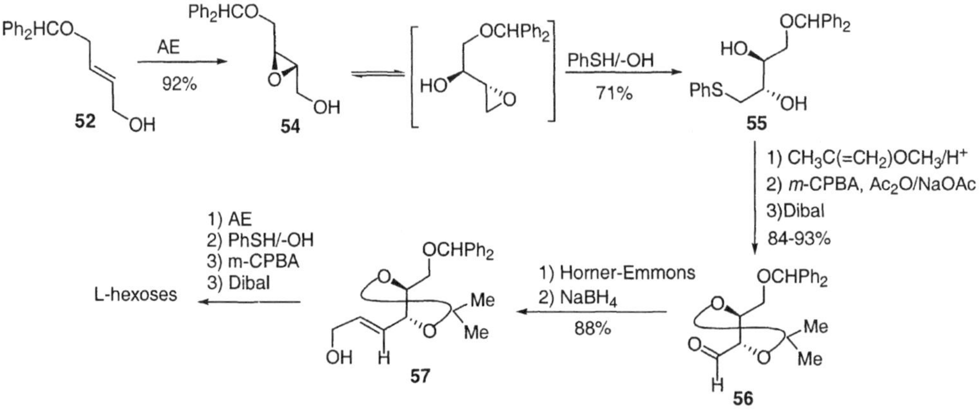

Desymmetrization of meso-bis-allylic alcohols is an effective method for the preparation of chiral functionalized intermediates from meso-substrates. Schreiber *et al* has shown that divinyl carbonyl **58** is epoxidized in good enantioselectivity.[51] However, because the product epoxy alcohols **59** and **60** also contain a reactive allylic alcohol that are diastereomeric in nature, a second epoxidation would occur at different rates and thus affect the observed ee for the first AE reaction and the overall de. Indeed, the major diastereomeric product epoxide **59** resulting from the first AE is less reactive in the second epoxidation. Thus, high de is easily obtainable since the second epoxidation removes the minor diastereomer.

Scheme 1.6.16

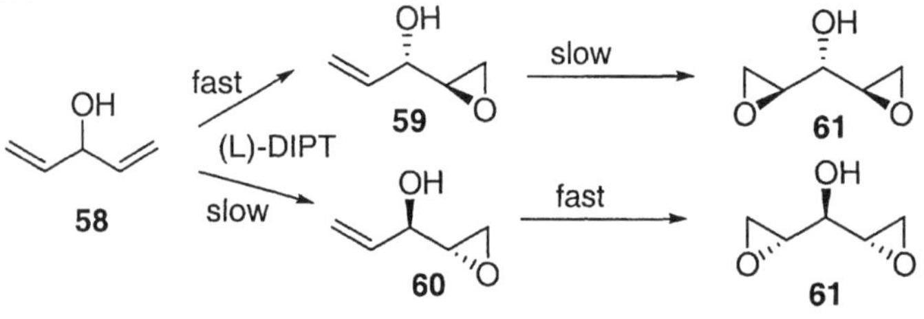

Time (h)	%ee	% de
3	84	92
24	93	99.7
140	>97	>99.7

1.6.6 Experimental[12]

Me ～～～～OH → Ti(O-*i*Pr)$_4$, (+)-DET / *t*-BuOOH, CH$_2$Cl$_2$ / -20 °C → Me ～～～～⟨O⟩OH

62

63

crude: 88% yield, 92.3 % ee
recrystallization: 73% yield, > 98% ee

An oven dried 1-L three-necked round-bottomed flask equipped with a magnetic stir bar, pressure equalizing addition funnel, thermometer, nitrogen inlet and bubbler was charged with 3.0 g of 4A powdered, activated molecular sieves and 350 mL of dry CH$_2$Cl$_2$. The flask was cooled to –20°C. L-(+)-Diethyl tartrate (1.24 g, 6.0 mmol) and Ti(O-i-Pr)$_4$ (1.49 mL, 1.42 g, 5.00 mmol, via syringe) were added sequentially with stirring. The reaction mixture was stirred at –20°C as TBHP (39 mL, 200 mmol, 5.17 M in isooctane) was added through the addition funnel at a moderate rate (over 5 min). The resulting mixture was stirred at –20°C for 30 min. (E)-2-octenol (**62**, 12.82 g, 100 mmol) dissolved in 50 mL of CH$_2$Cl$_2$ was then added drop wise through the same addition funnel over a period of 20 min while maintaining the temperature at –15 to –20°C. The mixture was stirred for an additional 3.5 h at –15 to –20°C. The reaction was allowed to warm to room temperature and poured into a beaker containing ferrous sulfate heptahydrate (33 g, 120 mmol), citric acid monohydrate (11 g, 60 mmol) and 100 mL of deionized water. The two-phase mixture was stirred for 10 min and separated. The aqueous phase was extracted with two 30 mL portions of ether. The combined organic layers were treated with 10 mL of precooled (0°C) solution of 30% NaOH (w/v) in saturate brine and stirred for 1 h at 0°C. To this was added 50 mL of water and the aqueous phase extracted with 2 × 50 mL of ether. The combined organic layers were dried over sodium sulfate, filtered and concentrated. Crude product yield was 88% with a 92.3% ee. Two crystallizations in petroleum ether affords the product **63** in > 98% ee and in 73% isolated yield.

1.6.7 References

1. For excellent reviews see: [R] (a) Johnson, R. A.; Sharpless, K. B. In *Comprehensive Organic Synthesis*; Trost, B. M., Ed,; Pergamon Press: New York, 1991; Vol. 7, Chapter 3.2. [R] (b) Johson, R. A.; Sharpless, K. B. In *Catalytic Asymmetric Synthesis*; Ojima, I., Ed,; VCH: New York, 1993; Chapter 4.1, pp 103-158.
2. Katsuki, T.; Sharpless, K.B. *J. Am. Chem. Soc.* **1980**, *102*, 5974.
3. Martin, V. S.; Woodard, S. S.; Katsuki, T.; Yamada, Y.; Ikeda, M.; Sharpless, K. B. *J. Am. Chem. Soc.* **1981**, *103*, 6237.
4. (a) Hatakeyama, S.; Sakurai, K.; Takano, S.; *J. Chem.. Soc., Chem. Commun.* **1985**, 1759. (b) Hafele, B.; Schroter, d.; Jager, V. *Angew Chem., Int. Ed. Engl.* **1986**, *25*, 87. c) Schreiber, S. L.; Schreiber, T. S; Smith, D. B. *J. Am. Chem. Soc.* **1987**, *109*, 1525.
5. [R] (Finn, M. G.; Sharpless, B. M. In Asymmetric Synthesis; Morrison, J. D., Ed.; Academic Press: New York,Vol 5, Chapter 8.
6. (a) Woodard, S. S.; Finn, M. G.; Sharpless, B. M. *J. Am. Chem. Soc.* **1991**, *113*, 106. (b) Finn, M. G.; Sharpless, K. B. *J. Am. Chem. Soc.* **1991**, *113*, 113-126. c) Potvin, P. G.; Bianchet, S. *J. Org. Chem.* **1992**, *57*, 6629.
7. Wu, Y.-D.; Lai, D. K. W. *J. Am. Chem. Soc.* **1995**, *117*, 11327.
8. Sharpless, K. B.; Woodard, S. S.; Finn, M. G. *Pure & Appl. Chem.*. **1983**, *55*, 1823.
9. X-Ray crystallographic structure of a Ti$_2$(dibenzyltartamide)$_2$(OiPr)$_4$ species has been obtained.
10. Potvin, P. G.; Bianchet, S. *J. Org. Chem.* **1992**, *57*, 6629.
11. Corey, E. J. *J. Org. Chem.* **1990**, *55*, 1693.

12.   Gao, Y.; Hanson, R. M.; Klunder, J. M.; Ko, S. Y.; Masamune, H.; Sharpless, K. B. *J. Am. Chem. Soc.* **1987,** *109,* 5765.
13.   Lu, L. D.-L.; Johnson, R. A.; Finn, M. G.; Sharpless, K. B. *J. Org. Chem.* **1984,** *49,* 728.
14.   (a) Meister, C.; Scharf, H. D. *Justus Liebigs Ann. Chem..* **1983,** 913. (b) Tanner, D.; Somfai, P. *Tetrahedron* **1986,** *42,* 5985. c) Giese, B.; Rupaner, R.; *Justus Liebigs Ann. Chem..* **1987,** 231.
15.   Klunder, J. M.; Ko, S. Y.; Sharpless, B. M. *J. Org. Chem.* **1986,** *51,* 3710.
16.   For examples of *in situ* derivatization see: (a) Berger, D.; Overman, L. E.; Renhowe, R. A. *J. Am. Chem. Soc.* **1993,** *115,* 9305. (b) Uemura, I.; Yamada, K.; Sugiura, K.; Miyagawa, H.; Ueno, T. *Tetrahedron Asymm.* **2001,** *12,* 942. (c) DeGoey, D. A.; Chen, H.-J.; Flost, W. J.; Grampovnik, D. J.; Yeung, C. M.; Klein, L. L.; Kempf, D. J. *J. Org. Chem.* **2002,** *67,* 5445.
17.   Gao, Y.; Hanson, R. M.; Klunder, J. M.; Ko, S. Y.; Masamune, H.; Sharpless, K. B. *J. Am. Chem. Soc.* **1987,** *109,* 5765.
18.   Tanner, D.; Somfai, P. *Tetrahedron* **1986,** 42, 5985.
19.   McDonald, F. E.; Bravo, F.; Wang, X.; Wei, X.; Toganoh, M.; Rodriguez, J. R.; Do, B.; Neiwert, W. A.; Hardcastle, K. I. *J. Org. Chem.* **2002,** *67,* 2523.
20.   Goument, B.; Duhamel, L.; Mauge, R. *Tetrahedron* **1994,** *50,* 171.
21.   Chakraborty, R. K.; Purkait, S.; Das, S. *Tetrahedron* **2003,** *59,* 9127.
22.   Lindsay, K. B.; Pyne, S. G. *J. Org. Chem.* **2002,** *67,* 7774.
23.   Ginesta, X.; Pericas, M. A.; Riera, A. *Tetrahedron Lett.* **2002,** *43,* 779.
24.   Makino, K.; Suzuki, T.; Awane, S.; Hara, O.; Hamada, Y. *Tetrahedron Lett.* **2002,** *43,* 9391.
25.   Green, M. P.; Prodger, J. C.; Hayes, C. J. *Tetrahedron Lett.* **2002,** *43,* 6609.
26.   Wood, R. D.; Ganem, B.; *Tetrahedron Lett.* **1982,** *23,* 707.
27.   Adam, W.; Braun, M.; Griesbeck, A.; Lucchinni, V.; Staab, E.; Will, B.; *J. Am. Chem. Soc.* **1989,** *111,* 203.
28.   Katsuki, T.; Lee, A. W.; Ma, P.; Martin, V. S.; Masamune, S.; Sharpless, K. B. *J. Org. Chem.* **1982,** *47,* 1373.
29.   Denis, J.-N.; Greene, A. E.; Serra, A. A.; Luche, M.-J. *J. Org. Chem.* **1986,** *51,* 46.
30.   Kabat, M. M. *Tetrahedron Asymm.* **1993,** *4,* 1417.
31.   Satake, A.; Shimizu, I. *Tetrahedron Asymm.* **1993,** *4,* 1405.
32.   Aiai, M.; Robert, A.; Baudy-Floc'h, M.; Le Grel, P. *Tetrahedron Asymm.* **1995,** *6,* 2249.
33.   Morimoto, Y.; Oda, K.; Shirahama, H.; Matsumoto, T.; Omura, S. *Chemistry Lett.* **1998,** 909.
34.   Sharpless, K. B.; Behrens, C. H.; Katsuki, T.; Lee, A. W.; Martin, V. S.; Takatani, M.; Viti, S. M.; Walker, F. J.; Woodard, S. S. *Pure Appl. Chem.* **1983,** *55,* 589.
35.   Ghosh, A. K.; Lei, H. *Tetrahedron Asymm.* **2003,** *14,* 629.
36.   Gabarda, A. E.; Du, W.; Isarno, T.; Tangirala, R. S.; Curran, D. P. *Tetrahedron* **2002,** *58,* 6329.
37.   Diaz, S.; Cuesta, J.; Gonzalez, A.; Bonjoch, J. *J. Org. Chem. Soc.* **2003,** *68,* 7400.
38.   Uemura, I.; Yamada, K.; Sugiura, K.; Miyagawa, H.; Ueno, T. *Tetrahedron Asymm.* **2001,** *12,* 943.
39.   Mohr, P. J.; Halcomb, R. L. *Org. Lett.* **2002,** *4,* 2413.
40.   Erickson, T. J. *J. Org. Chem.* **1986,** *51,* 934.
41.   Abad, A.; Agullo, c.; Arno, M.; Cunat, A. C.; Garcia, M. T.; Zaragoza, R. *J. Org. Chem.* **1996,** *61,* 5916.
42.   Furuichi, N.; Hara, H.; Osaki, T.; Mori, H.; Katsumura, S. *Angew. Chem. Int. Ed,* **2002,** *41,* 1023.
43.   Martin, V. S.; Woodard, S. S.; Katsuki, T.; Yamada, Y.; Ikeda, M.; Sharpless, K. B. *J. Am. Chem. Soc.* **1981,** *103,* 6237.
44.   Note that the maximum yield for a kinetic resolution is 50%.
45.   (a) See reference 43. (b) Carlier, P. R.; Mungall, W. S.; Schroder, G.; Sharpless, K. B. *J. Am. Chem. Soc.* **1988,** *110,* 2978.
46.   Lafontaine, J. A.; Provencal, D. P.; Gardelli, C.; Leahy, J. W. *J. Org. Chem.* **2003,** *68,* 4215.
47.   Takayama, H.; Kurihara, M.; Kitajima, M.; Said, I. M.; Aimi, N. *J. Org. Chem. Soc.* **1999,** *64,* 1772.
48.   Pommier, A.; Stepanenko, V.; Jarowicki, K.; Kocienski, P. J. *J. Org. Chem.* **2003,** *68,* 4008.
49.   Neagu, C.; Hase, T. *Tetrahedron Lett.* **1993,** *34,* 1629.
50.   Ko, S. Y.; Lee, A. W. M.; Masamune, S.; Reed, L. A. III; Sharpless, K. B; Walker, F. J. *Science,* **1983,** *220,* 949.
51.   (a) Schreiber, S. L.; Schrieber, T. S.; Smith, D. B. *J. Am. Chem. Soc.* **1987,** *109,* 1525-1529. (b) Poss, C. S.; Shreiber, S. L. *Acc. Chem. Res.* **1994,** *27,* 9.

Michael Palucki

1.7   Wenker Aziridine Synthesis

1.7.1   Description

The Wenker aziridine synthesis entails the treatment of a β-amino alcohol **1** with sulfuric acid to give β-aminoethyl sulfate ester **2** which is subsequently treated with base to afford aziridine **3**.[1] Before the discovery of the Mitsunobu reaction, which transforms an amino alcohol into an aziridine in one step under very mild conditions, the Wenker reaction was one of the most convenient methods for aziridine synthesis. However, due to the involvement of strong acid and then strong base, its utility has been limited to substrates without labile functionalities.

A related aziridine synthesis is the Gabriel reaction (a.k.a. Gabriel–Cromwell reaction),[2,3] which involves an intramolecular S_N2 reaction of a β-amino halide. However, the reaction has become so common that the name Gabriel is not tightly related to the transformation.

1.7.2   Historical Perspective

In 1935, Wenker[4] first prepared β-aminoethyl sulfate ester (**4**, a solid) from thermal dehydration of monoethanolamine acid sulfate at 250°C according to Gabriel's procedure.[4] Subsequently, the mixture of **4** and 40% NaOH aqueous solution was distilled. Further fractional distillation of the distillate in the presence of KOH and then Na at 55–56°C led to pure aziridine in 26.5% yield.

1.7.3   Mechanism

The mechanism for the Wenker aziridine synthesis is very straightforward. As depicted by conversion **2→3**, the transformation is a simple case of intramolecular S_N2 displacement process, in which the sulfate ester is the leaving group.

1.7.4   Variations and Improvements

Due to the convenience of the Wenker aziridine formation from β-aminoethyl sulfate ester (**4**) and base, many improvements ensued. Leighton *et al.* improved the yield of the first step for the formation of sulfate ester **4**.[5] First of all, both ethanolamine and 95% sulfuric acid were diluted with half of their weight of water and then slowly mixed together at 0°C. Finally, by keeping the temperature below 145°C, sulfate ester **4** was obtained in 90–95% yield.

Another improvement of the Wenker reaction was utilization of flash distillation, which boosted the yield of **5** to 83% based on **4**.[6] In addition, a procedure involving the use of an aqueous reaction medium and the generation of the aziridine in solution was developed, which obviates the isolation and handling of anhydrous aziridine.[7]

Mesylates and tosylates may be used as variants of the *O*-sulfate ester. For instance, 55% of aziridine **7** was obtained from base-mediated cyclization of amino mesylate **6**.[8] In comparison, the classic Wenker protocol only gave 3% of **7**. In another instance,[9] *N*-tosyl amino alcohol **8** was tosylated to give **9**, which was transformed to aziridine **10** in 64% yield, along with 29% of the β-elimination product due to the presence of the ester moiety. Likewise, aziridine **12** was assembled from tosylate **11** in two steps and 60% yield.[10]

11   **12**

1.7.5 Synthetic Utility

As described in Section 1.7.1, the utility of the Wenker reaction is limited to substrates without labile functionalities because of the involvement of strong acid and then strong base. The Fanta group prepared a variety of aziridines by taking advantage of the Wenker reaction.[11–14] For example, 6-aza-bicyclo[3.1.0]hexane (**14**) was produced from the ring-closure of (±)-*trans*-2-aminocyclopentanol hydrochloride (**13**).[11] In a similar fashion, sulfate ester **16** was prepared from *N*-methyl *dl*-*trans*-3-amino-4-hydroxytetrahydrofuran (**15**). Subsequent treatment of sulfate ester **16** with NaOH then delivered aziridine **17**.[14] Additional examples of Wenker aziridine synthesis may also be found in references 15–17.

13   **14**

15   **16**   **17**

Due to the abundance of epoxides, they are ideal precursors for the preparation of β-amino alcohols. In one case, ring-opening of 2-methyl-oxirane (**18**) with methylamine resulted in 1-methylamino-propan-2-ol (**19**), which was transformed to 1,2-dimethyl-aziridine (**20**) in 30–35% yield using the Wenker protocol.[18] Interestingly, 1-amino-3-buten-2-ol sulfate ester (**23**) was prepared from 1-amino-3-buten-2-ol (**22**, a product of ammonia ring-opening of vinyl epoxide **21**) and *chlorosulfonic acid*. Treatment of sulfate ester **23** with NaOH then led to aziridine **24**.[19]

18   **19**   **20**

21 → **22** → **23** → **24**

1.7.6 Experimental

25 → **26**

(S)-2-Benzyl-aziridine (26)[17]

A cold mixture of sulfuric acid (98%, 4 g), and water (4 mL) was added to an amino-alcohol **25** (40 mmol) in water (2.4 mL) at 0–5°C. The mixture was heated to 120°C and then water was carefully distilled off *in vacuo*. The solid sulfate residue was treated with 6.2 M potassium hydroxide, and steam-distilled. The distillate was saturated with potassium hydroxide pellets and the upper organic layer, which separated, was fractionally distilled from potassium hydroxide through a short column to give a colorless oil aziridine **26** in 96% yield.

In addition, an *Organic Synthesis* procedure of preparing aziridine from β-amino alcohol exists.[20]

1.7.7 References

1.   (a) [R] Deyrup, J. A. In *Small Ring Heterocycles*, Part 1, Hassner, A., ed.; Wiley-Interscience: New York, **1983**, 1–214. (b) [R] Dermer, O. C.; Ham, G. E. *Ethyleneimine and Other Aziridines*, Academic Press: New York, 1969.

2.   (a) Filigheddu, S. N.; Masala, S.; Taddei, M. *Tetrahedron Lett.* **1999**, *40*, 6503. (b) Shustov, G. V.; Krutius, O.; Voznesenskii, V. N.; Chervin, I. I.; Eremeev, A. V.; Kostyanovskii, R. G.; Polyak, F. D. *Tetrahedron* **1990**, *46*, 6741. (c) Okada, I.; Ichimura, K.; Sudo, R. *Bull. Chem. Soc. Jpn.* **1970**, *43*, 1185. (d) Gabriel, S. *Ber.* **1895**, *28*, 2929.

3.   Gabriel, S. *Ber.* **1888**, *21*, 1049, 2667.

4.   Wenker, H. *J. Am. Chem. Soc.* **1935**, *57*, 2328.

5.   Leighton, P. A.; Perkins, W. A.; Renquist, M. I. *J. Am. Chem. Soc.* **1947**, *69*, 1540.

6.   Reeves, W. A.; Drake, G. L., Jr.; Hoffpauir, C. L. *J. Am. Chem. Soc.* **1951**, *73*, 3522.

7.   Wystrach, V. P.; Kaiser, D. W.; Schaefer, F. C. *J. Am. Chem. Soc.* **1955**, *77*, 5915.

8.   Gaertner, V. R. *J. Org. Chem.* **1970**, *35*, 3952.

9.   (a) Nakagawa, Y.; Tsuno, T.; Nakajima, K.; Iwai, M.; Kawai, H.; Okawa, K. *Bull. Chem. Soc. Jpn.* **1972**, *45*, 1162. (b) Okawa, K.; Kinutani, T.; Sakai, K. *Bull. Chem. Soc. Jpn.* **1968**, *41*, 1353.

10.   Park, J.-i.; Tian, G.; Kim, D. H. *J. Org. Chem.* **2001**, *66*, 3696.

11.   Fanta, P. E. *J. Chem. Soc.* **1957**, 1441.

12.   Fanta, P. E.; Walsh, E. N. *J. Org. Chem.* **1966**, *31*, 59.

13.   Kaschelikar, D. V.; Fanta, P. E. *J. Am. Chem. Soc.* **1960**, *82*, 4927.

14.   Kaschelikar, D. V.; Fanta, P. E. *J. Am. Chem. Soc.* **1960**, *82*, 4930.

15.   Brois, S. J. *J. Org. Chem.* **1962**, *27*, 3532.

16.   Brewster, K; Pinder, R. M. *J. Med. Chem.* **1972**, *15*, 1078.
17.   Xu, J. *Tetrahedron: Asymmetry* **2002**, *13*, 1129.
18.   Minoura, Y.; Takebayashi, M.; Price, C. C. *J. Am. Chem. Soc.* **1959**, *81*, 4689.
19.   Stogryn, E. L.; Brois, S. J. *J. Am. Chem. Soc.* **1967**, *89*, 605.
20.   Allen, C. F. H.; Spangler, F. W.; Webster, E. R. *Org. Synth., Coll. Vol. IV,* **1963**, 433.

Jie Jack Li

2.1 Barton-Zard Reaction

2.1.1 Description

The Barton-Zard reaction refers to the base-induced reaction of nitroalkenes **1** with alkyl α-isocyanoacetates **2** to afford pyrroles **3**.[1,2] Solvents used are THF or alcohols (or mixtures) and the reaction often proceeds at room temperature.

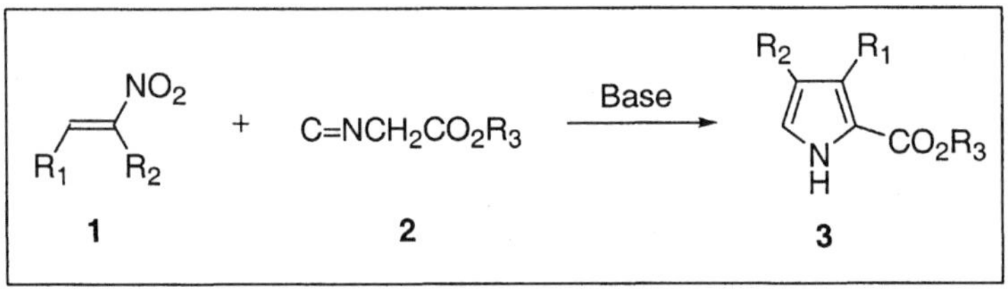

R$_1$ = H, alkyl, aryl
R$_2$ = H, alkyl
R$_3$ = Me, Et, *t*-Bu
Base = KO*t*-Bu, DBU, guanidine bases

The Barton-Zard (BZ) pyrrole synthesis is similar both to the van Leusen pyrrole synthesis that uses Michael acceptors and TosMIC[3] (Section 6.7) and the Montforts pyrrole synthesis using α,β-unsaturated sulfones and alkyl α-isocyanoacetates.[4] An alternative to the use of the reactive nitroalkenes **1** is their *in situ* generation from β-acetoxy nitroalkanes, which are readily prepared via the Henry reaction between an aldehyde and a nitroalkane followed by acetylation. Examples are shown later.

2.1.2 Historical Perspective

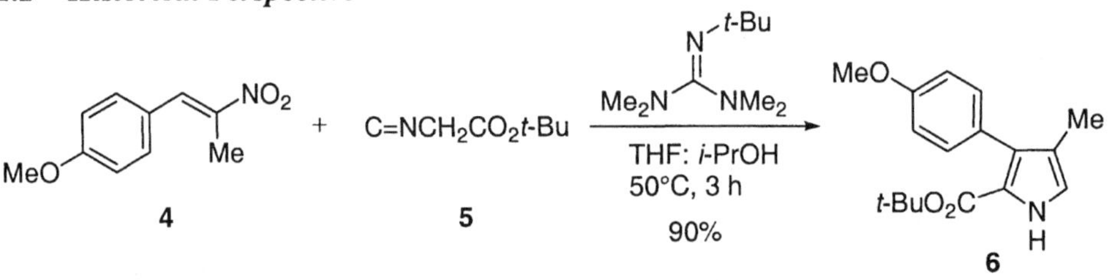

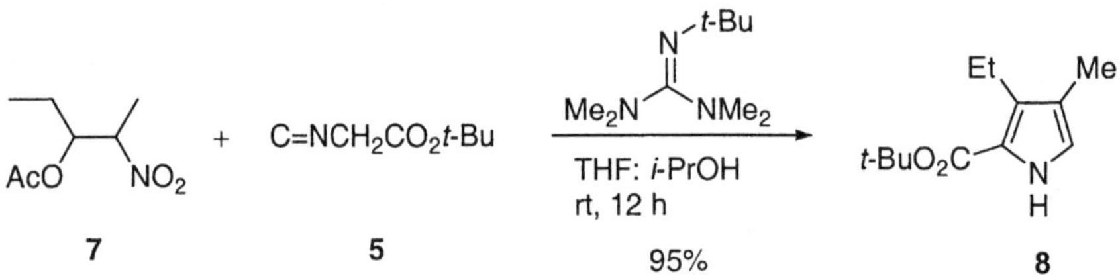

In 1985, in the course of their interest in nitroalkane chemistry, Barton and Zard reported the base-catalyzed reaction of nitroalkenes with α-isocyanoacetates leading to pyrrole esters having an ideal substitution pattern for the synthesis of porphyrins and bile pigments.[1] Indeed, this particular theme has been the most important application of the BZ reaction in synthesis. Two examples are shown.[1]

2.1.3 Mechanism

The mechanism is presumed to involve a pathway related to those proposed for other base-catalyzed reactions of isocyanoacetates with Michael acceptors. Thus base-induced formation of enolate **9** is followed by Michael addition to the nitroalkene and cyclization of nitronate **10** to furnish **11** after protonation. Loss of nitrous acid and aromatization affords pyrrole ester **12**.

2.1.4 Variations and Improvements

The use of stronger bases than, for example, DBU, such as proazaphosphatrane **13**[5,6] or phosphazene superbase **14**[6–8] has afforded pyrroles where weaker bases (i.e. DBU) fail or give non-pyrrole products.[6]   For example, whereas the reaction of **15** and ethyl isocyanoacetate (**16**) in the presence of **14** gives pyrrole **17**, the same reaction with DBU as base affords only pyrimidine *N*-oxide **18**.[6] The formation of the latter is frequently seen as a side product in the BZ reaction and a mechanism has been proposed.[9]

In some cases, potassium carbonate is superior to DBU and tetramethylguanidine (TMG) in the BZ reaction.[10] Thus **7** reacts with TosMIC (**19**) in the presence of K_2CO_3 to afford pyrrole **20** in excellent yield. The yield of **20** using DBU is 62%.

Improved syntheses of benzyl isocyanoacetate,[11] *p*-toluenesulfonylmethyl isocyanide (TosMIC),[12] and other isocyanides[13] are available. The BZ reaction has been modified to synthesize 2-cyanopyrroles **21** using isocyanoacetonitrile,[14] pyrrole-2-phosphonates **22** using isocyanomethylphosphonates,[15] and pyrrole esters **24** using imidothiolates **23** in place of isocyanoacetates.[16] This latter modification affords access to 5-substituted pyrrole esters, unlike the standard BZ reaction. Depending on the nitroalkene precursor, *N*-oxygenated pyrazoles can be obtained.[17] A solid-phase BZ reaction has been reported to generate an array of trisubstituted pyrroles.[18] In this study, the polymer supported guanidine base 1,5,7-triazabicyclo[4.4.0]dec-5-ene (TBD) was employed.

2.1.5 Synthetic Utility

Ono and Lash have been the two pioneers in applying the BZ reaction to the synthesis of pyrroles and, particularly, with applications to the synthesis of novel fused and other porphyrins. Although the concept was recognized by Barton and Zard,[1] Ono and Lash independently discovered the conversion of 2-pyrrolecarboxylates, prepared by the BZ reaction, into porphyrins by what is now a standard protocol (1. LiAlH₄; 2. H⁺; 3. oxidation), although Ono was first to publish;[19,20] for example, 25 to 27.[20]

Ono has employed this strategy for the synthesis of porphyrins with long-chain alkyl groups,[21] aryl groups,[22] and trifluoromethyl groups,[23] dodecaarylporphyrins,[19,24] water-soluble sugar-containing porphyrins,[25] benzoporphyrins,[26] and diporphyrins.[27] The BZ method for the synthesis of the requisite pyrrole esters for use in porphyrin construction is far superior to the classical Knorr method.

Lash, who simultaneously with Ono discovered this efficient route to porphyrins,[28] has been exceptionally clever in subsequent synthetic applications, with a pronounced focus on highly conjugated porphyrins and "geoporphyrins".[29] Thus, the BZ reaction and subsequent elaboration of the resulting pyrroles afford tetrahydrobenzoporphyrins,[30] dinaphthoporphyrins,[31] other highly conjugated porphyrins such as phenanthroporphyrins,[32] tetraacenaphthoporphyrins,[33] phenanthrolinoporphyrins,[34] and fluoroanthoporphyrins,[35] and other porphyrins.[36–40] In their work, both Lash and Ono discovered that many nitroarenes react under BZ conditions to give fused pyrroles,[8,32–40,41] such as 28 and 29.[8,41]

29

Other workers have employed a BZ reaction and subsequent chemistry to synthesize porphyrins.[5,7,42–47] Likewise, the BZ reaction has been extended to the preparation of cycloalkano-oligopyrroles,[48] novel polypyrroles,[49] bilirubin analogues,[12,50] phytochrome analogues,[51] porphobilinogen,[14b] and deoxypyrrolodine.[52] In addition to providing a powerful route to porphyrins, the BZ reaction offers a versatile synthesis of simple pyrroles, such as 3-arylpyrroles **30**,[53] pyrrole *C*-nucleosides,[54] 2,3-disubstituted 4-ethynylpyrroles, **31**,[55] benzyl pyrrole-2-carboxylates,[56] other alkyl pyrrole-2-carboxylates,[57] 3-alkanoylpyrroles,[58] phenylpyrrolylpyrroles,[59] axially dissymmetric pyrroles as new catalysts for the enantiomeric addition of dialkylzincs to arylaldehydes,[60] pyrrolnitrin analogues,[61] and the natural pyrrole pyrrolostatin.[62]

R_1 = H, NO_2, OMe, Cl
R_2 = Me, Et

30

R = Ph, 4-MePh, 2-MePh, 4-MeOPh, 1-naphthyl, 2-thienyl, cyclohexyl

31

The often inaccessible and labile isoindoles can be accessed by the BZ reaction,[63] as can be heteroisoindoles,[9,64,65] such as **32**.[64] Novel pyrroles fused to rigid bicyclic skeleta are readily crafted via a BZ reaction.[66] Certain nitroheterocycles undergo the BZ reaction, such as 3-nitroquinoline (**33**),[64] 3-nitrobenzothiophene (**34**),[41] and 3-nitroindole **35**.[67] Interestingly, whereas **35** gives the expected BZ pyrrolo[3,4-*b*]indole **36**, 3-nitroindole **37** affords the "abnormal" BZ pyrrolo[2,3-*b*]indole **38**, the product of an indole ring fragmentation-ring closure sequence, as promoted by the *N*-phenylsulfonyl group.[67,68] Some reactions of nitroheterocycles afford fused pyrimidine *N*-oxides, such as **39** to **40**.[9]

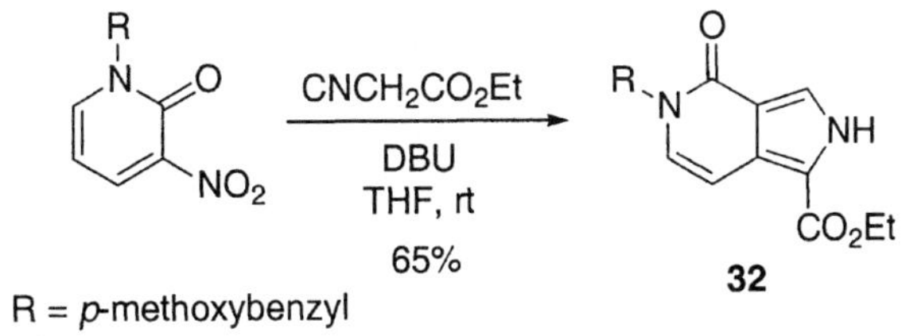

CNCH₂CO₂Et

R = *p*-methoxybenzyl

R—N ... CNCH₂CO₂Et / DBU / THF, rt / 65% → **32** CO₂Et

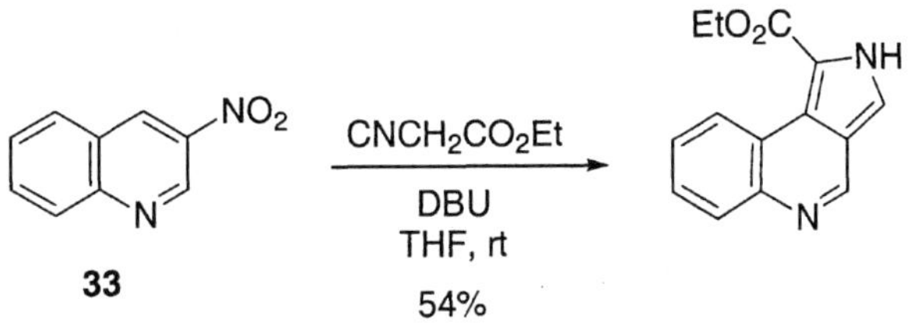

33 CNCH₂CO₂Et / DBU / THF, rt / 54% → EtO₂C

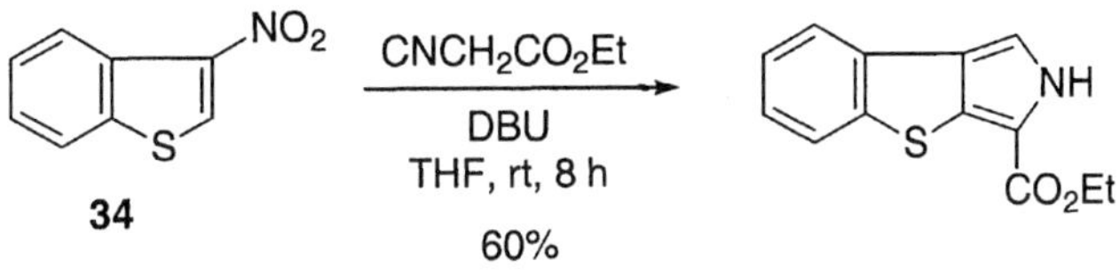

34 CNCH₂CO₂Et / DBU / THF, rt, 8 h / 60% → CO₂Et

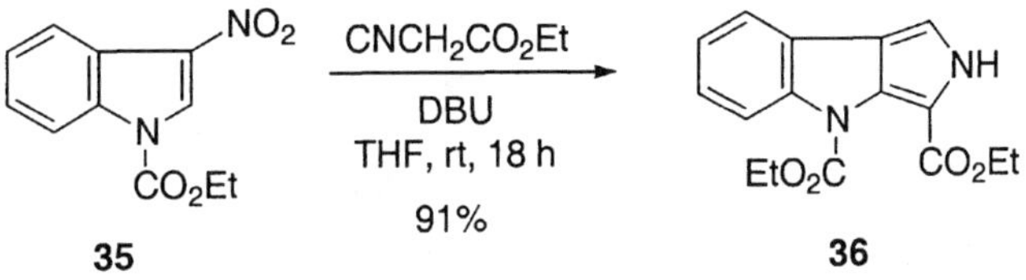

35 CNCH₂CO₂Et / DBU / THF, rt, 18 h / 91% → **36** EtO₂C CO₂Et

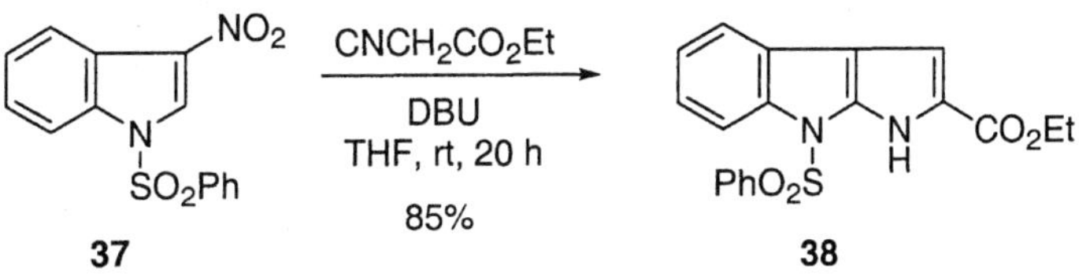

37 CNCH₂CO₂Et / DBU / THF, rt, 20 h / 85% → **38** PhO₂S CO₂Et

39 CNCH₂CO₂Et / DBU / THF, rt, 24 h / 40% → **40** CO₂Et

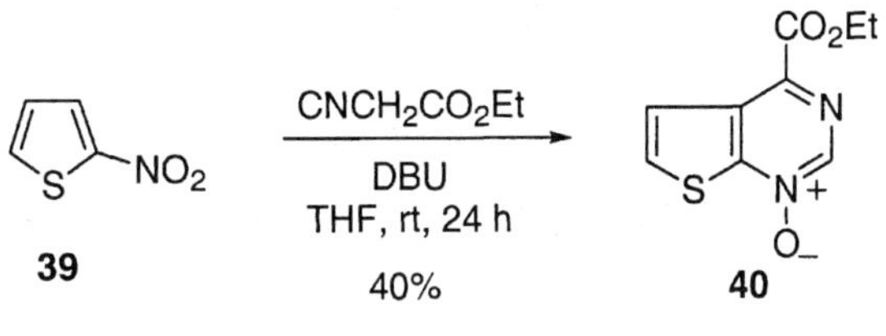

2.1.6   Experimental

The reader is referred to the synthesis of ethyl 3,4-diethylpyrrole-2-carboxylate published in *Organic Syntheses*.[69]

t-Butyl 3-(*p*-methoxyphenyl)-4-methylpyrrole-2-carboxylate:[1]

To a solution of nitroolefin **4** (200 mg) and isocyanide **16** (169 mg) in a 1:1 mixture of THF and isopropanol (5 mL) was added the guanidine base *N-t*-BuTMG (180 mg). The resulting solution was heated to 50°C for 3 h, poured into water, and extracted with dichloromethane. The organic layer was dried over sodium sulfate and filtered through a short column of silica gel (eluent: dichloromethane). Evaporation under vacuum of the solvent gave the desired pyrrole as a pale crystalline solid (272 mg, 90%): mp 142–144°C (from CCl$_4$-pentane); IR 3250, 1640 cm^{-1}; ^{1}H NMR (CDCl$_3$) δ 10.0 (1H, broad), 7.45 (2H, d, J = 9 Hz), 7.10 (2H d, J = 9 Hz), 6.90 (1H, d, J = 2 Hz), 3.90 (3H, s), 2.00 (3H, s), 1.40 (9H, s). Anal. Found: C, 70.77; H, 7.49; N, 4.62. Calcd for C$_{17}$H$_{21}$NO$_3$: C, 71.06; H, 7.37; N, 4.87.

2-Ethoxycarbonyl-3,4-dimethyl-1*H*-pyrrole:[50c]

In a 2-l round-bottomed flask equipped with a magnetic stirrer was charged **16** (148 g, 1.3 mol) and tetramethylguanidine (300 g, 2.54 mol). The stirrer was started and the flask was cooled in ice-water. To the mixture a solution of 3-acetoxy-2-nitrobutane (200 g, 1.24 mol) in dry THF (200 mL) and isopropanol (200 mL) was added dropwise at 0°C over a period of 30 min. The mixture was stirred at rt for 20 h after the addition was complete. The resulting mixture was concentrated under vacuum. The oil residue was dissolved in dichloromethane (2500 mL) and washed with water (3 × 400 mL), 5% aqueous hydrochloric acid (3 × 400 mL), water (400 mL), aqueous saturated sodium bicarbonate (400 mL), and brine (400 mL). After drying over anhydrous sodium sulfate and removing solvent under vacuum, the residue oil was crystallized from dichloromethane-hexane to give the expected product 110 g (53%): mp 92–94°C; ^{1}H NMR (CDCl$_3$) δ 1.3 (t, 3H, J = 7.2 Hz), 2.0 (s, 3H), 2.3 (s, 3H), 4.3 (q, 2H, J = 7.2 Hz), 6.7 (d, 1H, J = 2.7 Hz), 8.9 (br s, 1H); ^{13}C NMR (CDCl$_3$) δ 9.8, 10.2, 14.5, 59.7, 120.5, 120.9, 121.0, 126.9, 161.7.

2-Cyano-4-ethyl-3-methylpyrrole:[14a]

In a dry 25 mL single-necked round bottom flask equipped with magnetic stir bar and nitrogen inlet was placed *N*-formylamino acetonitrile (0.21 g, 3.0 mmol) in dry CH$_2$Cl$_2$ (5 mL) with added triethyl amine (0.75 mL, 5.4 mmol, 1.8 equiv), and then cooled to –25 °C. Phosphorus oxychloride (0.28 mL, 3.0 mmol) was added via syringe slowly over 2 min. The mixture was stirred for additional 10 min, the cooling bath was removed and the reaction mixture was allowed to warm to rt. The mixture was diluted with CH$_2$Cl$_2$ (20 mL) and poured into 20% Na$_2$CO$_3$ soln (4 mL). The organic layer was separated and washed with 20% Na$_2$CO$_3$ soln (4 mL), water (5 mL), dried (MgSO$_4$), and the solvent was removed using a rotary evaporator at low temperature (between 0–5°C). The crude isocyanoacetonitrile (0.24 g) was dissolved in THF (4 mL) and cooled to 0°C, and 2-acetoxy-3-nitropentane (0.175 g, 1.0 mmol) in THF (3 mL) was added via double ended needle followed by DBU (0.53 mL, 3.2 mmol) with syringe. After stirring the mixture for 30 min at 0°C, the cooling bath was removed, the mixture was allowed to warm to rt and stirred for 2 h. The resulting orange-red color precipitate was quenched with water (3 mL) and extracted with ethyl acetate (3 × 25 mL). The combined organic layer was washed with brine (5 mL), dried (MgSO$_4$), and the solvent was removed via rotary a evaporator. The crude compound was purified by silica gel column chromatography (20% ethyl acetate in hexane) to afford 0.12 g of the title compound in 90% yield as a colorless solid: mp 63–64°C; ^{1}H NMR (CDCl$_3$) δ 1.17 (t, 3H, J = 7.5 Hz), 2.17 (s, 3H), 2.41 (q, 2H, J = 7.5 Hz), 6.6 (d, 1H, J = 3.3 Hz), 8.49 (br s, 1H); ^{13}C NMR (CDCl$_3$) δ 9.8, 14.3, 18.2, 99.4, 115.0, 120.6, 126.5, 130.3; exact mass calc'd for C$_8$H$_{10}$N$_2$: 134.0844; found: 134.0842.

2.1.7 References

1. (a) Barton, D. H. R.; Zard, S. Z. *J. Chem. Soc., Chem. Commun.* **1985**, 1098. (b) Barton, D. H. R.; Kervagoret, J.; Zard, S. Z. *Tetrahedron* **1990**, *46*, 7587.
2. [R] Ferreira, V. F.; de Souza, M. C. B. V.; Cunha, A. C.; Pereira, L. O. R.; Ferreira, M. L. G. *Org. Prep. Proc. Int.* **2001**, *33*, 411.
3. van Leusen, A. M.; Siderius, H.; Hoogenboom, B. E.; van Leusen, D. *Tetrahedron Lett.* **1972**, 5337.
4. Haake, G.; Struve, D.; Montforts, F.-P. *Tetrahedron Lett.* **1994**, *35*, 9703.
5. Tang, J.; Verkade, J. G. *J. Org. Chem.* **1994**, *59*, 7793.
6. Murashima, T.; Tamai, R.; Fujita, K.; Uno, H.; Ono, N. *Tetrahedron Lett.* **1996**, *37*, 8391.
7. Bag, N.; Chern, S.-S.; Peng, S.-M.; Chang, C. K. *Tetrahedron Lett.* **1995**, *36*, 6409.
8. Lash, T. D.; Thompson, M. L.; Werner, T. M.; Spence, J. D. *Synlett* **2000**, 213.
9. Murashima, T.; Fujita, K.; Ono, K.; Ogawa, T.; Uno, H.; Ono, N. *J. Chem. Soc., Perkin Trans. 1* **1996**, 1403.
10. Bobál, P.; Lightner, D. A. *J. Heterocycl. Chem.* **2001**, *38*, 527.
11. Burns, D. H.; Jabara, C. S.; Burden, M. W. *Syn. Commun.* **1995**, *25*, 379.
12. Chen, Q.; Huggins, M. T.; Lightner, D. A.; Norona, W.; McDonagh, A. F. *J. Am. Chem. Soc.* **1999**, *121*, 9253.
13. Obrecht, R.; Herrmann, R.; Ugi, I. *Synthesis* **1985**, 400.
14. (a) Adamczyk, M.; Reddy, R. E. *Tetrahedron Lett.* **1995**, *36*, 7983. (b) Adamczyk, M.; Reddy, R. E. *Tetrahedron Lett.* **1995**, *36*, 9121.
15. (a) Yuan, C.; Huang, W. *Synthesis* **1993**, 473. (b) Huang, W.-S.; Zhang, Y.-X.; Yuan, C. *J. Chem. Soc., Perkin Trans. 1* **1996**, 1893.
16. Yokoyama, M.; Menjo, Y.; Wei, H.; Togo, H. *Bull. Chem. Soc. Japan* **1995**, *68*, 2735.
17. Uno, H.; Kinoshita, T.; Matsumoto, K.; Murashima, T.; Ogawa, T.; Ono, N. *J. Chem. Res. (S)* **1996**, 76.
18. Caldarelli, M.; Habermann, J.; Ley, S. V. *J. Chem. Soc., Perkin Trans. 1,* **1999**, 107.
19. Ono, N.; Maruyama, K. *Chem. Lett.* **1988**, 1511.
20. Ono, N.; Kawamura, H.; Bougauchi, M.; Maruyama, K. *Tetrahedron* **1990**, *46*, 7483.
21. Ono, N.; Maruyama, K. *Bull. Chem. Soc. Japan* **1988**, *61*, 4470.
22. (a) Ono, N.; Kawamura, H.; Bougauchi, M.; Maruyama, K. *J. Chem. Soc., Chem. Commun.* **1989**, 1580. (b) Ono, N.; Maruyama, K. *Chem. Lett.* **1989**, 1237.
23. Ono, N.; Kawamura, H.; Maruyama, K. *Bull. Chem. Soc. Japan* **1989**, *62*, 3386.
24. Ono, N.; Miyagawa, H.; Ueta, T.; Ogawa, T.; Tani, H. *J. Chem. Soc., Perkin Trans. 1* **1998**, 1595.
25. Ono, N.; Bougauchi, M.; Maruyama, K. *Tetrahedron Lett.* **1992**, *33*, 1629.
26. Ito, S.; Ochi, N.; Murashima, T.; Uno, H.; Ono, N. *Heterocycles* **2000**, *52*, 399.
27. Fumoto, Y.; Uno, H.; Tanaka, K.; Tanaka, M.; Murashima, T.; Ono, N. *Synthesis* **2001**, 399.
28. Results presented, in part, at the 23rd Midwest Regional ACS Meeting, University of Iowa, Iowa City, IA, Nov 1988; May, D. A., Jr.; Lash, T. D. *Program and Abstracts*, 192; and the 197th National Meeting of the American Chemical Society, Dallas, TX, April 1989; Lash, T. D.; Balasubramaniam, R. P.; May, D. A., Jr. *Book of Abstracts*, ORGN 257.
29. Freemantle, M. *Chem. Eng. News* **1994**, *72* (39), 25.
30. May, D. A., Jr.; Lash. T. D. *J. Org. Chem.* **1992**, *57*, 4820.
31. Lash, T. D.; Roper T. J. *Tetrahedron Lett.* **1994**, *35*, 7715.
32. (a) Lash, T. D.; Novak, B. H. *Tetrahedron Lett.* **1995**, *36*, 4381. (b) Lash, T. D.; Novak, B. H. *Angew. Chem. Int. Ed. Engl.* **1995**, *34*, 683. (c) Lash, T. D.; Novak, B. H.; Lin, Y. *Tetrahedron Lett.* **1994**, *35*, 2493. (d) Novak, B. H.; Lash, T. D. *J. Org. Chem.* **1998**, *63*, 3998.
33. (a) Lash, T. D.; Chandrasekar, P. *J. Am. Chem. Soc.* **1996**, *118*, 8767. (b) Spence, J. D.; Lash, T. D. *J. Org. Chem.* **2000**, *65*, 1530.
34. Lin, Y.; Lash, T. D. *Tetrahedron Lett.* **1995**, *36*, 9441.

35.   Lash, T. D.; Werner, T. M.; Thompson, M. L.; Manley, J. M. *J. Org. Chem.* **2001**, *66*, 3152.
36.   Chandrasekar, P.; Lash, T. D. *Tetrahedron Lett.* **1996**, *37*, 4873.
37.   Lash, T. D.; Wijesinghe, C.; Osuma, A. T.; Patel, J. R. *Tetrahedron Lett.* **1997**, *38*, 2031.
38.   Chen, S.; Lash, T. D. *J. Heterocycl. Chem.* **1997**, *34*, 273.
39.   Lash. T. D.; Chandrasekar, P.; Osuma, A. T.; Chaney, S. T.; Spence, J. D. *J. Org. Chem.* **1998**, *63*, 8455.
40.   Lash, T. D.; Gandhi, V. *J. Org. Chem.* **2000**, *65*, 8020.
41.   Ono, N.; Hironaga, H.; Ono, K.; Kaneko, S.; Murashima, T.; Ueda, T.; Tsukamura, C.; Ogawa, T. *J. Chem. Soc., Perkin Trans. 1* **1996**, 417.
42.   (a) Medforth, C. J.; Berber, M. D.; Smith, K. M.; Shelnutt, J. A. *Tetrahedron Lett.* **1990**, *31*, 3719. (b) Jaquinod, L.; Gros, C.; Olmstead, M. M.; Antolovich, M.; Smith, K. M. *Chem. Commun.* **1996**, 1475.
43.   Sessler, J. L.; Johnson, M. R.; Creager, S. E.; Fettinger, J. C.; Ibers, J. A. *J. Am. Chem. Soc.* **1990**, *112*, 9310.
44.   Bauder, C.; Ocampo, R.; Callot, H. J. *Tetrahedron* **1992**, *48*, 5135.
45.   Nakajima, S.; Osuka, A. *Tetrahedron Lett.* **1995**, *36*, 8457.
46.   Endisch, C.; Fuhrhop, J.-H.; Buschmann, J.; Luger, P.; Siggel, U. *J. Am. Chem. Soc.* **1996**, *118*, 6671.
47.   Czarnecki, K.; Proniewicz, L. M.; Fujii, H.; Kincaid, J. R. *J. Am. Chem. Soc.* **1996**, *118*, 4690.
48.   Fumoto, Y.; Uno, H.; Ito, S.; Tsugumi, Y.; Sasaki, M.; Kitawaki, Y.; Ono, N. *J. Chem. Soc., Perkin Trans. 1* **2000**, 2977.
49.   (a) Ono, N.; Hironaga, H.; Simizu, K.; Ono, K.; Kuwano, K.; Ogawa, T. *J. Chem. Soc., Chem. Commun.* **1994**, 1019. (b) Ono, N.; Tsukamura, C.; Nomura, Y.; Hironaga, H.; Murashima, T.; Ogawa, T. *Adv. Mater.* **1997**, *9*, 149. (c) Lee, D.; Swager, T. M. *J. Am. Chem. Soc.* **2003**, *125*, 6870.
50.   (a) Tipton, A. K.; Lightner, D. A. *Monatsh. Chem.* **1999**, *130*, 425. (b) Brower, J. O.; Lightner, D. A.; McDonagh, A. F. *Tetrahedron* **2001**, *57*, 7813. (c) Chen, Q.; Wang, T.; Zhang, Y.; Wang, Q.; Ma, J. *Synth. Commun.* **2002**, *32*, 1031.
51.   (a) Jacobi, P. A.; DeSimone, R. W. *Tetrahedron Lett.* **1992**, *33*, 6239. (b) Jacobi, P. A.; Guo, J.; Rajeswari, S.; Zheng, W. *J. Org. Chem.* **1997**, *62*, 2907. (c) Drinan, M. A.; Lash, T. D. *J. Heterocycl. Chem.* **1994**, *31*, 255.
52.   Adamczyk, M.; Johnson, D. D.; Reddy, R. E. *J. Org. Chem.* **2001**, *66*, 11.
53.   Alazard, J. P.; Boyé, O.; Gillet, B.; Guénard, D.; Beloeil, J. C.; Thal, C. *Bull. Soc. Chim. Fr.* **1993**, *130*, 779.
54.   Krishna, P. R.; Reddy, V. V. R.; Sharma, G. V. M. *Synlett* **2003**, 1619.
55.   Dell'Erba, C.; Giglio, A.; Mugnoli, A.; Novi, M.; Petrillo, G.; Stagnaro, P. *Tetrahedron* **1995**, *51*, 5181.
56.   (a) Lash, T. D.; Bellettini, J. R.; Bastian, J. A.; Couch, K. B. *Synthesis* **1994**, 170. (b) Ono, N.; Katayama, H.; Nisyiyama, S.; Ogawa, T. *J. Heterocycl. Chem.* **1994**, *31*, 707.
57.   Murata, Y.; Kinoshita, H.; Inomata, K. *Bull. Chem. Soc. Japan* **1996**, *69*, 3339.
58.   Boëlle, J.; Schneider, R.; Gérardin, P.; Loubinoux, B. *Synthesis* **1997**, 1451.
59.   (a) Dumoulin, H.; Rault, S.; Robba, M. *J. Heterocycl. Chem.* **1995**, *32*, 1703. (b) Dumoulin, H.; Rault, S.; Robba, M. *J. Heterocycl. Chem.* **1996**, *33*, 255. (c) Dumoulin, H.; Rault, S.; Robba, M. *J. Heterocycl. Chem.* **1997**, *34*, 13.
60.   Furusho, Y.; Tsunoda, A.; Aida, T. *J. Chem. Soc., Perkin Trans. 1* **1996**, 183.
61.   Santo, R. D.; Costi, R.; Artico, M.; Massa. S.; Lampis, G.; Deidda, D.; Pompei, R. *Bioorg. Med. Chem. Lett.* **1998**, *8*, 2931.
62.   Fumoto, Y.; Eguchi, T.; Uno, H.; Ono, N. *J. Org. Chem.* **1999**, *64*, 6518.
63.   Murashima, T.; Tamai, R.; Nishi, K.; Nomura, K.; Fujita, K.; Uno, H.; Ono, N. *J. Chem. Soc., Perkin Trans. 1* **2000**, 995.
64.   Murashima, T.; Nishi, K.; Nakamoto, K.; Kato, A.; Tamai, R.; Uno, H.; Ono, N. *Heterocycles* **2002**, *58*, 301.
65.   Murashima, T.; Shiga, D.; Nishi, K.; Uno, H.; Ono, N. *J. Chem. Soc., Perkin Trans. 1* **2000**, 2671.
66.   (a) Ito, S.; Murashima, T.; Ono, N. *J. Chem. Soc., Perkin Trans. 1* **1997**, 3161. (b) Uno, H.; Ito, S.; Wada, M.; Watanabe, H.; Nagai, M.; Hayashi, A.; Murashima, T.; Ono, N. *J. Chem. Soc., Perkin Trans. 1* **2000**, 4347.
67.   (a) Pelkey, E. T.; Gribble, G. W. *Chem. Commun.* **1997**, 1873. (b) Pelkey, E. T.; Gribble, G. W. *Synthesis* **1999**, 1117.
68.   Pelkey, E. T.; Chang, L.; Gribble, G. W. *Chem. Commun.* **1996**, 1909.
69.   Sessler, J. L.; Mozaffari, A.; Johnson, M. R. *Org. Syn.* **1991**, *70*, 68.

Gordon W. Gribble

2.2 Knorr and Paal–Knorr pyrrole syntheses

2.1.1 Description
Discovered more than a century ago, the Knorr and Paal–Knorr (PK) pyrrole syntheses are similar intermolecular condensations of amines with carbonyl compounds to give pyrroles.

2.2.1.1 Knorr pyrrole synthesis
The Knorr pyrrole synthesis involves the reaction between an α-amino ketone **1** and a second carbonyl compound **2**, having a reactive α-methylene group, to give a pyrrole **3**.[1-5] The amine **1** is often generated *in situ* by reduction of an oximino group.

R_1-R_4 = various alkyl, acyl, aryl groups

2.2.1.2 Paal–Knorr pyrrole synthesis
The Paal–Knorr pyrrole synthesis is the condensation of a primary amine **4** (or ammonia) with a 1,4-diketone **5** (or 1,4-dialdehyde) to give a pyrrole **6**.[1-5]

R_1-R_3 = H, alkyl, aryl

2.2.2 Historical Perspective
2.2.2.1 Knorr pyrrole synthesis
Knorr discovered that treatment of ethyl α-oximinoacetoacetate (**7**) and ethyl acetoacetate (**8**) with zinc and acetic acid affords 2,4-dicarboethoxy-3,5-dimethylpyrrole (**9**).[6] Extensive modifications of this reaction over the past 100 years have elevated the Knorr pyrrole synthesis to one of exceptional generality and versatility.

7   **8**   **9**

2.2.2.2   Paal–Knorr pyrrole synthesis

Paal and Knorr independently discovered the straightforward reaction of primary amines (or ammonia) with 1,4-diketones to give pyrroles following loss of water.[7] Like the Knorr pyrrole synthesis, the PK method is a powerful and widely used method of constructing pyrroles (*vide infra*).

2.2.3   Mechanism

2.2.3.1   Knorr pyrrole synthesis

The mechanism of the original Knorr pyrrole synthesis entails *in situ* reduction of the oxime moiety to an amine, condensation with the second carbonyl compound, and cyclization with loss of a second molecule of water to give a pyrrole; for example, **10** + **11** to **12**.[8] Several studies have demonstrated that different pathways and pyrrole products obtain depending on the substrates.[1,2,9–12]

10   **11**

12

2.2.3.2   Paal–Knorr pyrrole synthesis

Despite its apparent simplicity, the PK pyrrole synthesis has retained its mystique since being discovered. Several investigations into the PK mechanism have been reported,[1,2,13–17] including a gas phase study.[16] Current evidence (intermediate isolation, kinetics, isotope effects) suggests the following (abbreviated) mechanism for the formation of pyrrole **13**.[17] However, the specific PK mechanism is often dependent on pH, solvent, and amine and dicarbonyl structure, especially with regard to the ring-closing step.

13

2.2.4 Variations and Improvements

2.2.4.1 Knorr pyrrole synthesis

The major development in the Knorr pyrrole synthesis has been access to the amine component. For example, use of preformed diethyl aminomalonate with 1,3-diketones affords much higher yields of pyrroles **14**.[18] Reaction of ß-dicarbonyl compounds with hydroxylamine *O*-sulfonic acid gives pyrroles **15** in one step.[19] Weinreb α-aminoamides have found use in the Knorr pyrrole synthesis of a wide variety of pyrroles **16**.[20]

14

R$_1$ = Me, Et
R$_2$ = H, Me, Et, Pr
R$_3$ = Me, Et, Pr

15

R = OEt, OMe, Me

16

R$_1$ = H, Me, *i*-Pr
R$_2$ = H, Me, Et, Bu, Ph
R$_3$ = CN, CO$_2$Et, Ac, Bz

A zinc-free alternative to the Knorr pyrrole synthesis employs catalytic hydrogenation, as for **17** + **18** to **19**.[21] Oximes such as **17** are readily prepared by nitrosation (NaNO$_2$, HOAc) of the active methylene group.

17   **18**   **19**

2.2.4.2 Paal–Knorr pyrrole synthesis

The use of 2,5-dimethoxytetrahydrofuran (**20**) as a succinaldehyde equivalent has expanded the PK synthesis to include unsubstituted pyrroles, for example, **21**.[22] A novel synthesis of monosubstituted succinaldehydes is also available for the PK pyrrole synthesis.[23]

20   **21**

46–100%

R = Me, Bn, Ph, Ar, CH$_2$CO$_2$Me, Ts, Bz, CO(CH$_2$)$_{16}$CH$_3$

Newer catalysts for the PK synthesis include montmorillonite clays,[24,25] alumina,[24,26] zirconium salts,[27] and titanium reagents.[28] Some pyrroles synthesized from 2,5-hexanedione and primary amines using montmorillonite KSF clay include **22–26**.[25c]

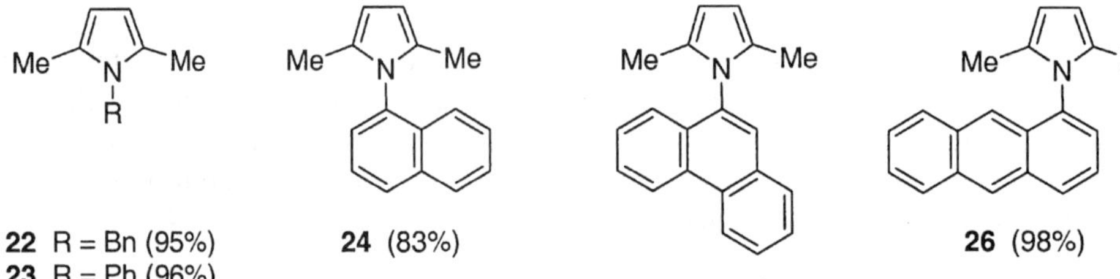

22 R = Bn (95%)   **24** (83%)   **26** (98%)
23 R = Ph (96%)

25 (94%)

Other PK variations include microwave conditions,[29] solid-phase synthesis,[30] and the fixation of atmospheric nitrogen as the nitrogen source (**27**→**28**).[31] Hexamethyldisilazane (HMDS) is also an excellent ammonia equivalent in the PK synthesis.[32]   For example, 2,5-hexanedione and HMDS on alumina gives 2,5-dimethylpyrrole in 81% yield at room temperature. Ammonium formate can be used as a nitrogen source in the PK synthesis of pyrroles from 1,4-diaryl-2-butene-1,4-diones under Pd-catalyzed transfer hydrogenation conditions.[33]

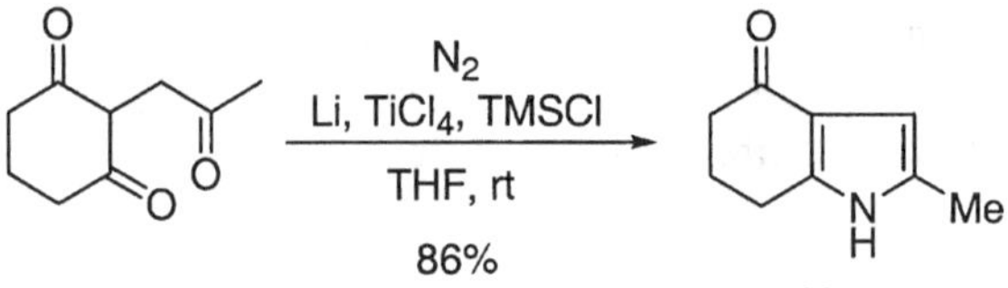

27   **28**

Several novel syntheses of 1,4-dicarbonyls have been recently developed *en route* to pyrroles by the PK method,[34] for example, **29** to **30**[34a] and **31** to **32/33**.[34c]

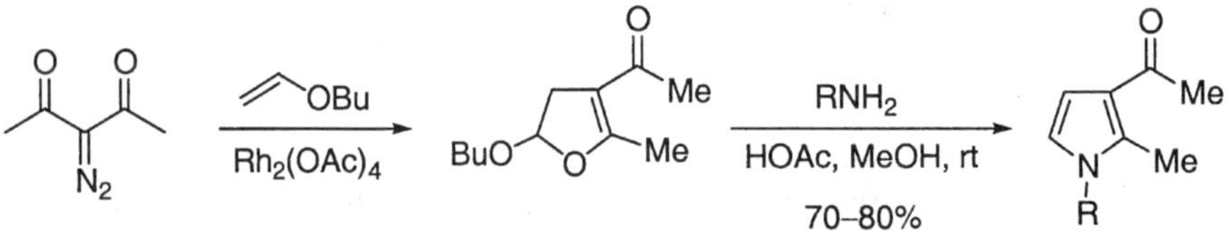

29   **30**

R = Bn, Bu, decyl, cyclohexyl

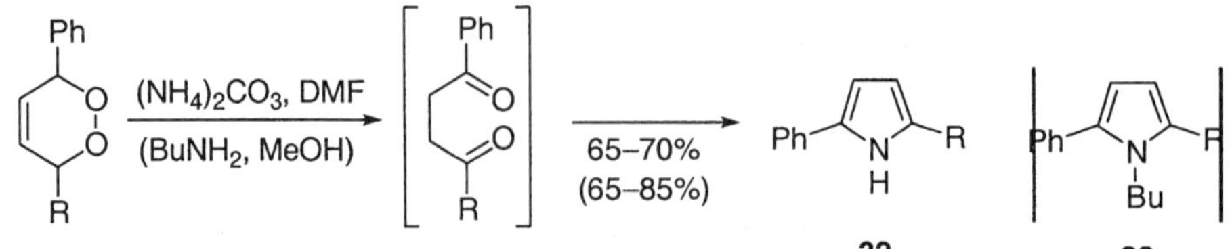

31   **32**   **33**

R = Ph, 4-Br-Ph, Me, H

2.2.5   Synthetic Utility
2.2.5.1   Knorr pyrrole synthesis

The major application of the Knorr pyrrole synthesis is in the construction of porphyrins, and many examples exist,[35,36] particularly from the work of Lash,[35] who also demonstrated the formation of novel pyrroles, such as **34**.[35h] Cyanopyrroles are available using a Knorr synthesis, **35**→**36**,[37] and the method has been adapted to the preparation of pyrroles for the analysis of GSA (glutamate-1-semialdehyde) in biological media,[38] to pyrroles related to pyrrolnitrin,[39] to novel tricyclic antiinflammatory pyrroles,[40] and to an antipsychotic pyrrolo[2,3-g]isoquinoline.[41]

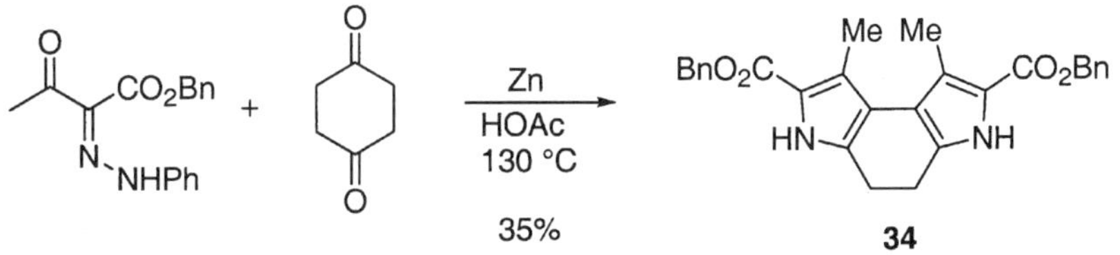

34

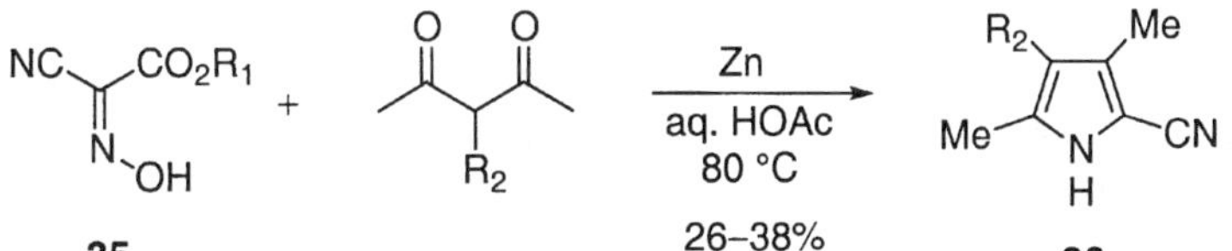

35

R_1 = Me, Et
R_2 = H, $CH_2CH_2CO_2Me$

36

An important extension of the Knorr pyrrole synthesis developed by Cushman utilizes ketone enolates and BOC-protected α-amino aldehydes and ketones.[42]   Two examples (**37**, **38**) are shown.

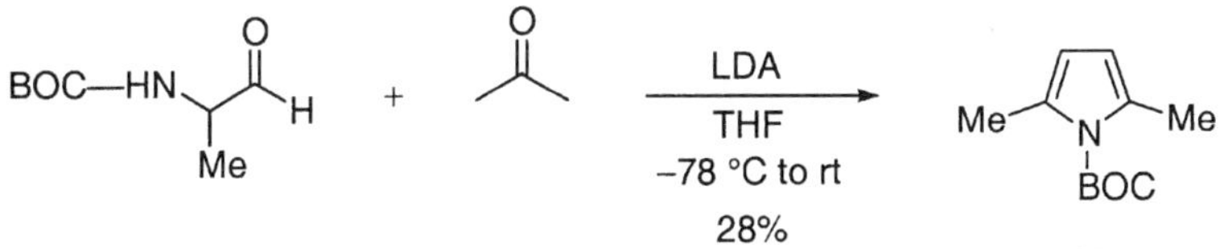

37

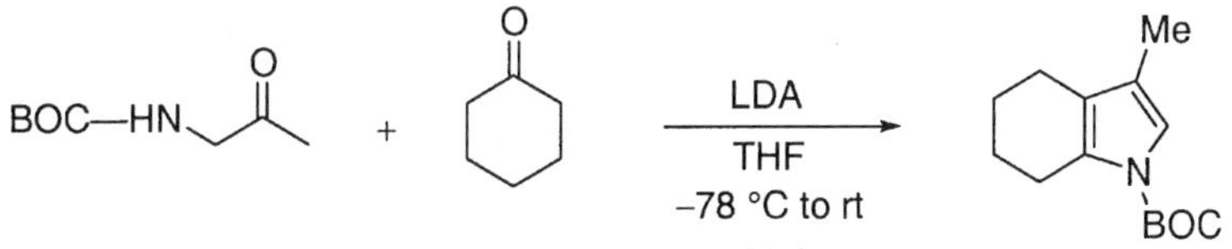

38

2.2.5.2   Paal–Knorr pyrrole synthesis

The simplicity of the PK pyrrole synthesis often makes it the method of choice.   The preparation of 2,5-dimethyl-1-phenylpyrrole from 2,5-hexanedione and aniline is an undergraduate laboratory experiment,[43] and applications of the PK reaction in research are extensive. Diarylpyrrole **39** was crafted for use in novel crown ethers,[44] and several 2,3,5-triarylpyrroles, for example, **40**, are p38 kinase inhibitors.[45]

Like thiophenes, pyrroles are important π-components of conducting polymers and the PK method was used to synthesize polymer **41** and higher homologues.[46] The DNA-interactive precursor bipyrrole **42** was constructed via a PK reaction.[47]

The pyrrole ring in numerous natural products has been constructed using a PK synthesis. Examples include lamellarin L,[48] funebrine,[49] magnolamide,[50] and roseophilin.[51] Thus, in the first total synthesis of magnolamide, pyrrole **43** was obtained using a titanium(IV) isopropoxide mediated PK synthesis.[50] In the course of one of several roseophilin syntheses, pyrrole **44** was prepared.[51c]

The wide applicability of the PK reaction is apparent in the synthesis of pyrroles, for example, **45**, *en route* to novel chiral guanidine bases,[52] levuglandin-derived pyrrole **46**,[53] lipoxygenase inhibitor precursors such as **47**,[54] pyrrole-containing zirconium complexes,[55] and *N*-aminopyrroles **48** from 1,4-dicarbonyl compounds and hydrazine derivatives.[56] The latter study also utilized Yb(OTf)$_3$ and acetic acid as pyrrole-forming catalysts, in addition to pyridinium *p*-toluenesulfonate (PPTS).[56b]

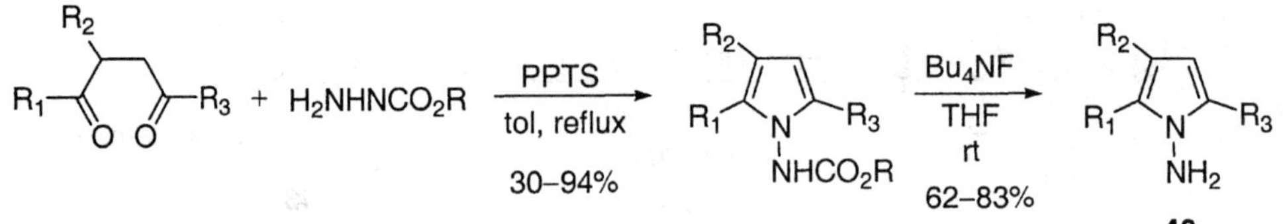

R$_1$ = Me, Ph, 2-furyl
R$_2$ = H, Ph
R$_3$ = H, Me, Ph
R = CH$_2$CH$_2$SiMe$_3$

A sequence of an ozonolysis-PK reaction has been used to convert functionalized cyclohexenes to pyrroles (for example **49** and **50**) that are important precursors to natural tetrapyrroles, hemes, and porphyrins.[57]

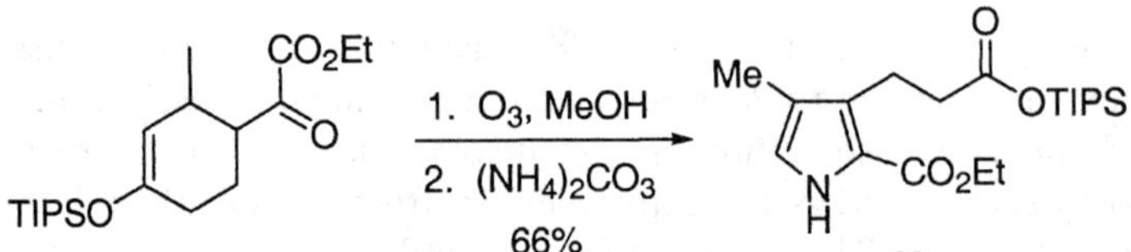

2.2.6   Experimental

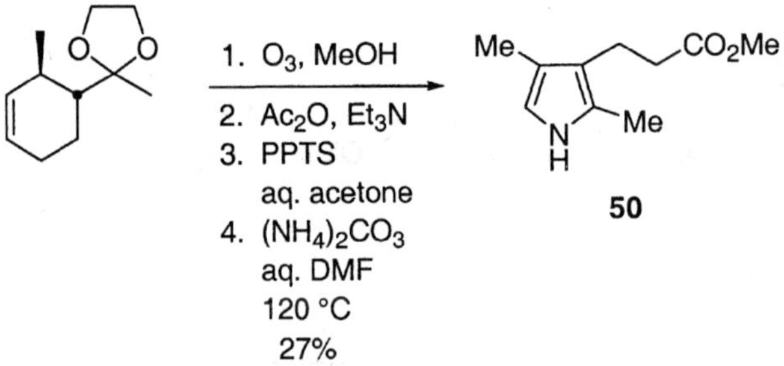

2.2.6.1   Knorr pyrrole synthesis
The reader is referred to the synthesis of 2,4-dicarbethoxy-3,5-dimethylpyrrole (**9**) published in *Organic Syntheses*.[58]

Dibenzyl 4,5-Dihydro-1,8-dimethyl-3H,6H-pyrrolo[3,2-e]indole-2,7-dicarboxylate (34):[35h]
In a 500 mL Erlenmeyer flask, 1,4-cyclohexanedione (1.12 g) and anhydrous sodium acetate (22 g) were dissolved in glacial acetic acid (60 mL), and the mixture was heated to 115 °C. Benzyl 2,3-dioxobutanoate-2-phenylhydrazone (5.92 g) was mixed with 20 g of zinc dust, and was added slowly to the stirred mixture keeping the temperature between 120–130 °C. Once the addition was complete, the solution was stirred at 100 °C for 30 min. The solution was cooled to 80 °C and poured into 700 mL of ice water with stirring. The residual zinc was washed several times with hot acetic acid and combined with the ice water. The precipitate was filtered and recrystallized from ethanol to give **34**

as off-white crystals (1.48 g, 35%), mp 218–222°; IR (Nujol): 3258, 1652 cm^{-1}; ^{1}H NMR (DMSO-d$_6$): δ 2.51 (6H, s), 2.75 (4H, s), 5.30 (4H, s), 7.30–7.60 (10H, m), 11.58 (2H, s). Anal. Calcd for $C_{28}H_{26}N_2O_4$: C, 73.99; H, 5.77; N, 6.16. Found: C, 73.71; H, 5.88; N, 6.06.

2.2.6.2   Paal–Knorr pyrrole synthesis
The reader is referred to the simple undergraduate laboratory synthesis of 2,5-dimethyl-1-phenylpyrrole in the *Journal of Chemical Education*.[43]

2,5-Dimethylpyrrole (13):[32]
Hexane-2,5-dione (342 mg, 3 mmol) was thoroughly mixed with alumina (1 g) before HMDS (1 mL, 4.8 mmol) was added, and the mixture was heated at 100–110 °C until the hexamethyldisiloxane formed was completely evaporated (about 20 min).   Once the mixture cooled down to rt, the product was eluted with CH_2Cl_2 and the oil obtained after evaporation of the solvent was purified by distillation; yield: 231 mg (81%), bp 68 °C/18 Torr.

2.2.7   References

1.   [R] Corwin, A. H. "Heterocyclic Compounds," Vol. 1, Wiley, NY, 1950; Chapter 6.
2.   [R] Jones, R. A.; Bean, G. P. "The Chemistry of Pyrroles," Academic Press, London, 1977, pp 51-57, 74-79.
3.   [R] Patterson, J. M. *Synthesis* **1976**, 281.
4.   [R] Ferreira, V. F.; de Souza, M. C. B. V.; Cunha, A. C.; Pereira, L. O. R.; Ferreira, M. L. G. *Org. Prep. Proc. Int.* **2001**, *33*, 411.
5.   [R] Sundberg, R. J. "Comprehensive Heterocyclic Chemistry II", Ed. Bird, C. W., Vol. 2, Pergamon, Oxford, 1996, Chapter 2.03.
6.   (a) Knorr, L. *Ber. Dtsch. Chem. Ges.* **1884**, *17*, 1635. (b) L. Knorr, *Justus Liebigs Ann. Chem.* **1886**, *236*, 290.
7.   (a) Paal, C. *Ber. Dtsch. Chem. Ges.* **1885**, *18*, 367. (b) Knorr, L. *Ber. Dtsch. Chem. Ges.* **1885**, *18*, 299.
8.   Bullock, E.; Johnson, A. W.; Markham, E.; Shaw, K. B. *J. Chem. Soc.* **1958**, 1430.
9.   Kleinspehn, G. G. *J. Am. Chem. Soc.* **1955**, *77*, 1546.
10.   Harbuck, J. W.; Rapoport, H. *J. Org. Chem.* **1971**, *36*, 853.
11.   Paine, J. B., III; Brough, J. R.; Buller, K. K.; Erikson, E. E. *J. Org. Chem.* **1987**, *52*, 3986.
12.   Yaylayan, V. A.; Keyhani, A. *Food Chem.* **2001**, *74*, 1.
13.   Katritzky, A. R.; Yousaf, T. I.; Chen, B. C.; Guang-Zhi, Z. *Tetrahedron* **1986**, *42*, 623.
14.   [R] Katritzky, A. R.; Ostercamp, D. L.; Yousaf, T. I. *Tetrahedron* **1987**, *43*, 5171.
15.   (a) Chiu, P.-K.; Sammes, M. P. *Tetrahedron* **1988**, *44*, 3531. (b) Chiu, P.-K.; Sammes, M. P. *Tetrahedron* **1990**, *46*, 3439.
16.   Gur, E. H.; de Koning, L. J.; Nibbering, N. M. M. *Int. J. Mass Spec. Ion Proc.* **1997**, *167/168*, 135.
17.   Amarnath, V.; Anthony, D. C.; Amarnath, K.; Valentine, W. M.; Wetterau, L. A.; Graham, D. G. *J. Org. Chem.* **1991**, *56*, 6924.
18.   Paine, J. B., III; Dolphin, D. *J. Org. Chem.* **1985**, *50*, 5598.
19.   Tamura, Y.; Kato, S.; Ikeda, M. *Chem. Ind.* **1971**, 767.
20.   (a) Alberola, A.; Ortega, A. G.; Sábada, M. L.; Sañudo, C. *Tetrahedron* **1999**, *55*, 6555. (b) Calvo, L.; González-Ortega, A.; Sañudo, M. C. *Synthesis* **2002**, 2450.
21.   Manley, J. M.; Kalman, M. J.; Conway, B. G.; Ball, C. C.; Havens, J. L.; Vaidyanathan, R. *J. Org. Chem.* **2003**, *68*, 6447.
22.   (a) Fang, Y.; Leysen, D.; Ottenheijm, H. C. J. *Synth. Commun.* **1995**, *25*, 1857. (b) Haubmann, C.; Hübner, H.; Gmeiner, P. *Bioorg. Med. Chem. Lett.* **1999**, *9*, 3143.
23.   Méndez, J. M.; Flores, B.; León, F.; Martínez, M. E.; Vázquez, A.; García, G. A.; Salmón, M. *Tetrahedron Lett.* **1996**, *37*, 4099.
24.   Texier-Boullet, F.; Klein, B.; Hamelin, J. *Synthesis* **1986**, 409.
25.   (a) Ruault, P.; Pilard, J.-F.; Touaux, B.; Texier-Boullet, F.; Hamelin, J. *Synlett* **1994**, 935. (b) Samajdar, S.; Becker, F. F.; Banik, B. K. *Heterocycles* **2001**, *55*, 1019. (c) Banik, B. K.; Samajdar, S.; Banik, I. *J. Org. Chem.* **2004**, *69*, 213.

26.   Ballini, R.; Barboni, L.; Bosica, G.; Petrini, M. *Synlett* **2000**, 391.
27.   Curini, M.; Montanari, F.; Rosati, O.; Lioy, E.; Margarita, R. *Tetrahedron Lett.* **2003**, *44*, 3923.
28.   (a) Broadbent, H. S.; Burnham, W. S.; Olsen, R. K.; Sheeley, R. M. *J. Heterocyl. Chem.* **1968**, *5*, 757. (b) Yu, S.-X.; Le Quesne, P. W. *Tetrahedron Lett.* **1995**, *36*, 6205.
29.   Danks, T. N. *Tetrahedron Lett.* **1999**, *40*, 3957.
30.   Raghavan, S.; Anuradha, K. *Synlett* **2003**, 711.
31.   Mori, M.; Hori, K.; Akashi, M.; Hori, M.; Sato, Y.; Nishida, M. *Angew. Chem. Int. Ed.* **1998**, *37*, 636.
32.   Rousseau, B.; Nydegger, F.; Gossauer, A.; Bennua-Skalmowski, B.; Vorbrüggen, H. *Synthesis* **1996**, 1336.
33.   Rao, H. S. P.; Jothilingam, S. *Tetrahedron Lett.* **2001**, *42*, 6595.
34.   (a) Cunha, A. C.; Pereira, L. O. R.; de Souza, R. O. P.; de Souza, M. C. B. V.; Ferreira, V. F. *Synth. Commun.* **2000**, *30*, 3215. (b) Braun, R. U.; Zeitler, K.; Müller, T. J. J. *Org. Lett.* **2001**, *3*, 3297. (c) Hewton, C. E.; Kimber, M. C.; Taylor, D. K. *Tetrahedron Lett.* **2002**, *43*, 3199. (d) Ferreira, P. M. T.; Maia, H. L. S.; Monteiro, L. S. *Tetrahedron Lett.* **2002**, *43*, 4491. (e) Quiclet-Sire, B.; Quintero, L.; Sanchez-Jimenez, G.; Zard, S. Z. *Synlett* **2003**, 75. (f) Yuguchi, M.; Tokuda, M.; Orito, K. *J. Org. Chem.* **2004**, *69*, 908.
35.   (a) Lash, T. D.; Bladel, K. A.; Johnson, M. C. *Tetrahedron Lett.* **1987**, *28*, 1135. (b) Lash, T. D.; Perun, T. J., Jr. *Tetrahedron Lett.* **1987**, *28*, 6265. (c) Lash, T. D. *Org. Geochem.* **1989**, *14*, 213. (d) Lash, T. D.; Balasubramaniam, R. P.; Catarello, J. J.; Johnson, M. C.; May, D. A., Jr.; Bladel, K. A.; Feeley, J. M.; Hoehner, M. C.; Marron, T. G.; Nguyen, T. H.; Perun, T. J., Jr.; Quizon, D. M.; Shiner, C. M.; Watson, A. *Energy & Fuels* **1990**, *4*, 668. (e) Lash, T. D.; Bladel, K. A.; Shiner, C. M.; Zajeski, D. L.; Balasubramaniam, R. P. *J. Org. Chem.* **1992**, *57*, 4809. (f) May, D. A., Jr.; Lash, T. D. *J. Org. Chem.* **1992**, *57*, 4820. (g) Quizon-Colquitt, D. M.; Lash, T. D. *J. Heterocycl. Chem.* **1993**, *30*, 477. (h) Lash, T. D.; Bellettini, J. R.; Voiles, S. *J. J. Heterocycl. Chem.* **1993**, *30*, 525. (i) Lash, T. D. *Tetrahedron* **1998**, *54*, 359. (j) Jiao, W.; Lash, T. D. *J. Org. Chem.* **2003**, *68*, 3896.
36.   (a) Paine, J. B. III; Dolphin, D. *J. Heterocycl. Chem.* **1975**, *12*, 1317. (b) Wang, C.-B.; Chang, C. K. *Synthesis* **1979**, 548. (c) Fujii, H.; Yoshimura, T.; Kamada, H. *Tetrahedron Lett.* **1997**, *38*, 1427. (d) Singh, V.; Baqi, A. *J. Indian Chem. Soc.* **2002**, *79*, 701. (e) Singh, V.; Baqi, A. *J. Indian Chem. Soc.* **2002**, *79*, 908.
37.   Cheng, L.; Lightner, D. A. *Synthesis* **1999**, 46.
38.   Liddell, P. A.; Forsyth, T. P.; Senge, M. O.; Smith, K. M. *Tetrahedron* **1993**, *49*, 1343.
39.   Umio, S.; Kariyone, K.; Tanaka, K.; Nakamura, H. *Chem. Pharm. Bull.* **1969**, *17*, 559.
40.   Goudie, A. C.; Rosenberg, H. E.; Ward, R. W. *J. Heterocycl. Chem.* **1983**, *20*, 1027.
41.   Coffen, D. L.; Hengartner, U.; Katonak, D. A.; Mulligan, M. E.; Burdick, D. C.; Olson, G. L.; Todaro, L. J. *J. Org. Chem.* **1984**, *49*, 5109.
42.   Nagafuji, P.; Cushman, M. *J. Org. Chem.* **1996**, *61*, 4999.
43.   Shaw, D. J.; Wood, W. F. *J. Chem. Ed.* **1992**, *69*, A313.
44.   Zhu, L.; Shen, D. *Synthesis* **1987**, 1019.
45.   de Laszlo, S. E.; Visco, D.; Agarwal, L.; Chang, L.; Chin, J.; Croft, G.; Forsyth, A.; Fletcher, D.; Frantz, B.; Hacker, C.; Hanlon, W.; Harper, C.; Kostura, M.; Li, B.; Luell, S.; MacCoss, M.; Mantlo, N.; O'Neill, E. A.; Orevillo, C.; Pang, M.; Parsons, J.; Rolando, A.; Sahly, Y.; Sidler, K.; Widmer, W. R.; O'Keefe, S. J. *Bioorg. Med. Chem. Lett.* **1998**, *8*, 2689.
46.   Ogura, K.; Yanai, H.; Miokawa, M.; Akazome, M. *Tetrahedron Lett.* **1999**, *40*, 8887.
47.   Mingoia, F. *Tetrahedron* **2001**, *57*, 10147.
48.   Peschko, C.; Winklhofer, C.; Steglich, W. *Chem. Eur. J.* **2000**, *6*, 1147.
49.   Dong, Y.; Pai, N. N.; Ablaza, S. L.; Yu, S.-X.; Bolvig, S.; Forsyth, D. A.; Le Quesne, P. W. *J. Org. Chem.* **1999**, *64*, 2657.
50.   Dong, Y.; Le Quesne, P. W. *Heterocycles* **2002**, *56*, 221.
51.   (a) Robertson, J.; Hatley, R. J. D. *Chem. Commun.* **1999**, 1455. (b) Robertson, J.; Hatley, R. J. D.; Watkin, D. J. *J. Chem. Soc., Perkin Trans. 1* **2000**, 3389. (c) Trost, B. M.; Doherty, G. A. *J. Am. Chem. Soc.* **2000**, *122*, 3801.
52.   Boyle, P. H.; Davis, A. P.; Dempsey, K. J.; Hosken, G. D. *J. Chem. Soc., Chem. Commun.* **1994**, 1875.
53.   Iyer, R. S.; Kobierski, M. E.; Salomon, R. G. *J. Org. Chem.* **1994**, *59*, 6038.
54.   Görlitzer, K.; Fabian, J.; Frohberg, P.; Drutkowski, G. *Pharmazie* **2002**, *57*, 243.
55.   Dreier, T.; Erker, G.; Fröhlich, R.; Wibbeling, B. *Organometallics* **2000**, *19*, 4095.
56.   (a) Bolós, J.; Pérez, A.; Gubert, S.; Anglada, L.; Sacristán, A.; Ortiz, J. A. *J. Org. Chem.* **1992**, *57*, 3535. (b) McLeod, M.; Boudreault, N.; Leblanc, Y. *J. Org. Chem.* **1996**, *61*, 1180.
57.   (a) Jacobi, P. A.; Cai, G. *Tetrahedron Lett.* **1991**, *32*, 1765. (b) Taber, D. F.; Nakajima, K. *J. Org. Chem.* **2001**, *66*, 2515.
58.   Fischer, H. *Org. Syn. Coll. Vol. II* **1943**, 202.

Gordon W. Gribble

2.3   Hofmann–Löffler–Freytag reaction

2.3.1   Description

The Hofmann–Löffler–Freytag reaction represents formation of pyrrolidines or piperidines by thermal or photochemical decomposition of protonated *N*-haloamines in the presence of strong acid such as sulfuric acid or trifluoroacetic acid.[1–3] The Hofmann–Löffler–Freytag reaction may also be carried out in milder conditions, for example, PhI(OAc)$_2$, I$_2$, *hv* as shown in section 2.3.4.

2.3.2   Historical Perspective

In 1878, Hofmann reported that treatment of *D*-1-bromo-2-propylpiperidine (**3**) with hot sulfuric acid gave rise to a tertiary amine **4**, *D*-octahydroindolizine.[4,5] In the ensuing decade, Löffler and Freytag extended the reaction to simple secondary amines and found it to be a general way to synthesize pyrrolidines as exemplified by transformation of *N*-bromo-*N*-methyl-2-butylaminylpyridine **5** to nicotine (**6**).[6–8] The Hofmann–Löffler–Freytag reaction is sometimes referred to as Löffler's method, Hofmann–Löffler reaction, Löffler–Hofmann reaction, as well as Löffler–Freytag reaction.

2.3.3   Mechanism

Wawzonek *et al.* first investigated the mechanism of the cyclization of *N*-haloamines and correctly proposed the free radical chain reaction pathway that was substantiated by experimental data.[9–11] Subsequently, Corey and Hertler examined the stereochemistry, hydrogen isotope effect, initiation, catalysis, intermediates, and selectivity of hydrogen transfer.[12] Their results pointed conclusively to a free radical chain mechanism involving intramolecular hydrogen transfer as one of the propagation steps. Accordingly, the mechanism of the Hofmann–Löffler–Freytag reaction can be delineated as follows: Treatment of *N*-chloroamine **7** with acid gives rise to *N*-chloroammonium salt **8**. The free radical chain reaction is initiated by heating **8** or via radiation with UV light to form nitrogen radical cation **9**, which is characterized as a *protonated aminyl radical*. An intramolecular 1,5-hydrogen atom transfer of **9** then results in intermediate **10**, which abstracts a chlorine atom from the starting material **7** to afford chloride **11**. Upon treatment of **11** with a base, S_N2 substitution takes place and fashions pyrrolidine **13** via intermediate **12**. The Hofmann–Löffler–Freytag reaction has an excellent regioselectivity to form 5-membered pyrrolidine.

2.3.4   Variations and Improvements

The starting material for the Hofmann–Löffler–Freytag reaction could be *N*-chloro-, *N*-bromo-, and *N*-iodoamines. When the initiation was carried out thermally, the *N*-chloroamines gave better yields for pyrrolidines because *N*-bromoamines are less stable thermally than the corresponding *N*-chloroamines.[13] In contrary, when the initiation was carried out by irradiation, the *N*-bromoamines gave higher yields for pyrrolidines.[9] As illustrated by transformation of **14** to **16**,[14] irradiation of *N*-bromoamide **14** gave rise to bromomethyl-cyclohexane-amide **15**. Upon treatment of **15** with base *in situ* then formed iminolactone **16** in 92% yield.

14   *hv*   **15**

92%   **16**   R = H or *t*-Bu

Another variation of the Hofmann–Löffler–Freytag reaction involves sulfonamides in place of *N*-chloroamines. For instance, in the presence of Na$_2$S$_2$O$_8$ and CuCl$_2$, butylsulfonamide (**17**) was transformed to 4-chlorobutylsulfonamide (**18**) and 3-chlorobutylsulfonamide (**18**) in the absence of an acid.[15]

A well-known modification of the Hofmann–Löffler–Freytag reaction is the so-called Suarez modification. In 1983, Suarez *et al.* described a process using neutral conditions for the Hofmann–Löffler–Freytag reaction of *N*-nitroamides, *N*-cyanamides or *N*-phosphoramidates.[16–18] For example, in the presence of lead tetraacetate or iodobenzene diacetate and iodine, irradiation of a solution of diethyl phosphoramidate **20** in cyclohexane at reflux with two 100-W tungsten-filament lamps for 2 h led to pyrrolidine derivative **21** in quantitative yield.[16] The reaction was carried out under neutral conditions thus tolerating even the acetate functionality which would not have survived the original strong acid then strong base conditions. Likewise, nitroamine **22** was converted to pyrrolidine nitroamine **23**[17] and primary amide **24** was transformed to lactam **25** as a mixture of the protyl and iodo-derivative in a 75% combined yield under similar conditions.

20   PhI(OAc)$_2$ or Pb(OAc)$_4$ I$_2$, *hv*, 99%   **21**

The Suarez modification of the Hofmann–Löffler–Freytag reaction was review by Togo in 1991.[19]

2.3.5 Synthetic Utility

Not surprising, the most prevalent synthetic utility is the assembly of the pyrrolidine ring. *N*-Chloroamine **27** was obtained by treatment of *N*-methyl-2-cyclopentylethylamine (**26**) with *N*-chlorosuccinimide. Under classic Hofmann–Löffler–Freytag reaction conditions, **27** was rearranged either thermally or by UV irradiation in sulfuric acid to bicyclic amine **28** (1-methyl-octahydro-cyclopenta[*b*]pyrrole) in 20% yield.[20] Similarly, oxazolidine **30** was formed from *N*-chloroamine **29** in moderate yield.[21]

Due to the radical nature of the Hofmann–Löffler–Freytag reaction, a deviation was observed when there was a pendant terminal olefin on the substrate. When the aminyl radical from *N*-chloroamine **31** had a choice between addition to the double bond and 1,5-hydrogen transfer via proton abstraction, the former choice prevailed to afford 1-butyl-2-chloromethyl-pyrrolidine (**32**).[22]

31   **32**

Utilizing the methodology developed by Lavergne's group,[23] Williams *et al.* constructed ^{13}C-labeled methylproline (**36**) as a tool to study the biosynthesis of paraherquamide.[24]   Treatment of [1-^{13}C]-L-isoleucine ethyl ester (**33**) with *t*-butylhypochlorite led to the *N*-chlorinated derivative **34** which, after exposure to a mercury lamp in 85% H_2SO_4, gave the Hofmann–Löffler–Freytag intermediate **35**. Upon neutralization, the intermediate suffered spontaneous cyclization to give the desired labeled β-methylproline ethyl ester. The whole process gave **36** in 53% total yield after treatment of the reaction mixture with di-*t*-butyl dicarbonate to facilitate the separation.

33   **34**

35   **36**

When formation of either the five- or six-membered ring was possible for *N*-chloroamine **37**, only the five-membered ring was conducive under the Hofmann–Löffler–Freytag reaction conditions, forming exclusively 6-ethyl-6-aza-bicyclo[3.2.1]-octane (**38**).[25] No 2-ethyl-2-aza-bicyclo[2.2.2]-octane (**39**) was observed. On the other hand, 2-methyl-2-aza-bicyclo[2.2.2]octan-6-one (**41**) was installed by UV irradiation of a solution of *N*-chloroamine **40** in TFA.[26] Ironically, when the ketone functionality on **40** was protected as its ethylene ketal group, the resultant steric interactions completely prohibited the classic Hofmann–Löffler–Freytag reaction.

39   **37**   **38**

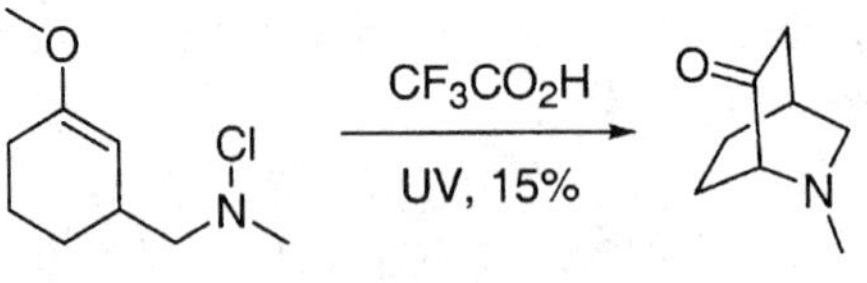

40 → **41**

CF$_3$CO$_2$H, UV, 15%

Ingenious utility was found in a synthesis of diaza-2,6-adamantane **44** by taking advantage of the Hofmann–Löffler–Freytag reaction.[27] *N*-Bromoamine **43**, prepared from 9-methyl-9-aza-bicyclo[3.3.1]nonan-3-one (**42**) in 2 steps, was dissolved in 84% sulfuric acid and heated at 65°C for 30 min to fashion diaza-2,6-adamantane **44** in 25% yield. A similar operation rendered *N*-chloroamine **45** to a longifolane-based tetracyclic *N*-methylpyrrolidine **46** although in only 10% yield.[28] Under neutral conditions, 2-methyl-4-piperidin-2-ylmethyl-quinolizidine (**47**) was treated with *N*-chlorosuccinamide and the resulting *N*-chloroamine was irradiated with 300 W high pressure mercury lamp to assemble (±)-dihydrodeoxyepiallocernuine (**48**).[29]

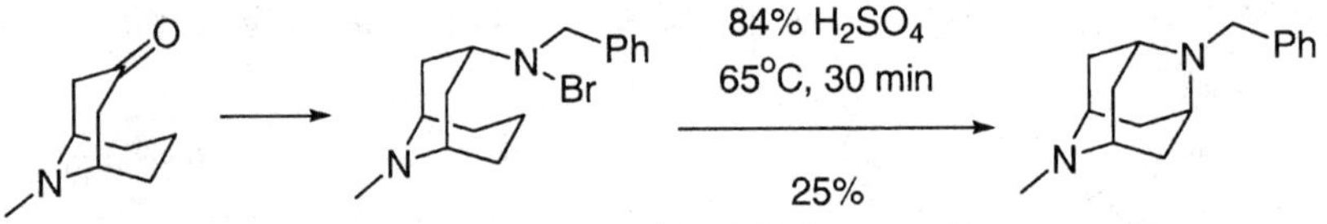

42 → **43**

84% H$_2$SO$_4$, 65°C, 30 min, 25% → **44**

45

84% H$_2$SO$_4$, 65°C, 30 min, 10% → **46**

47

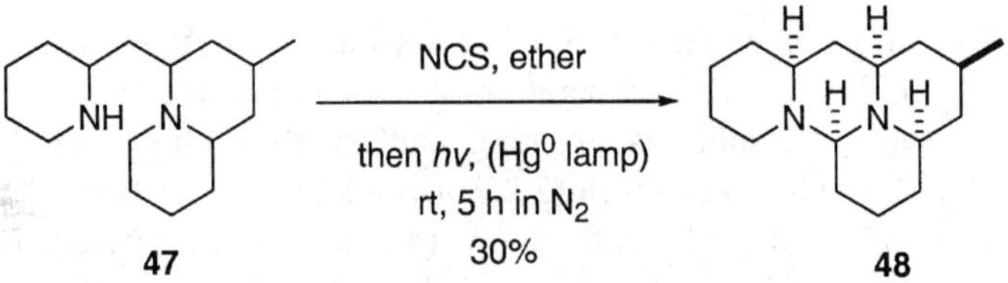

NCS, ether, then *hv*, (Hg^0 lamp), rt, 5 h in N$_2$, 30% → **48**

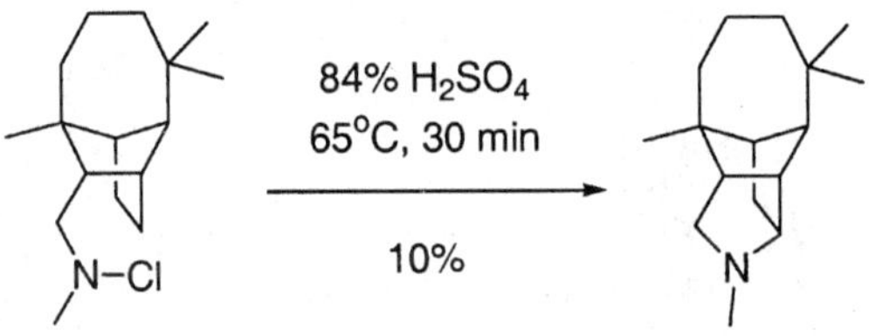

En route to the total synthesis of *cinchona* alkaloid meroquinene, a Hoffmann-La Roche group took advantage of the Hofmann–Löffler–Freytag reaction to functionalize the ethyl side chain in piperidine **49** to give chloroethylpiperidine **51** via the intermediacy of protonated aminyl radical **50**.[30]

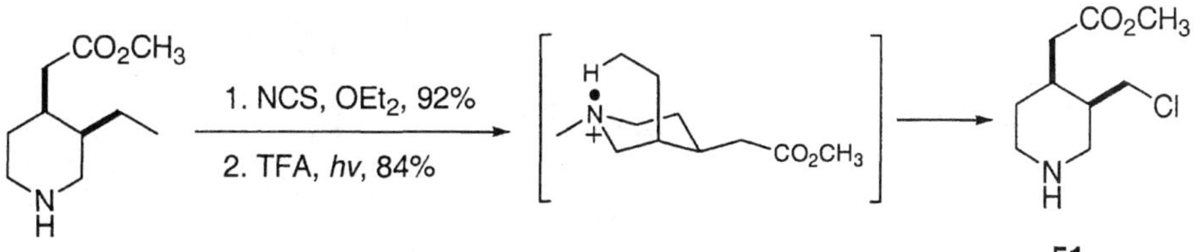

49 **50** **51**

The mechanism of the Hofmann–Löffler–Freytag reaction is analogous to that of the Barton nitrite photolysis reaction. Not surprisingly, both reactions have found synthetic utilities in the functionalization of steroids in a "remote control" fashion. In 1962, the Wolff group converted 20α-methylaminopregn-4-en-3-one (**52**) to the pyrrolidine analog **53** utilizing the Hofmann–Löffler–Freytag reaction.[32] In a similar fashion, aminosteroid **54** was transformed to its corresponding pyrrolidine analog **55**.[33]

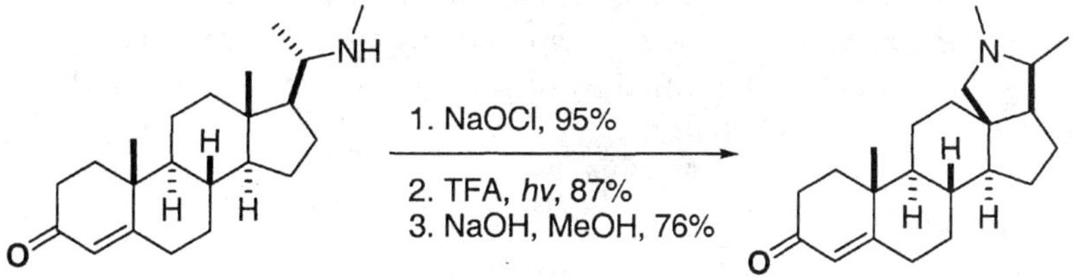

52 **53**

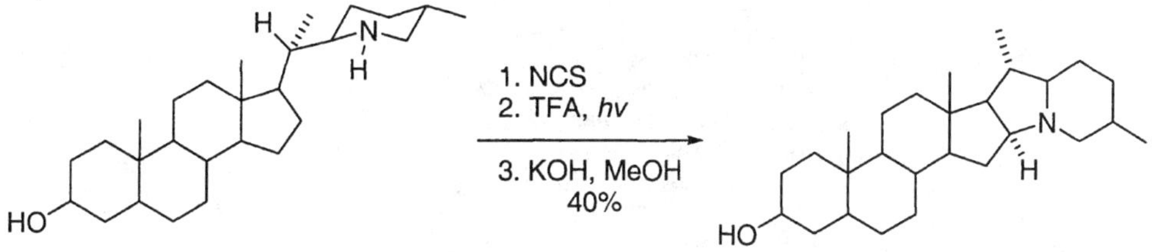

54 **55**

In conclusion, the Hofmann–Löffler–Freytag reaction tends to give moderate and sometimes poor yields for the preparation pyrrolidines under the classic conditions. Nonetheless, the utility of this reaction to functionalize molecules via the aminyl radical mechanism plays an unique role in the tool box for the organic chemist, enabling transformations not easily achievable using other means. Furthermore, milder conditions and better yields can be achieved by taking advantages of the newer developments such as the Suarez modification.

2.3.6 Experimental

Perhydro-dipyrido[1,2-*a*][1',2'-*c*]-pyrimidine[29b]

A solution of the starting material **56** (1 g, 5.1 mmol) and *N*-chlorosuccinamide (1.34 g, 10 mmol) in ether (200 mL) was stirred in an ice bath for 5 min, to which mixture was added triethylamine (2 g, 20 mmol). The mixture was immediately irradiated with the mercury lamp in a current of nitrogen in an ice bath fro 3.5 h. The precipitate was filtered off and the solvent and the excess of triethylamine were removed *in vacuo* to leave a residual oil, which was purified by distillation to give **57** as a colorless oil; yield: 980 mg, ca. 100%. b.p. 140°C/20 torr.

2.3.7 References

1.   [R] (a) Wolff, M. E. *Chem. Rev.* **1963**, *63*, 55. (b) Stella, L. *Angew. Chem., Int. Ed. Engl.* **1983**, *22*, 337–422.
2.   [R] Majetich, G.; Wheless, K. *Tetrahedron* **1995**, *51*, 7095.
3.   [R] Pellissier, H.; Santelli, M. *Org. Prep. Proced. Int.* **2001**, *33*, 455.
4.   Hofmann, A. W. *Ber. Dtsch. Chem. Ges.* **1879**, *12*, 984.
5.   Hofmann, A. W. *Ber. Dtsch. Chem. Ges.* **1883**, *16*, 558.
6.   Löffler, K.; Freytag, C. *Ber. Dtsch. Chem. Ges.* **1909**, *42*, 3427.
7.   Löffler, K.; Kober, S. *Ber. Dtsch. Chem. Ges.* **1909**, *42*, 3431.
8.   Löffler, K. *Ber. Dtsch. Chem. Ges.* **1910**, *43*, 2035.
9.   Wawzonek, S.; Thelan, P. J. *J. Am. Chem. Soc.* **1950**, *72*, 2118.
10.   Wawzonek, S.; Thelan, M. F., Jr.; Thelan, P. J. *J. Am. Chem. Soc.* **1951**, *73*, 2806.
11.   Wawzonek, S.; Culbertson, T. P. *J. Am. Chem. Soc.* **1959**, *81*, 3367.
12.   Corey, E. J.; Hertler, W. R. *J. Am. Chem. Soc.* **1960**, *82*, 1657.
13.   Coleman, G. H.; Goheen, G. E. *J. Am. Chem. Soc.* **1938**, *60*, 730.
14.   Chow, Y. L.; Mojelsky, T. W.; Magdzinski, L. J. *Can. J. Chem.* **1985**, *63*, 2197.
15.   Nikishin, G. I.; Troyansky, E. I.; Lazareva, M. I. *Tetrahedron Lett.* **1985**, *26*, 3743.
16.   Betancor, C.; Concepcion, J. I.; Hernandez, R.; Salazar, J. A.; Suarez, E. *J. Org. Chem.* **1983**, *48*, 4430.
17.   deArmas, P.; Carrau, R.; Concepcion, J. I.; Francisco, C. G.; Hernandez, R.; Salazar, J. A.; Suarez, E. *Tetrahedron Lett.* **1985**, *26*, 2493.
18.   (a) Hernandez, R.; Medina, M. C.; Salazar, J. A.; Suarez, E. *Tetrahedron Lett.* **1987**, *28*, 2533. (b) Dorta, R. L.; Francisco, C. G.; Suarez, E. *Chem. Commun.* **1989**, 1168.
19.   [R] Togo, H.; Katohgi, M. *Synlett* **2001**, 565.
20.   Kessar, S. V.; Rampal, A. L.; Mahajan, K. P. M. *J. Chem. Soc.* **1962**, 4703.
21.   Partch, R. *Tetrahedron Lett.* **1966**, 1361.
22.   Surzur, J.-M.; Stella, L.; Tordo, P. *Bull. Soc. Chim. Fr.* **1975**, 1429.
23.   Titouani, S. L.; Laverhne, J.-P.; Viallefont, P. *Tetrahedron* **1981**, *36*, 2961.
24.   Stocking, E. M.; Sanz,-Cervera, J. F.; Unkefer, C. J.; Williams, R. M. *Tetrahedron* **2001**, *57*, 5303.
25.   (a) Gassman, P. G.; Heckert, D. C. *Tetrahedron* **1965**, *21*, 2725. (b) Gassman, P. G.; Fox, B. L. *J. Org. Chem.* **1967**, *32*, 3679.
26.   (a) Furstoss, R.; Teissier, P.; Waegell, B. *Tetrahedron Lett.* **1970**, 1263. (b) Esposito, P. G.; Furstoss, R.; Waegell, B. *Tetrahedron Lett.* **1971**, 895. (c) Tadayoni, R.; Lacrampe, J.; Heumann, A.; Furstoss, R.; Waegell, B. *Tetrahedron Lett.* **1975**, 735.

27.   Dupeyre, R.-M.; Rassat, A. *Tetrahedron Lett.* **1973**, 2699.

28.   Deshpande, R. P.; Nayak, U. R. *Indian J. Chem., Sect. B* **1979**, *17B*, 310.

29.   (a) Ban, Y.; Kimura, M.; Oishi, T. *Chem. Pharm. Bull.* **1976**, *24*, 1490. (b) Kimura, M.; Ban, Y. *Synthesis* **1976**, 201.

30.   Uskokovic, M. R.; Reese, C.; Lee, H. L.; Grethe, G.; Gutzwiller, J. *J. Am. Chem. Soc.* **1971**, *93*, 5902.

31.   Uskokovic, M. R.; Henderson, T.; Reese, C.; Lee, H. L.; Grethe, G.; Gutzwiller, J. *J. Am. Chem. Soc.* **1978**, *100*, 571.

32.   Kerwin, J. F.; Wolff, M. E.; Owings, F. F.; Lewis, B. B.; Blank, B.; Magnani, A.; Karash, C.; Georgian, V. *J. Am. Chem. Soc.* **1962**, *27*, 3628.

33.   (a) Adam, G.; Schreiber, K. *Tetrahedron Lett.* **1963**, 943. (b) Adam, G.; Schreiber, K. *Tetrahedron* **1964**, *20*, 1719.

Jie Jack Li

Chapter 3   Indoles 99

3.1   Bartoli Indole Synthesis

3.1.1   Description

The reaction of the *o*-substituted nitrobenzene **1** with 3 moles of the vinyl magnesium halide **2** gives the 7-substituted indole **3** as the major product.

3.1.2   Historical Perspective

In 1989, Bartoli reported that vinylmagnesium bromide reacted with 2-nitrotoluene (**4**) at −40°C in THF to furnish 7-methylindole (**5**) in 67% yield. The reaction process also proceeded well with other 2-substituted nitrobenzenes. However, the 3- or 4-substituted nitrobenzenes provided either no indole products or indoles in poor yield.[1]

3.1.3   Mechanism

The proposed mechanism of the Bartoli reaction was postulated by Bartoli *et al* based on experimental evidence.[2] Nitrotoluene (**4**) is attacked at the oxygen atom by the Grignard reagent and is reduced to nitrosotoluene (**7**) *via* enolate elimination from intermediate **6**. An inverse 1,2 addition of the second mole of Grignard reagent to the N=O double bond provides the enolate **8**. The *N*-aryl-*O*-vinylhydroxylamino magnesium salt **8** undergoes a [3,3]-sigmatropic rearrangement and this is followed by a rapid ring closure to form the bicyclic intermediate **10**. The third mole of Grignard reagent behaves as a base on this bicyclic intermediate **10** to rearomatize the six-membered ring to provide **11**. Elimination of water from **11** ultimately results in formation of the indole **5**.

3.1.4 Synthetic Utility

The Bartoli process has been employed effectively in the synthesis of 7-substituted indoles including 7-alkoxy (**15**),[3] 7-alkyl (**18**)[4], and 7-formal indoles (**21**).[5] Although the yields are only moderate, this process does provide a simple entry into indoles which were heretofore difficult to obtain.

Dobbs then modified the approach to prepare indoles devoid of substituents at position 7 by employing a blocking group. Treatment of a range of *o*-bromonitrobenzenes (see **22**) under Bartoli conditions with various vinyl Grignard reagents (the *o*-bromine atom is employed to direct the cyclization), followed by reductive removal of the bromine substituent provides indoles such as **24**.[6]

Recently, the Bartoli indole synthesis was extended to solid supports.[7] In contrast to the earlier reports in the liquid phase, *o,o´*-unsubstituted nitro analogs (see **25**) prove to be useful substrates. In addition, fluoro/chloro substituted nitro derivatives are well tolerated, which typically undergo nucleophilic substitution under Bartoli conditions in the liquid phase.

R = H, 82% (purity), 15% (overall yield)
R = F, 80% (purity), 18% (overall yield)

3.1.5   *Experimental*

Preparation of 7-methylindole[1]
Vinylmagnesium bromide (15 mmol) was quickly added to a stirred solution (10 mL/mmol) of 2-nitrotoluene (**4**) (5 mmol) in THF cooled at –40°C under nitrogen. The reaction mixture was stirred for 20 minutes and then poured into a saturated aqueous

solution of ammonium chloride. The mixture was then extracted with ether and dried over sodium sulfate. After chromatographic purification on silica gel, 7-methylindole (**5**) was obtained in 67% yield.

3.1.6   References

1. Bartoli, G.; Palmieri, G.; Bosco, M.; Dalpozzo, R. *Tetrahedron Lett.* **1989**, *30*, 2129.
2. Bosco, M.; Dalpozzo, R.; Bartoli, G.; Palmieri, G.; Petrini, M. *J. Chem. Soc. Perkin Trans. 2* **1991**, 657.
3. Dobson, D.; Todd, A.; Gilmore, J. *Synth. Commun.* **1991**, *21*, 611.
4. Pirrung, M. C.; Wedel, M.; Zhao, Y. *Synlett* **2002**, 143.
5. Dobson, D.; Gilmore, J.; Long, D. A. *Synlett* **1992**, 79.
6. Dobbs, A. *J. Org. Chem.* **2001**, *66*, 638.
7. Brase, S.; Knepper, K. *Org. Lett.* **2003**, *5*, 2829.

Jin Li and James M. Cook

3.2 Batcho–Leimgruber Indole Synthesis

3.2.1 Description

The Batcho–Leimgruber indole synthesis involves the condensation of *o*-nitrotoluene derivatives **1** with formamide acetals **2**, followed by reduction of the *trans*-β-dimethylamino-2-nitrostyrene **3** which results to furnish indole derivatives represented by **4**.[1]

3.2.2 Historical Perspective

In 1971, Batcho and Leimgruber introduced a new method for the synthesis of indoles. For example, condensation of *o*-nitrotoluene (**5**) with *N,N*-dimethylformamide dimethyl acetal (**6**) (DMFDMA) was followed by reduction of the *trans*-β-dimethylamino-2-nitrostyrene (**7**) which resulted to provide the indole (**8**).[2,3]

3.2.3 Mechanism
There are no detailed reports on the mechanism of the Batcho-Leimgruber process. The mechanism proposed here is based on similar types of reactions reported in the literature.[4,5]

3.2.4 Synthetic Utility
Several types of substituted formamide acetals have been utilized in the Batcho-Leimgruber reaction, including *N,N*-dimethylformamide dimethyl and diethyl acetals, *N*-formylpyrrolidine dimethyl acetal, and *N*-formylpiperidine dimethyl acetal.[2,4] Other reagents including tris(*N,N*-dimethylamino)methane and Gold's reagent have been employed for the condensation as well.[2,6,7] Of these, the readily available *N,N*-dimethylformamide dimethyl acetal (**6**) (DMFDMA) has found the widest application. This condensation is regiospecific as shown which is an important advantage.

14	15	16
X = 6-Cl,	89%,	63%
X = 5-Cl,	88%,	78%
X = 5-F,	92%,	51%

The condensation reaction can be facilitated by addition of an amine, such as pyrrolidine or triethylamine which permits use of a lower temperature.[8,9] Many methods have been employed for the reductive cyclization of the β-dialkylamino-2-nitrostyrenes to the indoles.[2,10–14] These methods includes H$_2$/Pd on carbon, H$_2$/Ni, Fe/AcOH, Na$_2$S$_2$O$_4$, FeSO$_4$/NH$_4$OH, TiCl$_3$, Zn/HCl and Ni/NH$_2$NH$_2$, and so forth.

In the synthesis of chuangxinmycin, Kozikowski *et al.* employed the potassium salt of the acid **23** to condense with DMFDMA **6** to obtain the β-nitrostyrene **24** with the desired regiochemistry. This material was later converted into indole **25** by reductive cyclization. When the ester **26** was used, the condensation occurred at the more acidic methylene group to provide **27** as the major product.[10]

Somei *et al.* have demonstrated the utility of TiCl$_3$ as the reducing agent in the Batcho-Leimgruber synthesis. The product distribution is sometimes dependent on the amount of TiCl$_3$ used as well as the choice of the solvent employed in the reaction.[14–16]

Garcia *et al.* has extended the Batcho-Leimgruber procedure to the synthesis of 2-substituted indoles.[17] Treatment of **36** with *o*-fluorobenzoyl chloride **37**, followed by *in situ* hydrolysis and deformylation gave ketone **38**. Reduction of nitroarylketone **38** with sodium hydrosulfite then furnished indole **39**. Similarly, bromoacetylation of **36** gave an acylenamine which was converted into the phthalimido derivative **40**. Hydrolysis and deformylation gave phthalimidoketone **41** which underwent reductive cyclization to furnish indole **42**.

TiCl₃ (mol eq)	Yield	32	33	34	35
12		83	0	0	0
8		42	22	6	1
6		6	36	8	8
4		3	57	13	16

Coe *et al.* reported an efficient modification for the preparation of *N*-substituted indole analogs for biology screening[18] in good yield. The intermediate β-nitrostyrene **44**, prepared from the condensation of **43** with DMFDMA, underwent methanolysis and reduction to provide the aniline acetal intermediate **45**. Alkylation of amine **45** was carried out employing standard conditions of reductive alkylation to provide *N*-alkyl analogs represented by **46**. The indole **47** was generated by formation of the oxonium ion (from **46**) under acidic conditions, followed by cyclization, accompanied by loss of methanol.

3.2.5   Experimental

Preparation of Methyl indole-4-carboxylate (30)[19]

A. *Methyl trans-2-[β-(dimethylamino)vinyl]-3-nitrobenzoate(29)*. A solution of methyl 2-methyl-3-nitrobenzoate (**28**) (9.75g, 0.05 mol) and DMFDMA (**6**) (17.85 g, 0.15 mol) in 50 mL of dry DMF was heated to 130°C for 6 h. The DMF was then removed under reduced pressure. The residue was distilled (120–130°C, 0.2 mm) from bulb-to-bulb to yield the title compound **29** [10.7 g (86%)].

B. *Methyl indole-4-carboxylate (30)*. A mixture of 7.0 g (28 mmol) of methyl *trans*-2-[β-(dimethylamino)vinyl]-3-nitrobenzoate(**29**) in 140 mL of dry benzene which contained 1.4 g of 10% Pd/C was shaken in a Parr apparatus under H$_2$ (50 psi) for 1.5 h. The catalyst was removed by filtration, and the benzene solution was washed with 30 mL of 5% aq. HCl, brine and dried over MgSO$_4$. After removal of the solvent under reduced pressure, the residue was purified *via* chromatography on silica gel to furnish 6.9 g (82%) of methyl indole-4-carboxylate (**30**).

3.2.6   *References*

1.   [R] Clark, R. D.; Repke, D. B. *Heterocycles* **1984**, *22*, 195.
2.   Leimgruber, W.; Batcho, A. D. *Third International Congress of Heterocyclic Chemistry*: Japan, 1971.
3.   Leimgruber, W.; Batcho, A. D. *US 3732245* **1973**.
4.   Abdulla, R. F.; Brinkmeyer, R. S. *Tetrahedron* **1979**, *35*, 1675.
5.   [R] Sundberg, R. J. *The Chemistry of Indoles*; Academic Press: New York & London, 1970.
6.   Kruse, L. I. *Heterocycles* **1981**, *16*, 1119.
7.   Gupton, J. T.; Lizzi, M. J.; Polk, D. *Synth. Commun.* **1982**, *12*, 939.
8.   Repke, D. B.; Ferguson, W. J. *J. Heterocyclic Chem.* **1982**, *19*, 845.
9.   Feldman, P. L.; Rapoport, H. *Synthesis* **1986**, 735.
10. Kozikowski, A. P.; Greco, M. N.; Springer, J. P. *J. Am. Chem. Soc.* **1982**, *104*, 7622.
11. Ferguson, W. J. *J. Heterocyclic Chem.* **1982**, *19*, 845.
12. Maehr, H.; Smallheer, J. M. *J. Org. Chem.* **1981**, *46*, 1752.
13. Moyer, M. P.; Shiurba, J. F.; Rapoport, H. *J. Org. Chem.* **1986**, *51*, 5106.
14. Somei, M.; Tsuchiya, M. *Chem. Pharm. Bull.* **1981**, *29*, 3145.
15. Somei, M.; Inoue, S.; Tokutake, S.; Yamada, F.; Kaneko, C. *Chem. Pharm. Bull.* **1981**, *29*, 726.
16. Somei, M.; Shoda, T. *Heterocycles* **1982**, *17*, 417.
17. Garcia, E. E.; Fryer, R. I. *J. Heterocyclic Chem.* **1974**, *11*, 219.
18. Coe, J. W.; Vetelino, M. G.; Bradlee, M. J. *Tetrahedron Lett.* **1996**, *37*, 6045.
19. Imuta, M.; Kawai, K.-I.; Ziffer, H. *J. Org. Chem.* **1980**, *45*, 3352.

Jin Li and James M. Cook

3.3 Bucherer Carbazole Synthesis

3.3.1 Description

The Bucherer carbazole synthesis[1-4] involves the treatment of a naphthyl alcohol (**1** or **4**) or a naphthylamine[5,6] (**2** or **5**) with a phenylhydrazine **3** in the presence of aqueous sodium bisulfite to afford, after acidic work-up, either a benzo[a]carbazole **4** or benzo[c]carbazole **6**.

3.3.2 Historical Perspective

The Bucherer carbazole synthesis was first demonstrated when **7** was heated in the presence of phenylhydrazines **8**, sodium hydroxide and sodium bisulfite; after acidic work-up, the benzocarbazole product **9** was isolated (~70% yield). When 2-naphthol was used the reaction was significantly slower with the yield of benzocarbazole being only 46% after several days at 130 °C.[7] Bucherer and co-workers investigated this reaction extensively concluding, incorrectly, that intermediate products were probably carbazole-*N*-sulfonic acids due to the ease with which they lost the sulfonic acid residues to yield benzocarbazoles.

3.3.3 Mechanism

Whereas the similarity between the Bucherer carbazole synthesis and the Fischer indolisation was noted by Bucherer and Seyde in their pioneering work,[1] it was only later

that the reaction mechanism was established.   Through investigation of reaction
intermediates, Seeboth *et al.* established that sodium bisulfite reacted with naphthols, not
via bisulfite addition to the keto form of the naphthols as originally thought,[8] but rather
via an intermediate tetralonesulfonic acid derivative.[3] Thus, 2-naphthol **10** forms **13**; this
adduct reacts with phenylhydrazine to give arylhydrazone **14**, which then undergoes
Fischer indolisation (see Chapter 3.4) and finally loss of sodium bisulfite to form the
benzocarbazole **20**. Diamine **21** has been isolated from reaction mixtures and may itself
be converted into **20** in 81% yield when heated with either sodium bisulfite or sulfurous
acid,[9] suggesting that an alternative mechanism *via* **22** may also be operative. Bucherer
carbazole synthesis using 1-naphthol proceeds analogously, although the 1-
tetralonephenylhydrazone-3-sulfonic acid intermediates (*cf.* **14**) tend to be stable and only
form benzocarbazoles on treatment with strong acids.

     When naphthylamines (*e.g.* **23**) are used in the Bucherer carbazole synthesis, they
are converted by the catalytic action of aqueous bisulfite into tetralonesulfonic acid
derivative **13** by the Bucherer reaction.[5] Addition of $NaHSO_3$ gives an enamine, which
tautomerises to the imine **24**; **24** is hydrolysed to keto form **13** and subsequent Bucherer
carbazole synthesis follows to afford the benzocarbazole product **20**.[3]

3.3.4   Variations and Improvements
The Bucherer carbazole synthesis is, to a certain extent, limited by the availabilty of the
starting naphthol and naphthylamine reagents.[10] However, a variety of substituents have
been used and these are illustrated for the conversion of **25** to **27**. The use of 2-
aminoanthracene in the Bucherer carbazole synthesis has also been demonstrated.[11]

R = 4-OH, 6-OH, 6-SO_3H,
7-OH, 7-OMe

     A number of *p*-substituted phenylhydrazines (*e.g.* **28**, R^1 = Me, Br; R^2 = H) have
been used to yield benzocarbazole derivatives (*e.g.* **29**, R^1 = Me, Br; R^2 = H). Reaction

of N_α-alkylphenylhydrazines (*e.g.* **28**, R^1 = H, R^2 = Me, Et, ^nBu) has been shown to give *N*-alkyl benzocarbazole derivatives (*e.g.* **29**, R^1 = H, R^2 = Me, Et, ^nBu).[12,13]

10   **28**   **29**

Both 1- and 2-naphthylhydrazine have been shown to react in good yield with 2-hydroxy-3-naphthoic acid in the presence of sodium bisulfite to give, after acidic work-up, dibenzocarbazole **30** and **31**, respectively.[7] When either 1- or 2-naphthylhydrazine is heated with sodium bisulfite, dibenzocarbazoles **32** and **31**, respectively, are isolated after acidic work-up.[7] It is suggested that loss of the hydrazine residue to form a bisulfite addition compound of the parent naphthol occurs initially; further reaction of this adduct with naphthylhydrazine then affords, after work-up, the products.

30   **31**   **32**

Japp-Maitland Condensation
In a related reaction, a number of aryloic compounds, behaving as ketonic moieties in the Fischer indolisation, have been shown to afford carbazoles in the Japp-Maitland condensation. For example, when either 1-naphthol,[14] 2-naphthol,[14] 6-alkyl-2-naphthol,[15] 9-hydroxyphenanthrene **33a**[14] or 6-hydroxychrysene **33b**[16] was heated with dry phenylhydrazine and its hydrochloride, the corresponding carbazole **34** was obtained.

33a R = H
   b R-R = -(CH=CH-)$_2$-

34a R = H
   b R-R = -(CH=CH-)$_2$-

4-Methylphenylhydrazine and both 1- and 2-naphthylhydrazines are also reported to react similarly. Phenols, in general, do not undergo this reaction, which is favoured by compounds exhibiting keto-enol tautomerism.[14,16]

3.3.5   Synthetic Utility

The Bucherer carbazole synthesis was pivotal in the preparation of the first hexahelicene **37a**.[8]   Reaction of 2,7-dihydroxynaphthalene **35** with phenylhydrazine and sodium bisulfite afforded helicene **37a** although in low yield. More recently, the synthesis was extended to the preparation of **37b** using 2,5-dimethylphenylhydrazine **36b**.[17]

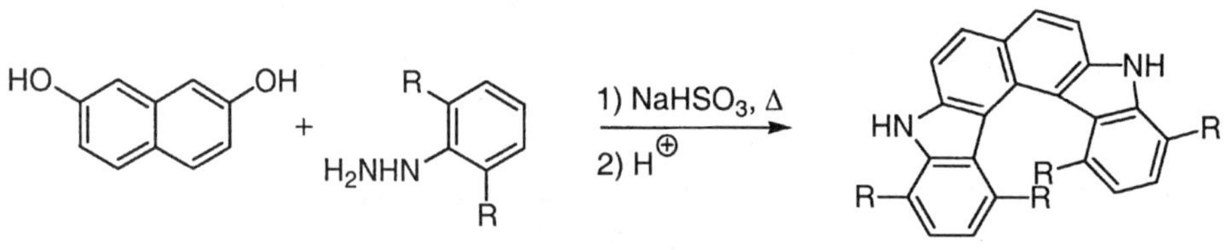

35	**36a** R = H	**37a** R = H
	b R = Me	**b** R = Me

The AMAPs (2-[{arylmethyl}amino]-1,3-propanediols) are a class of planar polycyclic aromatic derivatives, which contain polar side-chains. They are known to be DNA intercalators and possess broad spectrum antitumour activity. An approach to ^{14}C-radiolabelled AMAP derivative **40** used the Bucherer reaction as an initial starting reaction.[18]   2-Naphthol was reacted with 4-bromophenylhydrazine **38** in the presence of sodium metabisulfite and HCl to afford **39**. Subsequent derivatisation of **39** afforded **40**.

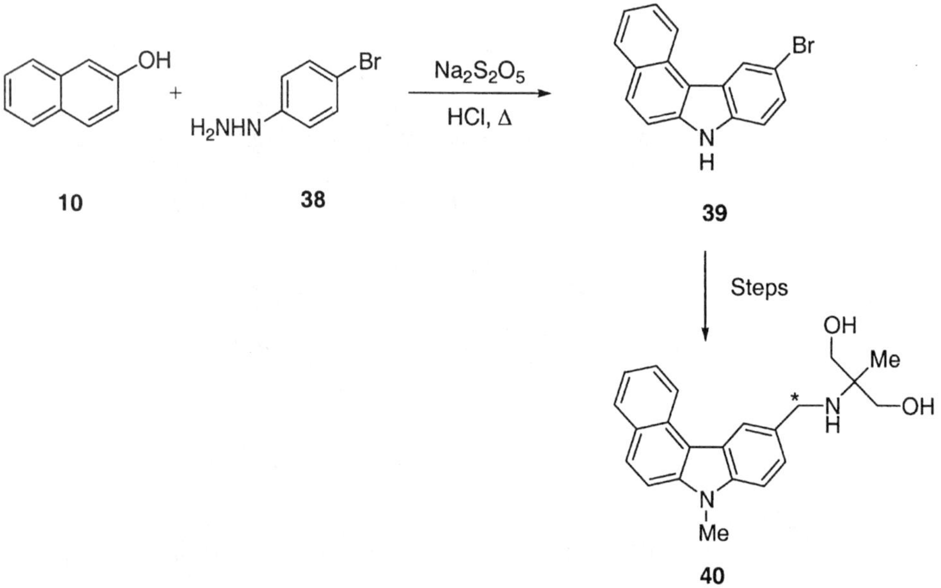

10	**38**	**39**

Steps

40

3.3.6 Experimental

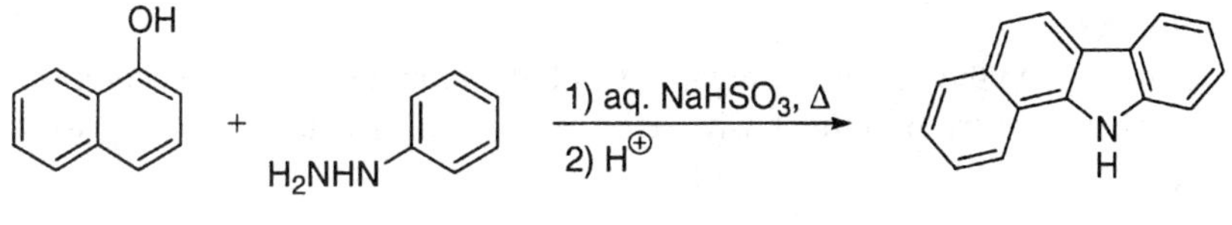

41   **42**   **43**

11*H*-benzo[a]carbazole (43)[19]:

A mixture of α-naphthol **41** (15.0 g, 0.1 mol), phenylhydrazine **42** (11.0 g, 0.1 mol) and sodium bisulfite solution (36 %, 250 g) was heated at reflux for 15 h. A further 4 g of phenylhydrazine was added and heating continued for 15 h, after which time the majority of the α-naphthol was consumed. After cooling, the mixture was extracted with ether. The oily, ether and aqueous insoluble residue was warmed with conc. HCl until a dark crystalline mass developed. After cooling, the mixture was extracted with ether. The organic extract was dried and concentrated to afford a crystalline residue which was purified by recrystallisation from ethanol to afford the product **43** as a white crystalline solid, mp 225 °C. A reaction yield is not given.

3.3.7 References

1.   Bucherer, H. T.; Seyde, F. *J. Prakt. Chem.* **1908**, *77*, 403.
2.   [R] Drake, N. L. *Org. React.* **1942**, *1*, 105.
3.   [R] Seeboth, H. *Angew. Chem., Int. Ed. Engl.* **1967**, *6*, 307.
4.   Robinson, B. *The Fischer Indole Synthesis*, Wiley-Interscience, New York, **1982**.
5.   The Bucherer reaction is the reversible conversion of a naphthylamine to a naphthol by the action of aqueous sulfite or bisulfite, see references 2 and 3.
6.   (a) Bucherer, H. T. *J. Prakt. Chem.* **1904**, *69*, 49; (b) Lepetit, R. *Bull. Soc. Ind. Mulhouse* **1903**, 326.
7.   Bucherer, H. T.; Schmidt, M. *J. Prakt. Chem.* **1909**, *79*, 369.
8.   Fuchs, W.; Niszel, F. *Chem. Ber.* **1927**, *60*, 209.
9.   Rieche, A.; Seeboth, H. *Liebigs Ann. Chem.* **1960**, *638*, 81.
10.   Buu-Hoï, N. P.; Hoán, N.; Khôi, N. H. *J. Org. Chem.* **1949**, *14*, 492.
11.   Zander, M.; Franke, W. *Chem. Ber.* **1963**, *96*, 699.
12.   (a) Friedländer, P. *Chem. Ztg.* **1916**, *40*, 918; (b) Friedländer, P. *Chem. Ber.* **1921**, *54*, 620; (c) Seeboth, H.; Bärwolff, D.; Becker, B. *Liebigs Ann. Chem.* **1965**, *683*, 85.
13.   (a) Farbenind, I. G. DE548,819, **1930**; (b) Gen. Aniline Works, US Patent 1,911,699, **1931**; (c) Gen. Aniline Works, US Patent 1,948,923, **1931**; (d) Farbenind, I. G. CH156,005, **1931**.
14.   Japp, F. R.; Maitland, W. *Proc. Chem. Soc.* **1901**, *17*, 176; (b) Japp, F. R.; Maitland, W. *J. Chem. Soc.* **1903**, *83*, 273.
15.   Buu-Hoï, N. P.; Royer, R.; Eckert, B.; Jacquignon, P. *J. Chem. Soc.* **1952**, 4867.
16.   Thang, D. C.; Can, C. X.; Buu-Hoï, N. P.; Jacquignon, P. *J. Chem. Soc., Perkin Trans. 1* **1972**, 1932.
17.   Pischel, I.; Grimme, S.; Kotila, S.; Nieger, M.; Vögtle, F. *Tetrahedron:Asymmetry* **1996**, *7*, 109.
18.   Hill, J. A.; Eaddy, J. F. *J. Labelled. Compd. Radiopharm.* **1994**, *34*, 697.
19.   Bucherer, H. T.; Sonnenburg, E. F. *J. Prakt. Chem.* **1910**, *81*, 1.

Adrian J. Moore

3.4   Fischer Indole Synthesis

3.4.1   Description

The Fischer indole synthesis can be regarded as the cyclization of an arylhydrazone **1** of an aldehyde or ketone by treatment with acid catalyst or effected thermally to form the indole nucleus **2**.[1-5]

The synthesis is often carried out by subjecting an equimolar mixture of the aryl hydrazine and aldehyde or ketone directly to the indolization conditions without isolation of the hydrazone.[6,7] Similarly, arylhydrazones, prepared by reduction of the corresponding aryldiazonium salt or *N*-nitrosoarylalkylamine[8,9] or by a palladium mediated coupling reaction,[10] can be subjected to the indolization conditions directly in the presence of the carbonyl moiety without isolation of the arylhydrazone. Such methods are useful when the arylhydrazone intermediates are unstable or toxic.

3.4.2   Historical Perspective

The first indolization of an arylhydrazone was reported in 1983 by Fischer and Jourdan[11] by treatment of pyruvic acid 1-methylphenylhydrazone **3** with alcoholic hydrogen chloride. However, it was not until the following year that Fischer and Hess[12] identified the product from this reaction as 1-methyl indole-2-carboxylic acid **4**.

Over 100 years after the initial discovery, the Fischer indole synthesis remains the most commonly employed method for the preparation of indoles.[13]

3.4.3   Mechanism

A number of reaction pathways have been proposed for the Fischer indolization reaction. The mechanism proposed by Robinson and Robinson in 1918,[14] which was extended by Allen and Wilson in 1943[15] and interpreted in light of modern electronic theory by Carlin and Fischer in 1948[16] is now generally accepted. The mechanism consists of three stages: (I) hydrazone–ene-hydrazine equilibrium; (II) formation of the new C–C bond *via* a [3,3]-sigmatropic rearrangement; (III) generation of the indole nucleus by loss of ammonia. Illustrated below is the indolization of the hydrazone **5** formed from 1-methyl-phenylhydrazine and acetone.[17]

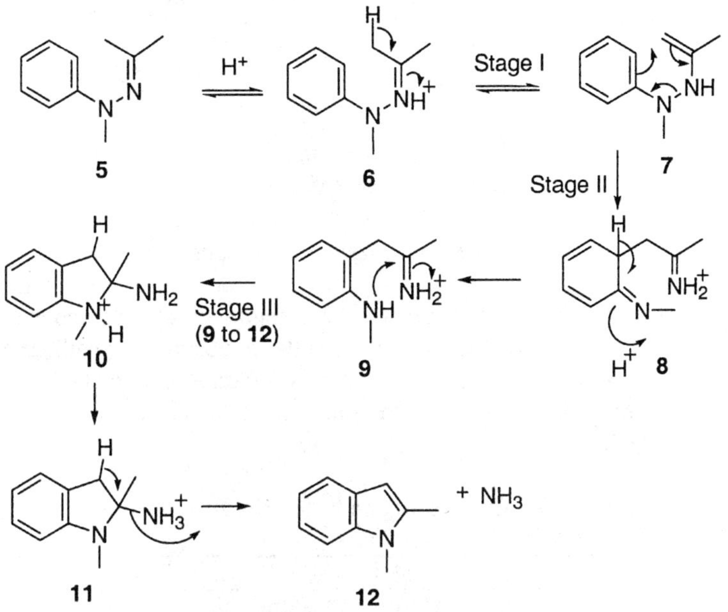

Under acidic conditions, the first step involves protonation of the imine nitrogen followed by tautomerization to form an ene-hydrazine intermediate (**7**). After the tautomerization, a [3,3]-sigmatropic rearrangement occurs, which provides intermediate **8**. Rearomatization then occurs *via* a proton shift to form the imine **9** which cyclizes to form the 5-membered ring **10**. Finally, loss of ammonia from **11** generates the indole nucleus in **12**.

3.4.4 Scope and Limitations
3.4.4.1. Catalytic/thermal indolization

A large number of Brønsted and Lewis acid catalysts have been employed in the Fischer indole synthesis. Only a few have been found to be sufficiently useful for general use. It is worth noting that some Fischer indolizations are unsuccessful simply due to the sensitivity of the reaction intermediates or products under acidic conditions. In many such cases the thermal indolization process may be of use if the reaction intermediates or products are thermally stable (*vide infra*). If the products (intermediates) are labile to either thermal or acidic conditions, the use of pyridine chloride[18] in pyridine or biphasic conditions are employed. The general mechanism for the acid catalyzed reaction is believed to be facilitated by the equilibrium between the aryl-hydrazone **13** (R' = H$^+$ or Lewis acid) and the ene-hydrazine tautomer **14**, presumably stabilizing the latter intermediate **14** by either protonation or complex formation (*i.e.* Lewis acid) at the more basic nitrogen atom (*i.e.* the 2-nitrogen atom in the arylhydrazone) is important.

It has been proposed[19] that protonation or complex formation at the 2-nitrogen atom of **14** would enhance the polarization of the 1',6'-π system and facilitate the rearrangement leading to new "C–C" bond formation. The equilibrium between the arylhydrazone and its ene-hydrazine tautomer is continuously promoted to the right by the irreversible rearomatization in stage II of the process. The indolization of arylhydrazones on heating in the presence of (or absence of) solvent under non-catalytic conditions can be rationalized by the formation of the transient intermediate **14** (R' = H). Under these thermal conditions, the equilibrium is continuously pushed to the right in favor of indole formation. Some commonly used catalysts in this process are summarized in Table 3.4.1.

Table 3.4.1. Commonly used catalysts [1,2] for the Fischer indole synthesis

Glacial acetic acid
Boron trifluoride in acetic acid
Copper (I) chloride
Formic acid
Conc. hydrochloric acid
Hydrogen chloride in ethanol
Hydrogen chloride in acetic acid
Polyphosphoric acid
Conc. sulfuric acid in ethanol
Conc. sulfuric acid in acetic acid
Zinc chloride
p-Toluenesulfonic acid

Although several examples of successful Fischer indolizations under thermal conditions in the absence of an acid catalyst had been reported, the general realization of thermally-mediated Fischer indole cyclizations did not become apparent until 1957.[20] They are particularly effective in indolizations which involve heteroaromatic hydrazones, such as pyridylhydrazones, pyrimidylhydrazones and quinolylhydrazones, that are difficult to cyclize under acidic conditions due to the deactivation of the heteronucleus by protonation of the heteroatom, as well as the inductive effects of the nitrogen atom.[2–4] Several recent reports have described the indolization of ketone **15** with a variety of arylhydrazines, including 2- and 3-hydrazinopyridine, 1- and 2-hydrazinonaphthalene, 2- and 3-hydrazinoquinoline, and 1-hydrazinophthalazine to provide the corresponding indoles *via* the thermal process. [21–24]

Bisagni *et al.*[25–29] described the indolization of a series of pyridone-based hydrazones **18** in 70–95% yield by refluxing them in diphenyl ether. Crooks and others reported successful thermal cyclizations as well.[30,31]

R = H, Me, Et
Y = CHOMe, NAc

18

19

3.4.4.2. Direction of indolization

Many aryhydrazones provide two or more isomers when subjected to the conditions of the Fischer indole cyclization. The product ratio and the direction of indolization can also be affected by different reaction conditions (*i.e.* catalysts and solvents), which is attributed, at least in part, to the relative stabilities of the two possible tautomeric ene-hydrazine intermediates. Generally, strongly acidic conditions favor formation of the least substituted ene-hydrazine, while cyclization carried out in weak acids favors the most substituted ene-hydrazine. Eaton's acid (10% P_2O_5 in $MeSO_3H$) has been demonstrated to be an effective catalyst for the preparation of 3-unsubstituted indoles from methyl ketones under strongly acidic conditions.[32] Many comprehensive reviews on this topic have appeared.[1–4]

3.4.4.2.1. Ketone/aldehyde components
A. Indolization of arylhydrazones of nonsymmetric ketones
The keto arylhydrazone **22** indolized to give only the 3-H substituted indole **23** upon treatment under the cyclization conditions. Indolization had occurred toward the more substituted carbon atom.[10,33]

The arylhydrazone **24** generally gives the 3-alkyl-2-methylindole **25** as major product. However, the indolization of ethyl methyl ketone has been reported to provide both 2,3-dimethyl indole and 2-ethyl indole.[34–36]

B. Indolization of aryhydrazones of cyclohexanone

The arylhydrazone **29** of a 2-substituted cyclohexanone gave a mixture of indolenine **30** and tetrahydrocarbazole **30**. It was reported that the relative amounts of **30** and **31** produced depended upon the catalyst employed.[37–39] For example, glacial acetic acid as catalyst provided largely **30**, whereas aqueous sulfuric acid gave **31** as the major product.

Catalyst	Solvent	% Yield	% Yield
AcOH	AcOH	61	6
H_2SO_4	H_2O	21	45

Similarly, it is expected that one or two of the 2- and 4-substituted isomers (**33** and **34**) would be formed by the indolization of phenylhydrazone **32**, depending on the reaction conditions as well as the nature of the substituent. [40–43]

32   **33**   **34**

35   **36**   **37**

In keeping with the thermodynamic stability of the ene-hydrazine, the mono-phenylhydrazone of cyclohexane-1,3-dione **38** provided only one of two possible isomers **39**, or at least, one dominant product **39** on indolization.[44,45]

38   **39**

However, when the reaction was carried out in a mixture of ethylene glycol and toluene and followed by hydrolysis the 2-keto derivative **42** was obtained in 54% yield, presumably *via* the more stable ene-hydrazine intermediate **41**.

40   **41**   **42**

In the same fashion, the phenylhydrazone of 2-tetralone **43** was found to provide only one isomer, carbazole **44**, when heated in dilute sulfuric acid.[46,47]

43   **44**

3.4.4.2.2. Hydrazine component
A.   Indolization of p-substituted phenylhydrazones

Indolization of the *p*-substituted phenylhydrazone **45** provides only one regioisomer as expected, the 5-substituted indole **46**. It is the most useful example of Fischer indole chemistry. An electron donating substituent on the phenyl ring in **45** enhances the rate of the indolization, whereas electron-withdrawing groups decrease the rate of cyclization.[4,48]

R_1=alkyl, alkoxy, halogen, nitro, etc.

45   **46**

PPA, 110°C, 1 h

78%

47   **48**

B. Indolization of o-substituted phenylhydrazones

Cyclization of an *o*-substituted arylhydrazone provides one isomer in many cases. However, the cyclization is more sluggish than the *m*- or *p*-substituted analogs. Sometimes the cyclization gives low yields of the desired indole products along with side products.[2]   Cyclization of the 2-substituted arylhydrazone can occur either to the unsubstituted side to provide the "normal" indole product (**50**), or to the substituted side, where other reactions take place (**53** and **56**).

Cyclization to the unsubstituted side

49   **50**

Cyclization to the substituted side (Nu = Cl⁻, if HCl is used as a catalyst)

Nu   **51**   **52**   **53**

In general, electron donating substituents tend to direct the cyclization to the substituted position instead of the unsubstituted position, especially when there is no steric hindrance to cyclization arising from the ketone portion of the hydrazone. In contrast, when an electron withdrawing group such as CF_3 or Cl is present in the *ortho*-position, indolization is then favored at the unsubstituted position (**58**).[5,49,50]

54 → [1,2]-shift → **55** → → **56**

57 $\xrightarrow[\text{EtOH, 33\%}]{\text{HCl}}$ **58** major product

C. Indolization of m-substituted phenylhydrazones

Indolization of 3-substituted phenylhydrazone **59** normally gives rise to two isomeric products, 4-substituted indole **60** and 6-substituted indole **61**.[2] The ratio of **60** to **61** from the cyclization depends on the nature of the R_1 group,[2] the structure of the carbonyl moiety of the hydrazone,[51] and the cyclization conditions (catalyst, solvent).[52,53]

59 R_1=alkyl, alkoxy, halogen   **60**   +   **61**

62   63:64=2:3   **63**   +   **64**

PCl$_3$, rt. 80%

There is a general trend resulting in a ratio of **60:61** < 1 when R_1 is an electron-donating group, whereas when R_1 is a *meta*- directing group **60:61** > 1.[52] Regardless of the electronic nature of the substituents, steric considerations will always favor the formation of the 6-substituted ring in preference to the 4-substituted indole.[51,52,54,55]

D. Indolization of Heterocyclic Hydrazines

Indolization of 3-pyridylhydrazone **65**(**67**) gives 4-azaindole **66**(**69**) as the major product.[8,56]

During indolization of the 3, 6 and 7-quinolylhydrazones, formation of the new C–C bond occurs between the appropriate carbon atom of the ketone/aldehyde moiety and the 4, 5 and 8 carbon atoms of the quinoline nucleus. It is consistent with the mechanism of formation of the C–C bond during indolization and the direction of electrophilic substitution in the quinoline nucleus.[57–60]

3.4.5   Synthetic Applications in the Pharmaceutical Industry
3.4.5.1. Synthesis of MK-677

MK-677, an orally active spiroindoline-based growth hormone secretagogue (GHS) agonist, discovered by Merck and currently in Phase II clinical studies, was synthesized with a Fischer indolization as a key step. The synthesis of this agonist involved a Fischer indole/reduction process and was achieved in 48% overall yield from the relatively cheap starting material, isonipecotic acid **72**.[61]

MK-677

one pot process from **73**, 93%

3.4.5.2 Synthesis of Imitrex® (Sumatriptan)

Imitrex® (Sumatriptan) **82** is the first selective 5-HT$_{1D}$ agonist developed by Glaxo for the treatment of migraine. The synthesis of Imitrex has been carried out by several different routes all of which involved a Fischer indolization reaction as the key step.[62-65] Outlined below is one of the synthetic routes.

Aniline **77** was converted into its diazonium salt with nitrous acid and this was followed by reduction with stannous chloride to afford the corresponding arylhydrazine **78**. Condensation of **78** with 3-cyanopropanal dimethylacetal **79** gave the arylhydrazone **80**. Treatment of **80** with PPE resulted in cyclization to indole **81**. The nitrile group was then reduced to the primary amine by catalytic hydrogenation. Reaction of the amine with excess formalin and sodium borohydride resulted in Imitrex (**82**).

3.4.6 Experimental

83   84

Preparation of 2-phenylindole[66] (84), a classic procedure.

In a tall 1 liter beaker was placed an intimate mixture of 53 g (0.25 mole) of freshly prepared acetophenone phenylhydrazone (**83**) and 250 g of powdered anhydrous zinc chloride. The beaker was immersed in an oil bath at 170°C, and the mixture was stirred vigorously by hand. The mass became liquid after 3–4 minutes, and evolution of white fumes began. The beaker was removed from the bath and the mixture was stirred for 5 minutes. In order to prevent solidification to a hard mass, 200 g of clean sand was thoroughly stirred into the reaction mixture. The zinc chloride was dissolved by digesting the mixture overnight on a steam cone with 800 mL of water and 25 mL of concentrated hydrochloric acid. The sand and crude 2-phenylindole were removed by filtration, and the solids were boiled with 600 mL of 95% ethanol. The hot mixture was decolorized with Norit and filtered through a hot 10-cm Büchner funnel. The sand and Norit were then washed with 75 mL of hot ethanol. After the combined filtrate was cooled to room temperature, the 2-phenylindole was collected on a 10-cm Büchner funnel and washed three times with small amounts (15–20 mL) of cold ethanol. The first crop was quite pure; after drying in a vacuum desiccator over calcium chloride it weighed 30–33 g and melted at 188–189°C (cor.). A little Norit was added to the combined filtrate and washings, which were then concentrated to a volume of 200 mL and filtered. The filtrate, on cooling, furnished a second crop of 5–6 g of impure product, which melted at 186–188°C. The total yield of 2-phenylindole (**84**) was 35–39 g (72–80%).

3.4.7 References

1.   [R] Hughes, D. L. *Org. Prep. Proced. Int.* **1993**, *25*, 609.
2.   [R] Robinson, B. *The Fischer Indole Synthesis*; J. Wiley & Sons: New York, 1982; 923pp.
3.   [R] Robinson, B. *Chem. Rev.* **1963**, *63*, 373.
4.   [R] Robinson, B. *Chem. Rev.* **1969**, *69*, 227.
5.   [R] Ishii, H. *Acc. Chem. Res.* **1981**, *14*, 275.
6.   Bosch, J.; Roca, T.; Armengol, M.; Fernández-Forner, D. *Tetrahedron* **2001**, *57*, 1041.
7.   Brodfuehrer, P. R.; Chen, B.-C.; Sattelberg, T. R., Sr.; Smith, P. R.; Reddy, J. P.; Stark, D. R.; Quinlan, S. L.; Reid, J. G.; Thottathil, J. K.; Wang, S.-P. *J. Org. Chem.* **1997**, *62*, 9192.
8.   Ficken, G. E.; Kendall, J. D. *J. Chem. Soc.* **1961**, 584.
9.   Clifton, P. V.; Plant, S. G. P. *J. Chem. Soc.* **1951**, 461.
10.   Wagaw, S.; Yang, B. H.; Buchwald, S. L. *J. Am. Chem. Soc.* **1999**, *121*, 10251.
11.   Fischer, E.; Jourdan, F. *Ber.* **1883**, *16*, 2241.
12.   Fischer, E.; Hess, O. *Ber.* **1884**, *17*, 559.
13.   Sundberg, R. J. *Best Synthetic Methods*; Academic Press: London, 1996.
14.   Robinson, G. M.; Robinson, R. *J. Chem. Soc.* **1918**, 639.
15.   Allen, C. F. H.; Wilson, C. V. *J. Am. Chem. Soc.* **1943**, *65*, 611.
16.   Carlin, R. B.; Fischer, E. E. *J. Am. Chem. Soc.* **1948**, *70*, 3421.
17.   Hughes, D. L.; Zhao, D. *J. Org. Chem.* **1993**, *58*, 228.
18.   Welch, W. M. *Synthesis* **1977**, 645.
19.   Arbuzov, A. E.; Kitaev, Y. P. *J. Gen. Chem. USSR.* **1957**, *27*, 2388.

20. Fitzpatrick, J. T.; Hiser, R. D. *J. Org. Chem.* **1957**, *22*, 1703.
21. Fukuda, N.; Trudell, M. L.; Johnson, B.; Cook, J. M. cook. *Tetrahedron Lett.* **1985**, *26*, 2139.
22. Trudell, M. L.; Basile, A. S.; Shannon, H. E.; Skolnick, P.; Cook, J. M. *J. Med. Chem.* **1987**, *30*, 456.
23. Tan, Y.; Trudell, M. L.; Cook, J. M. *Heterocycles* **1988**, *27*, 1607.
24. Martin, M. J.; Trudell, M. L.; Arauzo, H. D.; Allen, M. S.; LaLoggia, A. J.; Deng, L.; Schultz, C. A.; Tan, Y.; Bi, Y.; Narayanan, K.; Dorn, L. J.; Koehler, K. F.; Skolnick, P.; Cook, J. M. *J. Med. Chem.* **1992**, *35*, 4105.
25. Hung, N. C.; Bisagni, E. *Tetrahedron* **1986**, *42*, 2303.
26. Nguyen, C. H.; Bisagni, E. *Tetrahedron* **1987**, *43*, 527.
27. Bisagni, E.; Nguyen, C. H.; Pierre, A.; Pepin, O.; de Cointet, P.; Gros, P. *J. Med. Chem.* **1988**, *31*, 398.
28. Nguyen, C. H.; Lhoste, J.-M.; Lavelle, F.; Bissery, M.-C.; Bisagni, E. *J. Med. Chem.* **1990**, *33*, 1519.
29. Nyugen, C. H.; Lavelle, F.; Riou, J.-F.; Bissery, M.-C.; Huel, C.; Bisagni, E. *Anti-Cancer Drug Des.* **1992**, *7*, 235.
30. Crooks, P. A.; Robinson, B. *Chem. Ind.* **1967**, 547.
31. Crooks, P. A.; Robinson, B. *Can. J. Chem.* **1969**, *47*, 2061.
32. Zhao, D.; Hughes, D. L.; Bender, D. R.; DeMarco, A. M.; Reider, P. J. *J. Org. Chem.* **1991**, *56*, 3001.
33. Zimmermann, T. *J. Hetrocyclic. Chem.* **2000**, *37*, 1571.
34. Fischer, E. *Leibigs Ann. Chem.* **1886**, *236*, 116.
35. Korczynski, A.; Brydowna, W.; Kierzek, L. *Gazz. Chim. Ital.* **1926**, *56*, 903.
36. Zen, S.; Takeda, Y.; Yasuda, A.; Umezawa, S. *Bull. Chem. Soc. Jpn.* **1967**, *40*, 431.
37. Pausacker, K. H.; Schubert, C. I. *J. Chem. Soc.* **1949**, 1384.
38. Plancher, G. *Gazz. Chim. Ital.* **1900**, *30*, 558.
39. Buu-Hoï, N. P.; Jacquignon, P.; Loc, T. B. *J. Chem. Soc.* **1958**, 738.
40. Barclay, B. M.; Campbell, N. *J. Chem. Soc.* **1945**, 530.
41. Grammaticakis, P. *Compt. Rend. Acad. Sci.* **1940**, *210*, 569.
42. Borsche, W.; Witte, A.; Bothe, W. *Liebigs Ann. Chem.* **1908**, *359*, 49.
43. Ergün, Y.; Patir, S.; Okay, G. *J. Hetrocyclic Chem.* **2002**, *39*, 315.
44. Clemo, G. R.; Felton, D. G. I. *J. Chem. Soc.* **1951**, 700.
45. Rodriguez, J. G.; Temprano, F.; Esteban-Calderon, C.; Martinez-Ripoll, M. *J. Chem. Soc. Perkin Trans.* **1989**, 2117.
46. Ghigi, E. *Gazz. Chim. Ital.* **1931**, *61*, 43.
47. Katritzky, A. R.; Wang, Z.-Q. *J. Hetrocyclic Chem.* **1988**, *25*, 671.
48. Shono, T.; Matsumura, Y.; Kanazawa, T. *Tetrahedron Lett.* **1983**, 1259.
49. Luis, S. V.; Burguete, M. I. *Tetrahedron* **1991**, *47*, 1737.
50. Ishii, H.; Murakami, Y.; Ishikawa, T. *Chem. Pharm. Bull.* **1990**, *38*, 597.
51. Grandberg, I. I.; Belyaeva, L. D.; Dmitriev, L. B. *Chem. Het. Comps.* **1971**, *7*, 54.
52. Ockenden, D. W.; Schofield, K. *J. Chem. Soc.* **1957**, 3175.
53. Baccolini, G.; Marotta, E. *Tetrahedron* **1985**, *41*, 4615.
54. Grandberg, I. I.; Belyaeva, L. D.; Dmitriev, L. B. *Chem. Het. Comps.* **1971**, *7*, 1131.
55. Grandberg, I. I.; Belyaeva, L. D.; Dmitriev, L. B. *Chem. Het. Comps.* **1973**, *9*, 31.
56. Kelly, A. H.; Parrick, J. *J. Chem. Soc. C* **1970**, 303.
57. Clemo, G. R.; Felton, D. G. I. *J. Chem. Soc.* **1951**, 671.
58. Govindachari, T. R.; Rajappa, S.; Sudarsanam, V. *Tetrahedron* **1961**, *16*, 1.
59. Wieland, H.; Hornor, L. *Liebigs Ann. Chem.* **1938**, *536*, 89.
60. Buu-Hoï, N. P.; Périn, F.; Jacquignon, P. *J. Chem. Soc.* **1960**, 4500.
61. Maligres, P. E.; Houpis, I.; Rossen, K.; Molina, J. S.; Upadhyay, V.; Wells, K. M.; Reamer, R. A.; Lynch, J. E.; Askin, D.; Volante, R. P.; Reider, P. J. *Tetrahedron* **1997**, *53*, 10983.
62. Oxford, A. W. *German Offen DE3527648*, **1986**.
63. Pete, B.; Bitter, I.; Harsányi, K.; Töke, L. *Heterocycles* **2000**, *53*, 665.
64. Pete, B.; Bitter, I.; Szántay, C., Jr.; Schön; Töke, L. *Heterocycles* **1998**, *48*, 1139.
65. Dowle, M. D.; Coates, I. H. *U.S.4,816,470*, **Mar 28, 1989**.
66. Shriner, R. L.; Ashley, W. C.; Welch, E. *Org. Synth.* **1955**, *Coll. Vol. 3*, 725.

Jin Li and James M. Cook

3.5   Gassman indole synthesis

3.5.1   Description

The Gassman indole synthesis involves an one-pot process in which hypohalite, a β-carbonyl sulfide derivative **2**, and a base are added sequentially to an aniline or a substituted aniline **1** to provide 3-thioalkoxyindoles **3**.   Raney nickel-mediated desulfurization of **3** then produces the parent indole **4**.[1-5]

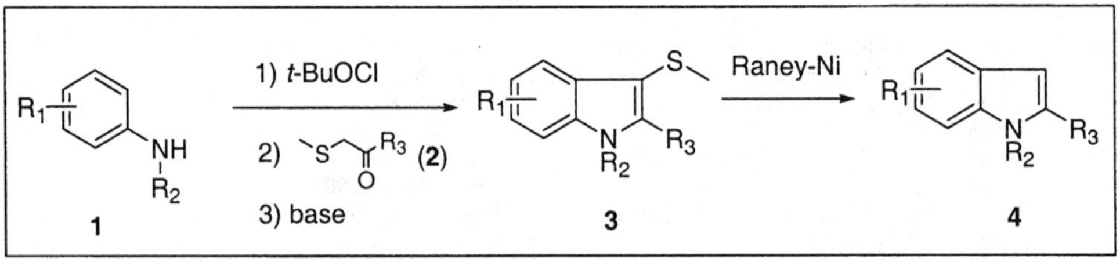

3.5.2   Historical Perspective

In 1974, Gassman *et al.* reported a general method for the synthesis of indoles.[1]   For example, aniline **5** was reacted sequentially with *t*-BuOCl, methylthio-2-propanone **6** and triethylamine to yield methylthioindole **7** in 69% yield.   The Raney-nickel mediated desulfurization of **7** then provided 2-methylindole **8** in 79% yield.[1]   The scope and mechanism of the process were discussed in the same report by Gassman and coworkers as well.

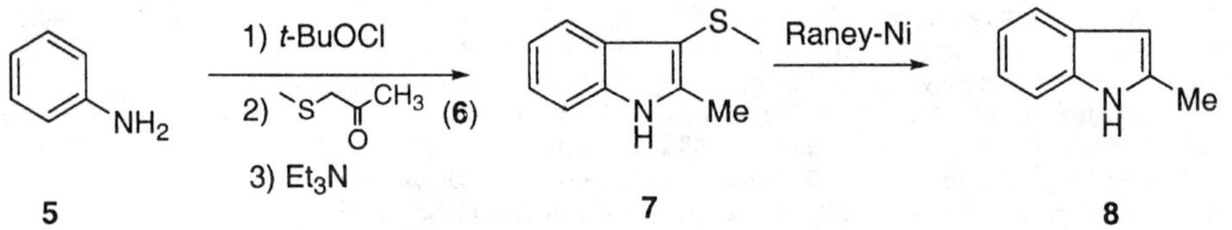

3.5.3   Mechanism

The mechanism of the indolization of aniline **5** with methylthio-2-propanone **6** is illustrated below.[1]   Aniline **5** reacts with *t*-BuOCl to provide *N*-chloroaniline **9**.   This chloroaniline **9** reacts with sulfide **6** to yield azasulfonium salt **10**.   Deprotonation of the carbon atom adjacent to the sulfur provides the ylide **11**.   Intramolecular attack of the nucleophilic portion of the ylide **11** in a Sommelet–Hauser type rearrangement produces **12**.   Proton transfer and re-aromatization leads to **13** after which intramolecular addition of the amine to the carbonyl function generates the carbinolamine **14**.   Dehydration of **14** by prototropic rearrangement eventually furnishes the indole **8**.

3.5.4 Synthetic Utility

The Gassman indole synthesis provides a single regioisomer when *ortho/para* substituted anilines are employed, the yields of which are quite good generally. This provides some advantage in the preparation of 7-substituted indoles compared to other methods which normally give low yields, that is, the Fischer indole process.[1]

R = *p*-OC(O)CH$_3$	68%	72%
R = *p*-CH$_3$	60%	80%
R = *p*-Cl	72%	74%
R = *p*-CO$_2$Et	58%	83%
R = *o*-CH$_3$	72%	73%

The *meta*-substituted anilines provide one or two isomeric products depending on the nature of the substituents (*vide infra*).

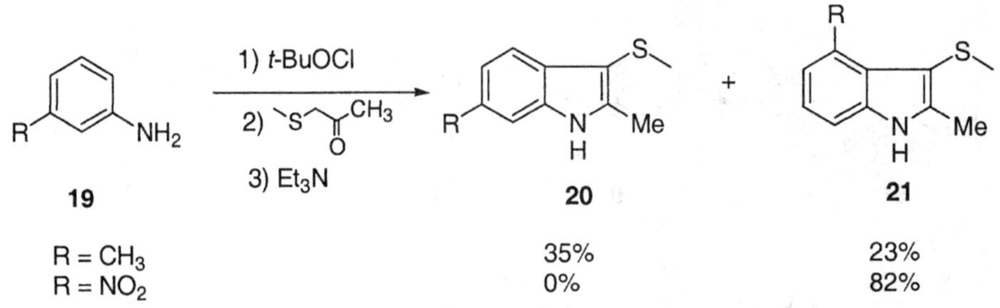

| R = CH$_3$ | 35% | 23% |
| R = NO$_2$ | 0% | 82% |

A variety of 2-substituted indoles can be prepared by the Gassman process. For example, when methyl phenacyl sulfide **22** was employed with aniline, the 2-phenyl indole was obtained in 81% yield as shown here.

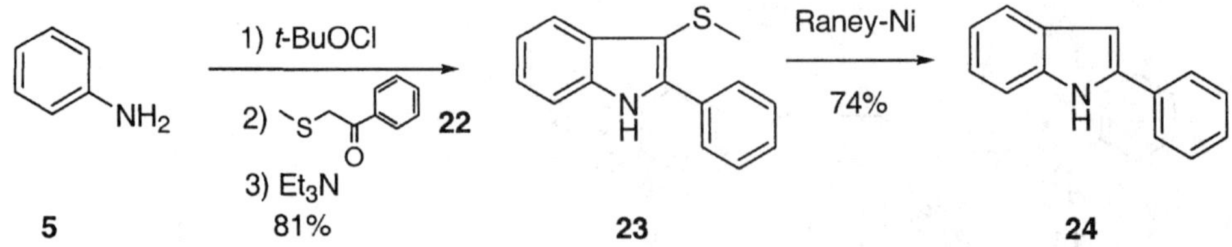

The Gassman process has also been applied to the preparation of indoles which are devoid of a substituent at the 2 position. For example, indolization of aniline **5** with methylthioacetaldehyde **25** or methylthioacetaldehyde dimethyl acetal **26** furnished indole **27**; however, a higher yield was obtained when the acetal **26** was employed.[1]

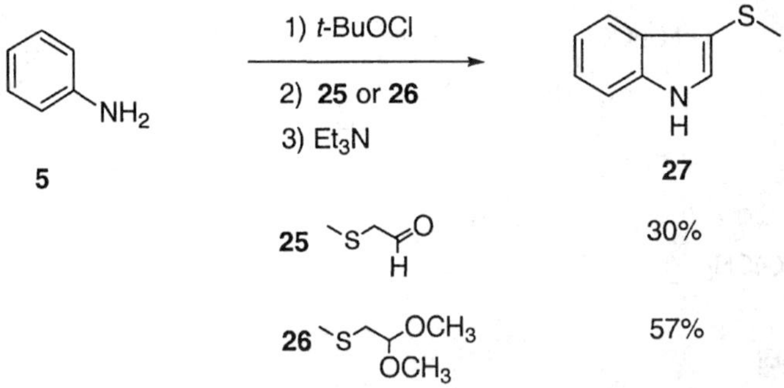

Many 3-substituted indoles have also been prepared with the use of α-alkyl or α-aryl-β-keto sulfides. Thus indolization of aniline **5** with 3-methylthio-2-butanone **27** furnished indolenine **28**, presumably *via* the same mechanism discussed earlier. The indolenine **28** was relatively unstable and reduced to the indole **29** without purification. Tetrahydrocarbazole **32** was prepared in 58% overall yield. Smith *et al.* made excellent use of the Gassman process in the total synthesis of (+)-paspalicine and (+)-paspalinine.[4,5]

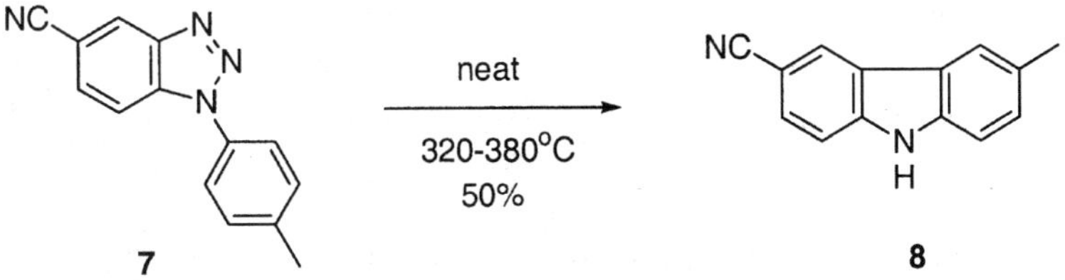

7 → **8**

neat

320–380°C

50%

The Graebe–Ullman carbazole synthesis has been employed in the preparation of substituted carbolines,[6,7] as well as indolo[2,3-b] quinolines,[8] which are often difficult to synthesize *via* other approaches, for example, the Fischer indole process.

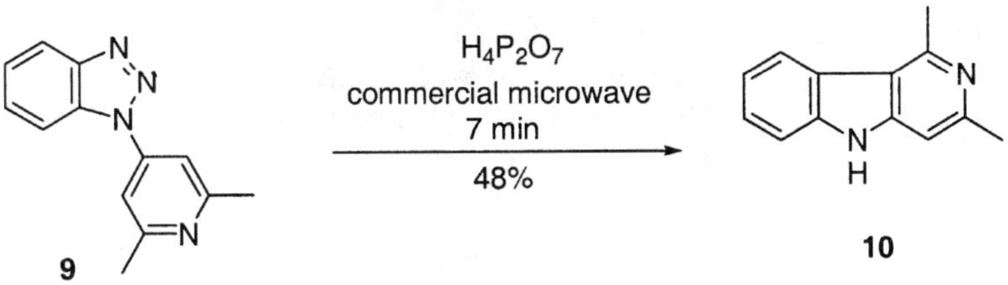

$H_4P_2O_7$

commercial microwave

7 min

48%

9 → **10**

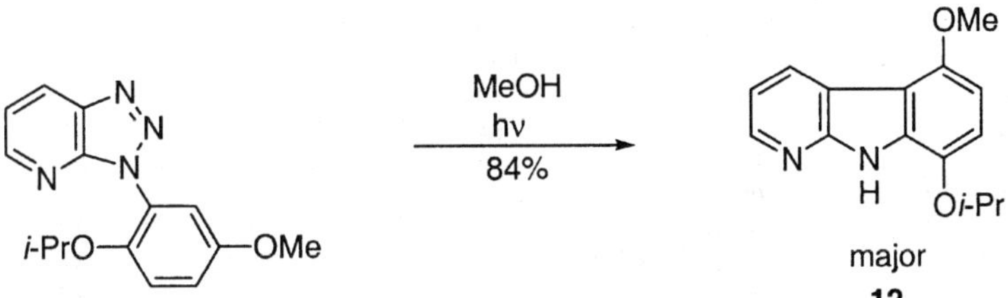

MeOH

hv

84%

11 → **12** (major)

Hagan *et al.* utilized the Graebe–Ullmann process to synthesize polycyclic acridines **14** which exhibit anti-tumor activity.[2]

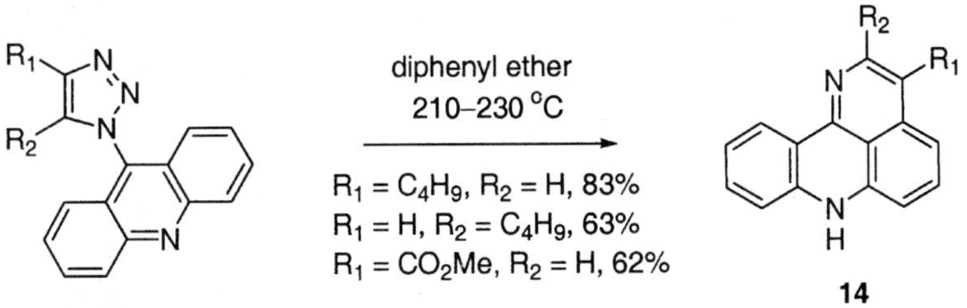

diphenyl ether

210–230 °C

$R_1 = C_4H_9, R_2 = H, 83\%$
$R_1 = H, R_2 = C_4H_9, 63\%$
$R_1 = CO_2Me, R_2 = H, 62\%$

13 → **14**

3.6.5 *Experimental*

Preparation of 6H-Indolo[2,3-b]quinoline 16[8]

Triazole **15** was mixed with polyphosphoric acid (150 mL) and the mixture was heated until gas evolution ceased (130–180°C) (**Caution: It is important to note that one mole of N_2 is evolved**). After cooling, the syrup was poured into ice water (2 L), and the precipitate was collected, washed with H_2O, heated on a steam bath with 25% ammonia (100 mL), and filtered. The product was then washed with H_2O and crystallized from pyridine to furnish **16** in 30% yield.

3.6.6 *References*

1.   Graebe, C.; Ullmann, F. *Ann.* **1896**, *291*, 16.
2.   Hagan, D. J.; Chan, D.; Schwalbe, C. H.; Stevens, M. F. G. *J. Chem. Soc., Perkin Trans I* **1998**, 915.
3.   Mitchell, G.; Rees, C. W. *J. Chem. Soc., Perkin Trans I* **1987**, 403.
4.   Ashton, B. W.; Suschitzky, H. *J. Chem. Soc.* **1957**, 4559.
5.   Coker, G. G.; Plant, S. G. P.; Turner, P. B. *J. Chem. Soc.* **1951**, 110.
6.   Molina, A.; Vaquero, J. J.; Garcia-Navio, J. L.; Alvarez-Builla, J.; Pascual-Teresa, B. D.; Gago, F.; Rodrigo,
        M. M.; Ballesteros, M. *J. Org. Chem.* **1996**, *61*, 5587.
7.   Mehta, L. K.; Parrick, J.; Payne, F. *J. Chem. Soc., Perkin Trans I* **1993**, 1261.
8.   Peczynska-Czoch, W.; Pognan, F.; Kaczmarek, L.; Boratynski, J. *J. Med. Chem.* **1994**, *37*, 3503.

Jin Li and James M. Cook

3.7 Hegedus Indole Synthesis

3.7.1 Description

The Hegedus indole synthesis involves one of the earlier (formal) examples of olefin hydroamination. An *ortho*-vinyl or *ortho*-allyl aniline derivative **1** is treated with palladium(II) to deliver an intermediate resulting from alkene aminopalladation. Subsequent reduction and/or isomerization steps then provide the indoline or indole unit **2**, respectively.

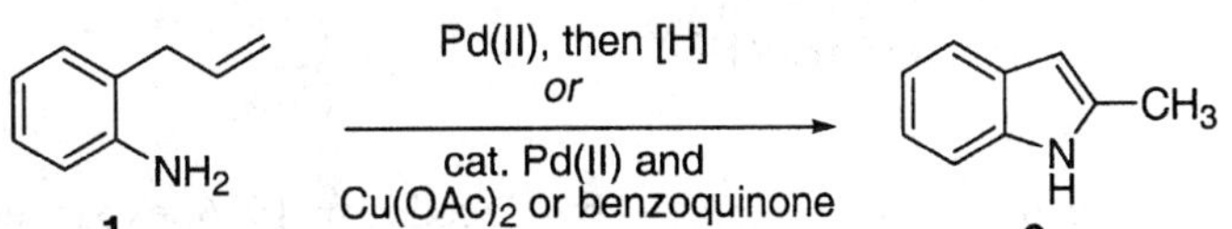

3.7.2 Historical Perspective

Metal-assisted heterocycle formation witnessed tremendous growth in the second half of the 20th century, largely patterned after mercury(II) and silver(I)-assisted olefin/heteroatom cyclizations. Using this intellectual springboard, Castro reported in 1966 that substituted indoles and benzofurans could be generated through the reaction of terminal alkynes and copper(I) salts with *ortho*-halo anilines and phenols, respectively.[1] Subsequent developments include the Mori-Ban indole synthesis (1976) through the use of an olefin insertion mechanism beginning from *ortho*-halo *N*-allylanilines such as **3** and using a stoichiometric amount of nickel(0) to give indole **4**.[2]

Olefin Oxymercuration

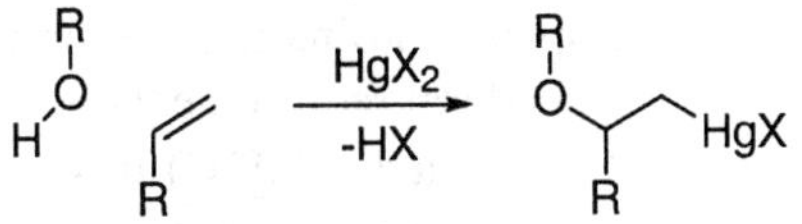

Castro

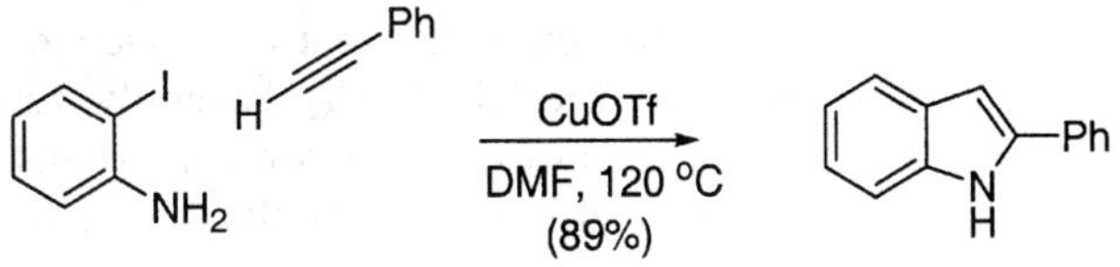

Mori-Ban

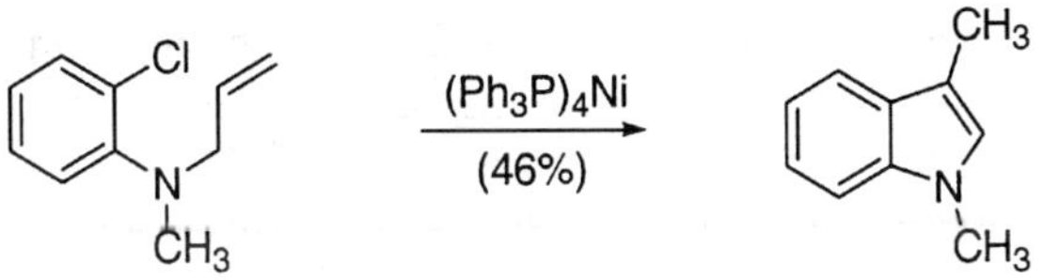

In 1974, Hegedus and coworkers reported the palladium(II)-promoted addition of secondary amines to α-olefins by analogy to the Wacker oxidation of terminal olefins and the platinum(II) promoted variant described earlier.[3] This transformation provided an early example of (formally) alkene hydroamination and a remarkably direct route to tertiary amines without the usual problems associated with the use of alkyl halide electrophiles.

The intramolecular variant, reported by Hegedus in 1976, expanded the scope to weakly nucleophilic amines.[4] *ortho*-Allylaniline **1** delivered 2-methyl indole in 84% yield using a stoichiometric amount of $(CH_3CN)_2PdCl_2$ and triethylamine in THF *via* the intermediacy of **5**.

3.7.3 Mechanism

The mechanistic information provided in the literature is both qualitative and often inspired by analogy to the behavior of palladium catalysts in the Heck reaction. The importance of maintaining an open coordination site on palladium(II) for olefin binding prior to amine results in two procedural requirements for the Hegedus synthesis. The first is the need to add the amine to a cooled (−50 °C) mixture of the palladium catalyst and olefin. At higher temperature, the amine displaces the bound olefin to give an aminopalladium complex that is not easily converted to the desired tertiary amine. Since formation of the amine-palladium(II) complex is not a productive pathway to product, aminopalladium catalysts such as $(Me_2NH)_2PdCl_2$ generally return only unreacted substrate. However, the optimal amount of amine is two equivalents relative to olefin. In the original work, the olefin component largely determined the efficiency of the process in the order: terminal > *trans*-disubstituted > *cis*-disubstituted >> cyclic disubstituted ~ trisubstituted with the latter providing less than 5% of the desired product. This trend is consistent with olefin-palladium(II) binding strength. Limitations on the amine component are considerably greater, being restricted to nonhindered secondary amines. However, unlike the analogous transformation with platinum, selectivity for olefin *mono*amination was uncorrupted.

Scheme 1. Proposed Mechanistic Steps Leading to Olefin Aminopalladation

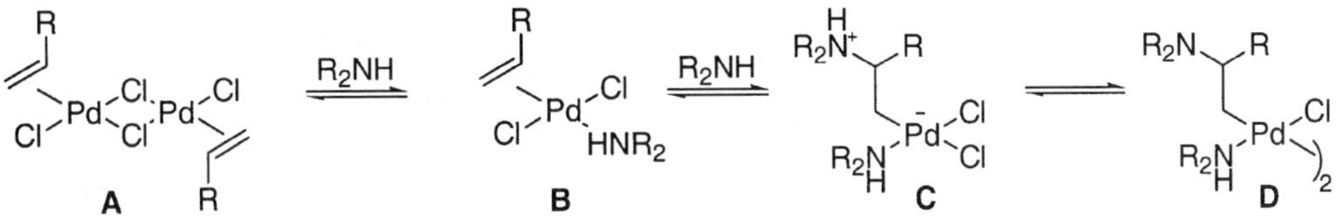

The characteristics of the intramolecular variant further supported the notion that a palladium(II)-olefin complex prone to amine attack on the complex in *anti*-fashion (**B** → **C**) is necessary for efficient nitrogen-carbon bond formation. Furthermore, the trend toward Markovnikov addition regioselection is driven by formation of the less substituted palladium alkyl intermediate (**D**). The intramolecular transformation is formally an isomerization, owing to reductive elimination of the alkylpalladium intermediate (to generate metallic palladium), whereas the increased stability of the analogous species in the intermolecular variant allowed its reduction in a subsequent step (H_2, $NaBH_4$, or HCl). Hegedus reported the variant substoichiometric in palladium(II) using either benzoquinone or copper(II) chloride as the stoichiometric oxidant.[5] Lithium chloride was found to be an additive beneficial to overall yield.

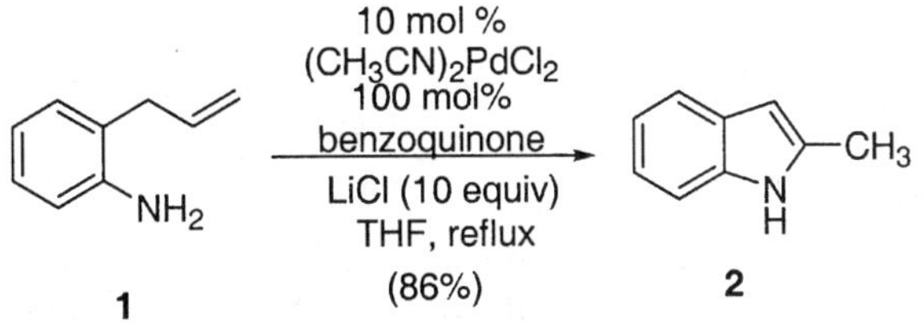

3.7.4 Synthetic Utility

3,4-Disubstituted indole represents a more prevalent synthon for target-oriented synthesis accessible via the Hegedus method. Synthetic access was gained to 4-bromo *N*-tosylindole (**8**) and 3-iodo-4-bromo indole (**6**) in six and eight steps, respectively, and good overall yield.[6] As one example of the general manner in which haloindoles such as these can be further elaborated,[7] the latter was further functionalized to the ergot alkaloid *rac*-aurantioclavine (**7**).[8]

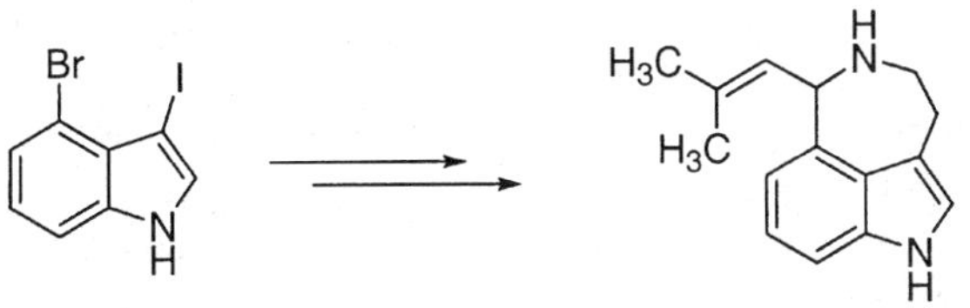

6

7, *rac*-aurantioclavine

8

9, arcyriacyanin A

10   **11**

4-Bromo-*N*-tosylindole (**8**) has been converted to arcyriacyanin A (**9**) by Steglich.[9] Similarly, Rapapport used 3,4-dibromoindole (**10**) to construct the ergot alkaloid tricyclic core (**11**),[10] and Murakami reported a relatively short route to costaclavine.[11] Similar uses in other natural product syntheses continue to appear in the literature.[12]

3.7.5   Experimental

2-Methylindole (2):[5]
2-(2-Propenyl)aniline (**1**, 1.0 g, 7.52 mmol), PdCl$_2$(CH$_3$CN)$_2$ (0.195 g, 0.75 mmol), benzoquinone (0.812 g, 7.52 mmol), and LiCl (3.158 g, 75.2 mmol) were combined in THF (95 mL). After 5 h at reflux, the solvent was removed and the residue was stirred with ether and decolorizing charcoal for approximately 20 min and filtered. The filtrate was washed five times with 50-mL portions of 1 M NaOH. The solvent was removed by vacuum, and the residue was placed on a silica gel column and eluted with 3:1 petroleum ether/ether. 2-Methylindole (**2**, 0.818 g, 86%) was collected as a white, crystalline solid, identical with authentic material.[13]

3.7.6   References

1   Castro, C. E.; Gaughan, E. J.; Owsley, E. C. *J. Org. Chem.* **1966**, *31*, 4071.
2   Mori, M.; Ban, Y. *Tetrahedron Lett.* **1976**, *17*, 1803.
3   Akermark, B.; Backvall, J. E.; Hegedus, L. S.; Zetterberg, K.; Siirala-Hansen, K.; Sjoberg, K. *J. Organomet. Chem.* **1974**, *782*, 127
4   Hegedus, L. S.; Allen, G. F.; Waterman, E. L. *J. Am. Chem. Soc.* **1976**, 98, 2674.
5   Hegedus, L. S.; Allen, G. F.; Bozell, J. J.; Waterman, E. L. *J. Am. Chem. Soc.* **1978**, *100*, 5800.
6   Harrington, P. J.; Hegedus, L. S. *J. Org. Chem.* **1984**, *49*, 2657.
7   (a) Harrington, P. J.; Hegedus, L. S. *J. Org. Chem.* **1984**, *49*, 2657. (b) Hegedus, L. S.; Sestrick, M. R.; Michaelson, E. T.; Harrington, P. J. *J. Org. Chem.* **1989**, 54, 4141.
8   Hegedus, L. S.; Toro, J. L.; Miles, W. H.; Harrington, P. J. *J. Org. Chem.* **1987**, *52*, 3319.
9   Brenner, M.; Mayer, G.; Terpin, A.; Steglich, W. *Chem. Eur. J.* **1997**, *3*, 70.
10   Hurt, C. R.; Lin, R.; Rapoport, H. *J. Org. Chem.* **1999**, *64*, 225.
11   Osanai, Y. Y.; Kondo, K.; Murakami, Y. *Chem. Pharm. Bull.* **1999**, *47*, 1587.

12   (a) Nicolaou, K. C.; Snyder, S. A.; Simonsen, K. B.; Koumbis, A. E. *Angew. Chem. Int. Ed. Engl.* **2000**, *39*, 3473. (b) Lee, K. L.; Goh, J. B.; Martin, S. F. *Tetrahedron Lett.* **2001**, *42*, 1635.
13   Kadin, S. B. *J. Org. Chem.* **1973**, *38*, 1348.

Jeffrey N. Johnston

3.8   Madelung Indole Synthesis

3.8.1   Description

The intramolecular cyclization of *N*-acylated-*o*-alkylanilines in the presence of a strong base at elevated temperatures is known as the Madelung indole synthesis.[1,2]

3.8.2   Historical Perspective

In 1912, Madelung reported that *o*-acetotoluidine **3** and *o*-benzotoluidine **5** provided the corresponding 2-methylindole **4** and 2-phenylindole **6** respectively when heated to 360–380°C with 2 molar equivalents of sodium ethoxide.[3]

3.8.3   Mechanism

The mechanism of the Madelung indole synthesis has not been fully established. An intramolecular Claisen type condensation is presumably involved in the process.[2,4,5]

3.8.4 Synthetic Utility

The most common conditions employed in the Madelung process are sodium/potassium alkoxide or sodium amide at elevated temperature (200–400°C).[1,2] The Madelung reaction could be effected at lower temperature when n-BuLi or LDA are employed as bases.[4] The useful scope of the synthesis is, therefore, limited to molecules which can survive strongly basic conditions. The process has been successfully applied to indoles bearing alkyl substituents.[6,7]

Houlihan et al. have described the successful indolization of N-acylated-o-alkylanilines by employing n-BuLi/LDA as bases at lower temperature.[4]

When the benzamide derivative of 3-picoline 17 was subjected to the cyclization conditions with n-BuLi, the reaction failed to yield the desired indole 18. However, when n-BuLi was replaced by LDA, the desired azaindole 18 was isolated in 22% yield.[4]

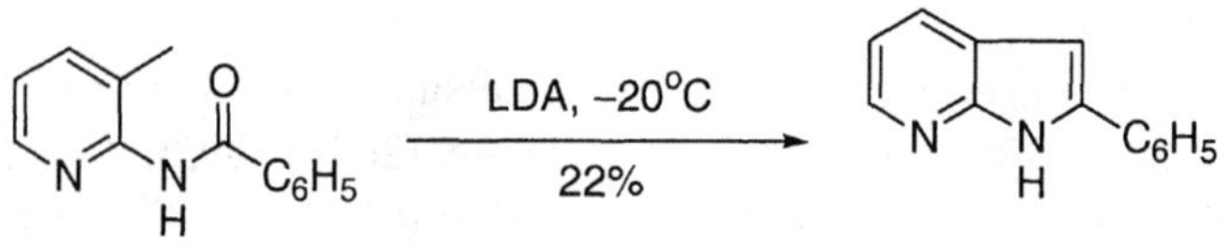

17 → **18**

LDA, –20°C
22%

The modification of the Madelung indole synthesis achieved by introduction of an electron withdrawing group (EWG) at the benzylic carbon atom of the *N*-acylated-*o*-alkylanilines has been quite successful.[8,9] Orlemans *et al.* reported that indoles were isolated in decent yields when the amides were treated with *t*-BuOK in THF for a period of 10 minutes at room temperature.[10]

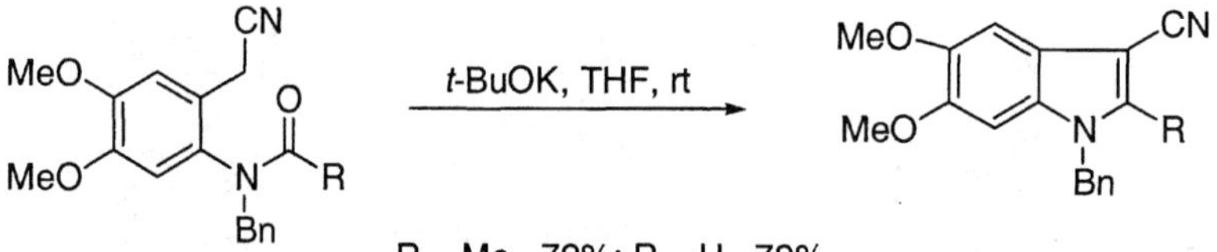

t-BuOK, THF, rt

R = Me, 79%; R = H, 79%
R = CF₃, 81%; R = Ph, 90%

19   **20**

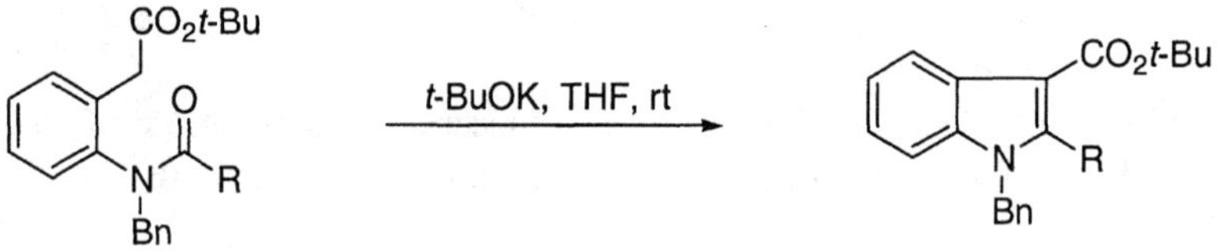

t-BuOK, THF, rt

R = Me, 79%; R = Ph, 83%

21   **22**

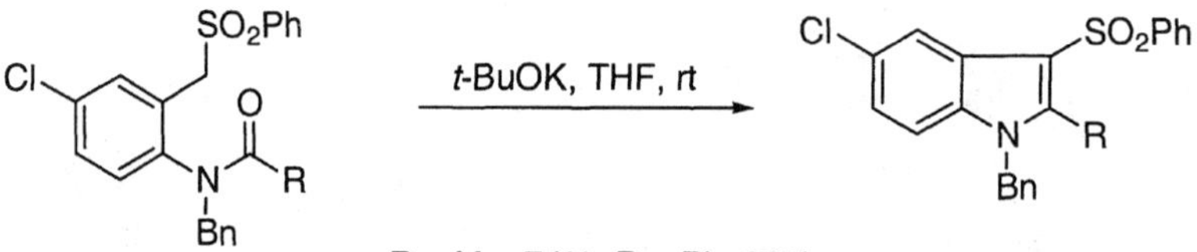

t-BuOK, THF, rt

R = Me, 74%; R = Ph, 83%

23   **24**

When *o*-dialkylanilides were treated with strong bases, the condensation took place at the methyl group in preference to the more substituted ethyl function as expected.[2]

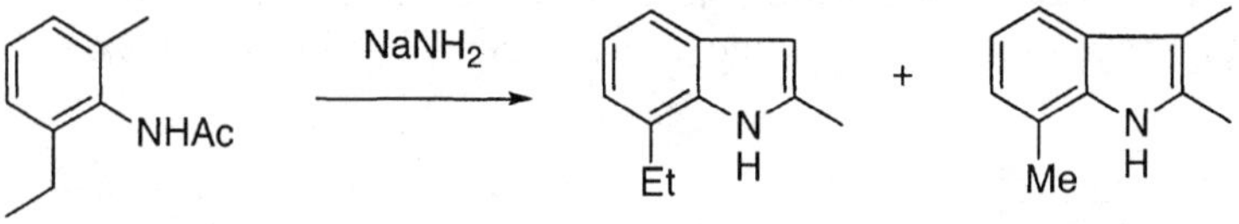

NaNH₂

25   26:27 = 35:1   **26**   **27**

The Madelung indole synthesis has been employed in the preparation of some complex indole systems. Uhle *et al.* reported the conversion of *N*-formyl-5,6,7,8-tetrahydronaphthylamine **28** into 1,3,4,5-tetrahydrobenz[*c,d*]indole **29** with *t*-BuOK in 11% yield in regard to synthesis of ergot alkaloids.[11]

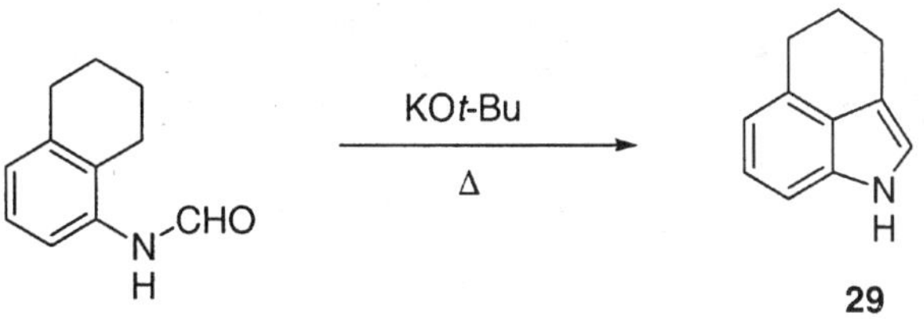

The spirocyclic cyclopenta[*g*]indole derivatives represented by **31** have also been prepared *via* the Madelung indole process.[12]

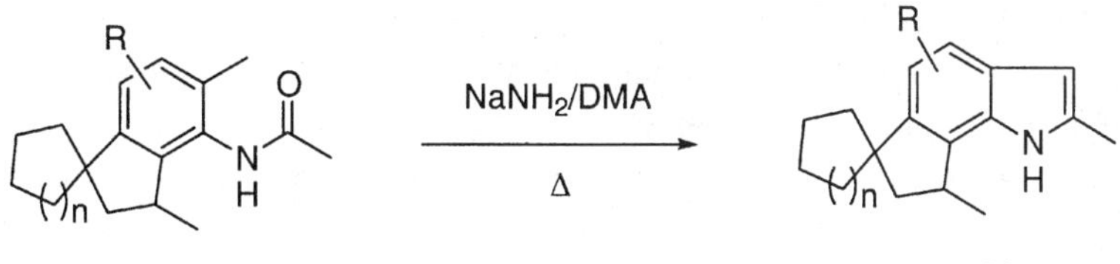

R = H, n = 1, 30%
R = H, n = 2, 46%
R = 4-Me, n = 2, 43%

R = H, n = 3, 32%
R = 5-Me, n = 2, 46%

Combinatorial chemistry has played an increasing role in drug discovery. Wacker *et al.* extended the Madelung indole process successfully to solid phase library synthesis for the preparation of 2,3-disubstituted indoles.[13] A number of examples follow in the table.

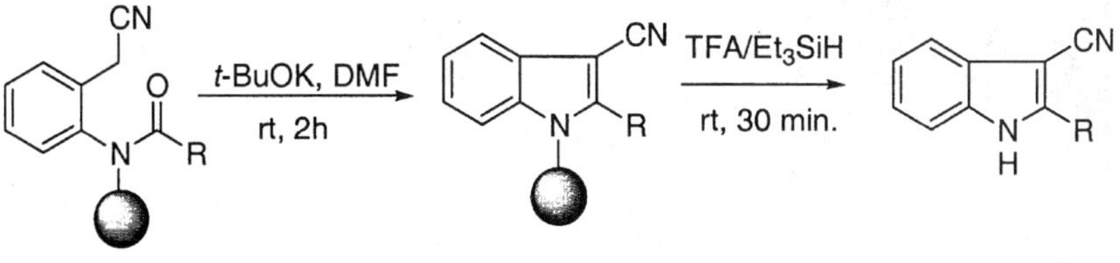

R	Yield of 34 (%)	Purity of 34 (%)	R	Yield (%)	Purity (%)
phenyl	88	95	*p*-MeO phenyl	84	95
p-CF₃ phenyl	83	94	*p*-NO₂ phenyl	75	87
o-Et phenyl	87	94	3-pyridyl	85	93
5-isoxazole	86	96	Et	86	95
t-butyl	84	95	PhCH₂	43	51

3.8.5   Experimental

Preparation of 2-Methylindole[14]

35 → **36**

In a 1 L flask was placed a mixture of 64 g of finely divided sodium amide and 100 g of *o*-acetyltoluidine **35**. Approximately 50 mL of dry ether was added and the apparatus was purged with dry nitrogen. Then, with a slow current of nitrogen passing through the mixture, the reaction flask was heated in a metal bath. The temperature was raised to 240–260°C over a 30 min period and was maintained in this range for 10 min. A vigorous evolution of gas occurred, the cessation of which indicated that the reaction was complete. The metal bath was removed, the flask was allowed to cool, and 50 mL of 95% ethanol and 250 mL of warm water (about 50°C) were added, successively, to the reaction mixture. The decomposition of the sodium derivative of 2-methylindole, and of any excess sodium amide, was completed by warming the mixture gently with a Bunsen burner. The cooled reaction mixture was extracted with two 200 mL portions of ether. The combined ether extracts were filtered, and the filtrate was concentrated to about 125 mL. The solution was then transferred to a 250 mL flask and distilled. The 2-methylindole **36** distilled at 119–126°C/3–4 mmHg as a liquid, which rapidly solidified in the receiver to a white crystalline mass. This product melted at 56–57°C. The yield was 70–72 g (80–83%).

3.8.6   References

1.   [R] Brown, R. K. In *Indoles, Part 1*; Houlihan, W. J. Ed.; Wiley: New York, 1972; pp 385-396.
2.   [R] Sundberg, R. J. *The Chemistry of Indoles*; Academic Press: New York & London, 1970.
3.   Madelung, W. *Ber.* **1912**, *45*, 1128.
4.   Houlihan, W. J.; Parrino, V. A.; Uike, Y. *J. Org. Chem.* **1981**, *46*, 4511.
5.   Fuhrer, W.; Gschwend, H. W. *J. Org. Chem.* **1979**, *44*, 1133.
6.   Tyson, B. F. *J. Am. Chem. Soc.* **1950**, *72*, 2801.
7.   Walton, E.; Stammer, C. H.; Nutt, R. F.; Jenkins, S. R.; Holly, F. W. *J. Med. Chem.* **1965**, *8*, 204.
8.   Schulenberg, J. W. *J. Am. Chem. Soc.* **1968**, *90*, 7008.
9.   Bergman, J.; Sand, P.; Tilstam, U. *Tetrahedron Lett.* **1983**, *24*, 3665.
10.   Orlemans, E. O. M.; Schreuder, A. H.; Conti, P. G. M.; Verboom, W.; Reinhoudt, D. N. *Tetrahedron* **1987**, *43*, 3817.
11.   Uhle, F. C.; Vernick, C. G.; Schmir, G. L. *J. Am. Chem. Soc.* **1955**, *77*, 3334.
12.   Kouznetsov, V.; Zubkov, F.; Palma, A.; Restrepo, G. *Tetrahedron Lett.* **2002**, *43*, 4707.
13.   Wacker, D. A.; Kasireddy, P. *Tetrahedron Lett.* **2002**, *43*, 5189.
14.   Allen, C. F. H.; VanAllan, J. *Org. Synth.* **1955**, *Coll. Vol. 3*, 597.

Jin Li and James M. Cook

3.9 Nenitzescu Indole Synthesis

3.9.1 Description

The Nenitzescu indole synthesis[1,2] involves the condensation of a quinone **1** and an enamine **2** to generate a hydroxyindole **3**.

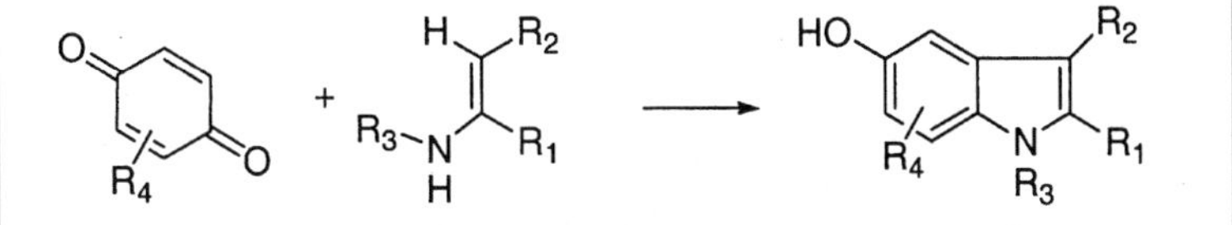

3.9.2 Historical Perspective

In 1929, Nenitzescu reported that *p*-benzoquinone (**4**) was treated with ethyl 3-aminocrotonate (**5**) in boiling acetone to yield ethyl 5-hydroxy-2-methylindole-3-carboxylate (**6**).[3]

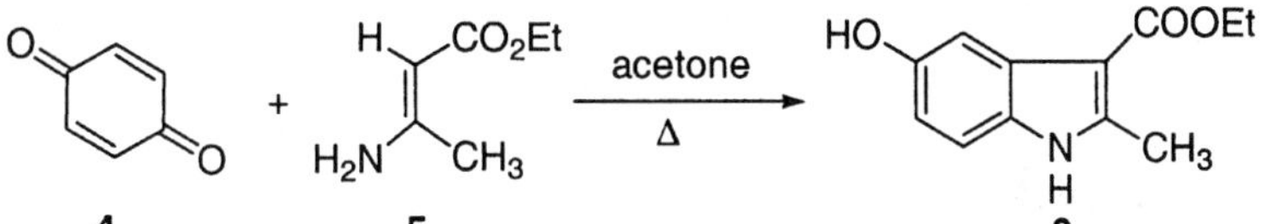

The procedure was largely ignored until the 1950s when interest in melanin-related substances and recognition of serotonin as a 5-hydroxy derivative stimulated exploration of the scope of the reaction. Nowadays, the Nenitzescu reaction is one of the most efficient processes for the preparation of 5-hydroxyindoles.[1,2,4–10]

3.9.3 Mechanism

At least two pathways have been proposed for the Nenitzescu reaction.[1,2] The mechanism outlined below is generally accepted.[11,12] Illustrated here is the indolization of the 1,4-benzoquinone (**4**) with ethyl 3-aminocrotonate (**5**). The mechanism consists of four stages: (I) Michael addition of the carbon terminal of the enamine **5** to quinone **4**; (II) Oxidation of the resulting hydroquinone **10** to the quinone **11** either by the starting quinone **4** or the quinonimmonium intermediate **13**, which is generated at a later stage; (III) Cyclization of the quinone adduct **11**, if in the *cis*-configuration, to the carbinolamine **12** or quinonimmonium intermediate **13**; (IV) Reduction of the intermediates **12** or **13** to the 5-hydroxyindole **6** by the initial hydroquinone adduct **7** (or **8**, **9**, **10**).

3.9.4 Scope and Limitations

The Nenitzescu reaction generally occurs under relatively mild reaction conditions. Moreover *mono-*, *di-*, and *tri-*substituted quinones react with equal facility. Many enamines including β-aminoacrylonitriles, β-aminoacrylamides, and β-amino-α,β-unsaturated ketones react with quinones to form indole nuclei as well. The mild reaction conditions and the availability of the starting material render it attractive even in those instances where the yield of the product is low.[1,2]

3.9.4.1 Structure of the quinone

Only *p*-quinones have been used in the Nenitzescu reaction, and the utility of 1,4-benzoquinone, 1,4-naphthoquinone, and substituted benzoquinones (*mono/di/tri*) in this procedure is well documented in literature.[1,2] Some of these substituted *p*-benzoquinones are: 2-hydroxy, 2-methoxy, 2-methyl, 2-chloro, 2-fluoro, or 2-bromo; 2-ethyl, 2-benzylthio, 2-trifluoromethyl, 2-carbethoxy, 2-acetyl, 2,3-dimethyl, 2-hydroxy-3,6-dimethyl, 2-chloro-5-trifluoromethyl, and 2-methoxy-5-trifluoromethyl.[1,2] The *mono-*substituted *p*-benzoquinone provides one or two products under Nenitzescu conditions depending on the nature of the substituents and steric requirements. It is known that electron-donating substituents at C-2 in the *p*-benzoquinone ring activate the C-5 and C-6 positions of the ring to nucleophilic agents while electron-withdrawing substituents at C-2 favor such attack at C-3.[13] It is worth noting that the enamino esters exert effects in determining the product profile in the Nenitzescu process as well (*vide infra*).[10,14–16]

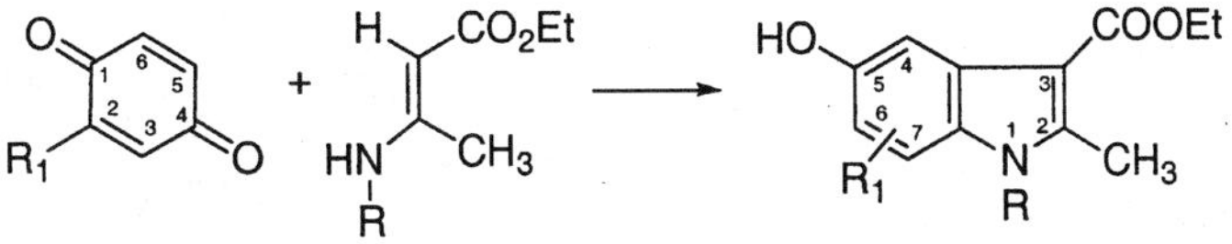

16 R_1 at 4 position
17 R_1 at 6 position
18 R_1 at 7 position

Substituents		Product, % yield		
R_1	R	**16**	**17**	**18**
CH_3	H	×	9	8
CH_3	CH_3	×	22	10
CH_3	C_2H_5	×	21	2
CH_3	n-C_3H_7	×	21	1
CH_3	n-C_4H_9	×	18	2
CH_3	i-C_3H_7	×	18	×
C_2H_5	H	×	30	×
C_2H_5	C_2H_5	×	14	0.9
CH_3O	H	×	19	×
F	H	×	12	×
Cl	H	×	20	4
Br	H	×	5	2
I	H	×	1	7
CF_3	H	54	×	×
CO_2CH_3	H	18	×	×

× — No product isolated

The more interesting situation arises in quinones which possess two dissimilar substituents. The site of initial carbon-to-carbon condensation is explicable in terms of the relative electronic effects. Thus condensation of 2-chloro-5-methylbenzoquinone (**19**) with *t*-butyl 3-aminocrotonate (**20**) in hot acetic acid furnished the 4-chloro-7-methylindole (**21**) in 51% yield.[17]

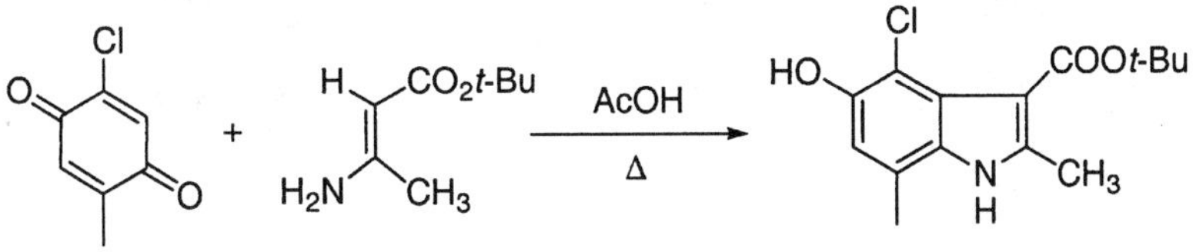

A trifluoromethyl moiety is a strong electron-withdrawing group which dominates the direction of indolization in the Nenitzescu process.[10]

The directing influence of the trifluoromethyl group also competes with that of the strongly electron-donating methoxyl group. Thus 2-methoxy-5-trifluoromethyl-1,4-benzoquinone (**26**) was treated with ethyl 3-aminocrotonate (**5**) to furnish 25% of each of the two possible isomeric indoles **27** and **28**.[10]

3.9.4.2 Structure of the enamine

Condensation of the *N*-substituted β-aminocrotonic acid ester **15** with *p*-benzoquinone (**4**) has been successfully carried out to furnish the 5-hydroxyindole **29** when the substituent R on the nitrogen of the aminocrotonic acid ester was methyl, ethyl, *n*-propyl, isopropyl, or *n*-butyl, *n*-hexyl, β-cyanoethyl, β-hydroxyethyl, carbethoxymethyl, benzyl, phenyl, *o*-tolyl, dimethylaminopropyl, γ-hydroxypropyl *etc.*[1,2]

Additional variations of the enamine moiety that have been satisfactorily condensed with a *p*-benzoquinone (**4**) to form 5-hydroxyindole **31** are given in the following table.

R_1	R_2	R_3	Ref.
C_2H_5	$CO_2C_2H_5$	C_2H_5	14
H	$CO_2C_2H_5$	OC_2H_5	18
H	$CO_2C_2H_5$	C_6H_5	19
alkyl	$CO_2C_2H_5$	C_6H_5	20,21
H	CN	C_6H_5	19
C_2H_5	$CONHC_6H_5$	CH_3	22
alkyl or aryl	$COCH_3$	CH_3	23,24
C_2H_5	$CO_2C_2H_5$	$CH_2CO_2C_2H_5$	25

3.9.5 Experimental Conditions

The choice of experimental conditions exerts a major influence on the course of Nenitzescu procedure and thereby determines the structure of the major product. The mechanism demonstrated to be operative for the method can be employed to understand the genesis of certain anomalous products and suggests ways to avoid them, thus increasing the efficiency of the synthesis of 5-hydroxyindoles.

3.9.5.1   Ratio of reactants

The best yields of 5-hydroxyindoles are obtained when equimolar amounts of the quinone and enamine are used. An excess of enamine gives rise to non-indolic products derived from reaction of two enamine units and one quinone unit or the product which results from the initial Michael addition of the enamine to the quinone. Use of excess quinone has been reported less frequently, but limited studies indicate no advantage.[26,27] When 2,5-dichloro-1,4-benzoquinone (32) was treated with a 50% excess of ethyl 3-

aminocrotonate (**5**), only a 1:2 adduct **34** was isolated. However, when quinone **32** was treated with an equimolar amount of aminocrotonate **5** the desired 5-hydroxyindole **33** was isolated in approximately 20% yield.[9] It is worth noting that the reactions were carried out in different solvents and a solvent effect could be involved as well.

3.9.5.2   *Effect of solvent*

The Nenitzescu process is presumed to involve an internal oxidation-reduction sequence. Since electron transfer processes, characterized by deep burgundy colored reaction mixtures, may be an important mechanistic aspect, the outcome should be sensitive to the reaction medium.   Many solvents have been employed in the Nenitzescu reaction including acetone, methanol, ethanol, benzene, methylene chloride, chloroform, and ethylene chloride; however, acetic acid and nitromethane are the most effective solvents for the process.[1,2] The utility of acetic acid[19] is likely the result of its ability to isomerize the olefinic intermediate (**9**) to the isomeric (**10**) capable of providing 5-hydroxyindole derivatives. The reaction of benzoquinone **4** with ethyl 3-aminocinnamate **35** illustrates this effect.[19]

The effectiveness of nitromethane can be attributed to its high dielectric constant, at least in part, which tends to promote reactions which involve electron-rich intermediates.   It may also result from the low solubility of the indole products in nitromethane since the indoles precipitate out of the reaction mixture in many cases.[28]

3.9.6   *Synthetic Applications in the Pharmaceutical Industry*

Synthesis of LY311727[29]

LY311727 is an indole acetic acid based selective inhibitor of human non-pancreatic secretory phospholipase A2 (hnpsPLA2) under development by Lilly as a potential treatment for sepsis. The synthesis of LY311727 involved a Nenitzescu indolization reaction as a key step.   The Nenitzescu condensation of quinone **4** with the β-aminoacrylate **39** was carried out in CH_3NO_2 to provide the desired 5-hydroxylindole **40** in 83% yield. Protection of the 5-hydroxyl moiety in indole **40** was accomplished in H_2O under phase transfer conditions in 80% yield. Lithium aluminum hydride mediated reduction of the ester functional group in **41** provided the alcohol **42** in 78% yield.

Addition of the alcohol **42** to a solution of BF$_3$•Et$_2$O/TMSCN in DCM provided the nitrile **43** in 83% yield. Hydrolysis of nitrile **43** then furnished amide **44** in 85% yield. Demethylation of the methoxyindole **44** with BBr$_3$ in DCM provided the hydroxyindole **45** in 80% yield. This was followed by alkylation of **45** with the bromide **46** under phase transfer conditions to provide the phosphonate ester **47** and subsequent cleavage of the methyl ester by TMS-I furnished trimethylsilyl phosphonic acid **48,** which upon alcoholic workup afforded **LY311727**.

3.9.7 Experimental

Preparation of ethyl 2,6-dimethyl-5-hydroxyindole-3-carboxylate(50) and ethyl 2,7-dimethyl-5-hydroxyindole-3-carboxylate(51).[2]

A solution of 3.61 kg (30.4 mol) of toluquinone **49** (practical grade) and 3.82 kg (30.4 mol) of ethyl 3-aminocrotonate (**5**) in 8.4 liters of acetone was heated on a steam bath with stirring. When the reflux temperature was attained, a vigorous exothermic reaction occurred which sustained boiling for 2 h despite the application of external cooling. The solution was maintained at reflux temperature for an additional 3 h, and then 3 L of acetone were removed by distillation. The concentrate was stored at 3–5°C for 16 h and filtered; the pink gray filter cake was washed thoroughly with ether and dried to give 3.425 kg of a solid. This material was stirred for 5 minutes with three successive 8 L portions of acetone and collected by filtration. The combined filtrates from the three acetone washes were concentrated to a volume of approximately 4 L and cooled to give 1.11 kg (15.5%) of ethyl 2,6-dimethyl-5-hydroxyindole-3-carboxylate (**50**), mp 225–228°C. The dry filter cake (1.85 kg) remaining from the acetone extraction was then extracted with 5 L of boiling acetone for 5 minutes, and the mixture was filtered hot to give 0.81 kg (11.5%) of ethyl 2,7-dimethyl-5-hydroxyindole-3-carboxylate (**51**), mp 192–196°C.

3.9.8 References

1.   [R] Brown, R. K. In *The Chemistry of Heterocyclic Compounds*; Houlihan, W. J. Ed.; Wiley: New York, 1972; pp 413.
2.   [R] Allen, G. R., Jr. *Org. React.* **1973**, *20*, 337.
3.   Nenitzescu, C. D. *Bull. Soc. Chim. Romania* **1929**, *11*, 37.
4.   Bernier, J. L.; Hénichart, J. P. *J. Org. Chem.* **1981**, *46*, 4197.
5.   Mukhanova, T. I.; Panisheva, E. K.; Lyubchanskaya, V. M.; Alekseeva, L. M.; Sheinker, Y. N.; Granik, V. G. *Tetrahedron* **1997**, *53*, 177.
6.   Kinugawa, M.; Arai, H.; Nishikawa, H.; Sakaguchi, A.; Ogasa, T.; Tomioka, S.; Kasai, M. *J. Chem. Soc., Perkin Trans. 1* **1995**, 2677.
7.   Bernier, J. L.; Hénichart, J. P.; Vaccher, C.; Houssin, R. *J. Org. Chem.* **1980**, *45*, 1493.
8.   Lyubchanskaya, V. M.; Alekseeva, L. M.; Granik, V. G. *Tetrahedron* **1997**, *53*, 177.
9.   Poletto, J. F.; Allen, G. R., Jr.; Sloboda, A. E.; Weiss, M. J. *J. Med. Chem.* **1973**, *16*, 757.
10.   Littell, R.; Allen, G. R., Jr. *J. Org. Chem.* **1968**, *33*, 2064.
11.   Allen, G. R., Jr.; Pidacks, C.; Weiss, M. J. *J. Am. Chem. Soc.* **1966**, *88*, 2536.
12.   Littell, R.; Morton, G. O.; Allen, G. R., Jr. *Chem. Commun.* **1969**, 1144.
13.   Wilgas III, H. S.; Frauenglass, E.; Jones, E. T.; Porter, R. F.; Gates, J. W. *J. Org. Chem.* **1964**, *29*, 594.
14.   Allen, G. R., Jr.; Pidacks, C.; Weiss, M. J. *Chem. Ind. (London)* **1965**, 2096.
15.   Allen, G. R., Jr.; Pidacks, C. D.; Weiss, M. J. *J. Am. Chem. Soc.* **1966**, *88*, 2536.
16.   Allen, G. R., Jr.; Weiss, M. J. *J. Org. Chem.* **1968**, *33*, 198.
17.   Poletto, J. F.; Weiss, M. J. *J. Org. Chem.* **1970**, *35*, 1190.
18.   Beer, R. S. J.; Davenport, H. F.; Robertson, A. *J. Chem. Soc.* **1953**, 1262.
19.   Raileanu, D.; Nenitzescu, C. D. *Rev. Roum. Chim.* **1965**, *10*, 339.

20.   Betkerur, S. N.; Siddappa, S. *J. Chem. Soc., C,* **1967**, 296.
21.   Betkerur, S. N.; Siddappa, S. *J. Chem. Soc., C,* **1968**, 1795.
22.   Grinev, A. N.; Ermakova, V. N.; Mel'nikova, I. A.; Terent'ev, A. P. *J. Gen. Chem. USSR* **1961**, *31*, 2146.
23.   Grinev, A. N.; Shvedov, V. I.; Sugrobova, I. P. *J. Gen. Chem. USSR* **1961**, *31*, 2140.
24.   Grinev, A. N.; Shvedov, V. I.; Panisheva, E. K. *Zh. Org. Khim* **1965**, *1*, 2051.
25.   Trofinov, F. A.; Nozdrich, V. I.; Grinev, A. N.; Shvedov, V. I. *Khim-Farm. Zh.* **1967**, *1*, 22.
26.   Yamada, Y.; Matsui, M. *Agr. Biol. Chem.* **1970**, *34*, 724.
27.   Yamada, Y.; Matsui, M. *Agr. Biol. Chem.* **1971**, *35*, 282.
28.   Patrick, J. B.; Saunders, E. K. *Tetrahedron Lett.* **1979**, *20*, 4009.
29.   Pawlak, J. M.; Khau, V. V.; Hutchison, D. R.; Martinelli, M. J. *J. Org. Chem.* **1996**, *61*, 9055.

Jin Li and James M. Cook

3.10 Reissert Indole Synthesis

3.10.1 Description

The Reissert procedure involves base-catalyzed condensation of an o-nitrotoluene derivative **1** with an ethyl oxalate (**2**) which is followed by reductive cyclization to an indole-2-carboxylic acid derivative **4**, as illustrated below[1,2].

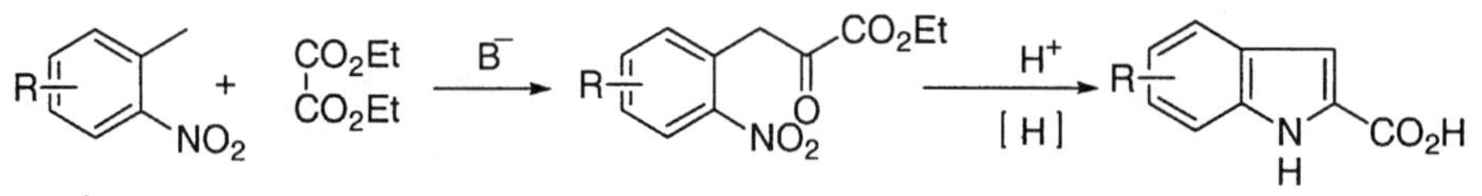

3.10.2 Historical Perspective

In 1897, Reissert reported the synthesis of a variety of substituted indoles from o-nitrotoluene derivatives.[3] Condensation of o-nitrotoluene (**5**) with diethyl oxalate (**2**) in the presense of sodium ethoxide afforded ethyl o-nitrophenylpyruvate (**6**). After hydrolysis of the ester, the free acid, o-nitrophenylpyruvic acid (**7**), was reduced with zinc in acetic acid to the intermediate, o-aminophenylpyruvic acid (**8**), which underwent cyclization with loss of water under the conditions of reduction to furnish the indole-2-carboxylic acid (**9**). When the indole-2-carboxylic acid (**9**) was heated above its melting point, carbon dioxide was evolved with concomitant formation of the indole (**10**).

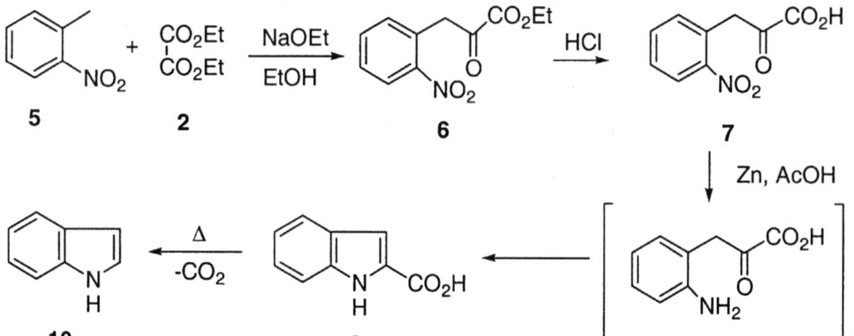

3.10.3 Mechanism

Under basic conditions, the o-nitrotoluene (**5**) undergoes condensation with ethyl oxalate (**2**) to provide the α-ketoester **6**. After hydrolysis of the ester functional group, the nitro moiety in **7** is then reduced to an amino function, which reacts with the carbonyl group to provide the cyclized intermediate **13**. Aromatization of **13** by loss of water gives the indole-2-carboxylic acid (**9**).

3.10.4   Synthetic Utility

Sodium alkoxides or potassium alkoxides generally serve as catalysts for the condensation of ethyl oxalate in the Reissert process. Activation of the methyl group by the electron withdrawal of the nitro group is required for successful base-catalyzed reaction of the o-nitrotoluene (**1**) with diethyl oxalate. For this reason electron-donor substituents in the o-nitrotoluene are expected to retard the reaction, whereas electron-withdrawing substituents should enhance the ease of proton removal to generate the carbanion. The reduction of the intermediate (see **7**), that is, o-nitrophenylpyruvic acid, can be carried out with Zn/HOAc, Zn/HOAc/Co(NO$_2$)$_2$, Zn-Hg/HCl, FeSO$_4$/NH$_4$OH, Fe/HCl, Fe/HOAc or Na$_2$S$_2$O$_4$, *etc.* The decomposition of the indole-2-carboxylic acid is achieved by heating it above its melting point alone, with calcium oxide, or in the presence of copper powder in quinoline. Many substituted indoles have been prepared by the Reissert indole synthesis.[2,4,5]

	yield		17		ratio		18
R= 6-OMe	80%		92	:			8
R= 5-OMe	82%		98	:			2
R= 3-Me	87%		90	:			6

The Reissert procedure has been employed for the preparation of 4- and 6-azaindoles (see **21**).[6]

In contrast to the facile condensation of *o*-nitrotoluene with diethyl oxalate, other α-alkyl nitrobenzenes are sluggish to react with diethyl oxalate or fail to react at all. It has been suggested that this is due both to steric and electronic factors effected by the alkyl group, which destabilizes the methylene group in regard to formation of the carbanion.[2]

Butin *et al.* reported that the indole derivative **29** was prepared by treatment of 2-tosylaminobenzylfuran **25** with ethanolic HCl in 78% yield. The furan ring served as the origin of a carbonyl group in this modification of the Reissert procedure.[7]

3.10.5 Experimental

Preparation of Ethyl indole-2-carboxylate 31[8]

A. *Potassium salt of ethyl o-nitrophenylpyruvate(30).*

Anhydrous ether (300 mL) was placed in a 5-L, three-neck, round-bottom flask fitted with a 500 mL dropping funnel, a motor-driven stirrer (with seal), and a reflux condenser protected with a calcium chloride drying tube. Freshly cut potassium (39.1 g, 1.00 g atom) was added. A slow stream of dry nitrogen was passed through the flask above the surface of the stirred liquid, and a mixture of 250 mL of commercial absolute ethanol and 200 mL of anhydrous ether was added from the dropping funnel just fast enough to maintain mild reflux. When all the potassium had dissolved, the nitrogen gas was shut off. The solution was allowed to cool to room temperature, and 2.5 L of anhydrous ether was added. Diethyl oxalate (146 g, 1.00 mole) was added with stirring, followed 10 minutes later by addition of 137 g (1.00 mole) of *o*-nitrotoluene (**5**). The stirring was discontinued after an additional 10 minutes, and the mixture was poured, with the aid of a connecting tube, into a 5L Erlenmeyer flask. The flask was stoppered and set aside for at least 24 hours. The lumpy deep-purple potassium salt of ethyl *o*-nitrophenylpyruvate (**30**) was separated by filtration and washed with anhydrous ether until the filtrate remained colorless. The yield of the air-dried salt was 204–215 g (74–78%).

B. *Ethyl indole-2-carboxylate(31)*.

Thirty grams (0.109 mole) of the potassium salt **30** was placed in a 400 mL hydrogenation bottle and dissolved by addition of 200 mL of glacial acetic acid, which produced a yellow, opaque solution. The platinum catalyst (0.20 g) was added and the bottle was placed on a Parr low-pressure hydrogenation apparatus. The system was flushed several times with hydrogen. With the initial reading on the pressure gauge at about 30 p.s.i., the bottle was shaken until hydrogen uptake ceases and continued shaking for an additional 1–2 hours. The catalyst was removed by filtration and washed with glacial acetic acid. The filtrate was placed in a 4-L beaker, and 3 L of water was added slowly with stirring. Ethyl indole-2-carboxylate precipitated as a yellow solid. It was separated by filtration, washed with five 100-mL portions of water, and dried over calcium chloride in a dessicator. The solid **31** weighed 13.2–13.6 g. (47–51% based on *o*-nitrotoluene); m.p. 118–124°C. The dried ester could be further purified by treatment with activated charcoal, followed by filtration, and recrystallization from a mixture of methylene chloride and light petroleum ether (b.p. 60–68°C). This gave 11.3–11.7 g (41–44% based on *o*-nitrotoluene) of ethyl indole-2-carboxylate **31** in the form of white needles, m.p. 122.5–124°C.

3.10.6 *References*

1.   [R] Julian, P. C.; Meyer, E. W.; Printy, S. C. *Heterocyclic Compounds*; John Wiley: New York, 1962; p 18.
2.   [R] Brown, R. K. In *Indoles, Part 1*; Houlihan, W. J. Ed.; Wiley: New York, 1972; pp 397-413.
3.   Reissert, A. *Ber.* **1897**, *30*, 1030.
4.   Leadbetter, G.; Fost, D. L.; Ekwuribe, N. N.; Remers, W. A. *J. Org. Chem.* **1974**, *39*, 3580.
5.   Suzuki, H.; Gyoutoku, H.; Yokoo, H.; Shinba, M.; Sato, Y.; Yamada, H.; Murakami, Y. *Synlett* **2000**, 1196.
6.   Frydman, B.; Despuy, M. E.; Rapoport, H. *J. Am. Chem. Soc.* **1965**, *87*, 3530.
7.   Butin, A. V.; Stroganova, T. A.; Lodina, I. V.; Krapivin, G. D. *Tetrahedron Lett.* **2001**, *42*, 2031.
8.   Noland, W. E.; Baude, F. J. *Org. Synth.*; Ed, Baumgarten, H. E.; John Wiley & Sons, 1973; p. 567.

Jin Li and James M. Cook

Chapter 4   Furans   **159**

4.1   Feist–Bénary Furan Synthesis

4.1.1. Description

The Feist–Bénary furan synthesis occurs when an α-halocarbonyl (**1**) reacts with a β-dicarbonyl (**2**) in the presence of a base. The resulting product (**3**) is a 3-furoate that incorporates substituents present in the two starting materials.[1-6]

Although bromo derivatives have been used, the two most common α-halocarbonyl compounds for this reaction are chloroacetaldehyde and chloroacetone. The dicarbonyl component is typically ethyl acetoacetate or one of its derivatives. A variety of bases including triethylamine and potassium hydroxide can promote the reaction; however, the most popular base is pyridine. Conversion to the furan takes place either at room temperature or upon heating to 50°C with reaction times varying from four hours to five days and yields ranging from 30–86%.

4.1.2. Historical Perspective

In 1902 Feist first described the combination of chloroacetone (**4**) and diethyl 3-oxoglutarate (**5**) in the presence of ammonia to yield trisubstituted furan **6**.[7]

In 1911 Bénary reported a modification of Feist's original procedure. He reacted chloroacetaldehyde (**8**), generated *in situ* from the ammonia promoted decomposition of 1,2-dichloroethyl ethyl ether (**7**), with ethyl acetoacetate (**9**) and ammonia to yield ethyl 2-methyl 3-furoate (**10**).[8]

4.1.3. Mechanism

The mechanism of the Feist–Bénary reaction involves an aldol reaction followed by an intramolecular *O*-alkylation and dehydration to yield the furan product.[5,9] In the example below, ethyl acetoacetate (**9**) is deprotonated by the base (**B**) to yield anion **10**; this carbanion reacts with chloroacetaldehyde (**8**) to furnish aldol adduct **11**. Protonation of the alkoxide anion followed by deprotonation of the β-dicarbonyl in **12** leads to formation of enolate **13**. Dihydrofuran **14** is produced by an intramolecular S_N2 reaction between the oxygen nucleophile and the carbon bearing the chloride leaving group. The desired furan **10** is then produced by dehydration of **14** via protonated intermediate **15**. In the so-called "interrupted" Feist–Bénary reaction the sequence does not proceed to yield a furan, but instead stops after the formation of dihydrofuran products like **14**.[10] It is also important to note that modifications of the standard experimental conditions can change the mechanism to favor alkylation in the first step. This produces a 1,4-dicarbonyl compound that then undergoes the Paal–Knorr furan synthesis (see section 4.2).[4,11]

4.1.4. Synthetic Utility
4.1.4.1.   Preparation of 2,3-disubstituted furans

The most common use of the Feist–Bénary furan synthesis is for the preparation of 2-substituted 3-furoates. Among these compounds, the most popular target is the furan originally prepared by Bénary,[8] namely, ethyl 2-methyl 3-furoate (**10**). The only improvement on Bénary's synthesis is the use of pyridine as the base to shorten the reaction time and increase the yield.[12] A common strategy for preparing a variety of these 2,3-disubstituted furans is to first synthesize a derivative of ethyl or methyl acetoacetate and then combine it with chloroacetaldehyde using the Feist–Bénary reaction. Furthermore, it is common to then convert the resulting ester into a carboxylic acid for use in a variety of subsequent transformations. For example, substituted ethyl acetoacetate **16** was combined with chloroacetaldehyde and pyridine to furnish furan **17** in 70% yield.[13] Saponification of the ester yielded acid **18** and decarboxylation produced 2-alkylfuran **19**. This example highlights a general strategy for the use of the Feist–Bénary reaction in the production of 2-substituted furans.

Several other 2-substituted 3-furoates have been prepared using the Feist–Bénary reaction. The examples below highlight the preparation of substrates containing an alkyl group (21),[14] aryl groups (23 and 25),[15,16] and an ester group in the 2-position (27).[17] Bisagni reported the syntheses of a variety of disubstituted furans starting with 1,3-diketones or 1,3-ketoesters using several different bases to furnish the target products in low to moderate yields (19–52%).[9a,18]

4.1.4.2.   Preparation of 2,3,4-trisubsituted furans

Syntheses of trisubstituted furans are much less common than the disubstituted derivatives; only one 2,4-disubstituted 3-furoate has been prepared using the Feist–Bénary reaction. Combination of chloroacetone (4) with ethyl acetoacetate (9) provides ethyl 2,4-dimethyl-3-furoate (28) in 54–57% yield.[9,19] The procedure for this transformation is different than standard Feist–Bénary conditions and involves stirring the reactants in cold hydrochloric acid followed by prolonged exposure to cold triethylamine.

4.1.4.3.   *Preparation of tetrasubstituted furans*

The only tetrasubstituted furans that have been prepared using the Feist–Bénary reaction are substituted tetrahydrobenzofurans and octahydrodibenzofurans. This strategy was pioneered by Stetter and Chatterjea and applied in a series of total syntheses by Magnus. Stetter demonstrated that 1,3-cyclohexanedione (**30**) can act as the β-dicarbonyl component and readily combines with either 3-bromo-2-ketobutyric (**29**) acid or ethyl 2-chloroacetoacetate (**32**) in the presence of potassium hydroxide to yield tetrahydrobenzofuran derivatives **31** and **33**, respectively.[20]

Chatterjea showed that cyclic α-halocarbonyls are acceptable substrates for the Feist–Bénary furan synthesis by combining 1-chlorocyclohexanone (**34**) with 1,3-cyclohexanedione (**30**) to yield octahydrodibenzofuran **35**.[21]

Magnus prepared tetrahydrobenzofuran **37** using a Feist–Bénary reaction of ethyl 2-chloroacetoacetate (**32**) and functionalized 1,3-cyclohexanedione **36**. Compound **37** was a key synthetic intermediate in Magnus's synthesis of linderalactone, isolinderalactone, and niolinderalactone.[22]

4.1.5. Variations

Several modifications of the Feist–Bénary furan synthesis have been reported and fall into two general classes: 1) reactions that yield furan products 2) reactions that yield dihydrofuran products. One variant that furnishes dihydrofurans uses substrates identical to the traditional Feist–Bénary furan synthesis with a slight modification of the reaction conditions. The other transformations covered in this section involve the combination of β-dicarbonyls with reagents that are not simple α-halocarbonyls. Several reactions incorporate α-halocarbonyl derivatives while others rely on completely different compounds.

4.1.5.1.   Preparation of furan derivatives

Several variations of the Feist–Bénary reaction furnish substituted furans as products. The following three examples provide synthetically useful alternatives to the standard reaction conditions.[23] One method is based on the reaction of a sulfonium salt with a β-dicarbonyl compound. For example, reaction of acetylacetone (39) with sulfonium salt 38 in the presence of sodium ethoxide yields 81% of trisubstituted furan 40.[24] This strategy provides a flexible method for the preparation of 2,3,4-trisubstituted furans.

A different procedure provides access to 2,3,5-trisubstituted furans. Deslongchamps discovered that simply heating a mixture of glyceraldehyde (41) and methyl acetoacetate (42) in DMF provides a high yield of furan 43.[25] Subsequent transformations enable selective substitution at the 2-position of the product.

Disubstituted furans are available from the combination of β-dicarbonyl compounds with bromoacetaldehyde diethyl acetal (44). For example, dibenzoylmethane (45) reacts with acetal 44 to furnish 2,3-disubstituted furan 46 in 77% yield.[26] This two-step reaction provides an excellent alternative to the traditional Feist–Bénary protocol for the synthesis of 2-substituted 3-furoates.

4.1.5.2.   Preparation of dihydrofuran derivatives

Three other modifications of the standard conditions provide synthetically useful strategies for the preparation of dihydrofurans. One method, called the interrupted Feist–Bénary reaction, utilizes milder reaction conditions to stop the final dehydration step. For example, Calter combined bromide **47** with dicarbonyl **48** to produce dihydrofuran **49** as a mixture of diastereomers.[10] He examined the scope and diastereoselectivity of this process[10] and applied this reaction toward the synthesis of the polycyclic core of the zaragozic acids.[27] A method principally designed to yield practical syntheses of cyclic ketodiesters also furnished a dihydrofuran via a variation of the interrupted Feist–Bénary reaction.[23a]

Another procedure relies on a domino Michael-*O*-alkylation reaction sequence to yield a variety of dihydrofurans. Combination of cyclohexanedione (**30**) with vinyl bromide **50** in the presence of 1,8-diazabicyclo[5.4.0]undec-7-ene (DBU) provides dihydrofuran **51** in 83% yield.[28] Numerous 1,3-dicarbonyls and vinyl bromides are amenable to this methodology, and thus a wide range of products like **51** are available via this strategy.

The final variation of the Feist–Bénary furan synthesis encompasses reactions of 1,3-dicarbonyls with 1,2-dibromoethyl acetate (**52**). For example, treatment of ethyl acetoacetate (**9**) with sodium hydride followed by addition of **52** at 50°C yields dihydrofuran **53**.[29] The product can be easily converted into the corresponding 2-methyl-3-furoate upon acid catalyzed elimination of the acetate, thus providing another strategy for the synthesis of 2,3-disubstituted furans.

4.1.6. Experimental

2-Methyl-3-carboethoxyfuran (10):[12a]

A solution containing 150 mL of pyridine and 70.2 g of ethyl acetoacetate (**9**, 0.539 mol) was treated dropwise with 100 g of a 45% aqueous chloroacetaldehyde (**8**, 0.573 mol) solution over a period of 15 min. The mixture was allowed to stir for 4 h during which time the temperature rose to 50°C and then slowly subsided. After stirring overnight, the reaction mixture was poured into water and extracted with ether. The ether layer was treated with a 10% aqueous hydrochloric acid solution to wash out the remaining pyridine, dried over magnesium sulfate, and concentrated under reduced pressure. Distillation gave 68 g (86%) of **10** as a colorless oil (bp 72–75°C (7.0 mm Hg)); IR (neat) 2980, 1720, 1605, 1430, 1415, 1300, 1235, 1190, 1160, 1130, 995, 945, 740 cm^{-1}; ^{1}H NMR (CDCl$_3$, 360 MHz) δ 1.35 (t, 3H, J = 7.1 Hz), 2.56 (s, 3H), 4.29 (q, 2H, J = 7.1 Hz), 6.64 (d, 1H, J = 1.9 Hz), 7.22 (d, 1H, J = 1.9 Hz).

4.1.7. References

1. [R] König, B. Product Class 9: Furans. In *Science of Synthesis: Houben-Weyl Methods of Molecular Transformations*; Maas, G., Ed.; Georg Thieme Verlag: New York, 2001; Cat. 2, Vol. 9, 183-278.
2. [R] Friedrichsen, W. Furans and Their Benzo Derivatives: Synthesis. In *Comprehensive Heterocyclic Chemistry II*; Katritzky, A. R., Rees, C. W., Scriven, E. F. V., Eds.; Pergamon: New York, 1996; Vol. 2, 351-393.
3. [R] Dean, F. M. Recent Advances in Furan Chemistry. Part I. In *Advances in Heterocyclic Chemistry*; Katritzky, A. R., Ed.; Academic Press: New York, 1982; Vol. 30, 167-238.
4. [R] Joule, J. A.; Mills, K. *Heterocyclic Chemistry*, 4th ed.; Blackwell Science: Cambridge, 2000; 310-311.
5. [R] Gupta, R. R.; Kumar, M.; Gupta, V. *Heterocyclic Chemistry*, Springer: New York, 1999; Vol. 2, 87-88.
6. [R] Gilchrist, T. L. *Heterocyclic Chemistry*, 3rd ed.; Longman: New York, 1997; 70, 211.
7. Feist, F. *Ber. Dtsch. Chem. Ges.* **1902**, *35*, 1545.
8. Bénary, E. *Ber. Dtsch. Chem. Ges.* **1911**, *44*, 493.
9. (a) Bisagni, É.; Marquet, J.-P.; Bourzat, J.-D.; Pepin, J.-J.; André-Louisfert, J. *Bull. Soc. Chim. Fr.* **1971**, 4041. (b) Alexander, E. R.; Baldwin, S. *J. Am. Chem. Soc.* **1951**, *73*, 356.
10. Calter, M. A.; Zhu, C. *Org. Lett.* **2002**, *4*, 205.
11. (a) Stauffer, F.; Neier, R. *Org. Lett.* **2000**, *2*, 3535. (b) Bambury, R. E.; Miller, L. F. *J. Heterocycl. Chem.* **1970**, *7*, 269. (c) Bambury, R. E.; Yaktin, H. K.; Wyckoff, K. K. *J. Heterocycl. Chem.* **1968**, *5*, 95. (d) Dann, O.; Distler, H.; Merkel, H. *Chem. Ber.* **1952**, *85*, 457.
12. (a) Padwa, A.; Gasdaska, J. R. *Tetrahedron* **1988**, *44*, 4147. (b) Elliott, J. D.; Hetmanski, M.; Stoodley, R. J.; Palfreyman, M. N. *J. Chem. Soc., Perkin Trans. 1* **1981**, 1782. (c) Winberg, H. E.; Fawcett, F. S.; Mochel, W. E.; Theobald, C. W. *J. Am. Chem. Soc.* **1960**, *82*, 1428.
13. Ho, T.-L.; Ho, M.-F. *J. Chem. Soc., Perkin Trans. 1* **1999**, 1823.
14. Trahanovsky, W. S.; Huang, Y.-C. J.; Leung, M.-K. *J. Org. Chem.* **1994**, *59*, 2594.
15. Baker, R.; Sims, R. J. *J. Chem. Soc., Perkin Trans. 1* **1981**, 3087.
16. Janda, M.; Šrogl, J.; Dvořáková, H.; Dvořák, D.; Stibor, I. *Collect. Czech. Chem. Commun.* **1981**, *46*, 906.
17. Tada, M.; Ohtsu, K.; Chiba, K. *Chem. Pharm. Bull.* **1994**, *42*, 2167.

18.   (a) Bisagni, É.; Rivalle, C. *Bull. Soc. Chim. Fr.* **1974**, 519. (b) Bisagni, É.; Marquet, J.-P.; André-Louisfert, J.; Cheutin, A.; Feinte, F. *Bull. Soc. Chim. Fr.* **1967**, 2796.

19.   Wenkert, E.; Khatuya, H.; Klein, P. S. *Tetrahedron Lett.* **1999**, *40*, 5171.

20.   Stetter, H.; Lauterbach, R. *Chem. Ber.* **1960**, *93*, 603.

21.   Chatterjea, J. N.; Ray, R. R. *Chem. Ber.* **1959**, *92*, 998.

22.   (a) Gopalan, A.; Magnus, P. *J. Org. Chem.* **1984**, *49*, 2317. (b) Gopalan, A.; Magnus, P. *J. Am. Chem. Soc.* **1980**, *102*, 1756.

23.   For variations of the Feist-Bénary reaction that yield mixtures of furans or furan byproducts, see: (a) Lavoisier-Gallo, T.; Rodriguez, J. *Synth. Commun.* **1998**, *28*, 2259. (b) Courtheyn, D.; Verhé, R.; De Kimpe, N.; De Buyck, L.; Schamp, N. *J. Org. Chem.* **1981**, *46*, 3226.

24.   (a) Howes, P. D.; Stirling, C. J. M. *Org. Synth., Coll. Vol.VI* **1988**, 31. (b) Batty, J. W.; Howes, P. D.; Stirling, C. J. M. *J. Chem. Soc., Perkin Trans. 1* **1973**, 65.

25.   Toró, A.; Deslongchamps, P. *Synth. Commun.* **1999**, *29*, 2317.

26.   Antonioletti, R.; Bonadies, F.; Scettri, A. *Gazz. Chim. Ital.* **1988**, *118*, 73.

27.   Calter, M. A.; Zhu, C.; Lachicotte, R. J. *Org. Lett.* **2002**, *4*, 209.

28.   Hagiwara, H.; Sato, K.; Nishino, D.; Hoshi, T.; Suzuki, T.; Ando, M. *J. Chem. Soc., Perkin Trans. 1* **2001**, 2946.

29.   Cambie, R. C.; Moratti, S. C.; Rutledge, P. S.; Woodgate, P. D. *Synth. Commun.* **1990**, *20*, 1923.

Kevin M. Shea

4.2   Paal–Knorr Furan Synthesis

4.2.1. Description

Treatment of 1,4-dicarbonyls (**1**) with catalytic acid yields substituted furans (**2**) and is called the Paal–Knorr furan synthesis. This method is used extensively to produce a variety of mono-, di-, tri-, and tetrasubstituted furans.[1–6]

A multitude of 1,4-dicarbonyls (**1**) undergo the Paal–Knorr reaction with R^2 and R^3 ranging from H to alkyl, aryl, carbonyl, nitrile, and phosphonate, while R^1 and R^4 vary between H, alkyl, aryl, trialkylsilyl, and O-alkyl. Protic acid catalysts are typically used with sulfuric, hydrochloric, and p-toluenesulfonic acids the most popular. Conversion to the furan takes place either at room temperature or upon heating with reaction times varying from five minutes to 24 hours and yields ranging from 17–100%.

4.2.2. Historical Perspective

In a series of papers in late 1884 and early 1885, Paal and Knorr demonstrated that several 1,4-dicarbonyls could be transformed into furans, pyrroles, and thiophenes. Paal first discovered this transformation and used it to prepare di-, tri-, and tetrasubstituted furans.[7] For example, dicarbonyl **3** yielded disubstituted furan **4** upon treatment with weak acid.

Shortly thereafter, Knorr reported that combining ammonia or primary amines with 1,4-dicarbonyls furnished substituted pyrroles (see Section 2.2),[8] and Paal produced thiophenes by addition of hydrogen sulfide with 1,4-dicarbonyls.[9]

4.2.3. Mechanism

The mechanism of the Paal–Knorr furan synthesis was investigated by Amarnath in 1995.[10] After detailed kinetic studies involving the d,l and meso diastereomers of 2,3-disubstituted 1,4-diketones (**5**) the following mechanistic pathway emerged. One of the carbonyls is reversibly protonated by the catalytic acid to yield **6**. Next, in the rate-determining step, water removes the proton adjacent to the unprotonated ketone which triggers simultaneous alkene formation and attack of the protonated ketone to form dihydrofuran **7**. Protonation of the alcohol in **7** to furnish cation **8** followed by elimination of water produces the furan product **9** and regenerates the acid catalyst. This is slightly different than the standard mechanistic proposal that involved a two-step conversion of **6** to **7** via an enol intermediate.[11] It is important to note that this initial hypothesis was made with little supporting evidence. Amarnath's data was inconsistent with the formation of the enol intermediate, thus the mechanism was revised to the version illustrated below.

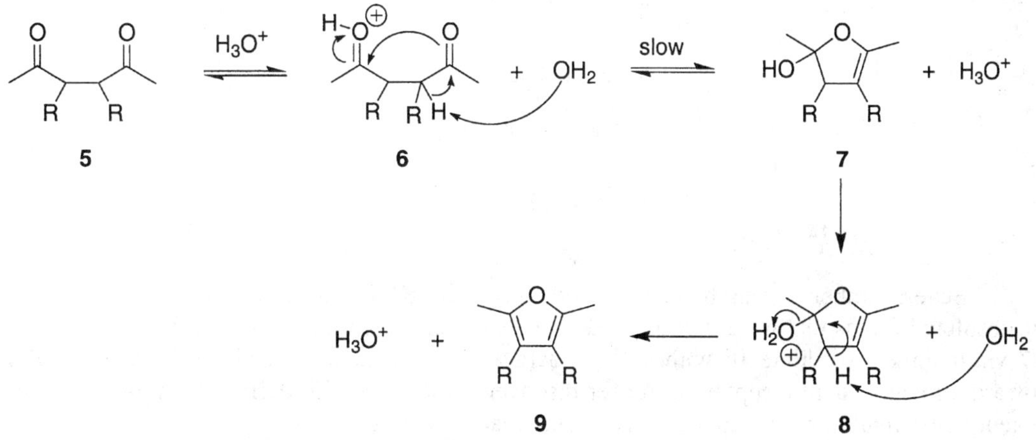

4.2.4. Synthetic Utility

4.2.4.1. Preparation of 2-substituted furans

Although nearly all Paal–Knorr condensations produce di-, tri-, or tetrasubstituted furans, it is possible to use this reaction to generate monosubstituted furans. Molander demonstrated the utility of this method with his synthesis of 2-(methyldiphenylsilyl)furan (**11**) from dicarbonyl **10**.[12]

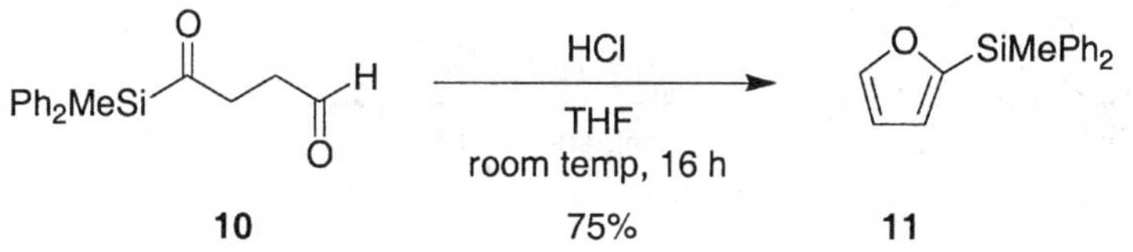

4.2.4.2. Preparation of 2,5-disubstituted furans

The most common use of the Paal–Knorr condensation begins with a 1,4-diketone and yields a 2,5-disubstituted furan. This method has been used to produce dialkyl and disilyl furans; however, the most popular use of this strategy is for the production of 2,5-diaryl furans. In addition to their utility as synthetic intermediates, these compounds are under investigation for novel electronic and pharmaceutical applications.

For example, treatment of dione **12** with hydrochloric acid yielded furan **13**, a key synthetic intermediate for the production of a variety of compounds that were recently investigated for anticancer activity.[13] Related inquiries by members of the same research team identified furans derived from **15** as potential treatments for RNA viruses. Furan **15** was prepared by condensation of dione **14** with catalytic sulfuric acid in refluxing acetic anhydride.[14]

Several other research teams used the Paal–Knorr condensation to prepare 2,5-disubstituted furans that were investigated as potential enzyme inhibitors. Nagai produced furan **17** via treatment of dione **16** with sulfuric acid and subsequently examined the activity of **17** toward a retenoic acid receptor.[15] Perrier discovered that furan **19**, derived from dione **18**, is a potent PDE4 inhibitor and may have anti-inflammatory activity.[16]

Juliá investigated the behavior of terfuran **22** and bis(thienyl)furan **23** by cyclic voltammetry as well as the EPR spectra of the radical cations derived from these two compounds. Condensation of the diketone **20** with sulfuric acid furnished furan **22** in 18% yield, while reaction of diketone **21** with hydrochloric acid produced **23** in 84% yield.[17] In a related report, Luo prepared oligomeric bis(thienyl)furans via similar methodology.[18]

Jones and Civcir prepared a variety of alternating oligomeric furan:pyridine compounds and studied their ^{13}C NMR chemical shifts as well as the pKa values for the corresponding conjugate acids. All of these compounds were synthesized by Paal–Knorr reactions of 1,4-diketones with hot polyphosphoric acid. A representative example is the conversion of **24** into **25** in 82% yield.[19]

Ibers used the Paal–Knorr furan synthesis to prepare a key intermediate for the synthesis of novel porphyrin-like aromatic macrocycles. Bis(pyrolyl)furan **27** was available in good yield via the acid catalyzed condensation of diketone **26**.[20]

The Paal–Knorr furan synthesis can also be used to prepare 2,5-arylalkylfurans, as illustrated in the following example. Salimbeni produced furan **29** from dione **28** and subsequently used the furan as an intermediate for the production of angiotensin II receptor antagonists.[21]

A variety of 2,5-dialkylfurans are available via the Paal–Knorr condensation; cyclization is possible for both hindered and unhindered 1,4-diketones. Fleming prepared 2-cyclohexyl-5-methylfuran (**31**) in 91% yield via treatment of dione **30** with catalytic p-toluenesulfonic acid in refluxing benzene.[22] Using the same methodology, Denisenko synthesized furan **33** in 35% yield from the corresponding dione (**32**).[23]

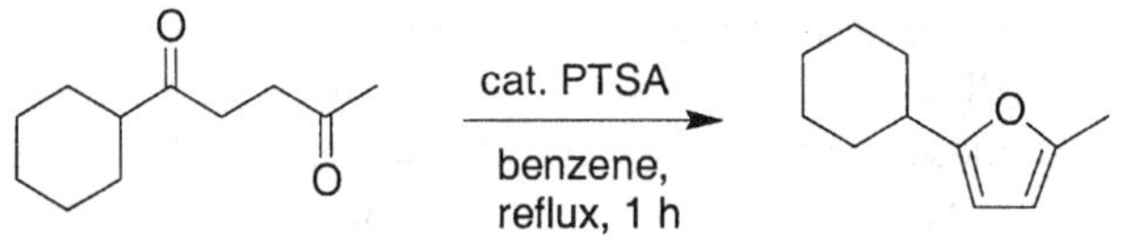
cat. PTSA
benzene,
reflux, 1 h

30 91 % **31**

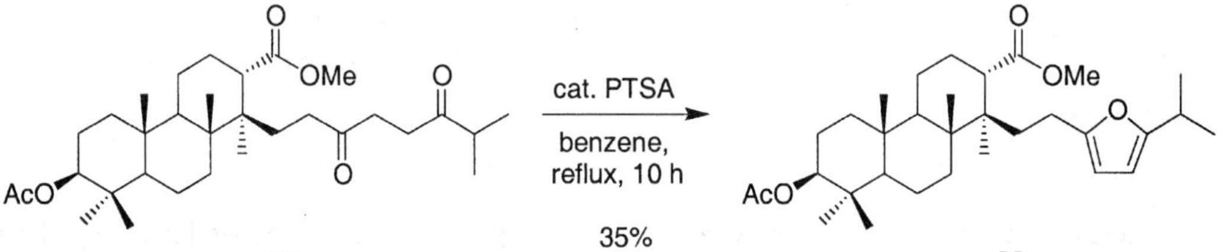
OMe
AcO
cat. PTSA
benzene,
reflux, 10 h
OMe
AcO

32 35% **33**

During studies of Horner–Wittig reactions, Warren prepared a furan incorporating a phosphine oxide using the Paal–Knorr reaction. Cyclization of dione **34** mediated by Amberlyst resin furnished furan **35** in excellent yield.[24] This strategy of using Amberlyst as the acid catalyst in the Paal–Knorr condensation originated with Scott and his report of a facile method for the production of 2,5-dimethylfuran (**37**) from 2,5-hexanedione (**36**).[25] Furan **37** has also been prepared from dione **36** using butyltin trichloride as the catalyst.[26]

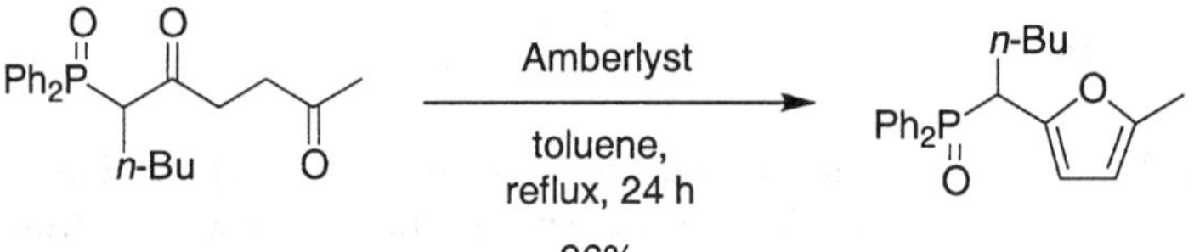
Ph2P
n-Bu
Amberlyst
toluene,
reflux, 24 h
n-Bu
Ph2P

34 96% **35**

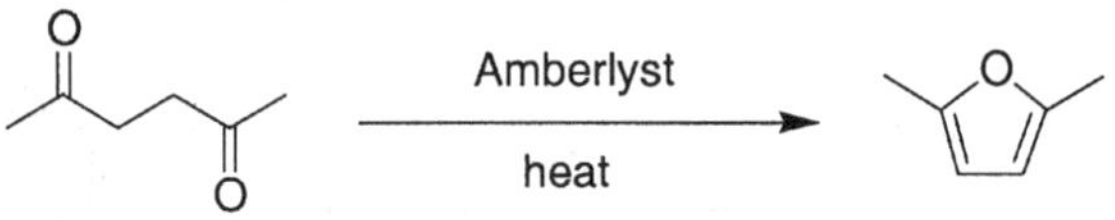
Amberlyst
heat

36 96% **37**

Portella reported the Paal–Knorr condensation of 1,4-bis(acylsilanes) **38** in the presence of p-toluenesulfonic acid to yield a variety of 2,5-disilylfurans **39**.[27] Presumably due to steric constraints, bis(acylsilanes) substituted in the 2-position failed to undergo the Paal–Knorr reaction to provide any of the expected trisubstituted furan products.

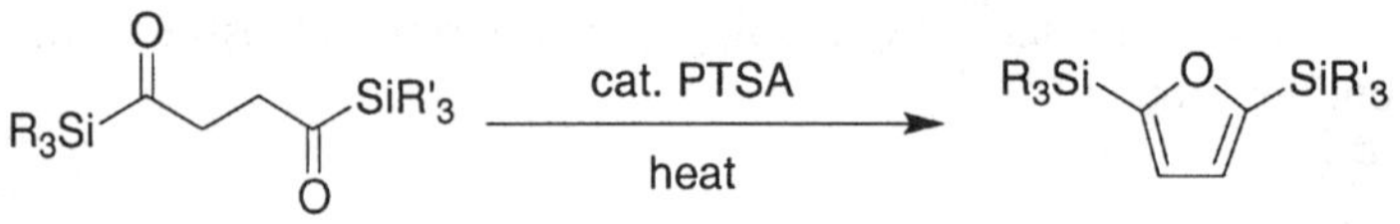
R3Si
SiR'3
cat. PTSA
heat
R3Si
SiR'3

38 61–71% **39**

4.2.4.3. Preparation of 2,3-disubstituted furans

Although it is far more common to synthesize these substrates using the Feist–Bénary reaction (Section 4.1), the Paal–Knorr reaction can also be used to prepare 2,3-disubstituted furans. In a recent example, Castagnoli converted 1,4-ketoaldehyde **40** into furan **41** in 97% yield upon exposure to hot sulfuric acid.[28]

4.2.4.4. Preparation of 2,4-disubstituted furans

Other less common products of the Paal–Knorr condensation include 2,4-disubstituted furans. An example of such a reaction is Molander's combination of dicarbonyl **42** with hydrochloric acid to furnish 4-methyl-2-(methyldiphenylsilyl)furan (**43**) in 87% yield.[12] It is important to note that this methodology can also be used to produce 2,5- and 2,3-disubstituted furans.[12]

4.2.4.5. Preparation of 2,3,5-trisubstituted furans

A multitude of 2,3,5-trisubstituted furans are available via the Paal–Knorr condensation. As with the synthesis of disubstituted furans, the scope of this version of the reaction is broad and includes incorporation of aryl, alkyl, ester, and phosphonate substituents.

Not surprisingly, based on the bioactivity of diaryl furans discussed in section 4.2.4.1, triaryl furans have also been investigated as enzyme inhibitors. De Laszlo prepared several 2,3,5-triarylfurans via the Paal–Knorr reactions (for example **44** to **45**) and tested these compounds for their activity toward P38 kinase.[29]

Diaryl bisfurans are available from two sequential Paal–Knorr reactions of tetraketones. For example, Barba converted **46** into 3,3′-bis-2,5-diphenylfuran (**47**) in good yield upon treatment with sulfuric acid and acetic anhydride.[30]

Ryder reported the preparation of an interesting alkyl diaryl furan that was subsequently polymerized and studied as a conducting polymer. The monomer furan **49** was available from the acid catalyzed cyclization of dione **48**.[31]

A recent report demonstrates that trisubstituted furans can be prepared on a solid support using the Paal–Knorr condensation. Raghavan synthesized a variety of triaryl and alkyl diaryl furans, one of which is highlighted below. Dione **50** was cyclized using p-toluenesulfonic acid in refluxing toluene followed by cleavage from the solid support to yield furan **51**.[32]

Several groups have used the Paal–Knorr reaction to synthesize trialkylfurans. Ballini prepared a series of 3-alkyl-2,5-dimethylfurans (**53**) in 60–94% yield by treatment of the corresponding diones (**52**) with tosic acid.[33] Weirsum produced the sterically congested tri-*t*-butylfuran **55** from dione **54**, but was unable to use the same strategy to furnish the tetra-*t*-butyl derivative.[34]

54   **55**

Neier recently developed two interesting furan syntheses based on a combination of the Feist–Bénary reaction (section 4.1) and the Paal–Knorr reaction. The first step involves the combination of *t*-butylacetoacetate (**56**) with an α-haloketone in the presence of sodium hydride. The resulting tricarbonyl (e.g., **57**) can either be treated with trifluoroacetic acid to yield a 3-acetoxy-2-hydroxyfuran or with another equivalent of sodium hydride followed by addition of a different bromoalkane to yield a more highly substituted dione ester (e.g., **58**) that cyclizes upon exposure to trifluoroacetic acid. For example, trialkylfuran **59** can be prepared in three steps from **56** with the final Paal–Knorr reaction proceeding in 67% yield, or tricarbonyl **57** can be converted directly into furan **60** in 67% yield.[35] As one would expect, trifluoroacetic acid catalyzes both the deprotection of the *t*-butyl ester and the Paal–Knorr cyclization. Successful formation of furan **59** also necessitates a decarboxylation reaction to allow complete unsaturation in the furan ring. Interestingly, treatment of **57** with TFA does not furnish a product resulting from reaction of the 1,4-diketone moiety. The furan product **60** is formed by attack of the carboxylic acid, available upon liberation of the *t*-butyl protecting group, on the only ketone present enabling formation of a five-membered ring.

Both 2,5-dialkyl-3-furoates and 2,5-dialkyl-3-phosphonofurans can be produced using the Paal–Knorr reaction. Methyl 2,5-diisopropyl-3-furoate (**62**) is available upon treatment of dione **61** with sulfuric acid.[36,37] Phosphonodiones **63** can be efficiently converted into 2-substituted-3-diethylphosphono-5-methylfurans **64** by exposure to Amberlyst in refluxing toluene.[38,39]

4.2.4.6. Preparation of tetrasubstituted furans

The Paal–Knorr reaction offers an excellent method for the preparation of tetrasubstituted furans; however, it does not work for some sterically congested substrates. Similar to di- and trisubstituted furans mentioned previously, tetrasubstituted furans have been investigated for biological acitivity. Katzenellenbogen has prepared numerous alkyl triarylfurans by the Paal–Knorr condensation (e.g. 65 to 66) and investigated their activity toward the estrogen receptor α.[40]

Several groups have employed the Paal–Knorr condensation for the preparation of disubstituted diarylfurans. Miyashita converted dione 67 into 3,4-disubstituted-2,5-diarylfuran 68 in good yield using standard Paal–Knorr conditions.[41] Lai demonstrated that 2,5-disubstituted-3,4-diarylfurans like 70 are available from dione 69 upon exposure to phosphorous pentoxide.[42]

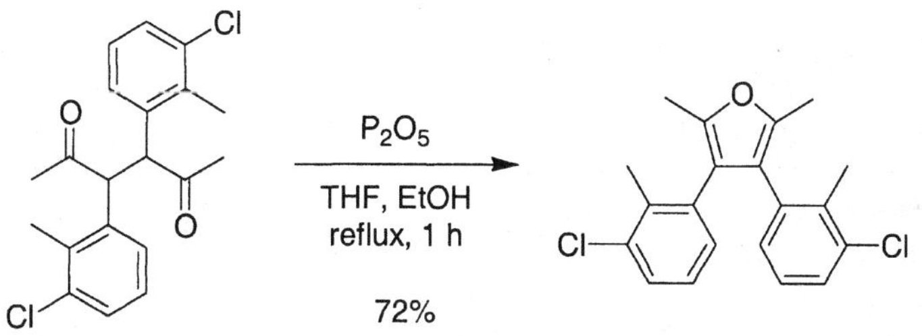

69 → **70**

P₂O₅

THF, EtOH
reflux, 1 h

72%

Tetraalkylfurans can be prepared via the Paal–Knorr reaction. For example, Muramatsu synthesized 2,5-diethyl-3,4-bis(trifluoromethyl)furan (**72**) in 94% yield by treatment of dione **71** with sulfuric acid.[43]

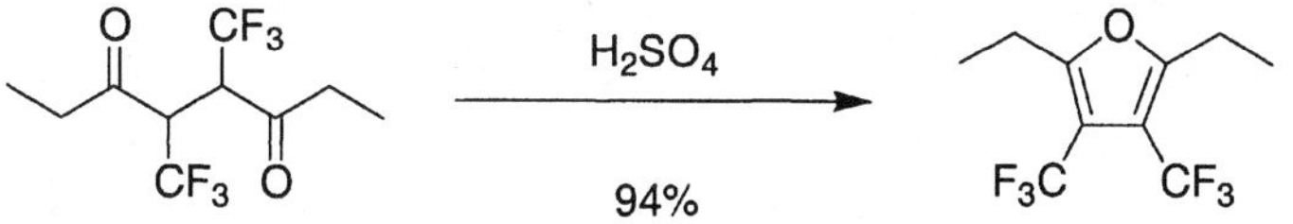

H_2SO_4

94%

71 → **72**

It is also possible to use the Paal–Knorr condensation to prepare 2,5-dialkyl-3,4-dicarbonyl substituted furans. For example, Zaleska converted diketone **73** into furan **74** in 92% yield,[44] and Pan treated tetracarbonyls **75** with p-toluenesulfonic acid to furnish furans **76** in excellent yields.[45]

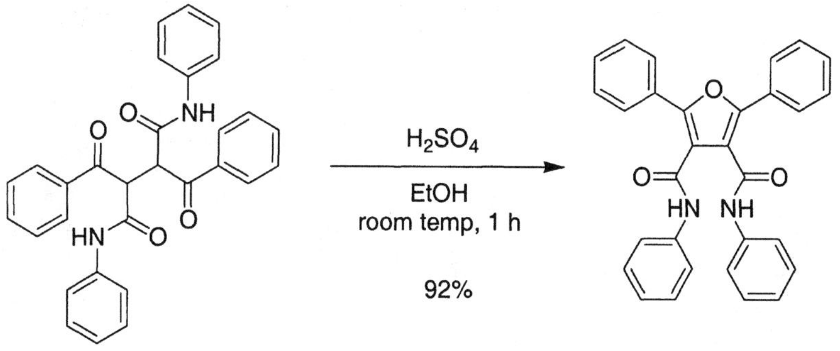

H_2SO_4

EtOH
room temp, 1 h

92%

73 → **74**

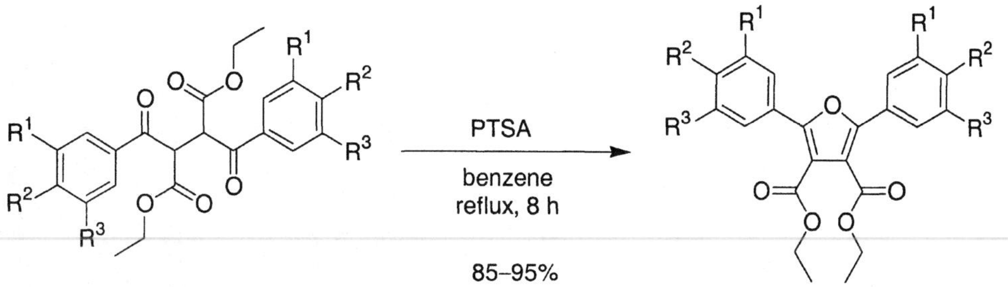

PTSA

benzene
reflux, 8 h

85–95%

75 → **76**

4.2.5. Variations

Numerous variations of the Paal–Knorr condensation are known. The most popular methods use starting materials that are converted to 1,4-dicarbonyls in situ and cyclize to yield furan products without isolating their dicarbonyl precursors. Other more specialized strategies have been developed for the preparation of heterosubstituted furans.

A common approach starts with a protected 1,4-dicarbonyl and unmasks the requisite carbonyl using acid, thus facilitating the Paal–Knorr reaction immediately upon deprotection. For example, Nagai used sulfuric acid to convert acetal **77** into 2,4-disubstituted furan **78** albeit in low yield.[15] Molander produced a different 2,4-disubstituted furan by a similar strategy. Thioketal acetal **79** was treated with mercury(II) chloride and furnished furan **80** in 71% yield.[12] Thus this strategy provides a useful approach for the synthesis of a variety of 2,4-disubsituted furans.

A less obvious method for the preparation of 2,4-disubstituted furans involves the treatment of epoxyketones like **81** with catalytic p-toluenesulfonic acid and their rearrangement to furans (for example **82**). Cormier developed this method, which presumably involves a 1,4-diketone intermediate, and works for a variety of epoxyketone derivatives to yield other disubstituted furan isomers as well as 2,3,5-trisubstituted furans.[46]

Another variation of the Paal–Knorr condensation involves starting with a derivative of 2-butene-1,4-dione and performing a reduction prior to the cyclization reaction. For example, Rao has recently reported that trisubstituted furans **84** can be produced in high yield upon treatment of diones **83** with formic acid, catalytic sulfuric acid, catalytic palladium on carbon, and exposure to microwave irradiation.[47] Haddadin used a similar strategy to produce several tri- and tetrasubstituted furans, however, he used triethylphosphite as the reductant.   A representative example involves the conversion of dione **85** into 2,3,5-triphenylfuran (**86**) in 66% yield.[48]

Perumal investigated a novel variation that involved the combination of a Vilsmeier reaction with a Paal–Knorr condensation.   Reaction of 3-benzoylpropionic acid (**87**) under Vilsmeier conditions furnished chloroformylfuran **88** in 75% yield, while reaction of acetonylacetone (**89**) provided formylfuran **90** in 60% yield.[49]

Ketoamides can participate in a variation of the Paal–Knorr condensation to yield 5-alkyl-2-aminofurans.   Boyd described the cyclization of 1,4-ketoamides **91** upon exposure to acetic anhydride and perchloric acid to yield imminium salts **92** that furnished aminofurans **93** after treatment with triethylamine.[50]

91 → **92** (29-97%) → **93** (87-100%)

4.2.6. Experimental

30 → **31** (91 %)

cat. PTSA
benzene,
reflux, 1 h

2-Cyclohexyl-5-methylfuran (31):[21]

p-Toluenesulfonic acid (70 mg) and 1-cyclohexylpentane-1,4-dione (0.64 g, 3.5 mmol) (**30**) were refluxed in benzene (100 mL) with a Soxhlet extractor containing molecular sieves (4 Å) for 1 h. The solvent was evaporated and the residue was dissolved in diethyl ether (50 mL), washed with aqueous sodium bicarbonate (saturated, 20 mL), dried over potassium carbonate, filtered through silica, and evaporated to yield the furan (0.53 g, 91%) (**31**); R_f (hexane/EtOAc, 6:1) 0.67; IR (neat) 3120, 2940, 2860, 1620, 1570 cm^{-1}; ^{1}H NMR (CDCl$_3$, 250 MHz) δ 1.17–1.43 (m, 5H), 1.58–1.78 (m, 3H), 1.82–2.02 (m, 2H), 2.24 (d, 3H, $J = 0.5$ Hz), 2.5 (m, 1H), 5.80 (d, 1H, $J = 3.1$ Hz), 5.83 (dd, 1H, $J = 1.0, 3.1$ Hz).

4.2.7   References

1.   [R] König, B. Product Class 9: Furans. In *Science of Synthesis: Houben-Weyl Methods of Molecular Transformations*; Maas, G., Ed.; Georg Thieme Verlag: New York, 2001; Cat. 2, Vol. 9, 183-278.
2.   [R] Friedrichsen, W. Furans and Their Benzo Derivatives: Synthesis. In *Comprehensive Heterocyclic Chemistry II*; Katritzky, A. R., Rees, C. W., Scriven, E. F. V., Eds.; Pergamon: New York, 1996; Vol. 2, 351-393.
3.   [R] Dean, F. M. Recent Advances in Furan Chemistry. Part I. In *Advances in Heterocyclic Chemistry*; Katritzky, A. R., Ed.; Academic Press: New York, 1982; Vol. 30, 167-238.
4.   [R] Joule, J. A.; Mills, K. *Heterocyclic Chemistry*, 4th ed.; Blackwell Science: Cambridge, 2000; 308-309.
5.   [R] Gupta, R. R.; Kumar, M.; Gupta, V. *Heterocyclic Chemistry*, Springer: New York, 1999; Vol. 2, 83-84.
6.   [R] Gilchrist, T. L. *Heterocyclic Chemistry*, 3rd ed.; Longman: New York, 1997; 211.
7.   Paal, C. *Ber. Dtsch. Chem. Ges.* **1884**, *17*, 2756.
8.   Knorr, L. *Ber. Dtsch. Chem. Ges.* **1885**, *18*, 299
9.   Paal, C. *Ber. Dtsch. Chem. Ges.* **1885**, *18*, 367.
10.   Amarnath, V.; Amarnath, K. *J. Org. Chem.* **1995**, *60*, 301.
11.   Drewes, S. E.; Hogan, C. J. *Synth. Commun.* **1989**, *19*, 2101.
12.   Siedem, C. S.; Molander, G. A. *J. Org. Chem.* **1996**, *61*, 1140.
13.   Lansiaux, A.; Dassonneville, L.; Facompré, M.; Kumar, A.; Stephens, C. E.; Bajic, M.; Tanious, F.; Wilson, W. D.; Boykin, D. W.; Bailly, C. *J. Med. Chem.* **2002**, *45*, 1994.
14.   Xiao, G.; Kumar, A.; Li, K.; Rigl, C. T.; Bajic, M.; Davis, T. M.; Boykin, D. W.; Wilson, W. D. *Bioorg. Med. Chem.* **2001**, *9*, 1097.
15.   Kikuchi, K.; Hibi, S.; Yoshimura, H.; Tokuhara, N.; Tai, K.; Hida, T.; Yamauchi, T.; Nagai, M. *J. Med. Chem.* **2000**, *43*, 409.
16.   Perrier, H.; Bayly, C.; Laliberté, F.; Huang, Z.; Rasori, R.; Robichaud, A.; Girard, Y.; Macdonald, D. *Bioorg. Med. Chem. Lett.* **1999**, *9*, 323.
17.   Fajarí, L.; Brillas, E.; Alemán, C.; Juliá, L. *J. Org. Chem.* **1998**, *63*, 5324.
18.   Chen, L.-H.; Wang, C.-Y.; Luo, T.-M. H. *Heterocycles* **1994**, *38*, 1393.
19.   Jones, R. A.; Civcir, P. U. *Tetrahedron* **1997**, *53*, 11529.
20.   Miller, D. C.; Johnson, M. R.; Becker, J. J.; Ibers, J. A. *J. Heterocycl. Chem.* **1993**, *30*, 1485.

21.   Salimbeni, A.; Canevotti, R.; Paleari, F.; Bonaccorsi, F.; Renzetti, A. R.; Belvisi, L.; Bravi, G.; Scolastico, C. *J. Med. Chem.* **1994**, *37*, 3928.
22.   Fleming, I.; Morgan, I. T.; Sarkar, A. K. *J. Chem. Soc., Perkin Trans. 1* **1998**, 2749.
23.   Denisenko, M. V.; Pokhilo, N. D.; Odinokova, L. E.; Denisenko, V. A.; Uvarova, N. I. *Tetrahedron Lett.* **1996**, *37*, 5187.
24.   Brown, P. S.; Greeves, N.; McElroy, A. B.; Warren, S. *J. Chem. Soc., Perkin Trans. 1* **1991**, 1485.
25.   Scott, L. T.; Naples, J. O. *Synthesis* **1973**, 209.
26.   Marton, D.; Slaviero, P.; Tagliavini, G. *Tetrahedron* **1989**, *45*, 7099.
27.   (a) Saleur, D.; Bouillon, J.-P.; Portella, C. *Tetrahedron Lett.* **2000**, *41*, 321. (b) Bouillon, J.-P.; Saleur, D.; Portella, C. *Synthesis* **2000**, 843.
28.   Castagnoli Jr., N.; Yu, J. *Bioorg. Med. Chem.* **1999**, *7*, 2835.
29.   de Laszlo, S. E.; Visco, D.; Agarwal, L.; Chang, L.; Chin, J.; Croft, G.; Forsyth, A.; Fletcher, D.; Frantz, B.; Hacker, C.; Hanlon, W.; Harper, C.; Kostura, M.; Li, B.; Luell, S.; MacCoss, M.; Mantlo, N.; O'Neill, E. A.; Orevillo, C.; Pang, M.; Parsons, J.; Rolando, A.; Sahly, Y.; Sidler, K.; Widmer, W. R.; O'Keefe, S. *J. Bioorg. Med. Chem. Lett.* **1998**, *8*, 2689.
30.   Barba, R.; de la Fuente, J. L. *J. Org. Chem.* **1993**, *58*, 7685.
31.   Schweiger, L. F.; Ryder, K. S.; Morris, D. G.; Glidle, A.; Cooper, J. M. *J. Mater. Chem.* **2000**, *10*, 107.
32.   Raghavan, S.; Anuradha, K. *Synlett* **2003**, 711.
33.   Ballini, R.; Bosica, G.; Fiorini, D.; Giarlo, G. *Synthesis* **2001**, 2003.
34.   Wynberg, H.; Wiersum, U. E. *J. Chem. Soc., Chem. Commun.* **1990**, 460.
35.   Stauffer, F.; Neier, R. *Org. Lett.* **2000**, *2*, 3535.
36.   Shono, T.; Soejima, T.; Takigawa, K.; Yamaguchi, Y.; Maekawa, H.; Kashimura, S. *Tetrahedron Lett.* **1994**, *35*, 4161.
37.   For a synthesis of ethyl 5-*t*-butyl-2-methyl-3-furoate, see Trahanovsky, W. S.; Chou, C.-H.; Cassady, T. J. *J. Org. Chem.* **1994**, *59*, 2613.
38.   Truel, I.; Mohamed-Hachi, A.; About-Jaudet, E.; Collignon, N. *Synth. Commun.* **1997**, *27*, 1165.
39.   For syntheses of other 2,5-dialkyl-3-phosphonofurans, see *Synlett* **2001**, 703.
40.   (a) Mortensen, D. S.; Rodriguez, A. L.; Carlson, K. E.; Sun, J.; Katzenellenbogen, B. S.; Katzenellenbogen, J. A. *J. Med. Chem.* **2001**, *44*, 3838. (b) Mortensen, D. S.; Rodriguez, A. L.; Sun, J.; Katzenellenbogen, B. S.; Katzenellenbogen, J. A. *Bioorg. Med. Chem. Lett.* **2001**, *11*, 2521.
41.   Miyashita, A.; Matsuoka, Y.; Numata, A.; Higashino, T. *Chem. Pharm. Bull.* **1996**, *44*, 448.
42.   Lai, Y.-H.; Chen, P. *J. Org. Chem.* **1996**, *61*, 935.
43.   Nishida, M.; Hayakawa, Y.; Matsui, M.; Shibata, K.; Muramatsu, H. *J. Heterocycl. Chem.* **1991**, *28*, 225.
44.   Zaleska, B.; Lis, S. *Synth. Commun.* **2001**, *31*, 189.
45.   Wu, A.; Wang, M.; Pan, X. *Synth. Commun.* **1997**, *27*, 2087.
46.   Cormier, R. A.; Francis, M. D. *Synth. Commun.* **1981**, *11*, 365.
47.   Rao, H. S. P.; Jothilingam, S. *J. Org. Chem.* **2003**, *68*, 5392.
48.   Haddadin, M. J.; Agha, B. J.; Tabri, R. F. *J. Org. Chem.* **1979**, *44*, 494.
49.   Venugopal, M.; Balasundaram, B.; Perumal, P. T. *Synth. Commun.* **1993**, *23*, 2593.
50.   Boyd, G. V.; Heatherington, K. *J. Chem. Soc., Perkin Trans. 1* **1973**, 2523.

Kevin M. Shea

Chapter 5 Thiophenes 183

5.1   Fiesselmann Thiophene Synthesis

5.1.1   Description

The Fiesselmann thiophene synthesis involves the condensation reaction of thioglycolic acid derivatives with α,β-acetylenic esters, which upon treatment with base results in the formation of 3-hydroxy-2-thiophenecarboxylic acid derivatives.

5.1.2   Historical Perspective

Soon after the 1946 report by Woodward on the condensation of thioglycolic acid and α,β-unsaturated esters in the presence of base to produce structures like 2-carbomethoxy-3-ketotetrahydrothiophene **4**,[1] Fiesselmann extended this reaction to α,β-acetylenic esters for the direct preparation of 3-hydroxy-2-thiophenecarboxylic acid derivatives.[2]

In subsequent years, Fiesselmann illustrated the applicability of this reaction for a number of related carbonyl systems. Thus, thiophenes have been produced from reactions of thioglycolic acid derivatives and β-keto esters, α,β-dihalo esters and α- and β-halovinyl esters, along with the corresponding nitriles, ketones and aldehydes. All of these starting materials display the same oxidation state at the β-carbon as demonstrated in the original studies using α,β-acetylenic esters. Furthermore, a variety of α-mercaptocarbonyl systems can be used in place of the thioglycolic acid derivatives, further extending the applicability of this reaction.

Unfortunately, much of Fiesselmann's work was documented only in patents and doctoral theses, allowing for the rediscovery of this classic reaction in recent years.[3] In fact, as late as 1997, the Fiesselmann reaction of **5** with methylthioglycolate was rediscovered as a novel, tandem Michael addition/intramolecular Knoevenagel approach to thiophenes such as **6**.[4]

1. **2**, THF, 0°C

2. CsCO$_3$/MgSO$_4$ (1:2)
   MeOH, 0°C – r.t.
   83% yield

5.1.3　Mechanism

The mechanism of the Fiesselmann reaction between methylthioglycolate and α,β-acetylenic esters proceeds via consecutive base-catalyzed 1,4-conjugate addition reactions to form thioacetal **7**.[2,5] Enolate formation, as a result of treatment with a stronger base, causes a Dieckmann condensation to occur providing ketone **8**. Elimination of methylthioglycolate and tautomerization driven by aromaticity provides the 3-hydroxy thiophene dicarboxylate **9**.

5.1.4 Synthetic Utility
5.1.4.1 With α,β-acetylenic carbonyl derivatives

Numerous examples exist in which α,β-acetylenic esters,[2,3,5-9] ketones,[4,10] aldehydes,[11] and nitriles[12] have been converted to the corresponding thiophene. One such notable example was utilized in conjunction with efforts to explore the biological activity of the designed antitumor agent golfomycin A (**10a**: R = H).[10] In order to evaluate the Michael acceptor capability of golfomycin A, **10b** (R = TBS) was treated with methyl thioglycolate and DBU. Nucleophilic addition of the thioglycolate was rapid, presumably eliminating considerable ring strain, followed by cyclization to provide **11** (55%) and thiophene **12** (20%). Dehydration to the thiophene is thought to reintroduce considerable ring strain in the formation of **12**.

The cyclization with α,β-unsaturated nitriles has proven effective for the synthesis of 3-aminothiophene **14**, a key intermediate for the synthesis of p38 kinase inhibitors.[12]

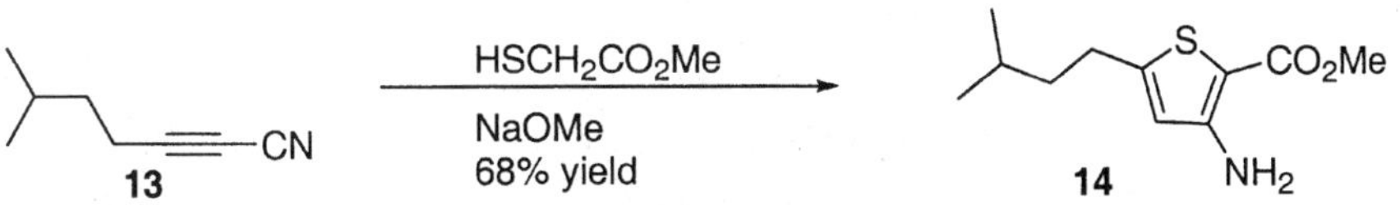

5.1.4.2 With β-halo-α,β-unsaturated carbonyl derivatives

The Fiesselmann reaction has been extensively used with β-halovinyl esters,[3,13] ketones,[3,14] aldehydes[3,15–21] and nitriles[3,22–31] as reaction partners for thioglycolic acid and its derivatives. This reaction with β-halovinyl aldehydes has been extensively explored as a result of the availability of β-chloro-α,β-unsaturated aldehydes via the Vilsmeier formylation of ketones.[32]   The preparation of **15**,[17] **16**,[15] **17**[21] and **18**[18] with β-halovinylaldehydes prepared by this two-step process speaks to its generality.

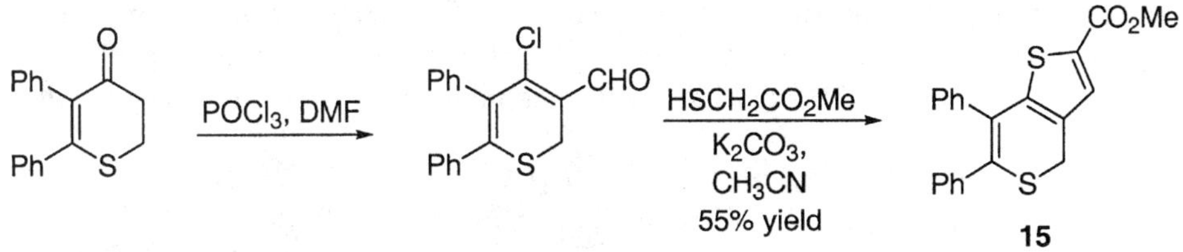

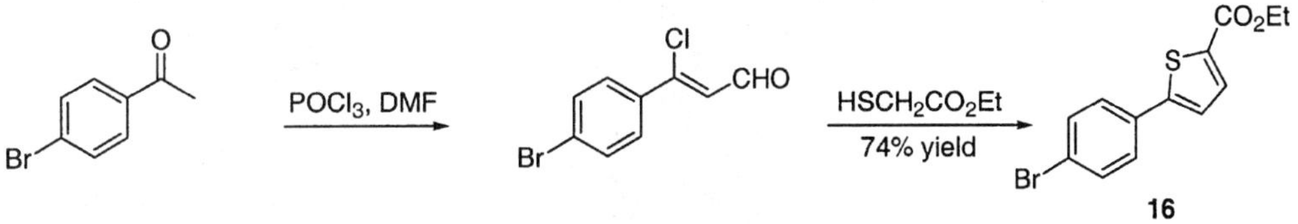

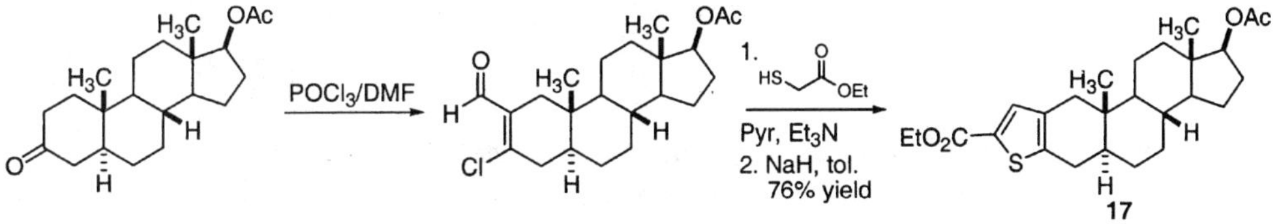

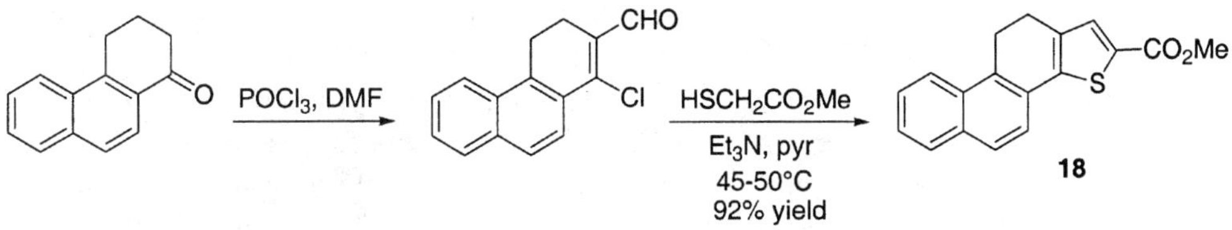

This reaction has recently been exploited for the synthesis of 2,3-diarylthiophenes.[33] Thus, β-chloro-α,β-unsaturated aldehyde **19** underwent reaction with ethylthioglycolate to produce **20**. The production of **20** by this method enabled the synthesis of a number of derivatives for investigations of their use as anti inflammatory agents.

Another frequently utilized variation of this theme involves the cyclization of thioglycolate esters with β-halovinylnitriles to efficiently provide 3-aminothiophenes in good yield. Examples of typical starting nitriles are especially prevalent in aromatic and heteroaromatic compounds where the halogen leaving group and the nitrile are disposed in an *ortho* relationship. Mechanistic considerations would suggest a process of nucleophilic aromatic substitution which would be facilitated by an electron deficient aromatic system for initial halogen displacement. The reaction has been exploited for the synthesis of heterocycles **21**[26] and **22**[27] with potential medicinal chemistry applications.

5.1.4.3 With α,β-dihalo- or α-halo-α,β-unsaturated carbonyl derivatives

A significant number of examples exist in which α,β-dihalogenated carbonyl derivatives undergo reactions with thioglycolates in the presence of base to produce thiophenes.[3,34] The reactions have been shown to occur through intermediate α-halo-α,β-unsaturated carbonyl derivatives produced by the elimination of HX.[3] Thus the use of α-halo-α,β-unsaturated carbonyl systems in place of the α,β-dihalocarbonyl compounds was found to efficiently provide thiophenes upon reaction with thioglycolates. In a modification of the work of Fiesselmann, readily accessible methyl-2-chloroacrylate **23** and 2-chloroacrylonitrile **24** have been used in this sense to provide **25** and **26**, respectively.[34]

5.1.4.4 With 1,3-dicarbonyl compounds

A significant number of examples exist in which 1,3-dicarbonyl derivatives undergo reaction with thioglycolates to produce thiophenes.[3,35–40] Such reactions are particularly effective when used in conjunction with β-ketoesters, as demonstrated by the preparation of **27**.[37]

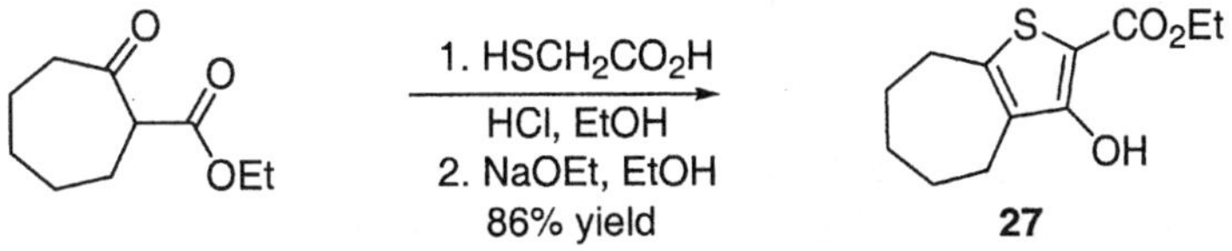

Complications often arise in the use of 1,3-diketones under the above reaction conditions. This is primarily due to the lack of regioselectivity with regard to formation of the intermediate thioacetal. However, when benzoyl acetone derivatives are employed, the thioketal forms preferentially with the aromatic ketone.[3]

An exceptional demonstration of this reaction is illustrated by the synthesis of a novel class of thienomorphinans, potent δ-opiod receptors ($K_i \sim 1.4 - 2.0$ nM) with increased lipophilicity over the corresponding pyrrolomorphinans.[39] Treatment of the diketone **28** under the conditions described by Lissavetsky[37] resulted in intermediate **29**, which upon exposure to base, cyclized to yield **30**. Notably, this reaction proceeded with complete regioselectivity in the formation of intermediate thioether **29**, and thus resulted in the formation of a single regioisomer **30**.

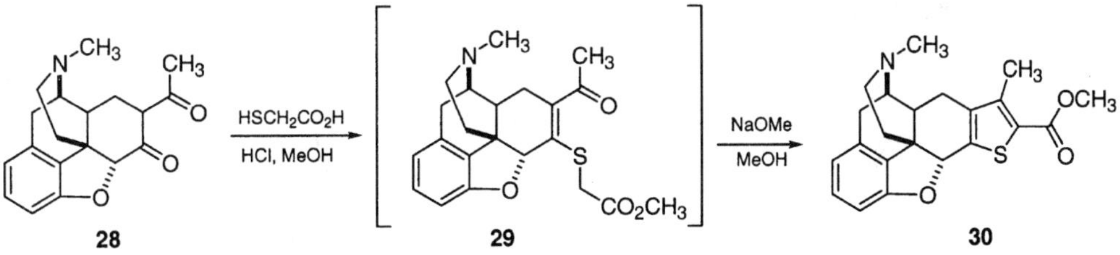

5.1.4.5 With other carbonyl derivatives

In a manner analogous to reactions with β-halo-α,β-unsaturated carbonyl derivatives, numerous other leaving groups can be utilized in place of the halogen. Among the substituents employed are dimethylamino,[41] nitro,[42–44] and ketones bearing β-triethylammonium,[14] and β-thio substituents.[45,46] A number of antileishmanial and antifungal agents similar to **31** were prepared in this manner.[46]

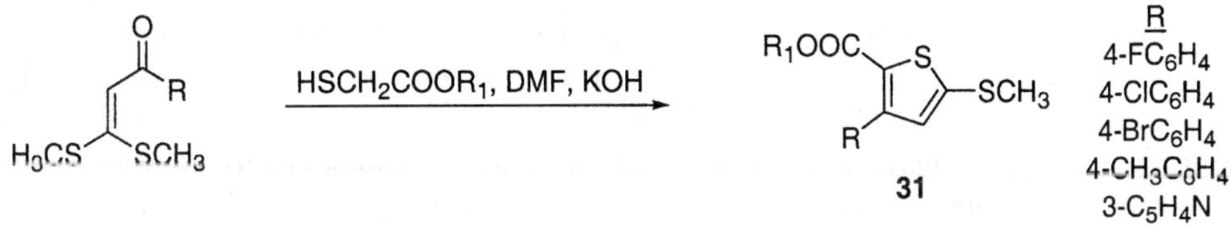

R
4-FC$_6$H$_4$
4-ClC$_6$H$_4$
4-BrC$_6$H$_4$
4-CH$_3$C$_0$H$_4$
3-C$_5$H$_4$N

This method has also been applied to the preparation of 4-fluorinated thiophenes such as **32**.[45]

5.1.4.6 With other α-mercapto derivatives

While the majority of examples of the Fiesselmann reactions found in the literature involve thioglycolic acid or the corresponding thioglycolates, the scope of this reaction has been expanded to a variety of other α-mercaptocarbonyl derivatives.[3] Some examples include α-mercapto aldehydes,[7,47] ketones,[7,14,47,48] and amides.[23,29] Notably, when compound **33** is treated with 2-mercaptoacetamide, compound **34** was produced in 48% yield.[29]

5.1.5 Experimental

Methyl 4,5,6,7-tetrahydrobenzo[c]thiophene-1-carboxylate (36):[38]

To a magnetically stirred mixture of 2-oxocyclohexanecarboxaldehyde **35** (3.6 g, 28 mmol) and methyl thioglycolate (6.0 g, 56 mmol) was added 3 drops of concentrated H_2SO_4. The resulting yellow solution was stirred at rt for 12 h, diluted with 25 mL of ice-water, and extracted with CH_2Cl_2 (25 mL). The aqueous phase was extracted with an additional 25 mL portion of CH_2Cl_2, and the combined organic phases were washed with 50 mL of a saturated aqueous NaCl solution and dried over Na_2SO_4. Removal of the drying agent by filtration followed by evaporation *in vacuo* gave a viscous yellow oil which was dissolved in 25 mL of MeOH and added dropwise over 1 h to a freshly-prepared solution of NaOMe (from 1.7 g, 2.5 equiv of sodium metal) in 100 mL of MeOH. The deep orange solution was allowed to stir overnight (12 h), concentrated to one-quarter volume *in vacuo*, and partitioned between CH_2Cl_2 (50 mL) and water (50 mL). The aqueous phase was extracted with an additional 25 mL portion of CH_2Cl_2, and the combined organic phases were washed with 50 mL of a saturated aqueous NaCl solution and dried over Na_2SO_4. The drying agent was removed by filtration and the filtrate concentrated *in vacuo* to afford a yellow oil which was purified by chromatography on silica using 5% EtOAc in hexanes as eluent to give 1.5 g (27%) of a clear, colorless liquid: ^1H NMR (CDCl$_3$, 300 MHz) δ 1.73 (m, 4H), 2.68 (t, 2H, J = 6.2 Hz), 3.02 (t, 2H, J = 6.2 Hz), 3.83 (s, 3H), 7.05 (s, 1H); ^{13}C NMR (CDCl$_3$, 75.6 MHz) δ 22.5, 22.5, 26.2, 26.6, 51.2, 125.1, 125.1, 139.8, 146.4, 162.9; IR (NaCl) 3088, 2935, 2848, 1698, 1538 cm^{-1}; MS *m/e* (relative intensity) 196 (99), 181 (15), 165 (55), 137 (92); HRMS calcd for $C_{10}H_{12}O_2S$: 196.0557, found: 196.0557.

37 **38**

Ethyl thieno[3,2-*b*]thiophene-2-carboxylate (38):[16]

3-Bromothiophene-2-carbaldehyde **37** (25.71 g, 134.0 mmol) was added to a stirred mixture of ethyl-2-sulfanyl acetate (14.8 mL, 16.22 g, 135.0 mmol), potassium carbonate (25.0 g) and *N*, *N*-dimethylformamide (250 mL) at ambient temperature and the resulting mixture was stirred for a further 72 h. The mixture was then poured into water (500 mL) and extracted with dichloromethane. The combined extracts were dried (MgSO$_4$), and filtered, and distillation of the solvents under reduced pressure gave the ester **38** (23.0 g, 81%): bp (Kugelrohr distillation) 120–125°C at 0.1 mm Hg; ν$_{max}$/cm^{-1} 1707 (CO); ^1H NMR δ$_H$ 1.37 (3H, t, J 7.0, Me), 4.34 (2H, q, J = 7.0, CH$_2$), 7.24 (1H, d $J_{6,5}$ = 5.0, H-6), 7.55 (1H, d, $J_{5,6}$ 5.0, H-5) and 7.97 (1H, s, H-3); MS: m/z 247 (M$^+$ + 1, 35%) Anal. Calcd for $C_9H_8S_2O_2$: C, 50.9; H, 3.8; S, 30.2. Found: C, 51.1; H, 3.9; S, 30.6.

5.1.6 References

1. Woodward, R. B.; Eastman, R. H. *J. Am. Chem. Soc.* **1946**, *68*, 2229.
2. Fiesselmann, H.; Schipprak, P. *Chem. Ber.* **1954**, *87*, 835.
3. For an outstanding review of the history and application of the Fiesselmann work through the 1960's, including references to his students' dissertations, see: [R] Gronowitz, S. In *Thiophene and Its Derivatives*, Part 1, Gronowitz, S., ed.; Wiley-Interscience: New York, 1985, 88–125.
4. Obrecht, D.; Gerber, F.; Sprenger, D.; Masquelin, T. *Helv. Chim. Acta* **1997**, *80*, 531.
5. Fiesselmann, H.; Schipprak, P.; Zeitler, L. *Chem. Ber.* **1954**, *87*, 841.
6. Courtin, A.; Class, E.; Erlenmeyer, H. *Helv. Chim. Acta* **1964**, *47*, 1748.
7. Bohlmann, F.; Bresinsky, E. *Chem. Ber.* **1964**, *97*, 2109.
8. Fiesselmann, H.; Schipprak, P. *Chem. Ber.* **1956**, *89*, 1897.
9. Hendrickson, J. B.; Rees, R.; Templeton, J. F. *J. Am. Chem. Soc.* **1964**, *86*, 107.
10. Nicolaou, K. C.; Skokotas, G.; Furuya, S.; Suemune, H.; Nicolaou, D. C. *Angew. Chem. Int. Ed. Engl.* **1990**, *29*, 1064.
11. Bohlmann, F.; Bornowski, H.; Kramer, D. *Chem. Ber.* **1967**, *100*, 107.
12. Redman, A. M.; Johnson, J. S.; Dally, R.; Swartz, S.; Wild, H.; Paulsen, H.; Caringal, Y.; Gunn, D.; Renick, J.; Osterhaut, M.; Kingery-Wood, J.; Smith, R. A.; Lee, W.; Dumas, J.; Wilhelm, S. M.; Housley, T. J.; Bhargava, A.; Ranges, G. E.; Shrikhande, A.; Young, D.; Bombara, M.; Scott, W. J. *Bioorg. Med. Chem. Lett.* **2001**, *11*, 9-12.
13. Dinescu, L.; Maly, K. E.; Lemieux, R. P. *J. Mater. Chem.* **1999**, *9*, 1679.
14. Alberola, A.; Andrés, J. M.; González, A.; Pedrosa, R.; Prádanos, P. *Synth. Comm.* **1990**, *20*, 2537.

15.   Hauptmann, S.; Weissenfels, M; Scholz, M.; Werner, E.-M.; Köhler, H. J.; Weisflog, J. *Tetrahedron Lett.* **1968**, *9*, 1317.
16.   Fuller, L. S.; Iddon, B.; Smith, K. A. *J. Chem. Soc. Perkin Trans. 1* **1997**, 3465.
17.   Vasumathi, N.; Ramana, D. V.; Ramadas, S. R. *Synth. Commun.* **1990**, *20*, 2749.
18.   Kar, G. K.; Karmakar, A. C.; Ray, J. K. *J. Heterocyclic Chem.* **1991**, *28*, 999.
19.   Athmani, S.; Farhat, M. F.; Iddon, B. *J. Chem. Soc. Perkin Trans. 1* **1992**, 973.
20.   Neidlein, R.; Schröder, G. *Helv. Chim. Acta* **1992**, *75*, 825.
21.   Kumar, V.; Daum, S. J.; Bell, M. R.; Alexander, M. A.; Christiansen, R. G.; Ackerman, J. H.; Krolski, M. E.; Pilling, G. M.; Herrmann, J. L.; Winneker, R. C.; Wagner, M. M. *Tetrahedron* **1991**, *47*, 5099.
22.   Tumkevicius, S.; Mickiene, J. *Org. Prep. Proc. Int.* **1991**, *23*, 413.
23.   Czech, K.; Haider, N.; Heinisch, G. *Monatsh fur Chemie* **1991**, *122*, 413.
24.   Dave, C. G.; Shah, P. R.; Shah, A. B. *Indian J. Chem.* **1992**, *31B*, 492.
25.   Peinador, C.; Veiga, C.; Vilar, J.; Quintela, J. M. *Heterocycles* **1994**, *38*, 1299.
26.   Showalter, H. D. H.; Bridges, A. J.; Zhou, H.; Sercel, A. D.; McMichael, A.; Fry, D. W. *J. Med. Chem.* **1999**, *42*, 5464.
27.   Hrib, N. J.; Jurcak, J. G.; Bregna, D. E.; Dunn, R. W.; Geyer, H. M.; Hartman, H. B.; Roehr, J. E.; Rogers, K. L.; Rush, D. K.; Szczepanik, A. M.; Szewczak, M. R.; Wilmot, C. A.; Conway, P. G. *J. Med. Chem.* **1992**, *35*, 2712.
28.   Migianu, E.; Kirsch, G. *Synthesis* **2002**, 1096.
29.   Athmani, S.; Iddon, B. *Tetrahedron*, **1992**, *48*, 7689.
30.   Mekheimer, R. A.; Ahmed, E. K.; El-Fahham, H. A.; Kamel, L. H. *Synthesis* **2001**, 97.
31.   Wang, Z.; Neidlein, R.; Krieger, C. *Synthesis* **2000**, 255.
32.   Arnold, Z.; Zemlicka, J. *Proc. Chem. Soc.* **1958**, 827.
33.   Tsuji, K.; Nakamura, K.; Ogino, T.; Konishi, N.; Tojo, T.; Ochi, T.; Seki, N.; Matsuo, M. *Chem. Pharm. Bull.* **1998**, *46*, 279.
34.   Huddleston, P. R.; Barker, J. M.; *Synth. Commun.* **1979**, *9*, 731.
35.   Fiesselmann, H.; Pfeiffer, G. *Chem. Ber.* **1954**, *87*, 848.
36.   Fiesselmann, H.; Thoma, F. *Chem. Ber.* **1956**, *89*, 1907.
37.   Donoso, R.; Jordán de Urríes, P.; Lissavetzky, J. *Synthesis* **1992**, 526.
38.   Taylor, E. C.; Dowling, J. E. *J. Org. Chem.* **1997**, *62*, 1599.
39.   Ronzoni, S.; Cerri, A.; Dondio, G.; Fronza, G.; Petrillo, P.; Raveglia, L. F.; Gatti, P. A. *Org. Lett.* **1999**, *1*, 513.
40.   Mullican, M. D.; Sorenson, R. J.; Connor, D. T.; Thueson, D. O.; Kennedy, J. A.; Conroy, M. C. *J. Med. Chem.* **1991**, *34*, 2186.
41.   Okada, E.; Masuda, R.; Hojo, M.; Imazaki, N.; Miya, H. *Heterocycles* **1992**, *34*, 103.
42.   Dal Piaz, V.; Ciciani, G.; Giovannoni, M. P. *Synthesis* **1994**, 669.
43.   Valderrama, J. A.; Valderrama, C. *Synth. Commun.* **1997**, *27*, 2143.
44.   Shkinyova, T. K.; Dalinger, I. L.; Molotov, S. I.; Shevelev, S. A. *Tetrahedron Lett.* **2000**, *41*, 4973.
45.   Andrès, D. F.; Laurent, E. G.; Marquet, B. S. *Tetrahedron Lett.* **1997**, *38*, 1049.
46.   Ram, V. J.; Goel, A.; Shukla, P. K.; Kapil, A. *Bioorg. Med. Chem. Lett.* **1997**, *7*, 3101.
47.   Bohlmann, F.; Bresinsky, E. *Chem. Ber.* **1967**, *100*, 107.
48.   Bohlmann, F.; Bornowski, H.; Kramer, D. *Chem. Ber.* **1963**, *96*, 584.

Richard J. Mullins and David R. Williams

5.2 Gewald Aminothiophene Synthesis

5.2.1 Description

The Gewald aminothiophene synthesis involves the condensation of aldehydes, ketones, or 1,3-dicarbonyl compounds **1** with activated nitriles such as malononitrile or cyanoacetic esters **2** and elemental sulfur in the presence of an amine to afford the corresponding 2-aminothiophene **3**.[1-3]

R_1, R_2 = H, alkyl, aryl, heteroaryl, CO_2R

X = CN, CO_2R, COPh, CO-heteroaryl, $CONH_2$

5.2.2 Historical Perspective

In 1966, German chemist Karl Gewald reported that aliphatic ketones, aldehydes, or 1,3-dicarbonyl compounds reacted with activated nitriles and sulfur in the presence of an amine at room temperature to give 2-aminothiophenes.[4-7] Cyclohexanone **4** and malononitrile **5** reacted to form 2-aminothiophene **6** in 86% yield. In an alternate procedure, the ketone and nitrile were combined to form the acrylonitrile compound **7**, which was then combined with sulfur and an amine to give thiophene **6** in 90% yield. Other aliphatic ketones and aldehydes gave similar results. In addition, several examples of aryl ketones and aldehydes were also reported. When 1,3-dicarbonyl compounds were investigated as starting materials, the corresponding 2-aminothiophenes were obtained in lower yields.

5.2.3 Mechanism

The first step in the Gewald reaction is a Knoevenagel condensation of an activated nitrile with a ketone or aldehyde to produce an acrylonitrile **8**, which is then thiolated at the methylene position with elemental sulfur. The sulfurated compound **9** initially decays to the mercaptide compound **10**, which then undergoes a cyclization reaction via mercaptide attack at the cyano group to provide **11**. Base-catalyzed tautomerization affords the 2-aminothiophene **3**.[8]

5.2.4   *Variations and Improvements*

Regioselectivity of the Gewald aminothiophene reaction is variable when an unsymmetrical ketone is used as a starting material, particularly when both the α- and α'-protons of the ketone are sterically accessible. Originally, Gewald circumvented this problem by first preparing α-mercaptoketones **12** which then underwent the cyclization reaction in the presence of base to give the corresponding aminothiophenes **13**.[9,10]

Because α-mercaptocarbonyl compounds are unstable and difficult to prepare, this method has been little used. Several groups[11–13] have demonstrated that ketones with a leaving group in the α-position undergo the Gewald reaction regioselectively to give a single isomer of the 2-aminothiophene. In these procedures, introduction of the sulfur atom occurs through nucleophilic displacement of the leaving group with sodium sulfide. In one example, condensation of 5-methyl-2-hexanone **14** with ethylcyanoacetate **15** in the presence of sulfur and an amine afforded a mixture of two isomeric aminothiophene compounds **16** and **17**. However, when 3-bromo-5-methyl-2-hexanone **18** was used as the starting material, condensation with ethylcyanoacetate and sodium sulfide in the presence of triethylamine afforded the aminothiophene **16** exclusively.

Enamines can be used in place of an aldehyde or ketone in the Gewald reaction. Compound **19** reacted with ethylcyanoacetate and sulfur in the presence of morpholine to give aminothiophene **20** in good yield.[14]

5.2.5   Synthetic Utility

A wide range of differentially substituted 2-aminothiophenes have been prepared using the Gewald aminothiophene reaction. A number of procedures have been developed for varying the substitution in the 5-position of the aminothiophene ring. 5-SO$_2$Ph-,[15-17] 5-NO$_2$-,[18] and 5-alkoxy-2-aminothiophenes[19] have been prepared in modest to good yields from readily available precursors. When the α-alkoxyketone **21** was combined with ethylcyanoacetate and sulfur in the presence of morpholine, the 5-alkoxy-2-aminothiophene **22** was the only product formed.[19] In these examples, sulfuration occured selectively at the methylene group adjacent to the alkoxy substitiuent. When R = t-BuMe$_2$Si-, desilylation of the product afforded the 5-hydroxythiophene which existed exclusively in the keto form.[19]

R = Me, -C$_5$H$_{11}$, -(CH$_2$)$_2$-aryl, -(CH$_2$)$_2$-heteroaryl, -SiMe$_2$$t$-Bu

2,4-Diaminothiophenes **24** were prepared in good yield from 3-amino-4,4-dicyano-3-butenoate **23**, sulfur, and diethylamine at room temperature.[20]

23   95%   **24**

When an aldehyde is used as the starting material in the Gewald reaction, a 5-alkyl-2-aminothiophene is the product isolated. For example, when 3-methylbutanal **25** was combined with ethylcyanoacetate **15** and sulfur in the presence of triethylamine, aminothiophene **26** is the only product observed in this reaction.[21] However, when an α-tosyloxy-ketone **27** and sodium sulfide were employed in the reaction, the corresponding 4-alkyl-2-aminothiophene **28** was the sole reaction product.[22]

Tetrasubstituted thiophenes obtained by the Gewald reaction serve as templates for structural diversification and semi-automated library synthesis.[23] Thiophene **31**, prepared from β-ketoester **29** and t-butylcyanoacetate **30**, could be selectively derivatized at three of the four substituents to maximize library diversity. This procedure represents an improvement over previously published methods[5] for utilizing 1,3-dicarbonyl compounds in the Gewald reaction.

The Gewald reaction was utilized to prepare glycopyranosyl thiophene C-nucleoside analogues.[24] Pyranosyl ethanal **32** was converted to the pyranosyl thiophene-3-carbonitrile **33** in good yield via initial condensation with malononitrile and subsequent reaction with sulfur and triethylamine. **33** was ultimately converted to thienopyrimidine **34**.

Recently, the Gewald protocol has been adapted to a solid-support reaction manifold which produces 2-aminothiophenes in good to excellent yields and high purity.[25,26] Acylation of Wang resin **35** with cyanoacetic acid under standard conditions gave the resin-bound cyanoacetic ester **36**. The resin was suspended in ethanol containing the ketone **37**, morpholine, and elemental sulfur to afford resin-bound aminothiophene **38**. Acylation of the amino group followed by resin cleavage produced thiophene **39**. The acylation step was necessary to prevent decomposition of the products during the cleavage step.[25]

92% yield
95% HPLC purity

5.2.6 Experimental

Ethyl 2-Amino-4-methyl-5-(*tert*-butoxycarbonyl)thiophene-3-carboxylate (42):[27]
A mixture of *tert*-butyl acetoacetate **40** (15.8 g, 100 mmol), ethyl cyanoacetate **15** (11.3 g, 100 mmol), sulfur (3.5 g, 110 mmol), and EtOH (25 mL) was stirred at 45°C. Morpholine (10 g, 115 mmol) was added drop-wise over 15 min. The mixture was stirred at 60°C for 5 h and filtered. The filtrate was diluted with water (50 mL) and cooled. The precipitate was collected by filtration, washed with 30% EtOH, and dried to obtain **42** (21.2 g, 74%): mp 116–117°C (cyclohexane/hexane); IR (KBr, cm⁻¹) 1670,

1588 (C=O); H^1 NMR (CDCl$_3$) δ 1.37 (t, J = 7.1 Hz, 3H), 1.53 (s, 9H), 2.67 (s, 3H), 4.31 (q, J = 7.1 Hz, 2H), 6.47 (s, 2H).

5.2.7   References

1.   [R] Sabnis, R. W.; Rangnekar, D. W.; Sonawane, N. D. *J. Heterocycl. Chem.* **1999**, *36*, 333.
2.   [R] Sabnis, R. W. *Sulfur Reports* **1994**, *16*, 1.
3.   [R] Mayer, R.; Gewald, K. *Angew. Chem., Int. Ed. Engl.* **1967**, *6*, 294.
4.   Gewald, K. *Z. Chem.* **1962**, *2*, 305.
5.   Gewald, K.; Schinke, E.; Böttcher, H. *Chem. Ber.* **1966**, *99*, 94.
6.   Gewald, K.; Neumann, G.; Böttcher, H. *Z. Chem.* **1966**, *6*, 261.
7.   Gewald, K.; Schinke, E. *Chem. Ber.* **1966**, *99*, 2712.
8.   Peet, N. P.; Sunder, S.; Barbuch, R. J.; Vinogradoff, A. P. *J. Heterocycl. Chem.* **1986**, *23*, 129.
9.   Gewald, K. *Angew. Chem.* **1961**, *73*, 114.
10.   Gewald, K. *Chem. Ber.* **1965**, *98*, 3571.
11.   Madding, G. D.; Thompson, M. D. *J. Heterocycl. Chem.* **1987**, *24*, 581.
12.   Hawksley, D.; Griffin, D. A.; Leeper, F. J. *J. Chem. Soc., Perkin Trans. 1* **2001**, 144.
13.   Buchstaller, H.-P.; Siebert, C. D.; Lyssy, R. F.; Frank, I.; Duran, A.; Gottschlich, R.; Noe, C. R. *Monatsh. Chem.* **2001**, *132*, 279.
14.   Bacon, E. R.; Daum, S. J. *J. Heterocycl. Chem.* **1991**, *28*, 1953.
15.   Sherif, S. M.; Hussein, A. M. *Monatsh. Chem.* **1997**, *128*, 687.
16.   Erian, A. W. *Synth. Commun.* **1998**, *28*, 3549.
17.   Erian, A. W.; Issac, Y. A.; Sherif, S. M. *Z. Naturforsch., B: Chem Sci.* **2000**, *55b*, 127.
18.   Schäfer, H.; Gewald, K. *Z. Chem.* **1983**, *23*, 179.
19.   Pinto, I. L.; Jarvest, R. L.; Serafinowska, H. T. *Tetrahedron Lett.* **2000**, *41*, 1597.
20.   Mittelbach, M.; Junek, H. *Liebigs Ann. Chem.* **1986**, 533.
21.   Stanetty, P.; Puschautz, E. *Monatsh. Chem.* **1989**, *120*, 65.
22.   Noe, C. R.; Buchstaller, H.-P.; Siebert, C. *Pharmazie* **1996**, *51*, 833.
23.   McKibben, B. P.; Cartwright, C. H.; Castelhano, A. L. *Tetrahedron Lett.* **1999**, *40*, 5471.
24.   Garcia, I.; Feist, H.; Cao, R.; Michalik, M.; Peseke, K. *J. Carbohydr. Chem.* **2001**, *20*, 681.
25.   Castenado, G. M.; Sutherlin, D. P. *Tetrahedron Lett.* **2001**, *42*, 7181.
26.   Hoener, A. P. F.; Henkel, B.; Gauvin, J.-C. *Synlett* **2003**, 63.
27.   Gütschow, M.; Kuerschner, L.; Neumann, U.; Pietsch, M.; Löser, R.; Koglin, N.; Eger, K. *J. Med. Chem.* **1999**, *42*, 5437.

Jennifer M. Tinsley

5.3 Hinsberg Synthesis of Thiophene Derivatives

5.3.1 Description

The Hinsberg synthesis of thiophene derivatives describes the original condensation of diethyl thiodiglycolate and α-diketones under basic conditions which provides 3,4-disubstituted-thiophene-2,5-dicarboxylic acids upon hydrolysis of the crude ester product with aqueous acid.[1]

5.3.2 Historical Perspective

In 1910, Hinsberg described the reaction between benzil and diethylthiodiacetate, resulting in the preparation of the thiophene ring system.[2] The reaction was run under Claisen condensation conditions, and after hydrolysis with aqueous acid at reflux, the free dicarboxylic acid **1** was produced.

5.3.3 Mechanism

For the fifty years following the discovery of the Hinsberg thiophene synthesis, it was incorrectly assumed that the product of the reaction was initially a diester, which, upon acid treatment, was hydrolyzed to the corresponding diacid. In 1965, Wynberg and Kooreman conclusively proved this assumption to be incorrect by utilizing an elegant ^{18}O-labelling study.[3] By condensing ^{18}O-enriched benzil, the researchers noted approximately 50% transfer of the [18]O label to carbon dioxide, following the thermal decarboxylation of **2**.

From this experiment, it was concluded that the reaction results in the production of a mono ester-carboxylate which underwent decarboxylation. The original Hinsberg conditions involved the hydrolysis of a single ethyl ester to the dicarboxylic acid **1**. This reaction mechanism parallels the Stobbe condensation reaction. Thus, similarities involve the condensation of the enolate of diethyl thiodiglycolate with benzil, and spontaneous lactonization to provide **3**. Base-induced, ring fragmentation in the elimination of the carboxylate and a subsequent Knoevenagel-type cyclization provides the mono-ester **4**. Reaction conditions must allow for the isomerization of the newly formed alkene via reversible conjugate addition reactions since only the Z-alkene geometry permits cyclization via a Claisen/elimination process to form the thiophene ring. Upon hydrolysis of **4** under acidic conditions, the thiophene dicarboxylic acid product **1** is obtained. This rationale explains the transfer and retention of ^{18}O from benzil to the carboxylate of **4** as indicated by decarboxylation and production of labeled CO_2.

5.3.4 *Synthetic Utility*

The Hinsberg thiophene synthesis has seen limited use owing to the potential for regioisomeric mixtures when unsymmetrical 1,2-dicarbonyls are condensed with unsymmetrical thiodiacetates. Thus, symmetrically substituted thiophenes are generally prepared in this manner.

A series of 3,4-disubstituted thiophene-2,5-dicarboxylates have been prepared using the Hinsberg reaction. The 3,4-disubstituted thiophene-2-carboxylates are also readily available via decarboxylation of the initially formed half-acid, half-esters which were earlier identified by the earlier mechanistic studies.[4] Concurrent with the elegant mechanistic elucidation of Wynberg and Zwanenburg, a route was sought to prepare **5** in order to study the effect of ring strain on the heteroaromatic properties of five-membered rings.[5] Treatment of biacetyl and diethyl thiodiacetate with potassium *t*-butoxide furnished the monoester **6** in good yield. Compound **5** was subsequently prepared in a series of four steps, including oxidation of the methyl substituents with *N*-bromosuccinimide.

A notable utilization of the Hinsberg procedure was executed in the synthesis of a thiophene analogue of porphyrin.[6] A previous synthesis of the tetrathioporphyrin dication of **7** found this material to be unsuitable for chemical exploration as a result of its limited accessibility and solubility.[7] Efforts by Vogel *et al.* were then directed toward the synthesis of the octaethyl derivative **8**.   The preparation of the α-hydroxymethylthiophene **9** was achieved by a Hinsberg reaction, decarboxylation, and reduction protocol.   Formation of the cyclic tetramer **8** occurred via a series of electrophilic aromatic substitution reactions from alcohol **9** (30% yield).   Compound **8** was converted to the perchlorate dication in a two-step process.

In addition to thiodiglycolic acid esters, the use of bis(cyanomethyl)sulfide in the Hinsberg reaction has facilitated the preparation of 5-cyano-thiophene-2-carboxamides.[8] Thus, the condensation of biacetyl with bis(cyanomethyl)sulfide resulted in the efficient preparation of **10** (94% yield).

Several related approaches to the thiophene ring system warrant inclusion here, although these are mechanistically distinct from the Hinsberg reaction.   The original publication demonstrating the Hinsberg reaction also reported diethyloxalate as a coupling partner in place of the usual 1,2-diketone.   Thiophene formation with dimethylthiodiglycolate produced a mixture of dimethyl- and diethylesters.[1]   This problem was alleviated when diethyl oxalate was used in place of dimethyl oxalate, producing 3,4-dihydroxy-2,5-dicarboxythiophene **11**.   Similar reactivity has been noted by others.[9]

The mechanism of this reaction is generally understood to consist of subsequent Claisen condensation reactions to produce an intermediate diketone **12**, which readily tautomerizes to the fully conjugated dihydroxythiophene **13**.[9]

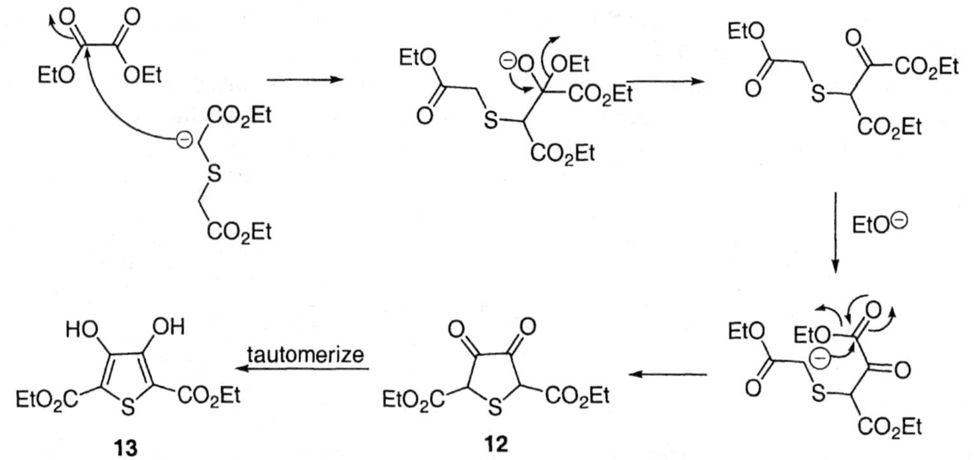

A notable exploitation of this reaction has been used for the preparation of potential anticancer agents. In an attempt to control the water solubility of thiophene analogs, a series of sodium salts of 2,5-dicarboethoxy-3,4-dihydroxythiophene have been produced by condensation between thiodiglycolate esters and diethyloxalate.[10] These condensation reactions consistently proceeded in good yield.

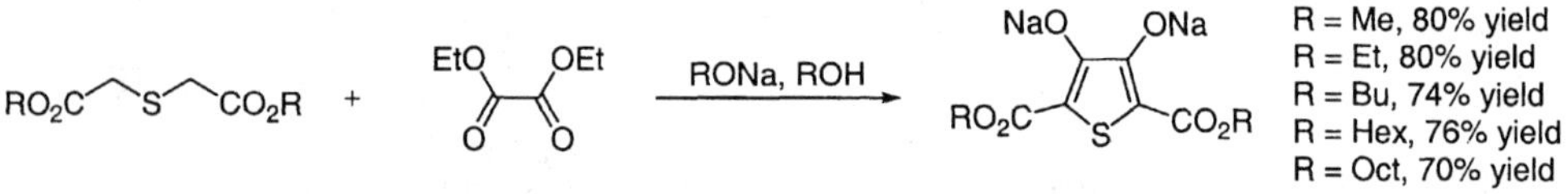

R = Me, 80% yield
R = Et, 80% yield
R = Bu, 74% yield
R = Hex, 76% yield
R = Oct, 70% yield

In this fashion, an α-ketoester was utilized as a means of studying the antiviral and antitumor activity of thiophene derivatives of pyrazofurin 14.[11] Reactions between dimethylthiodiglycolate and 15 provided intermediate 16, which was elaborated to 17 and 18 for comparisons with the antitumor activity of the C-nucleoside pyrazofurin 14.

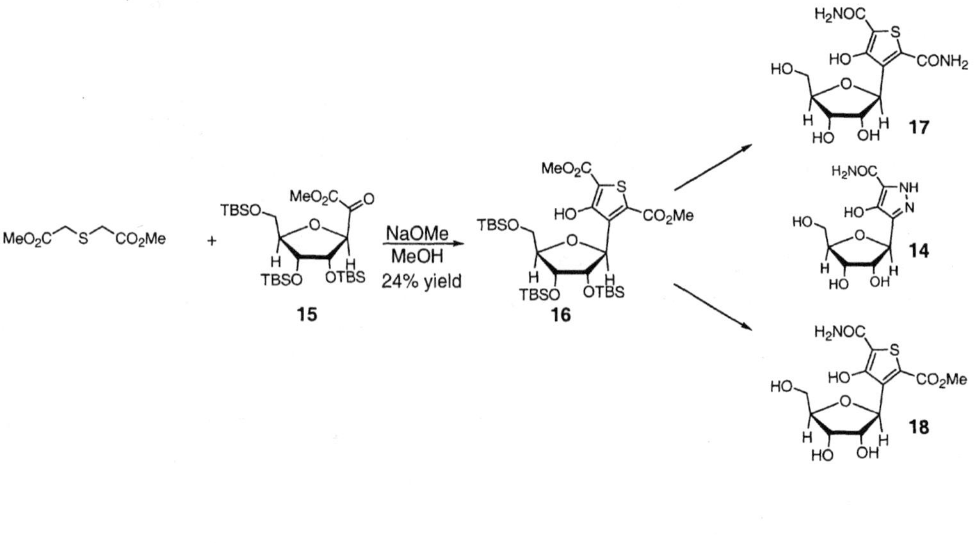

Another significant advance of the Hinsberg thiophene synthesis has explored the reactivity of diketosulfides in place of the use of diethylthiodiglycolate. This process has been extensively utilized for the preparation of novel thiophene containing systems. With glyoxal as condensation partner, the utility of this method has been pioneered Miyahara et al. in the synthesis of novel thiophenophanes **19**,[12] **20**,[13] and **21**.[14]

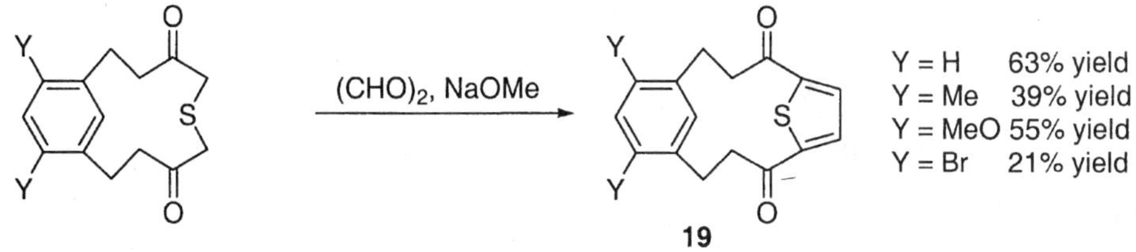

Y = H   63% yield
Y = Me   39% yield
Y = MeO 55% yield
Y = Br   21% yield

19

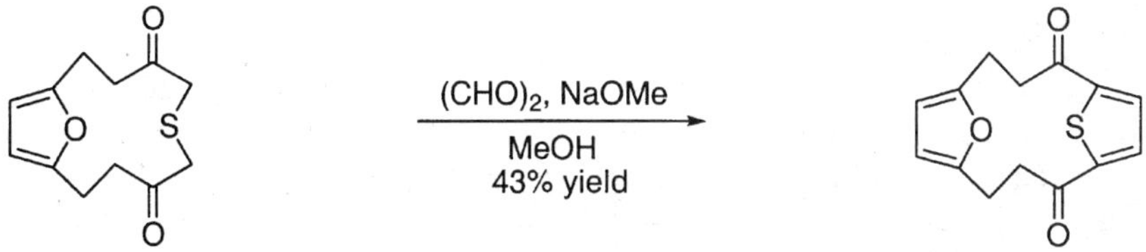

20

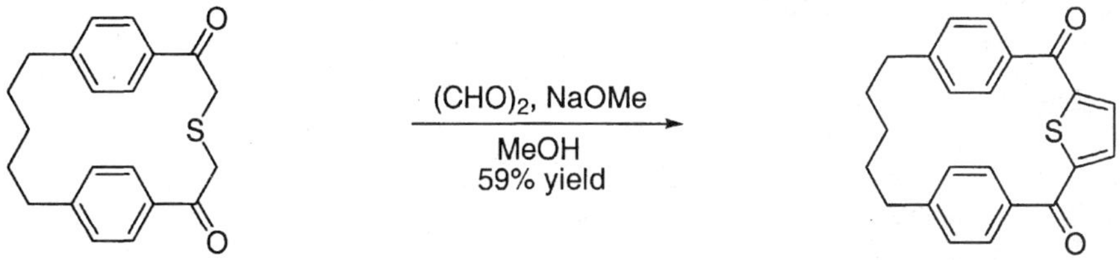

21

Mechanistically, the reaction of diketosulfides and glyoxal likely proceeds via an initial aldol reaction to provide **22**. A second intramolecular aldol reaction and the elimination of two equivalents of water produce the thiophene **23**. The timing of the elimination reactions and the ring-closing, carbonyl condensation reaction is not completely understood. However, 2,5-disubstituted thiophenes **23** are available in good yields via this process.

The reaction of diketosulfides with 1,2-dicarbonyl compounds other than glyoxal is often not efficient for the direct preparation of thiophenes. For example, the reaction of diketothiophene **24** and benzil or biacetyl reportedly gave only glycols as products.[15] The elimination of water from the β-hydroxy ketones was not as efficient as in the case of the glyoxal series. Fortunately, the mixture of diastereomers of compounds **25** and **26** could be converted to their corresponding thiophenes by an additional dehydration step with thionyl chloride and pyridine.

Application of this two-step technique has led to the synthesis of the novel thiophene **27**, albeit in low yield.[16]

Finally, the Hinsberg synthesis has been extended to the use of α-aryl-α-carboethoxydimethyl sulfide in conjunction with a series of 1,2-dicarbonyl compounds.[17] Specifically, the 4-nitroaryl substituent provides for sufficient activation of the α-proton to allow condensation and ring closure. These examples appear general and suggest future opportunities for the Hinsberg thiophene protocol.

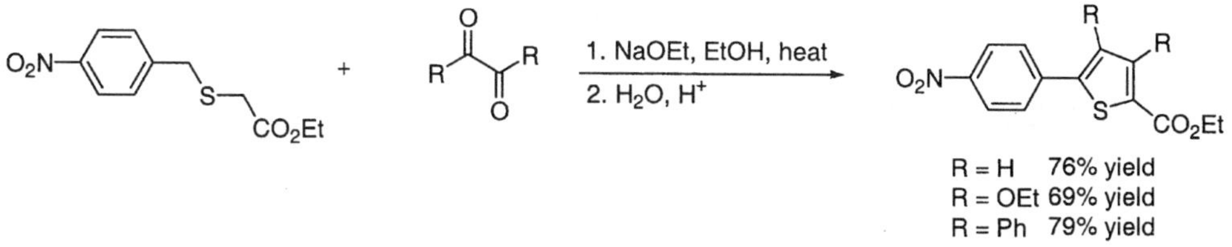

R = H   76% yield
R = OEt 69% yield
R = Ph   79% yield

5.3.5 Experimental

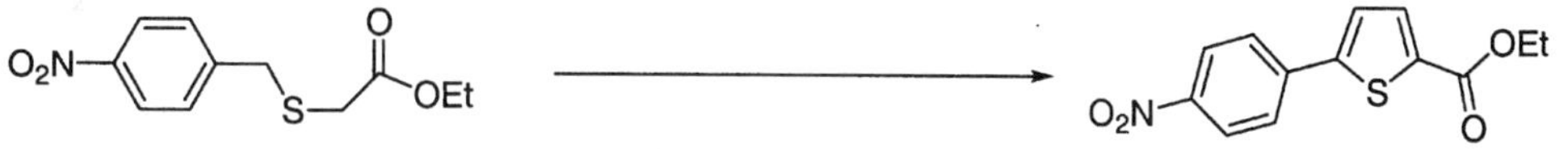

28   **29**

Ethyl 2-(4-nitrophenyl)thiophene-5-carboxylate 29:[17]

To an ethanolic solution of sodium ethoxide prepared by addition of 0.46 g (0.02 mole) of freshly cut sodium metal in 100 mL of absolute ethanol was slowly added 5.10 g (0.02 mole) of ethyl 4-nitrobenzylthioacetate **28** with stirring at 5°C. The mixture was refluxed for about 4 to 6 hours until the reaction was complete (monitored by tlc). The resultant mixture was allowed to cool to room temperature and then added into an ice-water mixture. The solution was neutralized with slow addition of dilute aqueous hydrochloric acid (10%). The precipitated solid was removed by filtration, washed with water, and recrystallized from a dimethylformamide-ethanol (1:1) mixture yielding 2.10 g (76 %) of **29** as a light brown crystalline solid, mp 227°C; ir (nujol): (neat (1710 cm^{-1}; ms: m/z 277 (M$^+$). *Anal*. Calcd. For $C_{13}H_{11}NO_4S$: C, 56.31; H, 3.97; N, 5.05; S, 11.55. Found: C, 56.36; H, 3.95; N, 5.01; S, 11.49.

5.3.6 References:

1.   [R] Gronowitz, S. In *Thiophene and Its Derivatives*, Part 1, Gronowitz, S., ed.; Wiley-Interscience: New York, 1985, 34-41.
2.   Hinsberg, O. *Ber.* **1910**, *43*, 901.
3.   Wynberg, H.; Kooreman, H. J. *J. Am. Chem. Soc.* **1965**, *87*, 1739.
4.   Chadwick, D. J.; Chambers, J.; Meakins, G. D.; Snowden, R. L. *J. Chem. Soc., Perkin I* **1972**, 2079.
5.   Wynberg, H.; Zwanenburg, D. J. *J. Org. Chem.* **1964**, *29*, 1919.
6.   Vogel, E.; Pohl, M.; Herrmann, A.; Wiss, T.; König, C.; Lex, J.; Gross, M.; Gisselbrecht, J. P. *Angew. Chem. Int. Ed. Engl.* **1996**, *35*, 1520.
7.   Vogel, E. *Pure Appl. Chem.* **1990**, *62*, 557.
8.   Beye, N.; Cava, M. P. *J. Org. Chem.* **1994**, *59*, 2223.
9.   Chadwick, D. J.; Chambers, J.; Meakins, G. D.; Snowden, R. L. *J. Chem. Soc., Perkin I* **1972**, 2079.
10.   Kumar, A.; Tilak, B. D. *Indian J. Chem.* **1986**, *25B*, 880.
11.   Huybrechts, L.; Buffel, D.; Freyne, E.; Hoornaert, G. *Tetrahedron* **1984**, *40*, 2479.
12.   Miyahara, Y.; Inazu, T.; Yoshino, T. *Tetrahedron Lett.* **1984**, *25*, 415.
13.   Miyahara, Y.; Inazu, T.; Yoshino, T. *Chem. Lett.* **1980**, 397.
14.   Miyahara, Y.; Inazu, T.; Yoshino, T. *J. Org. Chem.* **1984**, *49*, 1177.

15.   Miyahara, Y.; Inazu, T.; Yoshino, T. *Bull. Chem. Soc. Jpn.* **1980**, *53*, 1187.
16.   Christl, M.; Krimm, S.; Kraft, A. *Angew. Chem. Int. Ed. Engl.* **1990**, *29*, 675.
17.   Rangnekar, D. W.; Mavlankar, S. V. *J. Heterocyclic Chem.* **1991**, *28*, 1455.

Richard J. Mullins and David R. Williams

5.4 Paal Thiophene Synthesis

5.4.1 Description

The Paal thiophene synthesis involves the addition of a sulfur atom, typically from phosphorus pentasulfide, to 1,4-dicarbonyl compounds and subsequent dehydration.

5.4.2 Historical Perspective

In 1885, Paal reported the synthesis of 2-phenyl-5-methylthiophene **2** from **1**.[1]

Separately, Paal[1] and Knorr[2] described the initial examples of condensation reactions between 1,4-diketones and primary amines, which became known as the Paal-Knorr pyrrole synthesis. Paal also developed a furan synthesis in related studies.[3] The central theme of these reactions involves cyclizations of 1,4-diketones, either in the presence of a primary amine (Paal–Knorr pyrrole synthesis), in the presence of a sulfur(II) source (Paal thiophene synthesis), or by dehydration of the diketone itself (Paal furan synthesis).

5.4.3 Mechanism

Until 1952, it was postulated that the Paal thiophene synthesis proceeded via an initially formed furan via dehydration of the 1,4-diketone, followed by conversion of the furan to the thiophene. This hypothesis was considered because furans were often isolated as byproducts in the Paal thiophene synthesis. Parallel experiments conducted by Campaigne and coworkers provided the first evidence that furans were not likely intermediates on the pathway to thiophenes from 1,4-diketones. Direct comparisons were made between the reactions of acetonylacetone and 1,2-dibenzoylethane with P_2S_5 and the reactions of 2,5-dimethylfuran and 2,5-diphenylfuran under the Paal thiophene synthesis conditions. Reactions utilizing the diketones provided a greater yield of the thiophenes suggesting that the furan is not an essential intermediate in the reaction pathway, but rather a byproduct.[4] The reaction conditions facilitated an inefficient conversion of the 2,5-dimethylfuran to 2,5-dimethylthiophene. However, the more stable 2,5-diphenyl furan did not display this behavior.

$R = CH_3$	70% yield	13% yield
$R = Ph$	25% yield	0% yield

Based on these observations, it is likely that the mechanism involves initial formation of thione **3** (X = O or S), which is followed by tautomerization to **4** and cyclization to **5**. Aromaticity drives the facile elimination of either H_2O or H_2S resulting in the thiophene product.

5.4.4 Variations and Improvements

While the Paal thiophene synthesis was originally performed utilizing phosphorus pentasulfide as the sulfur atom source, a number of other reagents have since been developed for this purpose. As phosphorus pentasulfide can act as a dehydrating agent for 1,4-diketones, furans were often formed. Since hydrogen sulfide, in the presence of an acid catalyst, was found to be more efficient than phosphorus pentasulfide at converting ketones to thiones,[5] it was thought that this system might prove more effective in the Paal reaction. Indeed, as early as 1952, treatment of 1,2-di-*p*-bromobenzoylethane **6** with hydrogen sulfide, hydrogen chloride and stannic chloride, produced 2,5-di-*p*-bromophenylthiophene **7** in 73% yield.[4] The stannic chloride was found to function as a dehydrating agent as illustrated by the results of Table 5.4.1. Most notable was the fact that furans were not isolated under these conditions.

Table 5.4.1.

Dehydrating Agent	Yield, %	Recovered Diketone
None	0	80
Acetic anhydride	5	66
Zinc chloride	56	38
Stannic chloride	73	17

p-Methoxyphenylthionophosphine sulfide **8**, commonly known as Lawesson's reagent, was developed in 1978 and proved effective for the transformation of a number of carbonyl compounds to their thiocarbonyl derivatives.[6]

Thus, ketones,[6] amides,[7] and esters[8] all produce the corresponding thiocarbonyl derivative in nearly quantitative yield on treatment with **8**. Lawesson's reagent has proven to be an effective replacement for the traditional use of the P_4S_{10} reagent system for the conversion of 1,4-diketones to thiophenes. Both Lawesson's reagent and P_2S_5 resulted in mixtures of the desired thiophene **9** and the corresponding furan **10**. However, Lawesson's reagent proved far superior in terms of the yield of thiophene.[9]

	9	10
Phosphorus pentasulfide	20% yield	Yield of furan not reported
Lawesson's reagent	86% yield	

Another reagent system that has been recently employed in the Paal synthesis of thiophenes is the combination of bis(trialkyltin)- or bis(triaryltin) sulfides with boron trichloride. Known as the Steliou reagent,[10] it has been utilized in the transformation of 1,4-diketone **11** to thiophene **12**.[11] Higher yields are obtained in shorter reaction times in contrast to the use of Lawesson's reagent. Additionally, others have noted the relative ease of the work-up procedure using the Steliou conditions, and the fact that the tributyltinchloride byproduct of the reaction is reusable.[12] Similarly, the combination of the bis(trimethylsilyl)sulfide has been used in conjunction with trimethylsilyltriflate for the preparation of thiophenes in an analogous manner.[13]

5.4.5 Synthetic Utility
5.4.5.1 With 1,4-diketones and 1,4-dialdehydes

The Paal synthesis of thiophenes from 1,4-diketones, 4-ketoaldehydes and 1,4-dialdehydes has found great use in the synthesis of medicinally active compounds, polymers, liquid crystals and other important materials. Furthermore, the discovery of the catalyzed nucleophilic 1,4-conjugate addition of aldehydes, known as the Stetter reaction (Eq. 5.4.1), has enabled widespread use of the Paal thiophene synthesis, by providing 1,4-diketones from readily available starting materials.[14,15]

An interesting application of the Paal thiophene synthesis was documented for the synthesis of a polystyrene-oligothiophene-polystyrene copolymer.[16] In the Stetter reaction of aldehyde **13** and β-dimethylaminoketone **14**, *in situ* generation of the α,β-unsaturated ketone preceded nucleophilic 1,4-conjugate addition by the acyl anion equivalent of **13**. Treatment of bis-ketone **15** with an excess of Lawesson's reagent provided the triblock copolymer **16**.

Structure-activity studies of 5,6,7,8-tetrahdyro-5,5,8,8-tetramethyl-2-quinoxaline derivatives necessitated the preparation of thiophene-containing compound **17**.[17] Stetter conditions using thiazolium salt **20** as catalyst resulted in the preparation of 1,4-diketone **21** from **18** and **19**. Condensation of **21** with phosphorus pentasulfide followed by saponification resulted in **17**. In this fashion, the authors replaced the amide linker of parent compound **22** with the rigid thiophene moiety.

Novel compounds for use in liquid crystal displays have been prepared via the Stetter procedure followed by Paal thiophene formation with Lawesson's reagent. Thus aldehydes with a variety of alkyl chain lengths (n = 2–10) were reacted with α,β-unsaturated ketone **23**. Exposure of the diketone **24** to Lawesson's reagent gave **25**. The bromine in the aromatic ring was subsequently replaced by cyanide and homologues with n = 4 and 6 were found to be suitable components of mixtures with desired properties (viscosity ~ 30 cP, dielectric anisotropy ~ 11) for liquid crystal applications.[18]

23   n = 2-10   24   25

A notable variant of this scheme is useful for preparing symmetrical 1,4-diketones from a single aldehyde. Utilizing divinylsulfone **26** as the conjugate acceptor for the aldehyde-derived nucleophile, reactions in the presence of a catalytic amount of either cyanide or thiazolium salts give 1,4-diketones in good yields.[19] Following the initial conjugate addition to give **27**, elimination of vinylsulfinic acid produces α,β-unsaturated ketone **28** for the Stetter conjugate addition of a second equivalent of aldehyde. Thus production of symmetrical diketone **29** results in effective insertion of an ethylene unit between two aldehydes.

26   27   28   29

This strategy has been effectively used to prepare a number of 1,4-symmetrical diketones which have served as precursors to thiophenes such as **30**,[20] **31**,[21] and **32**[22] via the Paal synthesis.

Lawesson's Reagent

toluene, reflux
64% yield   **30**

Lawesson's Reagent

toluene, reflux
46% yield   **31**

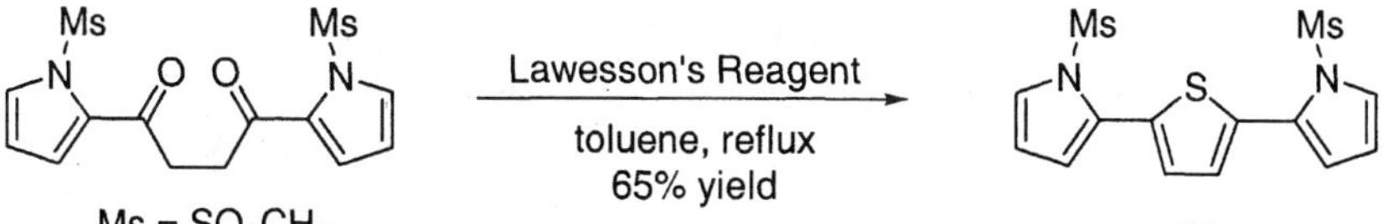

Ms = SO$_2$CH$_3$

32

A notable example of this strategy uses the divinylsulfone Stetter procedure for building complex molecules from simple substrates and was demonstrated in the synthesis of a novel 26π-aromatic macrocyclic ligand.[23] To that end, treatment of diketone **33** with Lawesson's reagent in refluxing toluene provided thiophene **34** which subsequently yielded macrocycle **35**. The latter demonstrated aromatic properties based on the inspection of its ^{1}H NMR spectrum which revealed extensive shielding of the internal pyrrole N−H signal and deshielding of the external hydrogens of the ring system.

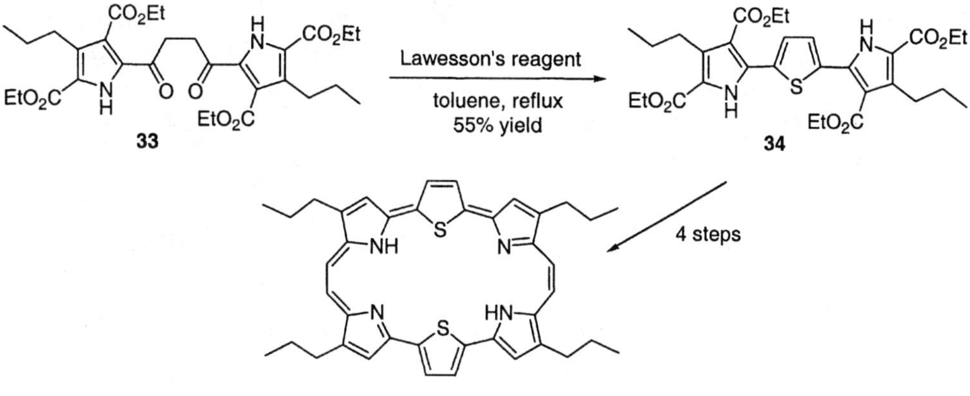

33   **34**

4 steps

35

An interesting variation on the Paal thiophene synthesis is observed when applied to an α,β-unsaturated γ-dialdehyde. Thus, treatment of o-phthaldehyde with Lawesson's reagent produced the dithiolactone **36**. When 2,3-napthalenedicarboxaldehyde reacted under similar conditions, dithio-2,3-naphthalide **37** was produced in excellent yield.[24]

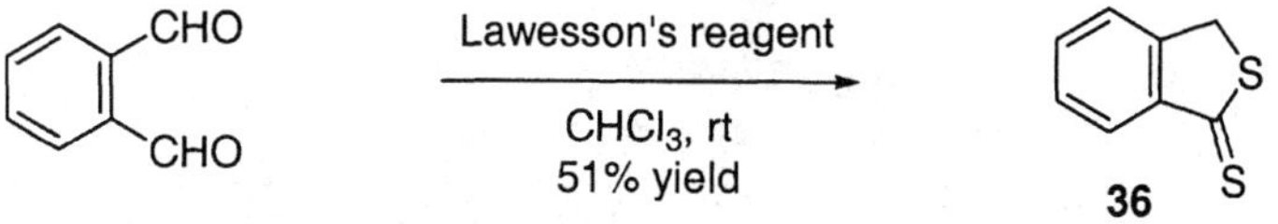

36

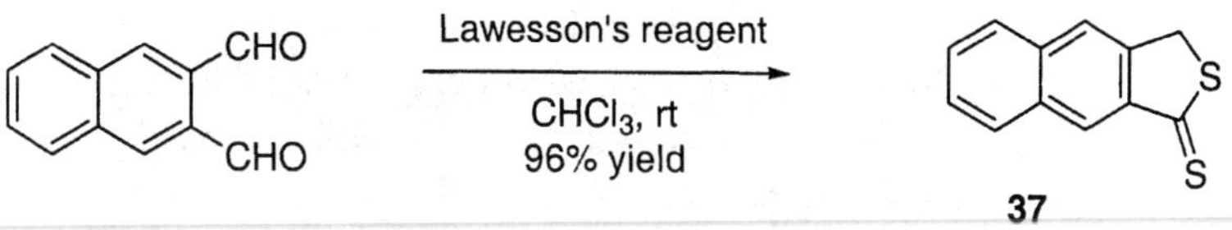

37

The mechanism for the redistribution in oxidation states begins similarly to that of the Paal thiophene synthesis. However, upon formation of dithione **38**, nucleophilic addition of one thiocarbonyl into the other produces the intermediate zwitterion **39**. A 1,3-tautomerization of hydrogen then gives **36**.[24]

This 1,3-migration of hydrogen was also observed when **40** reacted with Lawesson's reagent to produce the dithiolactone **41**.[25] However, when γ-hydroxy-α,β-unsaturated aldehyde **42** was reacted under similar conditions, thiophene **43** was prepared efficiently. These results are not surprising considering that the oxidation state of **42** is equivalent to the traditional saturated 1,4-dicarbonyl substrates of the Paal thiophene reaction via tautomerization of the double bond, and aromaticity is reestablished in the fully conjugated **43**.

5.4.5.2 With other 1,4-dicarbonyl compounds

While the Paal synthesis has been far-reaching in its application for the conversion of 1,4-diketones to thiophenes, there are surprisingly few examples which demonstrate its extension to related systems. The conversion of γ-ketoesters to 2-alkoxythiophenes has proven to be an extremely difficult transformation. Seed, et al. overcame this problem using Lawesson's reagent on route to prepare longer-chain alkoxythiophenes.[26] However, application of this method for the synthesis of short chain alkoxythiophenes proved futile and demanded further attention. To overcome this problem, the cyclization was effected using microwave irradiation in the absence of solvent. The procedure has proven somewhat general and resulted in the formation of 2-alkoxy-5-arylthiophenes with alkoxy groups of varying chain length.[27]

R = Et	90% yield
R = ^{n}Bu	89% yield
R = ^{n}Hex	94% yield
R = ^{n}Oct	89% yield
R = ^{n}Dec	93% yield
R = ^{n}Dod	87% yield

The duration of these reactions appears to be extremely important as longer reaction times resulted in deoxygenation of the thiophene derivative.

The preparation of 2-aminothiophenes has been demonstrated using the Paal synthesis.[28] For example, treatment of **44** with Lawesson's reagent for 10 min at 110°C resulted in the preparation of **45** in good yield.[29]

Interestingly, when utilizing a secondary amide, these reactions are limited by the simultaneous formation of the corresponding pyrrole as exemplified in the transformation of **46** to thiophene **47** and pyrrole **48**.[30]

5.4.6 Experimental

49   **50**

Thionation of 49.[31]

A mixture of **49** and (1g, 1.74 mmol) and Lawesson's reagent (1.1 g, 2.61 mmol) in chlorobenzene (40 mL) was refluxed under nitrogen for 36 h. The solvent was evaporated and to the residue was added a cold aqueous 10% NaOH solution (30 mL). The aqueous layer was extracted with CH_2Cl_2 (3 × 30 mL). The organic layers were combined, washed with water (3 × 20 mL), and dried over Na_2SO_4. Filtration to remove the drying agent, followed by concentration in vacuo provided the crude product which was purified by flash chromatography using hexanes:CH_2Cl_2 (5:1) as eluent, to yield a creamy white crystalline product **50** (0.68 g, 69%). An analytical sample was obtained by recrystallization in a hexanes:1,2-dichloroethane mixture: mp 289°C; UV (CHCl_3) λ_{max} (log ε) 295.5 (4.62): ^{1}H NMR δ 5.75 (s, 2H), 7.13 (dd, 2H, J_1 = 5.3 Hz, J_2 = 3.2 Hz), 7.32-7.38 (m, 20H), 7.43 (dd, 2H, J_1 = 5.3 Hz, J_2 = 3.2 Hz); ^{13}C NMR δ 44.3, 124.4, 126.2, 127.6, 128.5, 128.9, 132.9, 133.5, 142.9, 144.3; MS m/z (relative intensity) 570 (100, M$^+$), 537 (15), 449 (58), 372 (14), 187 (18), 121 (33). Anal. Calcd for $C_{40}H_{26}S_2$: C, 84.17; H 4.59. Found C, 84.29; H, 4.60.

51   **52**

2,5-Bis(4-chlorophenyl)-5-methylthiophene (52).[11]

To a 10-mL dry toluene solution of bis(triphenyltin) sulfide (697 mg, 0.95 mmol) and dione **51** (100 mg, 0.48 mmol) was injected a 1 M solution of BCl_3 in CH_2Cl_2 (0.64 mL, 0.64 mmol). The reaction solution was refluxed for 2h, cooled to 22–24°C, and added to 100 mL of 2:1 diethyl ether/ethyl acetate. The solution was washed with H_2O (3 × 100 mL), dried (MgSO_4), and concentrated. The residue was purified by chromatography using 100:1 hexanes/ethyl acetate solution to give 2,5-bis(4-chlorophenyl)-5-methylthiophene **52** as a white solid (97 mg, 98%), mp 161–162°C: ^{1}H NMR (300 MHz, CDCl_3) δ 7.23 (s, 2H), 7.32–7.53, (dd, 8H); ^{13}C NMR (75.5 MHz, CDCl_3) δ 124.39, 126.75, 129.07, 132.56, 133.37, 142.58.

5.4.7   References

1.   Paal, C. *Chem. Ber.* **1885**, *18*, 367.
2.   Knorr, L. *Chem. Ber.* **1885**, *18*, 299.
3.   Paal, C. *Chem. Ber.* **1884**, *17*, 2756.
4.   Campaigne, E.; Foye, W. O. *J. Am. Chem. Soc.* **1952**, *74*, 1405.
5.   [R] Campaigne, E. *Chem. Rev.* **1946**, *39*, 1.
6.   Pederson, B. S.; Scheibye, S.; Nilsson, N. H.; Lawesson, S.-O. *Bull. Soc. Chim. Belg.* **1978**, *87*, 223.
7.   Scheibye, S.; Pedersen, B. S.; Lawesson, S.-O. *Bull. Soc. Chim. Belg.* **1978**, *87*, 229.
8.   Pedersen, B. S.; Scheibye, S.; Clausen, K.; Lawesson, S.-O. *Bull. Soc. Chim. Belg.* **1978**, *87*, 293.
9.   Mitschke, U.; Osteritz, E. M.; Debaerdemaeker, T.; Sokolowski, M.; Bäuerle, P. *Chem. Eur. J.* **1998**, *4*, 2211.
10. Steliou, K.; Mrani, M. *J. Am. Chem. Soc.* **1982**, *104*, 3104.
11. Freeman, F.; Kim, D. S. H. L. *J. Org. Chem.* **1992**, *57*, 1722.
12. Kim, D. S. H. L.; Ashendel, C. L.; Zhou, Q.; Chang, C.-t.; Lee, E.-S.; Chang, C.-j. *Bioorg. Med. Chem.Lett.* **1998**, *8*, 2695.
13. Freeman, F.; Lee, M. Y.; Lu, H.; Wang, X. *J. Org. Chem.* **1994**, *59*, 3695.
14. Stetter, H.; Schreckenberg. H. *Angew. Chem. Int. Ed.* **1973**, *12*, 81.
15. [R] Stetter, H. *Angew. Chem. Int. Ed.* **1976**, *15*, 639.
16. Hempenius, M. A.; Langeveld-Voss, B. M. W.; van Haare, J. A. E. H.; Janssen, R. A. J.; Sheiko, S. S.; Spatz, J. P.; Möller, M.; Meijer, E. W. *J. Am. Chem. Soc.* **1998**, *120*, 2798.
17. Kikuchi, K.; Hibi, S.; Yoshimura, H.; Tokuhara, N.; Tai, K.; Hida, T.; Yamauchi, T.; Nagai, M. *J. Med. Chem.* **2000**, *43*, 409.
18. Brettle, R.; Dunmur, D. A.; Marson, C. M.; Piñol, M.; Toriyama, K. *Chem. Lett.* **1992**, 613.
19. Stetter, H.; Bender, H.-J. *Chem. Ber.* **1981**, *114*, 1226.
20. Kuroda, M.; Nakayama, J.; Hoshino, M.; Furusho, N.; Ohba, S. *Tetrahedron Lett.* **1994**, *35*, 3957.
21. Jones, R. A.; Civcir, P. U. *Tetrahedron* **1997**, *53*, 11529.
22. Merril, B. A.; LeGoff, E. *J. Org. Chem.* **1990**, *55*, 2904.
23. Johnson, M. R.; Miller, D. C.; Bush, K.; Becker, J. J.; Ibers, J. A. *J. Org. Chem.* **1992**, *57*, 4414.
24. Nugara, P. N.; Huang, N.-Z.; Lakshmikantham, M. V.; Cava, M. P. *Heterocycles* **1991**, *32*, 1559.
25. Lin, S.-C.; Yang, F.-D.; Shiue, J.-S.; Yang, S.-M.; Fang, J.-M. *J. Org. Chem.* **1998**, *63*, 2909.
26. Sonpatki, V. M.; Herbert, M. R.; Sandvoss, L. M.; Seed, A. J. *J. Org. Chem.* **2001**, *66*, 7283.
27. Kiryanov, A. A.; Sampson, P.; Seed, A. J. *J. Org. Chem.* **2001**, *66*, 7925.
28. Omar, M. T.; El-Aasar, N. K.; Saied, K. F. *Synthesis* **2001**, 413.
29. Thomsen, I.; Pedersen, U.; Rasmussen, P. B.; Yde, B.; Andersen, T. P.; Lawesson, S.-O. *Chem. Lett.* **1983**, 809.
30. Nishio, T. *Helv. Chim. Act.* **1998**, *81*, 1207.
31. Parakka, J. P.; Sadanandan, E. V.; Cava, M. P. *J. Org. Chem.* **1994**, *59*, 4308.

Richard J. Mullins and David R. Williams

Chapter 6   Oxazoles and Isoxazoles   **219**

6.1   Claisen Isoxazole Synthesis

6.1.1   Description

At the end of the nineteenth century, Claisen described the cyclization of β-keto esters with hydroxylamine to provide 3-hydroxyisoxazoles.[1] Substituents R_1 and R_2 in the β-keto ester make it possible to introduce substituents in the 4- and 5-position of the heterocyclic ring.

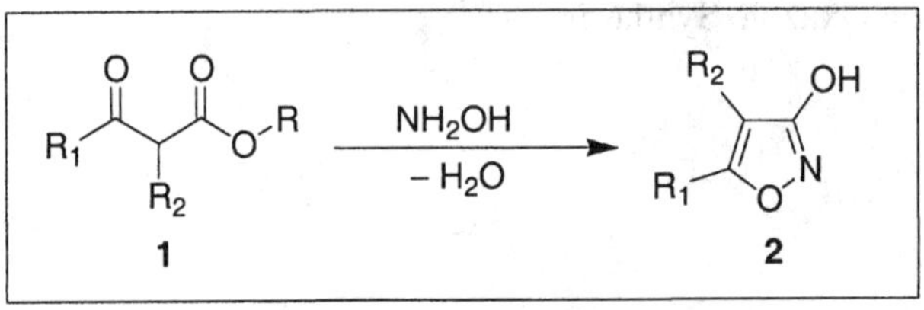

6.1.2   Historical Perspective and Mechanism

The condensation of β-keto esters with hydroxylamine can occur in two directions to give either isoxazolin-3-ones [which exist predominately as 3-hydroxyisoxazoles (2)] or isoxazolin-5-ones (3). Early work by Claisen,[1b] Hantzch,[2] and others[3] showed that the products from 2-unsubstituted β-keto esters were isoxazolin-5-ones. In the early 1960's, Katritzky found that 2-substituted analogues give 3-hydroxyisozaoles.[4] Jacquier later showed that both types of products could be produced from both types of keto esters depending on the precise pH variation during the reaction workup.[5]

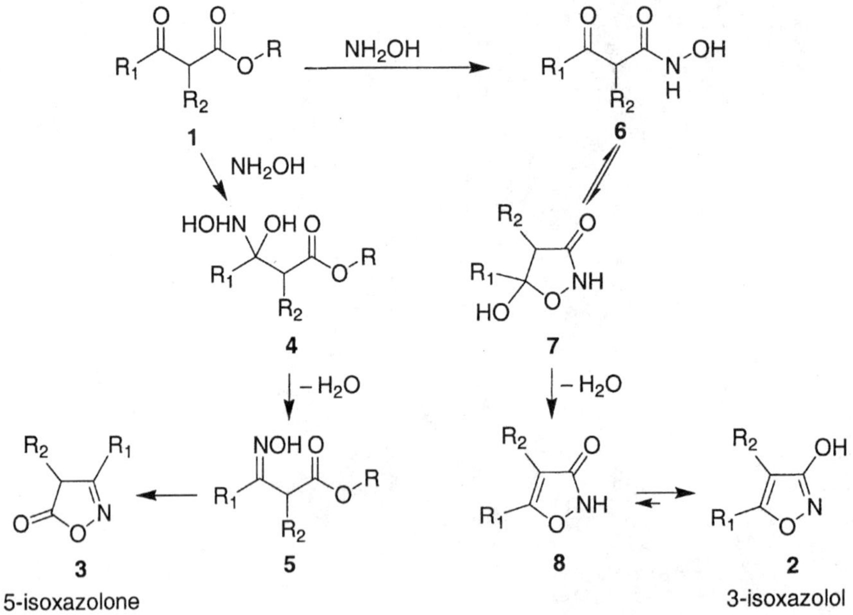

3
5-isoxazolone

2
3-isoxazolol

In the reaction of ethyl acetoacetate (1, R_1 = Me, R_2 = H, R = Et) with hydroxylamine at pH 6.5–8.5 (buffered), Cocivera et al. observed via ^{1}H NMR under stop-flow conditions that carbinolamine intermediate **4** dehydrates to form syn and anti oximes.[6] The syn isomer **5** cyclized within several minutes to **3**, while conversion of the anti form required several hours.

Jacobson conducted a detailed investigation of the effect of reaction pH on the direction of ring closure of a range of β-keto esters encompassing both 2-unsubstituted and 2-substituted compounds.[7] They found, for example, that the yield of the 3-hydroxyisoxazole was maximized at pH 10.0 ± 0.2. Katritzky and co-workers later carried out a detailed ^{13}C NMR study to rationalize the effects of the pH on the direction of ring closure.[8] The manner in which 3-hydroxyisoxazole **2** is formed was determined to occur through open chain hydroxamic acid **6**, which exists in dynamic equilibrium with **7**, followed by dehydration of 5-hydroxy-3-isoxazolidinone intermediate **8** under highly acidic conditions.

6.1.3 Variations and Improvements

Krogsgaard-Larsen and co-workers[9] have protected the β-keto functionality as a ketal as a modification to the traditional conditions so attack of hydroxylamine is directed towards the ester. They prepared hydroxamic acid **10** from ester **9** then cyclized with sulfuric acid to isoxazole **11**, in route to 4,5,6,7-tetrahydroisoxazolo[5,4-c]pyridin-3-ol (THIP), a selective GABA$_A$ receptor agonist studied clinically for insomnia.

In 2000, an efficient three-step procedure for the synthesis of 5-substituted 3-isoxazolols (without formation of undesired 5-isoxazolone byproduct) was published.[10] The method uses an activated carboxylic acid derivative to acylate Meldrum's acid,[11] which is treated with N,O-bis(tert-butoxycarbonyl)hydroxylamine to provide the N,O-di-Boc-protected β-keto hydroxamic acids **14**. Cyclization to the corresponding 5-substituted 3-isoxazolols **15** occurs upon treatment with hydrochloric acid in 76–99% yield.

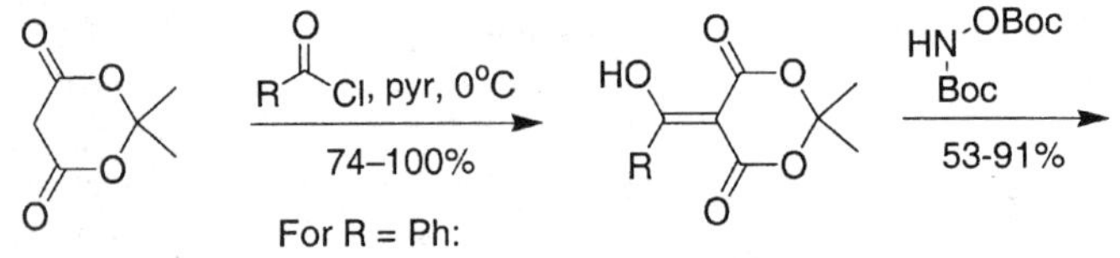

6.1.4 Synthetic Utility

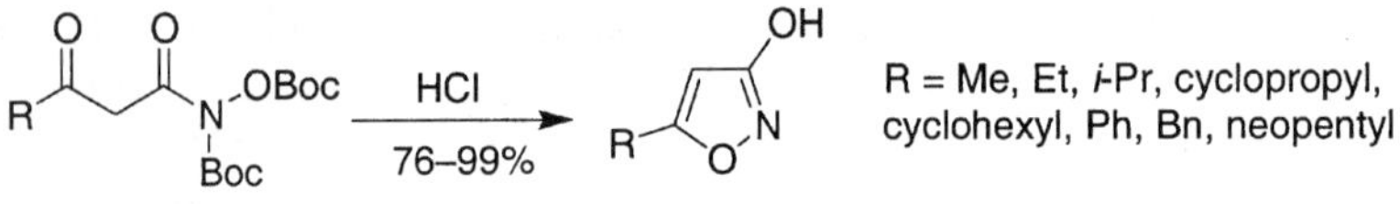

Krogsgaard-Larsen and co-workers continued to utilize Claisen isoxazole chemistry in the preparation of GABA$_A$ receptor antagonists reported in 2000.[12] In the synthesis of protected 3-isoxazolols **17a–f**, β-oxoesters **16a–f** were cyclized at –30°C followed by heating with concentrated hydrochloric acid at 80°C.

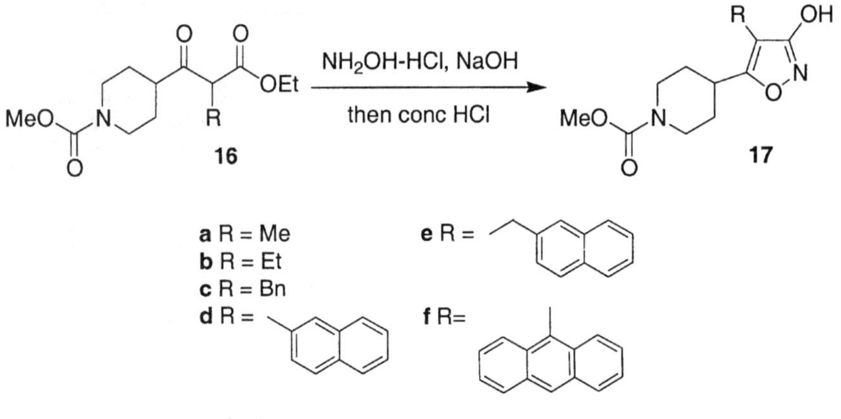

a R = Me   e R =
b R = Et
c R = Bn
d R =   f R=

These workers also prepared the thio analog of **17** (R = H)[13] by treating **16** (R = H) with aqueous ammonia to provide the β-oxoamide, which was converted into the corresponding enolized β-thioxoamide **18** by treatment with hydrogen sulfide and hydrogen chloride in ethanol. Compound **19** was synthesized by oxidation of **18** with iodine in ethanol under basic conditions.

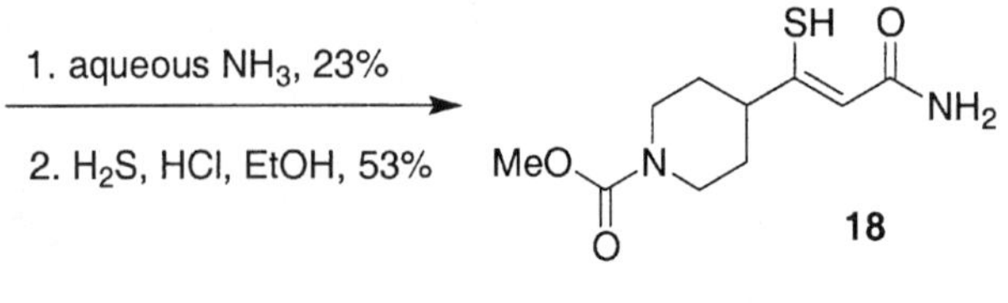

6.1.5 Experimental Procedure

5-Methyl-4-pentyl-3-isoxazolol (21).[14]

To a solution of $NH_2OH \cdot HCl$ (2.6 g, 42 mmol) in MeOH (10 mL) heated to 60°C was added a solution of NaOH (1.7 g, 42 mmol) in water (1 mL) and MeOH (10 mL). The mixture was cooled to –50°C. To a solution of **20** (3.72 g, 20 mmol) in MeOH (4 mL) was added a solution of NaOH (0.84 g, 21 mmol) in water (0.5 mL) and MeOH (4 mL). This mixture was cooled to –50°C, stirred for 10 min, and added to the above NH_2OH solution. The reaction mixture was stirred at –50°C for 2.5 h and acetone (2.1 g) was added. The reaction mixture was then quickly added to 4 M HCl (24 mL), heated to 85°C and stirred for 45 min at 85°C. The MeOH was evaporated and column chromatography (toluene-EtOAc-AcOH 20:2:1) gave **21** (1.2 g, 65%) as an oil: ^{1}H NMR (CDCl$_3$) δ 2.26 (t, J = 7.1 Hz, 2H), 2.24 (s, 3H), 1.52 (quintet, J = 7.1 Hz, 2H), 1.30 (m, 4H), 0.89 (t, J = 7.1 Hz, 3H).

6.1.6 References

1   (a) Claisen, L; Lowman, O.E. *Ber.* **1888**, *21*, 784. (b) Claisen, L.; Zedel, W. *Ber.* **1891**, *24*, 140.
2   Hantzsch *Ber.* **1891**, *24*, 495.
3   [R] Barnes, R.A. In *Heterocyclic Compounds*; Elderfield, R. C., Ed.; Wiley: New York, **1957**; Vol. 5, p 474 ff. (b) [R] Loudon, J. D. In *Chemistry of Carbon Compounds*; Rodd, E. H., Ed.; Elsevier: Amsterdam, **1957**; Vol. 4a, p. 345 ff.
4   (a) Boulton, A. J.; Katritzky, A.R. *Tetrahedron* **1961**, *12*, 41. (b) Boulton, A. J.; Katritzky, A.R. *Tetrahedron* **1961**, *12*, 51. (c) Katritzky, A.R.; Oksne, S.; Boulton, A. J. *Tetrahedron* **1962**, *18*, 777. (d) Katritzky, A. R.; Oksne, S. *Proc. Chem. Soc.* **1961**, 387.
5   Jacquier, R.; Petrus, C.; Petrus, F.; Verducci, J. *Bull. Soc. Chem. Fr.* **1970**, 2685.
6   Cocivera, M.; Effio, A.; Chen, H. E.; Vaish, S. *J. Am. Chem. Soc.* **1976**, *98*, 7362.
7   Jacobsen, N.; Kolind-Andersen, H.; Christensen, J. *Can. J. Chem.* **1984**, *62*, 1940.
8   Katritzky, A. R.; Barczynski, P.; Ostercamp, D. L.; Yousaf, T. I. *J. Org. Chem.* **1986**, *51*, 4037.
9   (a) Krogsgaard-Larsen, K. et al. *Org Prep. Proc. Int.* **2001**, *33*, 515. (b) P. H. Tygesen, presented at Optimizing Organic Reactions and Processes, June 17-19, 2002; Oslo, Norway.
10   Sorensen, U. S.; Falch, E.; Krogsgaard-Larsen, K. *J. Org. Chem.* **2000**, *65*, 1003.
11   (a) Chen, B.-C. *Heterocycles* **1991**, *32*, 529. (b) McNab, H. *Chem Soc. Rev.* **1978**, *7*, 345.
12   Frølund, B.; Tagmose, L.; Liljefors, T.; Stensbol, T.B.; Engblom, C.; Kristiansen, U.; Krogsgaard-Larsen, K. *J. Med. Chem.* **2000**, *43*, 4930.
13   Frølund, B.; Kristiansen, U.; Brehm, L.; Hansen, A.B.; Krogsgaard-Larsen, K.; Falch, E. *J. Med. Chem.* **1995**, *38*, 3287.

14   Madsen, U.; Bräuner-Osborne, H.; Frydenvang, K.; Hvene, L.; Johansen, T.N.; Nielsen, B.; Sánchez, C.; Stensbøl, T.B.; Bischoff, F.; Krogsgaard-Larsen, K. *J. Med. Chem.* **2001**, *44*, 1051.

Dawn A. Brooks

6.2 Cornforth Rearrangement

6.2.1 Description

The Cornforth rearrangement involves the thermal interconversion of 4-carbonyl substituted oxazoles, with "exchange" between the C–C–O side-chain and the C–C–O fragment of the oxazole ring.[1] These reactions generally involve compounds where a heteroatom (–OR, –SR, –Cl) is attached to the 5-position (R_2) of the starting oxazole.

6.2.2 Historical Perspective

Cornforth reported in 1949 that 2-phenyl-5-ethoxyoxazole-4-carboxamide (3) rearranged on heating to ethyl 2-phenyl-5-aminooxazole-4-carboxylate (4).[2]

Dewar and Turchi carried out similar rearrangements of secondary and tertiary alkyl and aryl oxazole-4-carboxamides (5a–e) to the corresponding secondary and tertiary 5-aminooxazoles (6a–e).[3] For example, they realized yields > 90% when the amide nitrogen is part of a heterocyclic ring system.

6.2.3 Mechanism

This rearrangement can be rationalized by postulating the dicarbonylnitrile ylide 8 as an intermediate. Dewar and Turchi obtained supporting evidence for the interconversion in

1975. They found that deuterium labeled 2-phenyl-5-methoxy-4-[(methoxy-d_3)-carbonyl]oxazole (**7**) scrambled on heating to give a 1:1 equilibrium mixture of **7** and the corresponding rearranged ester **9**.[3]

Dewar and Turchi studied the rates of Cornforth rearrangements and determined through variation of substituents in the 2-phenyl group that a small positive charge develops at the adjacent C-2 oxazole carbon on passing to the transition state.[4] They also found that in aprotic solvents, the rate changes little with the polarity of the solvent, suggesting that the transition state is not much more polar than the ground state. However, a substantial increase in rate was observed going from an aprotic ($PhNO_2$) to a protic solvent ($PhCH_2OH$) suggesting that a developing negative charge in the transition state is stabilized considerably by hydrogen bonding to the solvent. Recent *ab initio* and density functional calculations by Fabian *et al.* support their conclusions that the nitrile ylide intermediate is considerably less polar than a zwitterionic formula would suggest.[5] It is interesting to note that in experimental studies and MINDO/3 MO calculations by Turchi,[6] as well as *ab initio* and density functional calculations by Fabian *et al.*[5] that 4-(aminothiocarbonyl)oxazoles (**10**) rearrange to 5-aminothiazoles (**12**).

R_1 = Me, Ph
R_2, R_3 = H, Me, CH_2Ph, -$(CH_2)_n$-, -$(CH_2)_2$-O-$(CH_2)_2$-
n = 4, 5

78–92%

6.2.4 *Variations and Improvements*

Dewar and Turchi described the Cornforth rearrangement of 5-alkoxyoxazole-4-thiocarboxylates as a potentially general method for the synthesis of 5-thiooxazole-4-carboxylic esters. Specifically, they found that thiol ester **13** underwent thermal isomerization to the corresponding 5-thiooxazole **14** in 94% yield.

Williams and McClymont[7] have observed that acylation reactions of the dianion of 2-(5-oxazolyl)-1,3-dithiane (**15**) lead to formation of 4,5-disubstituted oxazole products through a Cornforth rearrangement pathway under base-induced, low-temperature conditions. For example, deprotonation of **15** with LiHMDS (3.0 equivalents) at –78°C, followed by addition of benzoyl chloride or p-chlorobenzoyl chloride and warming to 0°C, provided **16** in 74% and 47% yield, respectively.

Since neither direct acylation of the 2-position of oxazole **15** (H_a) nor acylation of the 1,3-dithianyl anion (H_b) was observed, the products were rationalized as arising through selective C-acylation of the ring-opened tautomer **15c**.

6.2.5 Synthetic Utility

L'abbe and coworkers studied the synthesis and thermolysis of 5-azido-4-formyl-oxazoles.[8] They found, for example, that azides **17a–b** rearrange at room temperature to yield isolable, isomeric oxazoles **18a–b**.

17a R = *i*-Pr
17b R = *t*-Bu

18a–b (60%)

6.2.6   Experimental procedure
4-Azidocarbonyl-2-tert-butyloxazole (18b).

Compound **17b** (1.5 g) was allowed to rearrange in chloroform (50 mL) at room temperature for 18 h. The reaction mixture was subjected to column chromatography on silica gel with hexane-ethyl acetate (4:1) as the eluent to give the acyl azide **18b** (0.9 g, 60%): m.p. 48°C; υ_{max}(KBr)/cm^{-1} 2150s (N$_3$) and 1700s, br (CO); ^{1}H NMR (400 MHz, CDCl$_3$) δ 1.34 (s, 9H), 8.2 (s, 1H); ^{13}C NMR (100 MHz, CDCl$_3$) δ 134.1, 144.6, 166.2, 172.8; MS m/z 194 (M$^+$) and 57 (*t*-Bu$^+$); Anal. Calcd for C$_8$H$_{10}$N$_4$O$_2$: C, 49.48; H, 5.15. Found: C, 49.3; H, 5.2.

6.2.7.   References

1   Turchi, I. J.; Dewar, M. J. S. *Chem Rev.* **1975**, 389.
2   [R] Cornforth, J. W. In *The Chemistry of Penicillin*, Princeton University Press: Princeton, N.J. **1949**, p 700.
3   Dewar, M. J. S.; Turchi, I. J. *J. Org. Chem.* **1975**, *40*, 1521.
4   a.) Dewar, M. J. S.; Spanninger, P. A.; Turchi, I. J. *Chem. Commun.* **1973**, 925. b) Dewar, M. J. S.; Turchi, I. J. *J. Am. Chem. Soc.* **1974**, *96*, 6148.
5   Fabian, W. M. F.; Kappe, C. O.; Bakulev, V. A. *J. Org. Chem.* **2000**, *65*, 47.
6   Corrao, S. L.; Macielag, M. J.; Turchi, I. J. *J. Org. Chem.* **1990**, *55*, 4484.
7   Williams, D. R.; McClymont, E. L. *Tetrahedron Lett.* **1993**, *34*, 7705.
8   L'Abbé, G.; Ilisiu, A.-M.; Dehaen, W.; Toppet, S. *J. Chem Soc, Perkin Trans I* **1993**, 2259.

Dawn A. Brooks

6.3   Erlenmeyer–Plöchl Azlactone Synthesis

6.3.1 Description

Formation of 5-oxazolones (or 'azlactones') (**2**) by intramolecular condensation of acylglycines (**1**) in the presence of acetic anhydride is known as the Erlenmeyer–Plöchl azlactone synthesis.[1]

6.3.2 Historical Perspective and Mechanism

The azlactones of α-benzoylaminocinnamic acids have traditionally been prepared by the action of hippuric acid (**1**, R_1 = Ph) and acetic anhydride upon aromatic aldehydes, usually in the presence of sodium acetate. The formation of the oxazolone (**2**) in Erlenmeyer–Plöchl synthesis is supported by good evidence.[2] The method is a way to important intermediate products used in the synthesis of α-amino acids, peptides and related compounds.[3] The aldol condensation reaction of azlactones (**2**) with carbonyl compounds is often followed by hydrolysis to provide unsaturated α-acylamino acid (**4**). Reduction yields the corresponding amino acid (**6**), while drastic hydrolysis gives the α-oxo acid (**5**).[4]

In 1959, Crawford and Little reported superior yields of **3** in reactions of aromatic aldehydes by using isolated, crystalline 2-phenyloxazol-5-one (**2**, R_1 = Ph) compared to direct reaction with hippuric acid (**1**, R_1 = Ph).[5] An early report by Boekelheide and Schramm on the use of ketones in the Erlenmeyer azlactone synthesis includes treatment of cyclohexane with hippuric acid and anhydrous sodium acetate in acetic anhydride to provide 49% yield of 2-phenyl-4-cyclohexylidene-5-oxazolone **7**.[6] Hydrolysis of **7** with concentrated hydrochloric acid gave cyclohexyloxoacetic acid (**8**) in 60% yield.

Cornforth has reviewed literature reports and independently studied the special cases of reaction of **1** with salicylaldehyde and with 2-acetoxybenzaldehyde.[7] Coumarins (**10**) are afforded in the condensation of **1** with salicylaldehyde or its imine,[8] whereas when 2-acetoxybenzaldehyde is used, acetoxy oxazolone **12** is the major product. The initial aldol condensation product between the oxazolone and 2-acetoxybenzaldehyde is the 4-(α-hydroxybenzyl)oxazolone **11**, in which base-catalyzed intramolecular trans-acetylation is envisioned. The product **9** (R = Ac) can either be acetylated on the phenolic hydroxy group, before or after loss of acetic acid, to yield the oxazolone **12**, or it can rearrange, by a second intramolecular process catalyzed by base and acid, to the hydrocoumarin, which loses acetic acid to yield **10**. When salicylaldehyde is the starting material, aldol intermediate **9** (R = H) can rearrange directly to a hydrocoumarin. Cornforth also accessed pure 4-(2′-hydroxyphenylmethylene)-2-phenyloxazol-5(4H)-one (**13**) through hydrolysis of **12** with 88% sulfuric acid.

6.3.3 Variations and Improvements

Armstrong and Combs have demonstrated an efficient one-pot Erlenmeyer–Plöchl reaction that provides a highly convergent and general route to dehydroamino acid derivatives through oxidative cleavage of *D*-mannitol diisopropylidine **14** to *D*-glyceraldehyde and condensation with hippuric acid.[9] The crude azlactone intermediate **15** was immediately reacted with *n*-butylamine to afford a 67% isolated yield of **16** as a 5.5:1 Z/E ratio of geometrical isomers.

Bismuth(III) acetate catalyzes the synthesis of azlactones (17) from aromatic aldehydes in moderate to good yields via the Erlenmeyer synthesis.[10] While the standard procedure for azlactone synthesis consists of using a stoichiometric amount of fused anhydrous sodium acetate, 10 mol% of Bi(OAc)$_3$ is sufficient to catalyze the reaction and the crude product is found to be > 98% pure.

Modification of the Erlenmeyer reaction has been developed using imines of the carbonyl compounds, obtained with aniline,[11,8b] benzylamine[12] or n-butylamine.[13] Ivanova has also shown that an N-methylketimine is an effective reagent in the Erlenmeyer azlactone synthesis.[14] Quantitative yield of 19 is generated by treatment of 3 equivalents of 2-phenyl-5(4H)-oxazolone (2) (freshly prepared in benzene) with 1 equivalent of N-methyl-diphenylmethanimine (18) in benzene. Products resulting from aminolysis (20), alkali-catalyzed hydrolysis (21), and alcoholysis (22) were also described.

6.3.4 Synthetic Utility

Monsanto's commercial route to the Parkinson's drug, L-DOPA (3,4-dihydroxyphenylalanine), utilizes an Erlenmeyer azlactone prepared from vanillin. The pioneering research in catalytic asymmetric hydrogenation by William Knowles as exemplified by his reduction of **24** to **25** in 95% *ee* with the DiPAMP diphosphine ligand was recognized with a Nobel Prize in Chemistry in 2001.[15]

Kirk and coworkers recently reported preparation of 6-flouro-meta-tyrosine **29** based on an Erlenmeyer–Plöchl azlactone strategy from 2-benzyloxy-5-fluorobenzaldehyde.[16]

6.3.5 Experimental

2-Phenyl-4-veratral-5(4)-oxazolone (30).[17]

A mixture of veratraldehyde (160 g, 0.96 mol), powdered, dry hippuric acid (192 g, 1.07 mol), powdered sodium acetate (80 g, 0.98 mol), and high-grade acetic anhydride (300 g, 278 mL, 2.9 mol) is heated 110°C, with constant stirring. The mixture becomes almost solid, and then, as the temperature rises, it gradually liquefies and turns deep yellow in color. After 2 h, the reaction is allowed to cool and then ethanol (400 mL) is added slowly to the contents of the flask. After allowing the reaction mixture to stand overnight, the yellow crystalline product is filtered and washed with ice-cold ethanol (2 × 100 mL) and finally with boiling water (2 × 100 mL). After drying, the product (30) weighs 205–215 g (69–73% yield) and melts at 149–150°C. This material is sufficiently pure for many purposes; it can be purified further by crystallization from hot benzene to obtain 180–190 g of the pure azlactone, melting at 151–152°C.

6.3.6   References

1   (a) Erlenmeyer, E. *Ann.* **1893**, 275, 1; (b) Plöchl, J. *Ber.* **1883**, *16*, 2815; *Ber.* **1884**, *17*, 1616.
2   (a) [R] Carter, H. E. In *Org. React.* Adams, R., Ed.; Wiley: New York, **1946**, Vol. 3, pp. 198-239; (b) Baltazzi, E. *Quart. Rev.* **1955**, *9*, 150.
3   (a) Steglich, W. *Fortschr. Chem. Forsch.* **1969**, *12*, 84; (b) [R] Filler, R.; Rao, Y. S. In *Adv. Heterocycl. Chem* Katritzky, A. R. and Boulton, A. Y., Eds; Academic Press, Inc: New York, **1977**, Vol. 21, pp. 175-206; (c) [R] Mukerjee, A. K. *Heterocycles* **1987**, *26*, 1077.
4   Schmidt, C.L.A *The Chemistry of the Amino Acids and Proteins* (Springfield, IL) 1944, p. 54
5   Crawford, M.; Little, W. T. *J. Chem. Soc.* **1959,** 729.
6   Boekelheide, V.; Schramm, L. M. *J. Org. Chem.* **1949**, *14*, 298.
7   Cornforth, J.; Ming-hui, D. *J. Chem. Soc., Perkin Trans. I* **1991**, 2183.
8   (a) Kumar, P.; Mukerjee, A. K. *Indian J. Chem.* **1980**, *19B*, 704; (b) Kumar, P.; Mukerjee, A. K. *Indian J. Chem.* **1981**, *20B (5)*, 418.
9   Combs, A. P.; Armstrong, R. W. *Tetrahedron Lett.* **1992**, *33*, 6419.
10   Monk, K. A.; Sarapa, D.; Mohan, R. S. *Synth. Commun.* **2000**, *30*, 3167.
11   (a) Rai, M.; Krishan, K.; Singh, A. *Indian J. Chem.* **1977**, *15B (9)*, 847; (b) Kumar, P.; Mishra, H. D.; Mukerjee, A. K. *Synthesis* **1980**, *10*, 836.
12   Kumar, S.; Rai, M.; Krishan, K.; Singh, A. *J. Indian Chem. Soc.* **1979**, *56 (4)*, 432.
13   [R] Mukerjee, A. K.; Kumar, P. *Heterocycles* **1981**, *16 (11)*, 1995.
14   Ivanova, G. G. *Tetrahedron* **1992**, *48*, 177.
15   (a) [R] Knowles, W. S. *Acc. Chem Res.* **1983**, *16*, 106; (b) Vineyard, B. D.; Knowles, W. S.; Sabacky, M. J. Bachman, G. L. Weinkauff, D. J. *J. Am. Chem. Soc.* **1977**, *99*, 5946.
16   Konkel, J. T.; Fan, J.; Jayachandran, B.; Kirk, K. L. *J. Fluorine Chem.* **2002**, *115*, 27.
17   (a) Kropp, W.; Decker, H. *Ber.* **1909**, *42*, 1184; (b) Buck, J. S.; Ide, W. S. *Organic Synthesis, CV 2*, 55.

Dawn A. Brooks

6.4   Fisher Oxazole Synthesis

6.4.1   Description

The Fisher oxazole synthesis involves condensation of equimolar amounts of aldehyde cyanohydrins (1) and aromatic aldehydes in dry ether in the presence of dry hydrochloric acid.[1]

6.4.2   Historical Perspective

In 1896, Emil Fisher found that 2,5-diphenyloxazole hydrochloride was precipitated by passing gaseous hydrogen chloride into an absolute ether solution of benzaldehyde and benzaldehyde cyanohydrin.[2] The oxazole hydrochloride can be converted to the free base by addition of water or by boiling with alcohol. Many different aromatic aldehydes and cyanohydrin combinations have been converted to 2,5-diaryloxazoles 4 by this procedure in 80% yield.[3]

6.4.3   Mechanism

Ingham proposed the following sequence to explain the formation of oxazole products following his study of the reaction of benzaldehyde with mandelonitrile and hydrogen chloride.[4] In the event, addition of hydrogen chloride to the cyanide is the first step providing the intermediate iminochloride 5 (Ar_1 = Ph), which upon reaction with benzaldehyde affords oxazole 2 (Ar_1, Ar_2 = Ph) via intermediate 6 (Ar_1, Ar_2 = Ph).

Ingham describes two by-products isolated in these reactions, (a) arylene-mandeloamides (**8**) formed in the presence of water and (b) diazines (**11**) resulting from dimerization of the chloroimine.   Reconsideration by Cornforth and Cornforth demonstrated that reaction of **5** with water actually produces oxazolidone **9**.[5]

6.4.4   Variations and Improvements

In 1949, Cornforth showed that preparation of 2,5-disubstituted oxazoles was not limited to diaryloxazoles through condensation of aldehydes (benzaldehyde, *n*-hept-aldehyde) with α-hydroxy-amides (lactamide). The intermediate oxazolidone **13** were converted into oxazoles **14** on warming with phosphoryl chloride.[5]

6.4.5   Synthetic Utility

Onaka demonstrated the utility of a modified Fisher method in the one-step synthesis of oxazole alkaloid Halfordinol (**16**) in higher overall yield[6] than previously reported by Robinson–Gabriel synthesis.[7]

6.4.6   Experimental
Halfordinol (16).

Dry HCl gas was saturated in an ice-cold solution of freshly prepared crystals of p-hydroxymandelonitrile[8] (**15**, 0.94 g) in 45 mL of anhydrous ether. After addition of $SOCl_2$ (1.12 g), the reaction mixture was stirred for 10 min. with external cooling. Addition of nicotinaldehyde (0.75 g) followed and the reaction mixture was saturated with dry HCl gas once again. After standing at room temperature for 2 days, the reaction mixture was poured into water and the separated organic layer was further extracted with aq. HCl. Neutralization of the combined aqueous layers with Na_2CO_3 resulted in the precipitation of halfordinol (**16**), which was collected by filtration and recrystallized from methanol to 248 mg (16.5%) as fine cream needles of mp 254–255 °C (lit. 255 °C).[9]

6.4.7   References

1   (a) [R] Wiley, R. H. *Chem. Rev.* **1945**, *37*, 401. (b) Cornforth, J. W.; Cornforth, R. H. *J. Chem. Soc.* **1949**, 1028. (c) [R] Cornforth, J. W. In *Heterocyclic Compounds, 5*; Wiley: New York, **1957**, pp 302-309.
2   Fischer, E. *Ber.* **1896**, **29**, 205.
3   Minovici, S.; Nenitzescu, C. D.; Angelescu, B. *Bull Soc. Chem. Romania* **1928**, *10*, 149; *Chem. Abstracts* **1929**, *23*, 2716.
4   Ingham, B. H. *J. Chem. Soc.* **1927**, 692.
5   Cornforth, J. W.; Cornforth, R. H. *J. Chem. Soc.* **1949**, 1028.
6   Onaka, T. *Tetrahedron Lett.* **1971**, 4393.
7   Brossi, A.; Wenis, E. *J. Heterocycl. Chem.* **1965**, *2*, 310.
8   Ladenburg, K.; Folkers, K.; Major, R. T. *J. Am. Chem. Soc.* **1936**, *58*, 1292.
9   Crow, W. D.; Hodgkin, J. H. *Tetrahedron Lett.* **1963**, *2*, 85; *Austr. J. Chem.* **1964**, *17*, 119.

Dawn A. Brooks

6.5 Meyers Oxazoline Method

6.5.1 Description

Chiral oxazolines developed by Albert I. Meyers and coworkers have been employed as activating groups and/or chiral auxiliaries in nucleophilic addition and substitution reactions that lead to the asymmetric construction of carbon-carbon bonds.[1-5] For example, metalation of chiral oxazoline **1** followed by alkylation and hydrolysis affords enantioenriched carboxylic acid **2**. Enantioenriched dihydronaphthalenes are produced via addition of alkyllithium reagents to 1-naphthyloxazoline **3** followed by alkylation of the resulting anion with an alkyl halide to give **4**, which is subjected to reductive cleavage of the oxazoline moiety to yield aldehyde **5**. Chiral oxazolines have also found numerous applications as ligands in asymmetric catalysis; these applications have been recently reviewed, and are not discussed in this chapter.[6-8]

6.5.2 Historical Perspective

The first synthesis and use of a chiral oxazoline was reported by Meyers in 1974. The chiral oxazoline **1** was prepared in two steps by condensation of (+)-1-phenyl-2-amino-1,3-propanediol (**6**) with the ethyl imidate of propionitrile[9] followed by O-methylation of the resulting alcohol **7** with NaH/MeI. Meyers demonstrated chiral oxazoline **1** could be alkylated by sequential treatment with LDA followed by ethyl iodide to afford **8**. Acidic hydrolysis of **8** provided enantioenriched carboxylic acid **9** in 95% yield and 67% optical purity. Since this report chiral oxazolines have become widely used as chiral auxiliaries and ligands in asymmetric synthesis.[1–8]

The first use of chiral oxazolines as activating groups for nucleophilic additions to arenes was described by Meyers in 1984.[10] Reaction of naphthyloxazoline **3** with phenyllithium followed by alkylation of the resulting anion with iodomethane afforded dihydronaphthalene **10** in 99% yield as an 83 : 17 mixture of separable diastereomers. Reductive cleavage of **10** by sequential treatment with methyl fluorosulfonate, NaBH$_4$, and aqueous oxalic acid afforded the corresponding enantiopure aldehyde **11** in 88% yield.

6.5.3   Mechanism of Asymmetric Alkylation

The mechanism of the asymmetric alkylation of chiral oxazolines is believed to occur through initial metalation of the oxazoline to afford a rapidly interconverting mixture of **12** and **13** with the methoxy group forming a chelate with the lithium cation.[11] Alkylation of the lithiooxazoline occurs on the less hindered face of the oxazoline **13** (opposite the bulky phenyl substituent) to provide **14**; the alkylation may proceed via complexation of the halide to the lithium cation. The fact that decreased enantioselectivity is observed with chiral oxazoline derivatives bearing substituents smaller than the phenyl group of **3** is consistent with this hypothesis.[12] Intermediate **13** is believed to react faster than **12** because the approach of the electrophile is impeded by the alkyl group in **12**.

Acidic hydrolysis of **14** occurs via protonation of the nitrogen followed by attack of water on the resulting cationic intermediate. Proton transfer followed by ring-opening affords cation **15**, which is trapped by a second equivalent of water. Another proton transfer followed by loss of the amino group affords protonated carboxylic acid **16**, which loses H$^+$ to provide the carboxylic acid product.

6.5.4   Mechanism of Asymmetric Addition to Naphthyl Oxazolines

The mechanism of organolithium addition to naphthyl oxazolines is believed to occur via initial complexation of the alkyllithium reagent to the oxazoline nitrogen atom and the methyl ether to form chelated intermediate **17**.[13] Addition of the alkyl group to the arene π-system affords azaenolate **18**, which undergoes reaction with an electrophile on the opposite face of the alkyl group to provide the observed product **4**. The chelating methyl ether group is not required for high asymmetric induction provided that a sufficiently bulky group is used in place of the hydroxymethyl substituent (*vide infra*).

Cleavage of the chiral auxiliary is effected in a three-step procedure commencing with quaternization of the nitrogen with methyl fluorosulfonate, methyl trifluoromethanesulfonate, or trimethyloxonium tetrafluoroborate. Reduction of the corresponding iminium salt **19** with NaBH₄ and acidic hydrolysis of the resulting product affords substituted aldehyde **5** without epimerization of either stereocenter.

6.5.5 *Variations and Improvements on Alkylations of Chiral Oxazolines*

Metalated chiral oxazolines can be trapped with a variety of different electrophiles including alkyl halides, aldehydes,[14] and epoxides to afford useful products. For example, treatment of oxazoline **20** with *n*-BuLi followed by addition of ethylene oxide and chlorotrimethylsilane yields silyl ether **21**. A second metalation/alkylation followed by acidic hydrolysis provides chiral lactone **22** in 54% yield and 86% *ee*.[15,16] A similar

strategy involving sequential alkylations was employed in the synthesis of 2-substituted butyrolactones. Alkylation of oxazoline **20** with γ-silyloxy alkyl bromide **23** then with benzyl bromide afforded **24**, which was subjected to acidic hydrolysis to provide **25** in 66% yield and 70% *ee*.[16]

Unsaturated chiral oxazolines have been employed in conjugate addition reactions for the asymmetric synthesis of 3-substituted carboxylic acids.[17] For example, metalation of oxazoline **20** with LDA followed by reaction with benzaldehyde and acidic workup leads to unsaturated oxazoline **26** via an aldol/dehydration process. Treatment of **26** with ethyllithium followed by acidic hydrolysis of the oxazoline afforded chiral carboxylic acid **27** in 66% yield and 97% *ee*. Chiral oxazolines derived from *tert*-leucine have also been employed in asymmetric conjugate addition reactions. As shown below, reaction of **28** with *t*-BuLi affords chiral aldehyde **29** in 74% yield and 94% *ee* upon cleavage of the oxazoline moiety.[18]

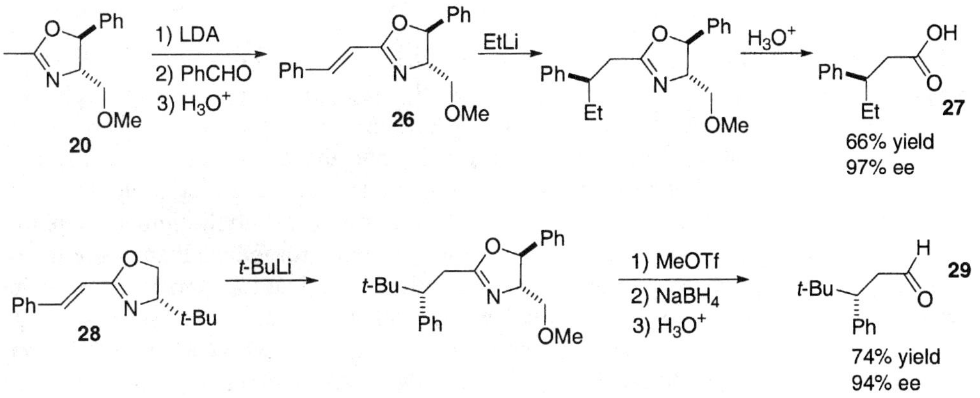

Chiral oxazolines have also been utilized for the synthesis of chiral ketones bearing quaternary carbon stereocenters. As shown below, reaction of substituted oxazoline **30** with 2 equiv PhLi followed by treatment with benzyl bromide gives ketone **33** upon acidic hydrolysis. This reaction is believed to proceed via addition of PhLi to keteneimine **31** to afford metalated enamine **32**, which undergoes alkylation at the nucleophilic carbon to provide **33** after aqueous workup.[19]

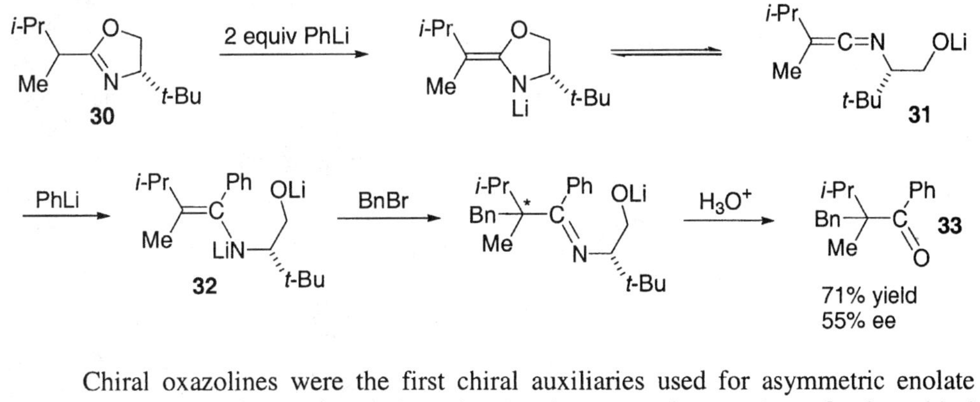

Chiral oxazolines were the first chiral auxiliaries used for asymmetric enolate alkylations. Subsequent studies led to the development of a number of other chiral auxiliaries (**34–38**) including those reported by Evans,[20,21] Myers,[22] Enders,[23,24] Schollkopf,[25] and others,[26,27] which are now widely used in asymmetric synthesis. Although these new auxiliaries frequently provide higher yields and enantioselectivities than the oxazolines originally developed by Meyers, the pioneering work of Meyers laid the groundwork for these later studies.

34   **35**   **36**   **37**   **38**

6.5.6 Variations and Improvements on Asymmetric Additions to Naphthyl Oxazolines

Meyers has demonstrated that chiral oxazolines derived from valine or *tert*-leucine are also effective auxiliaries for asymmetric additions to naphthalene. These chiral oxazolines (**39** and **40**) are more readily available than the methoxymethyl substituted compounds (**3**) described above but provide comparable yields and stereoselectivities in the tandem alkylation reactions. For example, addition of *n*-butyllithium to naphthyl oxazoline **39** followed by treatment of the resulting anion with iodomethane afforded **41** in 99% yield as a 99 : 1 mixture of diastereomers. The identical transformation of valine derived substrate **40** led to a 97% yield of **42** with 94% de.[28] As described above, sequential treatment of the oxazoline products **41** and **42** with MeOTf, NaBH$_4$ and aqueous oxalic acid afforded aldehydes **43** in > 98% *ee* and 90% *ee*, respectively. These experiments demonstrate that a chelating (methoxymethyl) group is not necessary for reactions to proceed with high asymmetric induction.

39 R = *t*-Bu   **41** R = *t*-Bu (94% yield, 98% de)   98% ee from **41**
40 R = *i*-Pr   **42** R = *i*-Pr (97% yield, 94% de)   90% ee from **42**

The application of this strategy to the synthesis of chiral cyclohexadienes has been demonstrated by Kündig. Addition of MeLi to the Cr(CO)$_3$-complexed chiral phenyl oxazoline **43** followed by reaction with allyl bromide produced cyclohexadiene **44** in 69% yield and >98% *de*.[29]

The asymmetric addition of naphthyl Grignard reagents to 1-methoxy-2-naphthyloxazolines has been used by Meyers[30] for the synthesis of nonracemic binaphthyl derivatives. These reactions are believed to occur via addition of the organolithium to naphthyloxazoline **45** followed by elimination of lithium methoxide to afford binaphthyl compound **46**; hydrolysis and reduction produced alcohol **47**. This methodology has been extended to the synthesis of chiral biaryl compounds **49** from 2-methoxyphenyl oxazolines **48**. A related enantioselective synthesis of binaphthyls has been reported by Cram, in which asymmetric induction was controlled by use of a chiral oxygen substituent and an achiral oxazoline.[31]

Meyers has also reported the use of chiral oxazolines in asymmetric copper-catalyzed Ullmann coupling reactions.[32] For example, treatment of bromooxazoline **50** with activated copper powder in refluxing DMF afforded binaphthyl oxazoline **51** as a 93 : 7 mixture of atropisomers; diastereomerically pure material was obtained in 57% yield after a single recrystallization. Reductive cleavage of the oxazoline groups as described above afforded diol **52** in 88% yield. This methodology has also been applied to the synthesis of biaryl derivatives.[33]

The chiral naphthyloxazoline substrates can also be employed in asymmetric carbon-heteroatom bond-forming reactions with lithium amides, which provide unusual β-amino acid products.[34] Treatment of oxazoline **53** with N-lithiopiperidine followed by alkylation with iodomethane affords aniline derivative **54** in 94% yield and 99% *de*. Hydrolysis of the oxazoline group provided amino acid **55** in 92% yield and >99% *ee*.

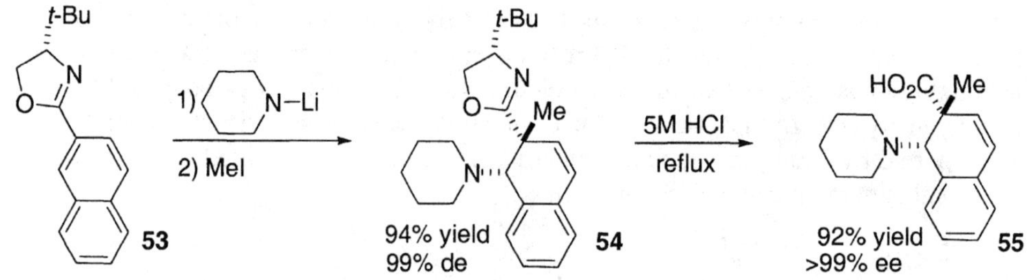

6.5.7 Applications in Natural Products Synthesis

Chiral oxazoline-based synthetic methods have been employed in the asymmetric synthesis of a large number of natural products. A few representative examples of these applications are shown below.

The sesquiterpenoid hydrocarbons (S)-α-curcumene (59) and (S)-xanthorrhizol (60) were prepared by asymmetric conjugate addition of the appropriate aryllithium reagent to unsaturated oxazoline 56 to afford alcohols 57 (66% yield, 96% ee) and 58 (57% yield, 96% ee) upon hydrolysis and reduction. The chiral alcohols were subsequently converted to the desired natural products.[35]

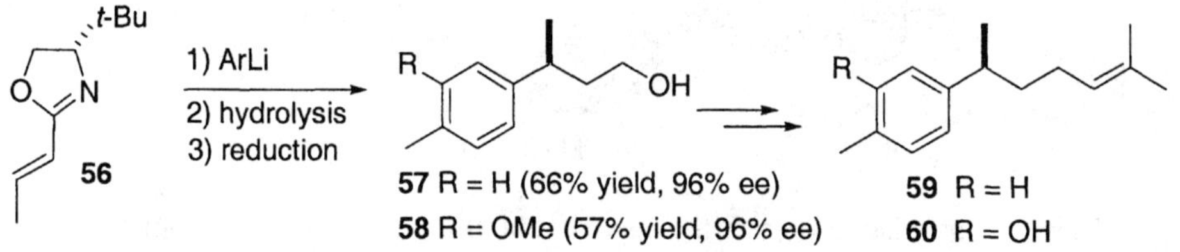

The asymmetric addition of organolithium reagents to aryloxazolines has been used to construct highly complex polycyclic terpene structures found in natural products. For example, the asymmetric addition of vinyllithium to chiral naphthyloxazoline 3 followed by treatment of the resulting anionic intermediate with iodoethyl dioxolane 61 afforded 62 in 99% yield as a single diastereomer.[36] This intermediate was converted into the terpenoid compound 63, which is structurally related to the natural product aphidicolin.

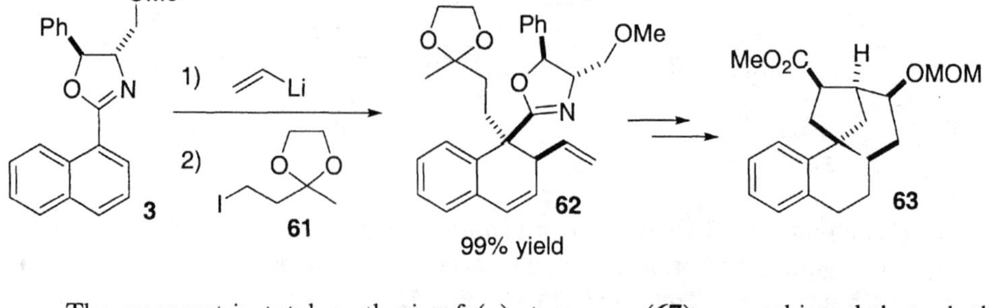

The asymmetric total synthesis of (–)-steganone (67) was achieved through the asymmetric Mg-mediated coupling of bromide 64 and oxazoline 65, which provided a

65% yield of biphenyl derivative **66**. Subsequent elaboration of this intermediate provided the natural product (**67**).[37]

The axially chiral natural product mastigophorene A (**70**) was synthesized via a copper-catalyzed asymmetric homocoupling of bromooxazoline **68**. Treatment of **68** with activated copper in DMF afforded **69** in 85% yield as a 3 : 1 mixture of atropisomers. The major atropisomer was converted into mastigophorene A (**70**); the minor regioisomer was transformed into the atropisomeric natural product mastigophorene B.[38]

6. 6   Experimental Procedures

Naphthyloxazoline 71.[32]

To a solution of 1-bromo-2-naphthoic acid (4.89g, 19.5 mmol) in CH_2Cl_2 (100 mL) was added oxalyl chloride (8.7 mL, 99 mmol) and DMF (6 drops). The mixture was stirred at rt overnight under an atmosphere of argon, then concentrated *in vacuo*. The crude material was dissolved in CH_2Cl_2 (50 mL) and was added to a solution of *tert*-leucinol (2.5 g, 21.5 mmol) and triethylamine (10 mL) in CH_2Cl_2 (100 mL) at 0°C. The mixture was stirred at rt overnight under argon and then diluted with water. The layers were separated and the organic layer was dried over anhydrous magnesium sulfate, filtered, and concentrated. The crude material was dissolved in CH_2Cl_2 (100 mL), $SOCl_2$ (10 mL) was added and the mixture was stirred at rt for 8 h. The mixture was cooled to 0°C and water and 4 M NaOH were added. The layers were separated and the organic phase was dried over anhydrous magnesium sulfate, filtered, and concentrated. The residue was dissolved in acetonitrile (300 mL) and water (25 mL) and solid K_2CO_3 were added. The mixture was heated to reflux for 3d, then cooled to rt and concentrated. The crude product was extracted with CH_2Cl_2, the organic extracts were concentrated, and the material was purified by flash chromatography on silica gel to afford 5.13 g (79%) of the title compound as a viscous oil. ^{1}H NMR (300 MHz, $CDCl_3$) δ 8.40 (d, *J* = 8.4 Hz, 1 H), 7.83–7.78 (m, 2 H), 7.63–7.50 (m, 3 H), 4.42 (dd, *J* = 8.6, 10.2 Hz, 1 H), 4.31 (t *J* = 8.3 Hz, 1 H), 4.15 (dd, *J* = 8.1, 10.2 Hz, 1 H), 1.02 (s, 9 H).

Binaphthyl bis(oxazoline) 72.

To a mixture of naphthyloxazoline **71** (4.31 g, 12.97 mmol) in pyridine (4 mL) was added activated copper (1.99 g). The mixture was heated to reflux for 24 h then was cooled to rt, diluted with CH_2Cl_2 and washed with aqueous ammonia until the copper had been completely removed. The organic phase was washed with water then dried over anhydrous magnesium sulfate, filtered, and concentrated to afford the title compound as a tan solid. This material was used without further purification.

Binaphthyl bis(methyl ester) 73.

To a solution of crude binaphthyl bis(oxazoline) **72** in THF (100 mL) was added water (5 mL), trifluoroacetic acid (11 mL), and sodium sulfate (55 g). The resulting suspension was stirred overnight at rt then filtered and concentrated *in vacuo*. The crude product was dissolved in CH$_2$Cl$_2$ (200 mL) and pyridine (12 mL) and acetic anhydride (20 mL) were added. The mixture was stirred overnight at rt and then washed with 1 M HCl (3 × 100 mL), and water (100 mL). The organic phase was dried over anhydrous magnesium sulfate, filtered, and concentrated *in vacuo*. The resulting brown solid was recrystallized from ethyl acetate and then purified by radial chromatography on silica gel to afford 2.3 g (57%) of an ester amide as a white solid.

To a solution of the ester amide (160 mg, 0.26 mmol) in methanol (3 mL) and THF (3 mL) was added a 1 M solution of NaOMe in methanol (5 mL). The mixture was stirred at rt for 1.5 d then neutralized with methanolic acetic acid and concentrated *in vacuo*. The crude material was partitioned between water and CH$_2$Cl$_2$. The organic phase was dried over anhydrous magnesium sulfate, filtered, and concentrated *in vacuo* to afford the bis(ester) **73** as a colorless solid, mp 154.4–155.5 °C, [α] –17° (c = 0.3, MeOH).

Dihydronaphthalene 41.[28]

A solution of naphthyloxazoline **39** (200 mg, 0.79 mmol) in THF was cooled to –78°C and a solution of *n*-butyllithium (0.79 mL, 1.5 M in hexanes, 1.19 mmol) was added dropwise. The mixture was stirred at –78°C for 2 h then iodomethane (1.21 mL, 2.37 mmol) was added. The mixture was warmed to rt, stirred for 1h, then quenched with saturated aqueous ammonium chloride (30 mL). The mixture was extracted with CH$_2$Cl$_2$ (3 × 30 mL) and the combined organic extracts were dried over anhydrous sodium sulfate, filtered, and concentrated *in vacuo*. The crude material was purified by flash chromatography on silica gel to afford 259 mg (100%) of the title compound as a colorless oil. ^{1}H NMR (300 MHz, CDCl$_3$) δ 7.3–7.0 (m, 4 H), 6.40 (d, *J* = 9.8 Hz, 1 H), 5.96 (dd, *J* = 4.3, 9.8 Hz, 1 H), 4.1–3.9 (m, 2 H), 3.83 (dd, *J* = 7.0, 10.0 Hz, 1 H), 2.4–2.3 (m, 1 H), 1.64 (s, 3 H), 1.6–1.2 (m, 6 H), 1.0–0.8 (m, 3 H), 0.87 (s, 9 H).

Aldehyde 43.

To a solution of dihydronaphthalene **41** (250 mg, 0.77 mmol) in CH$_2$Cl$_2$ (5 mL) was added methyl trifluoromethanesulfonate (227 mg, 1.38 mmol). The mixture was stirred at rt until the starting material had been completely consumed as judged by TLC analysis (3 h). The mixture was cooled to 0°C and a solution of NaBH$_4$ (111 mg, 2.92 mmol) in 4:1 MeOH:THF (3 mL) was slowly added. The mixture was warmed to rt then quenched with saturated aqueous ammonium chloride (50 mL). The resulting mixture was extracted with CH$_2$Cl$_2$ (3 × 50 mL) and the combined organic extracts were dried over anhydrous sodium sulfate, filtered, and concentrated *in vacuo*. The resulting material was dissolved in 4:1 THF/H$_2$O (5 mL) and oxalic acid (485 mg, 3.85 mmol) was added. The reaction mixture was stirred at rt for 12h then was quenched with saturated aqueous sodium bicarbonate (50 mL). The resulting mixture was extracted with CH_2Cl_2 and the combined organic extracts were dried over anhydrous sodium sulfate, filtered, and concentrated *in vacuo*. The crude product was purified by flash chromatography on silica gel to afford 134 mg (76%) of the title compound as a colorless oil. ^{1}H NMR (300 MHz, CDCl$_3$) δ 9.80 (s, 1 H), 7.2–7.1 (m, 4 H), 6.45 (dd, J = 2.1, 9.8 Hz, 1 H), 5.95 (dd, J = 3.8, 9.8 Hz, 1 H), 2.5–2.4 (m, 1 H), 1.4 (s, 9 H), 1.5–1.2 (m, 6 H), 0.87 (t, J = 7.1 Hz, 3 H).

6.7   *References*

1.   [R] Meyers, A. I. *J. Heterocycl. Chem.* **1998**, *35*, 991–1002.
2.   [R] Gant, T. G.; Meyers, A. I. *Tetrahedron* **1994**, *50*, 2297–2360.
3.   [R] Reuman, M.; Meyers, A. I. *Tetrahedron* **1985**, *41*, 837–860.
4.   [R] Meyers, A. I.; Mihelich, E. D. *Angew. Chem. Int. Ed.* **1976**, *15*, 270.
5.   [R] Meyers, A, I. *Acc. Chem. Res.* **1978**, *11*, 375–381.
6.   [R] Helmchen, G.; Pfaltz, A. *Acc. Chem. Res.* **2000**, *33*, 336–345.
7.   [R] Johnaon, J. S.; Evans, D. A. *Acc. Chem. Res.* **2000**, *33*, 325–335.
8.   [R] Rechavi, D.; Lemaire, M. *Chem. Rev.* **2002**, *102*, 3467–3494.
9.   Meyers, A. I.; Knaus, G.; Kamata, K. *J. Am. Chem. Soc.* **1974**, *96*, 268–270.
10.   Barner, B. A.; Meyers, A. I. *J. Am. Chem. Soc.* **1984**, *106*, 1865–1866.
11.   Meyers, A. I.; Knaus, G. *J. Am. Chem. Soc.* **1974**, *96*, 6508–6510.
12.   Meyers, A. I.; Mazzu, A.; Whitten, C. E. *Heterocycles* **1977**, *6*, 971–977.
13.   Meyers, A. I.; Barner, B. A. *J. Org. Chem.* **1986**, *51*, 120–122.
14.   Meyers, A. I.; Knaus, G. *Tetrahedron Lett.* **1974**, 1333–1336.
15.   Meyers, A. I.; Mihelich, E. D. *J. Org. Chem.* **1975**, *40* 1186–1187.
16.   Meyers, A. I.; Yamamoto, Y.; Mihelich, E. D.; Bell, R. A. *J. Org. Chem.* **1980**, *45*, 2792–2796.
17.   Meyers, A. I.; Whitten, C. E. *J. Am. Chem. Soc.* **1975**, *97*, 6266–6267.
18.   Meyers, A. I.; Shipman, M. *J. Org. Chem.* **1991**, *56*, 7098–7102.
19.   Dwyer, M. P.; Price, D. A.; Lamar, J. E.; Meyers, A. I. *Tetrahedron Lett.* **1999**, *40*, 4765–4768.
20.   Evans, D. A.; Takacs, J. M. *Tetrahedron Lett.* **1980**, *21*, 4233–4236.
21.   Evans, D. A.; Ennis, M. D.; Mathre, D. J. *J. Am. Chem. Soc.* **1982**, *104*, 1737–1739.
22.   Myers, A. G.; Yang, B. H.; Chen, H.; McKinstry, L.; Kopecky, D. J.; Gleason, J. L. *J. Am. Chem. Soc.* **1997**. *119*, 6496–6511.
23.   [R] Enders, D.; Klatt, M. *Synthesis* **1996**, 1403–1418.
24.   [R] Job, A.; Janeck, C. F.; Bettray, W. Peters, R.; Enders, D. *Tetrahedron* **2002**, *58*, 2253–2329.
25.   Schollkopf, U.; Hartwig, W.; Groth, U. *Angew. Chem. Int. Ed.* **1979**, *18*, 863–864.
26.   [R] Wirth, T. *Angew. Chem. Int. Ed. Engl.* **1997**, *36*, 225–227.
27.   [R] Ager, D. J.; Prakash, I.; Schaad, D. R. *Chem. Rev.* **1996**, *96*, 835–875.
28.   Rawson, D. J.; Meyers, A. I. *J. Org. Chem.* **1991**, *56*, 2292–2294.
29.   Kündig, E. P.; Ripa, A.; Bernardinelli, G. *Angew. Chem. Int. Ed.* **1992**, *31*, 1071–1073.
30.   Meyers, A. I.; Lutomski, K. A. *J. Am. Chem. Soc.* **1982**, *104*, 879–881.
31.   Wilson, J. M.; Cram, D. J. *J. Am. Chem. Soc.* **1982**, *104*, 881–884.
32.   Nelson, T. D.; Meyers, A. I. *J. Org. Chem.* **1994**, *59*, 2655–2658.
33.   Nelson, T. D.; Meyers, A. I. *Tetrahedron Lett.* **1993**, *34*, 3061–3062.
34.   Shimano, M.; Meyers, A. I. *J. Am. Chem. Soc.* **1994**, *116*, 6437–6438.
35.   Meyers, A. I.; Stoianova, D. *J. Org. Chem.* **1997**, *62*, 5219–5221.
36.   Robichaud, A. J.; Meyers, A. I. *J. Org. Chem.* **1991**, *56*, 2607–2609.
37.   Meyers, A. I.; Flisak, J. R.; Aitken, R. A. *J. Am. Chem. Soc.* **1987**, *109*, 5446–5452.
38.   Degnan, A. P.; Meyers, A. I. *J. Am. Chem. Soc.* **1999**, *121*, 2762–2769.

John P. Wolfe

6.6   Robinson–Gabriel Synthesis

6.6.1   Description

The Robinson–Gabriel cyclodehydration of 2-acylamidoketones **1** is one of the oldest yet most versatile synthesis of 2,5-di- and 2,4,5-trialkyl, aryl, heteroaryl-, and aralkyloxazoles **2**.[1]

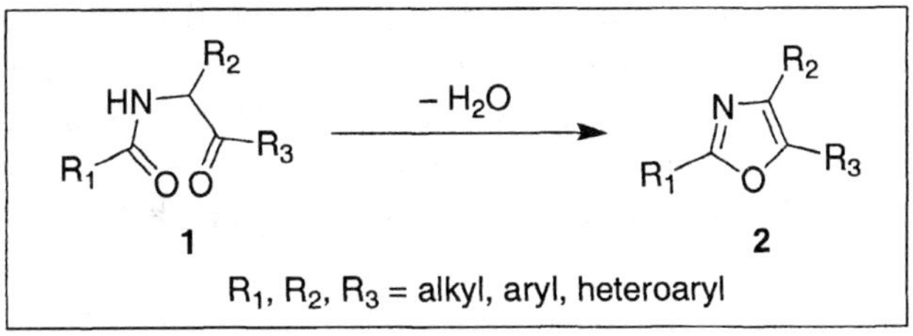

R₁, R₂, R₃ = alkyl, aryl, heteroaryl

6.6.2   Historical Perspective

In 1909, Robinson demonstrated the utility of acylamidoketones as intermediates to aryl- and benzyl-substituted 1,3-oxazoles through cyclization with sulfuric acid.[1a] Extension of sulfuric acid cyclization conditions to alkyl-substituted oxazoles can give low yields, for example 10–15% for 2,5-dimethyl-1,3-oxazole.[2] Wiegand and Rathburn found that polyphosphoric acid can provide alkyl-substituted oxazoles **4** in yields equal to or greater than those obtained with sulfuric acid.[3] Significantly better yields are seen in the preparation of aryl- and heteroaryl-substituted oxazoles. For example, reaction of ketoamides **5** with 98% phosphoric acid in acetic anhydride gives oxazoles **6** in 90–95% yield.[4]

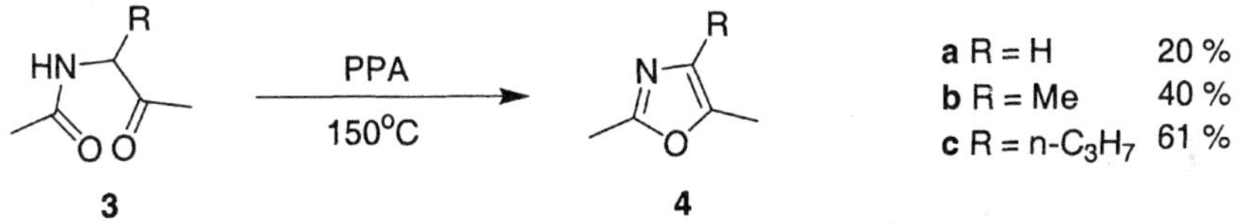

a R = H	20 %
b R = Me	40 %
c R = n-C₃H₇	61 %

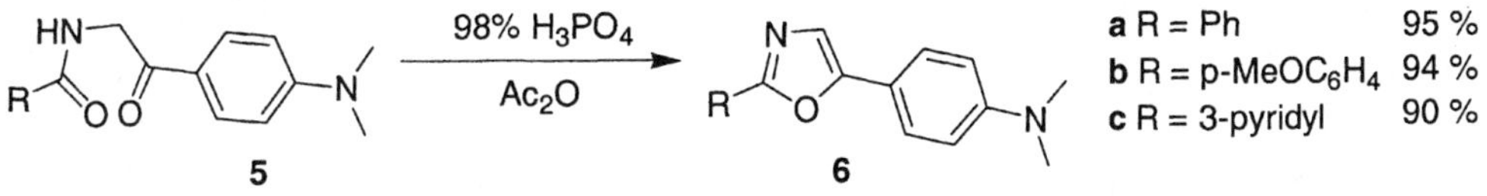

a R = Ph	95 %
b R = p-MeOC₆H₄	94 %
c R = 3-pyridyl	90 %

6.6.3   Mechanism

Wasserman demonstrated with ^{18}O labeling studies that the amide carbonyl oxygen is incorporated into the oxazole ring upon cyclization of ^{18}O-labeled 2-benzamidopropiophenone consistent with the mechanism shown below.[5]

6.6.4 Variations and Improvements

Wipf and Miller have reported side-chain oxidation of β-hydroxy amides with the Dess–Martin periodinane, followed by immediate cyclodehydration with triphenylphosphine-iodine, which provides a versatile extension of the Robinson–Gabriel method to substituted oxazoles.[6] Application of this method was used to prepare the oxazole fragment **10** in 55% overall yield from β-hydroxy amide **8**.

More recent examples have employed a milder reagent system, triphenyl-phosphine and dibromotetrachloroethane to generate a bromo-oxazoline, which is subsequently dehydrohalogenated. Wipf and Lim utilized their method to transform intermediate **11** into the 2,4-disubstituted system of (+)-Hennoxazole A.[7] Subsequently, Morwick and coworkers reported a generalized approach to 2,4-disubstituted oxazoles from amino acids using a similar reagent combination, triphenylphosphine and hexachloroethane.[8]

(+)-Hennoxazole A

Cyclodehydration of 2-acylamino carbonyl compounds with the Burgess reagent[9] has been shown by Brian and Paul to proceed rapidly and cleanly under monomode microwave conditions.[10] For example, irradiation of **1** for 2–4 min at 100 W afforded oxazoles **2** in excellent yield. It is also noteworthy that these conditions provided a high-yielding synthesis of 2-monosubstituted oxazoles **13**, which are historically problematic to prepare from 2-acylamino aldehydes.[11]

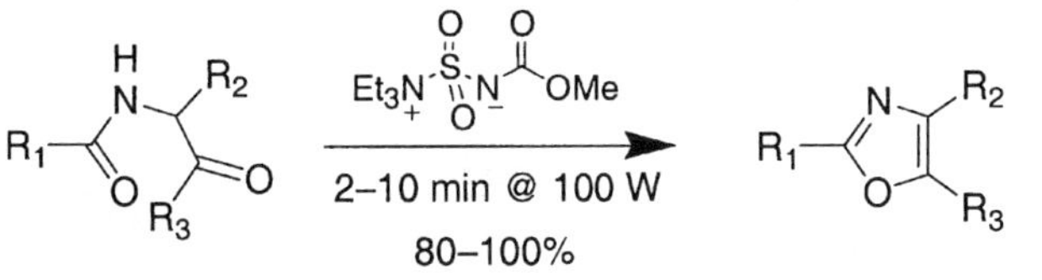

1

2

R_1 = Me or Ph
R_2 = H or Me
R_3 = Me or Ph

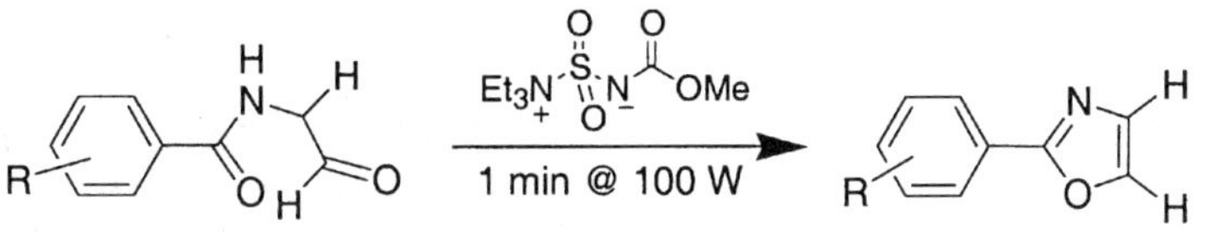

12

13

R = 3-OPh, 93%
R = 2,4-DiMe, 81%

Pulici and coworkers have reported a solid-phase variation of the Robinson-Gabriel for the production of parallel libraries of oxazole-containing molecules.[12] The preparation is based on a solid supported 2-acylamino ketone **16** that can be cleaved by means of a volatile anhydride and cyclized in solution to obtain a substituted oxazole ring (**17**) that does not contain traces of the linker moiety.

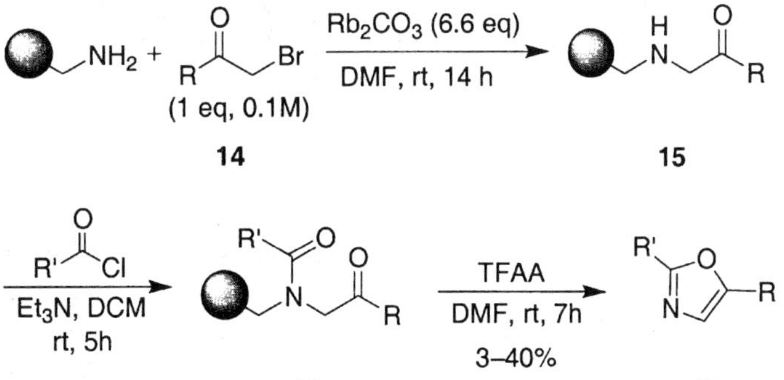

14

15

16

17

6.6.5 Synthetic Utility

Meguro and co-workers described the synthesis and hypoglycemic activity of 4-oxazoleacetic acid derivatives.[13] For example, cyclization of **18** was performed by using phosphorus oxychloride in refluxing toluene to provide 5-methyl-2 (1-methylcyclohexyl)-4-oxazoleacetate in 73% yield and the subsequent hydrolysis gave **19**.

Nicolaou and co-workers established the severely strained A-ring oxazole (**21**) in their total synthesis of antitumor agent diazonamide A through initial oxidation of the hindered alcohol of intermediate **20** with TPAP and subsequent Robinson–Gabriel cyclodehydration of the resultant ketoamide with a mixture of $POCl_3$ and pyridine (1:2) at $70\,^{\circ}C$.[14]

Workers at Lilly prepared the oxazole-containing, dual PPAR α/γ agonist **23**, through Robinson–Gabriel cyclodehydration of ketone **22** with acetic anhydride and sulfuric acid in refluxing ethyl acetate.[15]

6.6.6   Experimental

2-{4-[2-(2-Biphenyl-4-yl-5-methyl-oxazol-4-yl)-ethoxy]-phenoxy}-2-methyl-propionic acid (23) Ketone **22** (5.00 g, 10.51 mmol) was dissolved in 40 mL EtOAc. Acetic anhydride (3.22 g, 31.54 mmol) and 95–98% sulfuric acid (0.31 g, 3.16 mmol) in 2.5 mL EtOAc were added and the mixture was heated at reflux for 3 h. The reaction mixture was cooled and 5N NaOH (12.6 mL, 63 mmol) diluted to 25 mL with water was added.   The reaction was heated at reflux for 30 min, and then cooled to room temperature.   The resulting layers were separated. The organic layer was washed with 1N HCl and 10% brine, dried (Na_2SO_4), concentrated *in vacuo* to 26.5 g, and stirred overnight at room temperature.   The resulting slurry was diluted with heptane (24 mL) and cooled at 0°C for 1 hour. Filtration and drying yielded 4.21 g (88% yield) of **23** in 95% yield as a white solid: mp 141–143.5°C; ^{1}H NMR (300 MHz, DMSO-d$_6$) δ 7.99 (d, $J = 9.0$ Hz, 2H), 7.80 (d, $J = 6.0$ Hz, 2H), (d, $J = 9.0$ Hz, 2H), 7.49 (t, $J = 7.6$ Hz, 2H),

7.38 (t, J = 7.6 Hz, 1H), 6.91–6.79 (m, 4H), 4.17 (t, J = 6.4 Hz, 2H), 2.92 (t, J = 6.4 Hz, 2H), 2.36 (s, 3H), 1.44 (s, 6H); ^{13}C NMR (75 MHz, DMSO-d$_6$) δ 175.0, 158.1, 153.6, 148.8, 145.1, 141.4, 139.1, 132.8, 129.0, 127.9, 127.1, 126.6, 126.0, 121.0, 114.9, 78.9, 66.5, 25.7, 24.9; HRMS-FAB (m/z): [M+H]$^+$ calcd for $C_{28}H_{28}NO_5$, 458.1967; found, 458.1958; Anal. Calc'd for $C_{28}H_{27}NO_5$: C, 73.51; H, 5.95; N, 3.06. Found: C, 73.81; H, 6.16; N, 3.13.

6.6.7 References

1. (a) Robinson, R. *J. Chem Soc.* **1909**, *95*, 2167. (b) Gabriel, S. *Chem. Ber.* **1910**, *43*, 134, 1283. (c) [R] 2. Turchi, I. J. In *The Chemistry of Heterocyclic Compounds*, *45*; Wiley: New York, **1986**; pp 1-342.
2. Wiley, R. H.; Borum, O. H. *J. Am. Chem Soc.* **1948**, *70*, 2005.
3. Wiegand, E. E.; Rathburn, D. W. *Synthesis* **1970**, 649.
4. Kerr, V. H.; Hayes, F. N.; Ott, D. G.; Lier, R.; Hansbury, E. *J. Org. Chem.* **1959**, *24*, 1864.
5. Wasserman, H. H.; Vinick, F. J. *J. Org. Chem.* **1973**, *38*, 2407.
6. Wipf, P.; Miller, C. P. *J. Org. Chem.* **1993**, *58*, 3604.
7. (a) Wipf, P. Lim, S. *J. Am. Chem. Soc.* **1995**, *117*, 558; *Chimia* **1996**, *50*, 157. (b) Yokokawa, F.; Asano, T.; Shioiri, T. *Org Lett* **2000**, *2*, 4169; *Tetrahedron* **2001**, *57*, 6311.
8. Morwick, T.; Hrapchak, M.; DeTuri, M.; Campbell, S. *Org Lett* **2002**, *4*, 2665.
9. Atkins, G. M.; Burgess, E. M. *J. Am. Chem. Soc.* **1968**, *90*, 4744.
10. Brain, C. T.; Paul, J. M. *Synlett* **1999**, *10*, 1642.
11. Two known literature examples: Sen, P.K. Veal, C. J., Young, D. W. *J. Chem. Soc., Perkin Trans 1* **1981**, 3052 (8.5% using SOCl2). See also ref. 6 (17% overall).
12. Pulici, M.; Quartieri, F.; Felder, E. *Fifth Int. Electronic Conf. on Syn. Org. Chem.* (ECSOC), 1-30 Sept. 2001.
13. Meguro, K.; Tawada, H.; Sugiyama, Y.; Fujita, T.; Kawamatsu, Y. *Chem. Pharm. Bull.* **1986**, *34*, 2840.
14. Nicolaou, K.C.; Rao, P. B.; Hao, J.; Reddy, M. V.; Rassias, G.; Huang, X.; Chen, D. Y.-K.; Snyder, S. A. *Angew. Chem. Int. Ed.* **2003**, *42*, 1753.
15. Godfrey, A. G.; Brooks, D. A.; Hay, L. A.; Peters, M.; McCarthy, J. R.; Mitchell, D. *J. Org. Chem.* **2003**, *68*, 2623.

Dawn A. Brooks

6.7   van Leusen Oxazole Synthesis

6.7.1   Description

The van Leusen reaction forms 5-substituted oxazoles through the reaction of *p*-tolylsulfonylmethyl isocyanide (**1**, TosMIC)[1] with aldehydes in protic solvents at refluxing temperatures. Thus 5-phenyloxazole (**2**) is prepared in 91% yield by reacting equimolar quantities of TosMIC and benzaldehyde with potassium carbonate in refluxing methanol for 2 hrs.[2]

6.7.2   Historical Perspective

In 1972, van Leusen, Hoogenboom and Siderius introduced the utility of TosMIC for the synthesis of azoles (pyrroles, oxazoles, imidazoles, thiazoles, etc.) by delivering a C–N–C fragment to polarized double bonds. In addition to the synthesis of 5-phenyloxazole, they also described reaction of TosMIC with *p*-nitro- and *p*-chloro-benzaldehyde (**3**) to provide analogous oxazoles **4** in 91% and 57% yield, respectively. Reaction of TosMIC with acid chlorides, anhydrides, or esters leads to oxazoles in which the tosyl group is retained. For example, reaction of acetic anhydride and TosMIC furnish oxazole **5** in 73% yield.[2]

Van Leusen and co-workers also demonstrated the condensation of heteroaromatic aldehydes with TosMIC.[3] Table 6.7.1 shows the 5-heteroaryloxazoles **6** prepared in 47–88% yield in the presence of equimolar amounts of potassium carbonate in refluxing methanol.

Table 6.7.1. 5-Heteroaryloxazoles

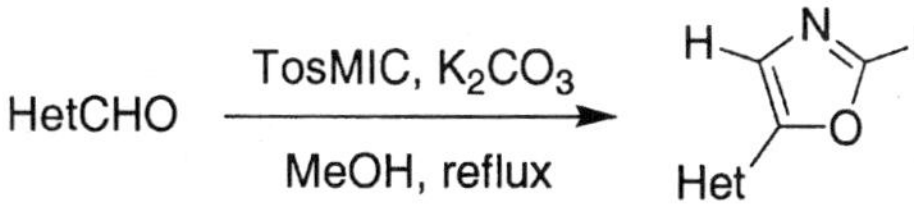

HetCHO $\xrightarrow[\text{MeOH, reflux}]{\text{TosMIC, K}_2\text{CO}_3}$ 6

Het =			Yield (%)
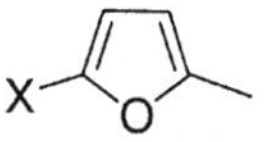 (furan)	X =	H	82
		NO$_2$	83
		CO$_2$Me	88
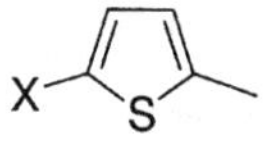 (thiophene)	X =	H	80
		NO$_2$	68
(pyrrole)			47
(pyridine)			82 (o-)
			80 (m-)
			67 (p-)

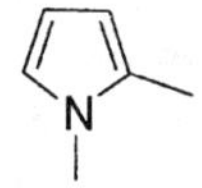

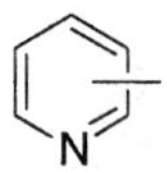

Van Leusen and Possel described the use of mono-substituted tosylmethyl isocyanides (TosCHRN=C; R = alkyl, benzyl, allyl) in the synthesis of 4,5-substituted oxazoles.[4] For example, 4-ethyl-5-phenyloxazole (8) was prepared in 82% yield by refluxing α-tosylpropyl isocyanide (7) and benzaldehyde for 1 hr with 1.5 equivalent of K$_2$CO$_3$ in MeOH.

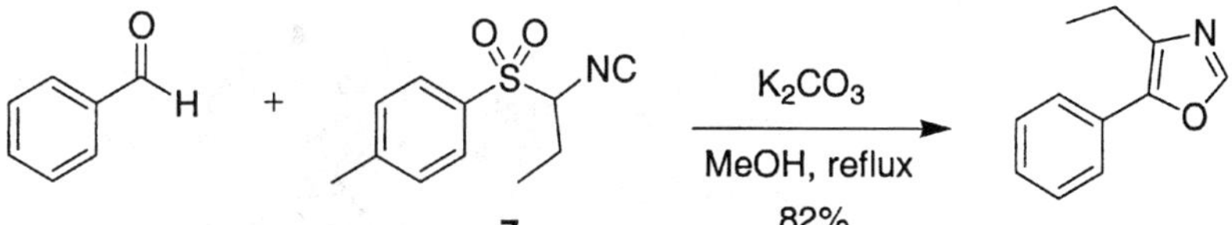

benzaldehyde + 7 $\xrightarrow[\substack{\text{MeOH, reflux} \\ 82\%}]{\text{K}_2\text{CO}_3}$ 8

6.7.3 Mechanism

The propensity of isocyanides to undergo nucleophilic α-additions at the terminal carbon, together with the presence of an activated methylene and a potential leaving group (i.e. tosyl), led van Leusen to suggest the following reaction path:

Thus attack of the TosMIC anion **9** on a carbonyl carbon is followed (or accompanied) by ring closure of the carbonyl oxygen to the electrophilic isocyano carbon to form an oxazoline (**12**). Loss of *p*-tolylsulfinic acid provides the 5-substituted oxazole **13**.[2]

6.7.4   *Variations and Improvements*

Van Leusen and co-workers also demonstrated the utility of dilithio-tosylmethyl isocyanide (dilithio-TosMIC) to extend the scope of the application.[5] Dilithio-TosMIC is readily formed from TosMIC and two equivalents of n-butyllithium (BuLi) in THF at –70°C. Dilithio-TosMIC converts ethyl benzoate to oxazole **14** in 70% yield whereas TosMIC monoanion does not react. In addition, unsaturated, conjugated esters (**15**) react with dilithio-TosMIC exclusively through the ester carbonyl to provide oxazoles (**16**). On the other hand, use of the softer TosMIC-monoanion provides pyrroles through reaction of the carbon-carbon double bond in the Michael acceptor.

Workers at SmithKline Beecham extended the synthetic access to interesting mono- and di-substituted oxazoles through an improved procedure for aryl-substituted TosMIC reagents.[6a] For example, glyoxylic acid ethyl ester undergoes cycloaddition with (2-naphthyl) tosylmethyl isonitrile (**17**) to produce oxazole **18** in good yield.[6b]

17   **18**

Recently, Ganesan and Kulkarni reported a solid-phase version of TosMIC and demonstrated its utility in the synthesis of 5-aryloxazoles.[7] They prepared the resin (PS-TosMIC) starting with polystyrene-SH through dehydration of N-(p-tolylsulfonylmethyl)-formamide (**20**) with Ph_3P/CCl_4. They observed that elimination of p-tolylsulfinic acid from the intermediate oxazoline takes place at room temperature and found quaternary ammonium hydroxide to be the optimum base. Yields of oxazoles **22**, following preparative TLC, ranged from 25–50% for ten demonstrated aromatic aldehydes.

20   **21**   **22**

Ganesen and Kulkarni also reported use of Ambersep 900 hydroxide resin as an ion-exchange base with TosMIC to prepare a variety of aromatic oxazoles in good isolated yields (54–85%) but moderate crude purities (57–94%).[8] Barrett and co-workers recently utilized ring-opening metathesis, polymer-supported (ROMPgel) TosMIC reagent **23**, for the conversion of a range of aromatic and heteroaromatic aldehydes to oxazoles.[9] Products were isolated without chromatography in excellent purity (> 95%) and in 68–90% yield. Electron rich aldehydes such as benzaldehyde and anisaldehyde could only be driven to 90 and 66% conversion, respectively, by this method.

23

6.7.5 Synthetic Utility

The development of the key intermediate, 5-(2-methoxy-4-nitrophenyl)oxazole (**25**), in the preparation of the hepatitis C drug candidate, **VX-497**, utilizes a van Leusen reaction of aldehyde **24** with TosMIC.[10]

VX-497

Workers at Lilly prepared the oxazole-containing partial ergot alkaloid, **27**, a 5-HT1A agonist, through van Leusen reaction of aldehyde **26**.[11]

6.7.6 Experimental procedure

6-(5-Oxazoyl)-*N,N*-dipropylbenz[*cd*]indole-4-amine (27).

To a solution of the aldehyde **26** (20.0 g, 70.4 mmol) in MeOH (200 mL) was added NaOMe (12.8 g, 237 mmol) as a solid portion-wise. After the solution was stirred for 5 min, tosylmethyl isocyanide (16.5 g, 84.5 mmol) was added as a solid portion-wise. The resulting solution was refluxed for 5 h, after which water (100 mL) was added to the hot reaction mixture. After cooling to rt, the mixture was cooled at 0°C and filtered. The solid was washed with cold 50% MeOH in water to afford 18.4 g (81%) of **27** as a tan solid. An analytically pure sample of **27** could be acquired by passing the material through a silica plug with EtOAc: mp 165–166°C; ^{1}H NMR (500 MHz, CDCl$_3$) δ 7.94 (s, 2 H), 7.48 (d, J = 8.4 Hz, 1 H), 7.22 (d, J = 8.4 Hz, 1 H), 7.19 (s, 1 H), 6.90 (s, 1 H), 3.25 (m, 2 H), 3.01 (m, 2 H), 2.82 (m, 1 H), 2.59 (m, 4 H), 1.49 (q, J = 7.2, 14.6 Hz, 4 H), 0.92 (t, J = 7.3 Hz, 6 H); ^{13}C NMR (75 MHz, CDCl$_3$) δ 151.8, 149.7, 133.6, 129.5, 127.1,

122.1, 121.8, 118.5, 115.9, 114.6, 108.8, 58.5, 53.1, 29.4, 23.8, 22.6, 11.9; MS m/z 323 (M^+). Anal. Calc'd for $C_{20}H_{25}N_3O$: C, 74.27; H, 7.79; N, 12.99. Found: C, 74.33; H, 7.90; N, 13.19.

6.7.7. *References*

1   [R] van Leusen, A. M.; van Leusen, D. In *Encyclopedia of Reagents of Organic Synthesis*; Paquette, L. A., Ed.; Wiley: New York, **1995**; Vol. 7, pp 4973-4979.

2   van Leusen, A. M.; Hoogenboom, B. E.; Siderius, H. *Tetrahedron Lett.* **1972,** *13,* 2369.

3   Saikachi, H.; Kitagawa, T.; Sasaki, H.; van Leusen, A. M. *Chem. Pharm. Bull.* **1979**, *27*, 793.

4   Possel, O.; van Leusen, A. M. *Heterocycles* **1977**, *7*, 77.

5   van Nispen, S. P. J. M.; Mensink, C.; van Leusen, A. M. *Tetrahedron Lett.* **1980**, *21*, 3723.

6   (a) Sisko, J.; Mellinger, M.; Sheldrake, P. W.; Baine, N. *Tetrahedron Lett.* **1996**, *37*, 8113; Sisko, J.; Mellinger, M.; Sheldrake, P. W.; Baine, N. *Org. Synth.* **2000**, *77*, 198. (b) Sisko, J.; Kassick, A. J.; Mellinger, M.; Filan, J. J.; Allen, A.; Olsen, M. A. *J. Org. Chem.* **2000**, *65*, 1516.

7   Kulkarni, B. A.; Ganesan, A. *Tetrahedron Lett.* **1999**, *40*, 5633

8   Kulkarni, B. A.; Ganesan, A. *Tetrahedron Lett.* **1999**, *40*, 5637.

9   Barrett, A. G.; Cramp, S. M.; Hennessy, A. J.; Procopiou, P. A.; Roberts, R. S. *Org. Lett.* **2001**, *3*, 271.

10   Herr, R. J.; Fairfax, D. J.; Meckler, H.; Wilson, J. D. *Org. Process Rec. Dev.* **2002**, *6*, 677.

11   Anderson, B. A.; Becke, L. M.; Booher, R. N.; Flaugh, M. E.; Harn, N. K.; Kress, T. J.; Varie, D. L.; Wepsiec, J. P. *J. Org. Chem.* **1997**, *62*, 8634.

Dawn A. Brooks

Chapter 7 Other Five-Membered Heterocycles 261

7.1   Auwers Flavone Synthesis

7.1.1   Description

The Auwers flavone synthesis consists of treatment of dibromo-coumarones **1** with alcoholic alkali to give the flavonols **2**.[1] It can also be described as the three-step sequence of **3 → 6**.

7.1.2   Historical Perspective

In 1908, while working at University of Heidelberg, Auwers and Müller described the transformation of 4-methyl-2-cumaranone (**3**) to flavanol **6**.[2] Thus aldol condensation of **3** with benzaldehyde gave benzylidene derivative **4**, which was brominated to give dibromide **5**. Subsequent treatment of **5** with alcoholic KOH then furnished 2-methylflavonol **6**. In the following years, Auwers published more extensively on the scope and limitations of this reaction.[3–5]

7.2.3   Mechanism

There is no published mechanistic study on the Auwers flavone synthesis. The mechanism may involve the nucleophilic addition of oxonium **7**, derived from **1**, with hydroxide to give **8**. Base-promoted ring opening of **8** could provide the putative intermediate **9**, which then could undergo an intramolecular Michael addition to form **10**. Expulsion of bromide ion from **10** would then give flavonol **2**.

7.1.4 Variations and Improvements

Auwers[3–5] and others[6] soon discovered that the transformation **3** → **6** did not consistently give flavonols such as **2**. For example, alcoholic alkali treatment of dibromide **11** produced 2-benzoyl-benzofuran-3-one **12** instead of the corresponding flavonol. The same observation was made by Robert Robinson in a failed attempt to make datiscetin in 1925.[7] It has reported that when there is a *meta* (to the coumarone ring oxygen) substituent such as methyl or methoxy, flavonol formation is hindered, whereas methyl, methoxy, and chlorine substituents at the *ortho* and *para* positions are conducive to flavonol formation.[1]

7.2.5 Synthetic Utility

Adopting Auwers' original method, Milton and Stephen prepared 2-chloroflavonol **16** from 4-chlorocoumaran-2-one **13** in 3 steps in 70% overall yield.[8] 4-Chloroflavonol was synthesized via the same sequence. The same group also carried out the bromination of 2-benzylidenedihydro-b-naphthafurano-1-one (**17**) and subsequently treated the dibromide with aqueous potassium hydroxide to give 5,6-benzflavonol **18**.[9] However, considerable difficulty was encountered in preparing the dibromides of the other arylidene compounds.

In summary, the Auwers flavone synthesis has seen only very limited utility in organic synthesis.

7.2.6   Experimental[8]

7-Chloro-2-benzylidene-coumaran-3-one (14)
A solution of coumaranone **13** (2.09, 12 mmol) and benzaldehyde (3.2 g, 30 mmol) in EtOH was heated at 60°C and 36% HCl (1 mL) was added slowly. On cooling, **14** crystallized. The filtered and dried product melted at 143°C.

7-Chloro-2-benzylidene-coumaran-3-one dibromide (15)
To a solution of **14** (5.0 g, 20 mmol) in CHCl₃ (10 mL) was added a solution of bromine (3.2 g, 20 mmol) in CHCl₃ (10 mL). After 24 h the solvent was removed at 20–25°C and the residue recrystallized from HOAc: **15**, mp 147°C.

8-Chlorotlavonol (16)
A solution of **15** (2.09, 5 mmol) in EtOH (150 mL) was treated with 0.1 N KOH (100 mL). The mixture was boiled for 10 min and the product was precipitated with water. Recrystallization from HOAc yielded 1 g of **16** (70%), mp 187°C.

7.1.7   References

1.   [R] Wawzonek, S. In *Heterocyclic Compounds* **1951**, *2*, 229–246.
2.   v. Auwers, K.; Müller, K. *Ber. Dtsch. Chem. Ges.* **1908**, *41*, 4233.
3.   v. Auwers, K.; Pohl, P. *Ann.* **1914**, *405*, 243.
4.   v. Auwers, K.; Pohl, P. *Ber. Dtsch. Chem. Ges.* **1915**, *48*, 85.
5.   v. Auwers, K. *Ber. Dtsch. Chem. Ges.* **1916**, *49*, 809.
6.   Dean, H. F.; Nierenstein, M. *J. Am. Chem. Soc.* **1925**, *47*, 1676.
7.   Kalff, J.; Robinson, R. *J. Chem. Soc.* **1925**, 1968.
8.   Minton, T. H.; Stephen, H. *J. Chem. Soc.* **1922**, *121*, 1598.
9.   Ingham, B. H.; Stephen, H.; Timpe, R. *J. Chem. Soc.* **1931**, 895.

Jie Jack Li

7.2   Bucherer–Bergs Reaction

7.2.1   Description

The formation of hydantoin (2) from carbonyl compound 1 with potassium cyanide and ammonium carbonate or from cyanohydrin 3 and ammonium carbonate is referred to as the Bucherer–Bergs reaction.[1,2]   It belongs to the category of multiple-component reactions (MCR).

7.2.2   Historical Perspective

In a German patent issued in 1929, Bergs described a synthesis of some 5-substituted hydantoins by treatment of aldehydes or ketones (1) with potassium cyanide, ammonium carbonate, and carbon dioxide under several atmospheres of pressure at 80°C.[3] In 1934, Bucherer et al. isolated a hydantoin derivative as a by-product in their preparation of cyanohydrin from cyclohexanone.[4–6]   They subsequently discovered that hydantoins could also be formed from the reaction of cyanohydrins (e.g. 3) and ammonium carbonate at room temperature or 60–70°C either in water or in benzene. The use of carbon dioxide under pressure was not necessary for the reaction to take place. Bucherer and Lieb later found that the reaction proceeded in 50% aqueous ethanol in excellent yields for ketones and good yields for aldehydes.[6]

7.2.3   Mechanism

The mechanism that Bucherer and Steiner proposed in 1934 has mostly withstood the test of time.[5] However, they erroneously suggested that 5-imino-oxazolidin-2-one **7** was transformed to **2** via a one-step intramolecular rearrangement directly. Ironically, it was not until 1980s when it was concluded that the conversion went through an isocyanate intermediate (e.g. **8**).[7-9] Thus the overall mechanism may be summarized as follows: Addition of ketone **1** with KCN gives rise to cyanohydrin **4**, which is followed by an S_N2 reaction with $(NH_4)_2CO_3$ to form aminonitrile **5**. Nucleophilic addition of **5** to carbon dioxide leads to cyano-carbamic acid **6**, which undergoes an intramolecular cyclization to 5-imino-oxazolidin-2-one **7**. Subsequently, **7** rearranges to hydantoin **2** via the intermediacy of isocyanate **8**.

7.2.4   Variations and Improvements

The first improvement of the Bucherer–Bergs reaction was the Bucherer–Lieb variation[6] using the diluted alcoholic solution as described at the end of section 7.2.2. The Bucherer–Lieb variation is possibly the most popular process for synthesizing hydantoins. Another notable variation is the Henze modification[10,11] using fusing acetamide as the solvent in place of water, benzene or 50% alcohol. Recently, ultrasound-promoted hydantoin synthesis has been reported to accelerate the reaction.[12,13]

Thiohydantoin **9** was obtained from the treatment of carbonyl **1** with carbon disulfide and ammonium cyanide in aqueous methanol.[14] The transformation could also be carried out step-wise, that is, treatment of **1** with ammonium cyanide to form aminonitrile **10** followed by reaction with carbon disulfide to produce thiohydantoin **9**. Alternatively, 5,5-disubstituted 4-thiohydantoins could be prepared by the reaction of ketones with ammonium monothiocarbamate and sodium cyanide.[15]

In at least one case, the standard Bucherer–Bergs conditions gave rise to oxazole rather hydantoin. Specifically, when 5-benzyloxy-pyridine-2-carbaldehyde (**11**) was treated with potassium cyanide, ammonium chloride, and ammonium carbonate in boiling ethanol/water, 5-amino-oxazol-2-ol **12** was obtained. Subsequent heating of oxazole **12** with acetic acid at reflux overnight then produced the Bucherer–Bergs product, hydantoin **13**.[16]

7.2.5 Synthetic Utility

Many hydantoins are endowed with significant pharmacological activities as highlighted by 5,5-diphenylhydantoin (Dilantin®), an anticonvulsant and antiepileptic discovered by Parke–Davis in 1940's. Despite the lapse of more than half a century, Dilantin® still plays an important role in modern medicine. Meanwhile, another anticonvulsant **15** was synthesized from 9,10-dimethoxy-1,3,4,6,7,11b-hexahydro-pyrido[2,1-*a*]isoquinolin-2-one (**14**) under the standard Bucherer–Lieb variation in a 2:1 water–ethanol solution (**15a:15b** = 8:1).[17]

In addition, Sarges *et al.* at Pfizer prepared spirohydantoins such as **17** by significantly modifying the standard Bucherer–Bergs conditions, which gave very low yields.[18,19] The best conditions were: use of 2 mol of KCN and 7 mol of (NH$_4$)$_2$CO$_3$ per mole of ketone (e.g. **16**), addition of 1 mole of NaHSO$_3$, use of formamide as the solvent (lower melting point than acetamide as employed in the Henze modification), and running the reaction at a maximum temperature of 50°C for 3 days. These conditions produced spirohydantoin **17** in 50% yield from ketone **16**.

The hydantoin moiety has been utilized as a biostere for the peptide linkage, transforming a peptide lead into an orally available drug candidate. Therefore, an Arg–Gly–Asp–Ser tetrapeptide (**18**) lead structure was modified to a non-peptide RGD mimetic as an orally active fibrinogen receptor antagonist **19**.[20,21]

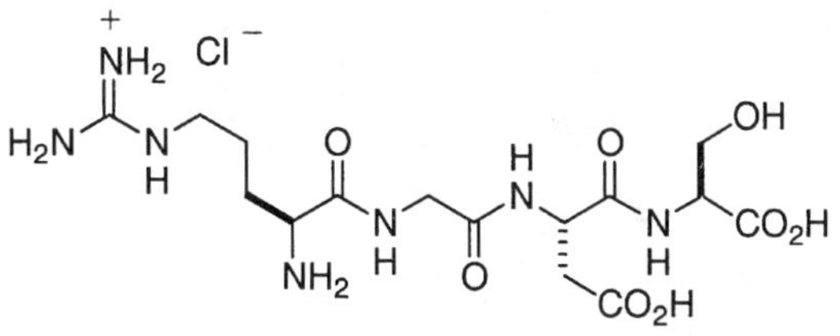

18

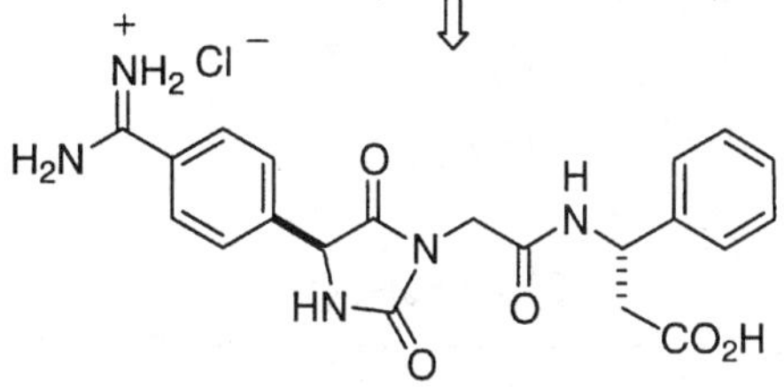

19

Notably, some substrates possess enough steric bias to exert sterospecificity for the Bucherer–Bergs reaction. For instance, ketone **21**, derived from enone **20** via a Corey–Chaykovsky reaction, underwent a Bucherer–Bergs reaction to fashion spirohydantoin **22** as a single isomer.[22]

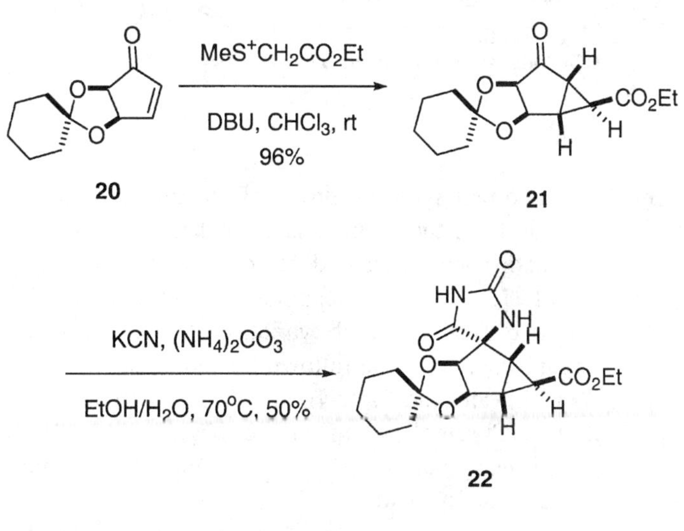

20

21

22

Furthermore, pharmacological active hydantoins **23–26** have been synthesized utilizing the Bucherer–Bergs conditions.[23–26]

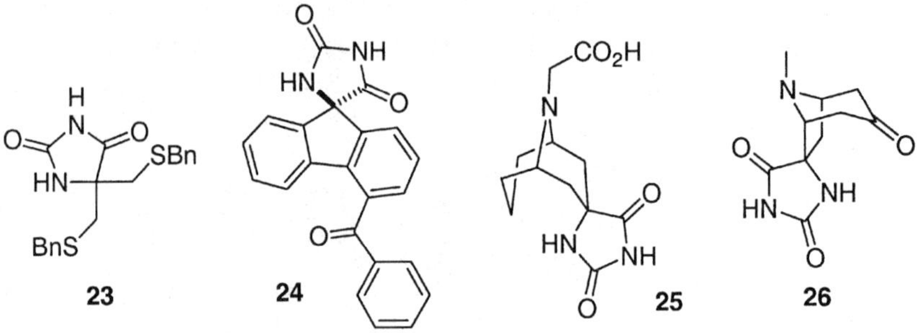

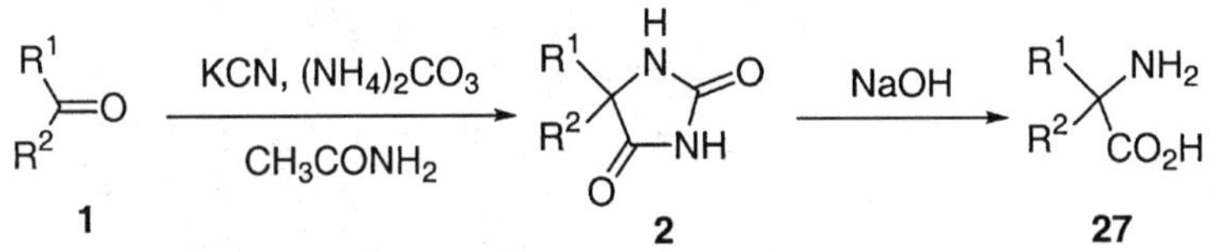

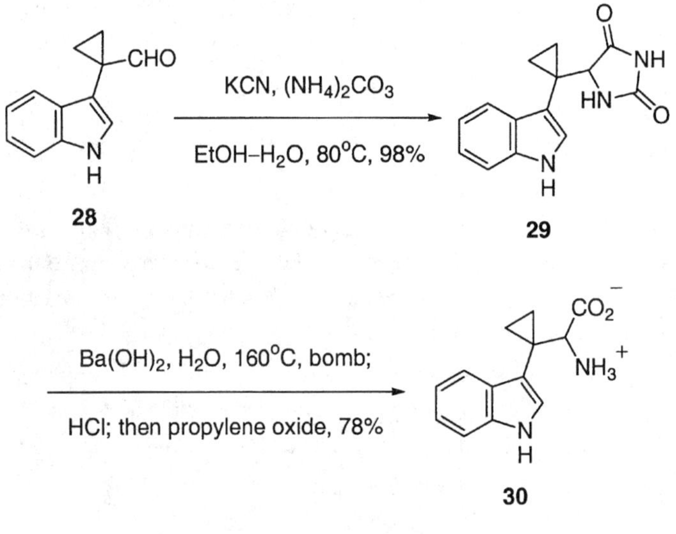

23   **24**   **25**   **26**

1   **2**   **27**

28   **29**

30

The other most important synthetic utility of the Bucherer–Bergs reaction is the preparation of amino acids from the hydrolysis of hydantoins. When carbonyl **1** was symmetrical, the Henze modification gave hydantoin **2**, which was then hydrolyzed to the corresponding amino acid **27**.[27] In another example, indolyl aldehyde **28** was converted to hydantoin **29** using the Bucherer–Lieb variation.[28] The difficult hydrolysis of hydantoin **29** was accomplished using the following conditions: the free amino acid was obtained by heating **29** with water and Ba(OH)$_2$ at 160°C in a bomb for 12 h. Converting the free amino acid to the HCl salt allowed for the easy separation of inorganic material. The resulting compound was then transformed to the zwitterionic species **30**, a conformationally constrained amino acid.

When a steric bias exists for the carbonyl substrate, a selectivity issue arises. There are cases where the substrates are sufficiently sterically biased so that the Bucherer–Bergs reaction occurs *specifically*. Using 3-*tert*-butyl-cyclohexanone (31) as an example, the Bucherer–Bergs reaction was followed by hydrolysis to deliver (R)-1-amino-3-*tert*-butyl-cyclohexanecarboxylic acid (32) exclusively.[29,30] Interestingly, the Strecker synthesis conditions transformed 28 to the enantiomer of 32, that is, (S)-1-amino-3-*tert*-butyl-cyclohexanecarboxylic acid (33), as a single stereoisomer. The Strecker synthesis was carried out by converting ketone 31 to the corresponding imine, which was treated with KCN to produce amino nitrile intermediate, which was then hydrolyzed to the final amino acid 33. Similar results have been observed with adamantan-2-one (34) as the substrate. The Bucherer–Lieb variation on 34 gave hydantoin 35 as a single stereoisomer. The difficult hydrolysis of hydantoin 35 was then carried out under high pressure and high temperature to afford the corresponding amino acid 36.[30]

At the other extreme of the stereoselectivity spectrum of the Bucherer–Bergs reaction, the steric bias is sometimes not powerful enough to exert any selectivity at all, as exemplified by the conversion of 37 → 38.[32–34] Amino acid 38 was produced as a 1:1 mixture of two diastereomers.

In most cases, however, many substrates give a mixture of stereoisomers with a certain degree of stereoselectivity. When ketone 39 was treated with potassium cyanide and ammonium carbonate in ethanol/water, a mixture of epimeric hydantoins 40 and 41 were isolated.[35] Similarly, the Bucherer–Bergs reaction of ketone 42 gave rise to a mixture of diastereomers **43** and **44** in a 87:13 ratio.[36] Finally, oxoproline **45** underwent the Bucherer–Bergs reaction to afford a mixture of diastereomers **46** and **47** in a 6:94 ratio. The major diastereomer **47** was further manipulated to deliver the natural product (–)-cucurbitine (**48**).[37]

In summary, the Bucherer–Bergs reaction converts aldehydes or ketones to the corresponding hydantoins. It is often carried out by treating the carbonyl compounds with potassium cyanide and ammonium carbonate in 50% aqueous ethanol. The resulting hydantoins, often of pharmacological importance, may also serve as the intermediates for amino acid synthesis.

7.2.6 Experimental

14 → KCN, (NH₄)₂CO₃, 48 h, 60°C, 83% → **15a** + **15b**

Spiro hydantoin 15[17]

Using the standard Bucherer–Lieb variation, a mixture of 9,10-dimethoxy-1,3,4,6,7,11b-hexahydro-pyrido[2,1-*a*]isoquinolin-2-one (**14**, 3 g, 12 mmol), potassium cyanide (1.17 g, 18 mmol) and ammonium carbonate (6.9 g) was dissolved in a 2:1 water–ethanol solution (45 mL). The reaction flask was sealed and heated for 48 h in an oven at 60°C. The cooled reaction mixture left a precipitate, which was filtered to yield 3.7 g (97%) of crude hydantoin **15**. Further analysis revealed that it was consisted of 83% of (±)-2*S*,11b*S* isomer **15a** and 9% of (±)-2*R*,11b*S* isomer **15b**.

7.2.7 References

1. [R] Ware, E. *Chem. Rev.* **1950**, *46*, 422.
2. [R] Wieland, H. *et al.*, in *Houben-Weyl's Methoden der organischen Chemie*, Vol. XI/2, p 371 (**1958**).
3. Bergs, H. DE 566,094, (**1929**); *C.A.*, **1933**, *27*, 1001.
4. Bucherer, H. T., Fischbeck, H. T. *J. Prakt. Chem.* **1934**, *140*, 69, 129, 151.
5. Mechanism, Bucherer, H. T., Steiner, W. *J. Prakt. Chem.* **1934**, *140*, 291.
6. Bucherer, H. T., Lieb, V. A. *J. Prakt. Chem.* **1934**, *141*, 5.
7. Mechanism: Rousset, A.; Laspéras, M.; Taillades, J.; Commeyras, A. *Tetrahedron* **1980**, *36*, 2649.
8. Mechanism: Bowness, W. G.; Howe, R.; Rao, B. S. *J. Chem. Soc., Perkin Trans. 1* **1983**, 2649.
9. Mechanism: Taillades, J.; Rousset, A.; Laspéras, M.; Commeyras, A. *Bull. Soc. Chim. Fr.* **1986**, 650.
10. Henze, H. R.; Long, L. M. *J. Org. Chem.* **1941**, *63*, 1936.
11. Henze, H. R.; Long, L. M. *J. Org. Chem.* **1941**, *63*, 1941.
12. Li, J.; Li, L.; Li, T.; Wang, J. *Indian J. Chem.* **1998**, *37B*, 298.
13. Thennarasu, S.; Perumal, P. T. *Indian J. Chem.* **2001**, *40B*, 1174.
14. Carrington, H. C. *J. Chem. Soc.* **1947**, 681.
15. Carrington, H. C.; Vasey, C. H.; Waring, W. S. *J. Chem. Soc.* **1959**, 396.
16. Herdeis, C.; Gebhard, R. *Heterocycles* **1986**, *24*, 1019.
17. Menéndez, J. C.; Díaz, M. P.; Bellver, C.; Söllhuber, M. M. *Eur. J. Med. Chem.* **1992**, *27*, 61.
18. Sarges, R.; Schnur, R. C.; Belletire, J. L.; Peterson, M. J. *J. Med. Chem.* **1988**, *31*, 230.
19. Sarges, R.; Goldstein, S. W.; Welch, W. M.; Swindell, A. C.; Siegel, T. W. *J. Med. Chem.* **1990**, *33*, 1859.
20. Stilz, H. U.; Beck, G.; Jablonka, B.; Just, M. *Bull. Soc. Chim. Belg.* **1996**, *105*, 711.
21. Stilz, H. U.; Guba, W.; Jablonka, B.; Just, M.; Klingler, O.; König, W.; Wehner, V.; Zoller, G. *J. Med. Chem.* **2001**, *44*, 1158.
22. Domínguez, C.; Ezquerra, A.; Prieto, L.; Espada, M.; Pedregal, C. *Tetrahedron: Asymmetry* **1997**, *8*, 511.
23. Comber, R. N.; Reynolds, R. C.; Friedrich, J. D.; Manguikian, R. A.; Buckheit, W. W., Jr.; Truss, J. W.; Shannon, W. M.; Secrist, J. A., III. *J. Med. Chem.* **1992**, *35*, 3567.
24. Bovy, P.; Lenaers, A.; Callaert, M.; Herickx, N.; Gillet, C.; Roba, J.; Dethy, J.-M.; Callaert-Deveen, B.; Janssens, M. *Eur. J. Med. Chem.* **1988**, *23*, 165.
25. Trigo, G. G.; Avendaño, C.; Ballesteros, P.; Sastre, A. *J. Heterocyclic Chem.* **1980**, *17*, 103.
26. Villacampa, M.; Martínez, M.; González-Trigo, G.; Söllhuber, M. M. *Heterocycles* **1992**, *34*, 1885.
27. Xiao, Z.; Timberlake, J. W. *J. Heterocyclic Chem.* **2000**, *37*, 773.

28. Horwell, D. C.; McKiernan, M. J.; Osborne, S. *Tetrahedron Lett.* **1998**, *39*, 8729.
29. Maki, Y.; Masugi, T. *Chem. Pharm. Bull.* **1973**, *21*, 685.
30. Maki, Y.; Masugi, T.; Ozeki, K. *Chem. Pharm. Bull.* **1973**, *21*, 2466.
31. Nagasawa, H. T.; Elberling, J. A.; Shirota, F. N. *J. Med. Chem.* **1973**, *16*, 823.
32. Collado, I.; Ezquerra, A.; Mazón, A.; Pedregal, C.; Yruretagoyena, B.; Kingston, A. E.; Tomlinson, R.; Wright, R. A.; Johnson, B. G.; Schoepp, D. D. *Bioorg. Med. Chem. Lett.* **1998**, *8*, 2849.
33. Ornstein, P.; Bleisch, T. J.; Arnold, M. B.; Wright, R. A.; Johnson, B. G.; Schoepp, D. D. *J. Med. Chem.* **1998**, *41*, 346.
34. Ornstein, P.; Bleisch, T. J.; Arnold, M. B.; Kennedy, J. H.; Wright, R. A.; Johnson, B. G.; Tizzano, J. P.; Helton, D. R.; Kallman, M. J.; Schoepp, D. D. *J. Med. Chem.* **1998**, *41*, 358.
35. Ezquerra, A.; Yruretagoyena, B.; Avendaño, C.; de la Cuesta, E.; González, R.; Prieto, L.; Pedregal, C.; Espada, M.; Prowse, W. *Tetrahedron* **1995**, *51*, 3271.
36. Monn, J. A.; Valli, M. J.; Massey, S. M.; Wright, R. A.; Salhoff, C. R.; Johnson, B. G.; Howe, T.; Alt, C. A.; Rhodes, G. A.; Robey, R. L.; Griffy, K. R.; Tizzano, J. P.; Kallman, M. J.; Helton, D. R.; Schoepp, D. D. *J. Med. Chem.* **1997**, *40*, 528.
37. Paik, S.; Kwak, H. S.; Park, T. H. *Bull. Korean Chem. Soc.* **2000**, *21*, 131.

Jie Jack Li

7.3   Cook–Heilbron 5-Amino-Thiazole Synthesis

7.3.1   Description

The Cook–Heilbron reaction involves the reaction of α-aminonitriles with salts and esters of dithioacids, carbon disulfide, carbon oxysulfide, and isothiocyanates under extremely mild conditions to form 5-aminothiazoles.

7.3.2   Historical Perspective

Prior to the 1947 report by Cook and Heilbron on their novel synthesis, 5-aminothiazoles were mostly unknown in the literature.   Previous syntheses included the Curtius degradation of ethyl thiazole-5-carboxylates which did not have general applicability; there was also difficultly in obtaining the necessary starting materials. During a study on penicillin,[1]   Cook   and   Heilbron   found   that   the   reaction   between   methyl dithiophenylacetate and ethyl aminocyanoacetate gave what was initially believed to be ethyl   phenylthionacetamidocyanoacetate **4**.   However   further   studies   proved   the compound to be 5-amino-4-carbethoxy-2-benzyl-thiazole **5**, which was basic.

Further investigation on this type of thiazole synthesis in subsequent years led to the preparation of 5-aminothiazoles in which the 2-position was varied through reaction of the aminonitrile with salts and esters of dithioacids, carbon disulfide, carbon oxysulfide, and isothiocyanates.

7.3.3 Mechanism

The mechanism of the Cook–Heilbron reaction between α-aminonitriles and dithioformic ester **6** proceeds via an acyclic intermediate **7**, as proven by its isolation in several cases. Nucleophilic attack of the amine function on the sulfur-bearing carbon leads to the elimination of hydrogen sulphide. Cyclization of the acyclic thiacetoamide results in a five membered ring which aromatises favourably to give 5-amino-2-benzylthiazole **8**.

7.3.4   Variations and Improvements

The reaction of α-aminonitriles and carbon disulphide was stated by Cook and Heilbron to give 5-amino-2-mercaptothiazoles; however, they later found that the same reaction with aminoacetonitrile was more complex.[2]   When aminoacetonitrile sulphate in ethanolic solution was treated with carbon disulphide, the dithiodicarbamate **9** was formed.   Benzylation was then carried out; treatment of the resulting ester **10** with phosphorus tribromide with subsequent loss of water gave 5-amino-2-benzylthiothiazole **11** in a quantitative fashion.   The rapid reaction was thought to be the first example of the formation of a 5-aminothiazole from an α-aminoamide.

The synthesis of 5-aminothiazoles via the reaction of isocyanate derivatives with aminomalononitrile p-toluenesulfonate (AMNT) has been investigated.[3]   It was found that AMNT **12** reacted with alkyl and aryl isothiocyanates in 1-methyl-2-pyrrolidine (NMP) to furnish 5-amino-2-(alkylamino)-4-cyanothiazoles (**13a**) and 5-amino-2-(arylamino)-4-cyanothiazoles (**13b–c**) in 44–81 % yields.[4–7]

13
13a Ar = C_4H_9 (44 %)
13b Ar = 4-$CH_3OC_6H_4$ (55 %)
13c Ar = C_6H_5 (66 %)
13d Ar = 4-ClC_6H_4 (58 %)
13e Ar = 4-$O_2NC_6H_4$ (81%)
13f Ar = 1-$C_{10}H_7$ (74 %)

14a Ar = C_6H_5
14b Ar = 4-$CH_3OC_6H_4$

These thiazoles are of specific interest in that they display exceptional pharmacological properties.[6,7] Additionally, the unsaturated 2-aminonitrile functionality of the above thiazoles is recognized for its versatile functionality and therefore for its ensuing significance in the synthesis of heterocycles.[6] The synthetic utility of thiazoles **13a–f** is illustrated by the reactions of the unsaturated 2-aminonitrile functionality in compounds **13b** and **13c** with formamidine acetate, resulting in the thiazolopyrimidines **14a** and **14c** respectively. The synthesis of this relatively rare family of heterocycles provides a route into structurally similar bioactive compounds.[8–11]

7.3.5 Synthetic Utility

7.3.5.1 With dithioacid

The synthesis of 5-amino-4-carbethoxy-2-benzylthiazole **17** via the reaction of ethyl aminocyanoacetate **15** with methyl dithiophenylacetate **16** provided the first general synthesis of the previously little known 5-aminothiazoles.[1] Similarly, the reaction between aminoacetonitrile **18** and sodium dithiophenylacetate **19** at room temperature gave 5-amino-2-benzylthiazole **20** in excellent yield.

Similar syntheses carried out with sodium dithioformate **21** instead of dithiophenylacetate gave 5-amino-4-carbethoxythiazole **22** in a facile reaction, where **22** is identical with the product obtained by heating the analogous 2-mercaptothiazole with Raney nickel.[12]

7.3.5.2 With carbon disulphide

Cook and Heilbron report the formation of highly crystalline Schiff bases via the reaction of 5-aminothiazoles and acetone, aldehydes such as cinnamaldehyde, or ketones such as acetophenone.[12] The reaction of α-aminobenzyl cyanide **23** with carbon disulphide **2** gave 2-mercaptothiazole **24** which was subsequently condensed with acetone to give the

Schiffs base **25**. It was also observed that the stability of 5-amino-2-mercaptothiazoles varied depending on the nature of the 4-position substituent.[2,13,14]

Several examples exist for the conversion of 5-aminothiazoles into the corresponding thiazolopyrimidines.[15]   Shaw and Butler[16] report the formation of aminothiazole thiocarboxyamide **27** from the thioamide **26** and carbon disulphide using Cook and Heilbron's procedure.[17] Methylation of **27** gave carboxythioimidate **28** which then reacted with sodium hydroxide to give amino-nitrile **29,** and with formic acid and acetic anhydride to give the thiazolopyrimidine **30**.

7.3.5.3 With carbon oxysulphide

The reaction of carbon oxysulphide with α-aminonitriles results in 5-amino-2-hydroxy thiazoles; these are structurally similar to the 2-mercaptothiazoles but are found to be less stable, readily undergoing cleavage or rearrangement to give 4-thiohydantoins.[18] Thus the reaction between ethyl aminocyanoacetate and carbon oxysulphide **31** in ether afforded 5-amino-2-hydroxy-4-carbethoxythiazole **32**, which in the presence of aqueous ammonia was converted into 5-carbethoxy-4-thiohydantoin **33**.   When using sodium carbonate in place of aqueous ammonia the monohydrated sodium salt of the thiohydantoin was isolated as an intermediate.

15   **31**   **32**   **33**

The synthesis of amino acids, amino mercapto acids and polypeptides has been made more facile with the use of 2-mercapto-5-thiazolone as a precursor.[2] However, it has been found that in order to synthesize these biologically important moieties, 2-mercapto-5-thiazolone does not undergo the necessary ring fissions or rearrangements as readily as desired. In order to improve on this activity the 2-hydroxy analogue was required and a route envisaged to its preparation. Accordingly, α-aminobenzylcyanide **34** and carbon oxysulphide were reacted in ethanol resulting in 5-amino-2-hydroxyphenylthiazole **35**. Similarly, reaction between ethyl aminocyanoacetate and ethereal carbon oxysulphide gave 5-amino-2-hydroxy-4-carbethoxythiazole **36**.

34   **35**

36

7.3.5.4 With isothiocyanate derivatives

The base-catalysed isomerisation of thiazoles to imidazoles initially reported by Cook *et al.*[19] has been further investigated more recently. One notable example reports the reaction of 2-amino-2-cyanoacetamide **37** with benzylisothiocyanate to give 5-aminothiozole **38**.[20] Base-catalysed isomerisation resulted in the corresponding imidazole **39**.

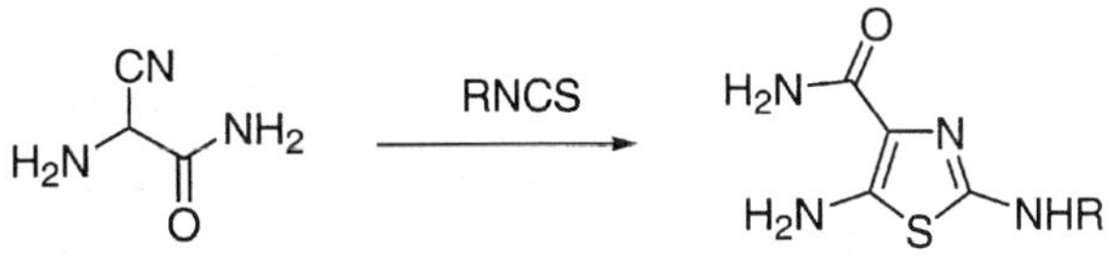

37 → **38**

38a R = C$_6$H$_5$CH$_2$
38b R = CH$_2$CHCH$_2$

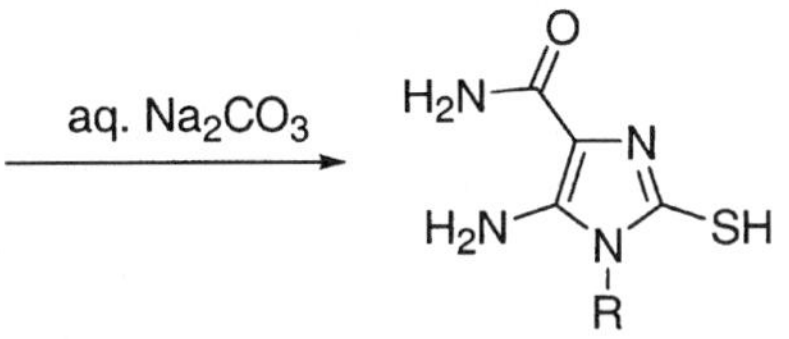

39

39a R = C$_6$H$_5$CH$_2$
39b R = CH$_2$CHCH$_2$

The formation of 5-benzamido-2-mercapto glyoxalines via the isomerisation of certain 5-amino-2-benzamidothiazoles in the presence of weak alkali has been reported.[21] Further to this, the reactions between α-aminonitriles and ethyl isothiocyanatoformate were investigated.[22] Therefore, α-aminopropionitrile **40** was reacted with ethyl isothiocyanatoformate **41** to give an unstable, colourless compound which isomerised on standing to give 5-aminothiazole **42**. Treatment of **42** with isocyanate **43** afforded the thiazole **44**. Subsequent heating of either **41** or **42** with aqueous sodium carbonate caused isomerization resulting in glyoxaline **45**. Elimination of the thiol group with Raney nickel then gave the glyoxoline **46**.

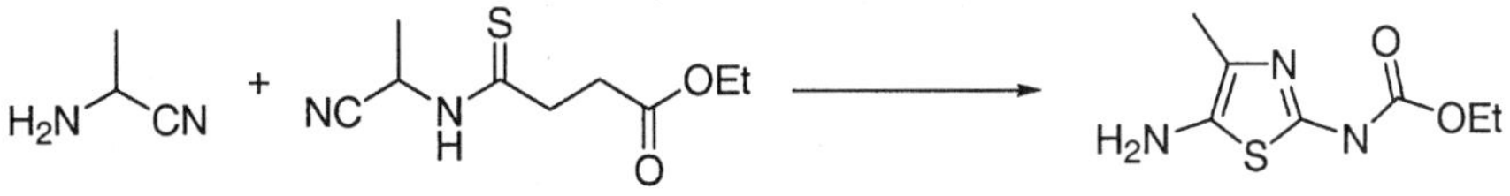

40   **41**   **42**

43

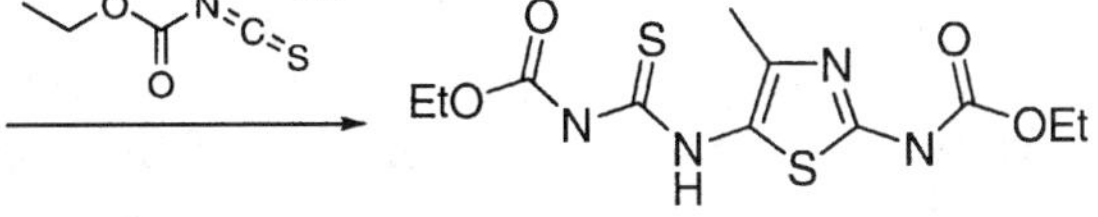

44

40 + **41** — Na$_2$CO$_3$ (aq) → **45** — Raney Ni → **46**

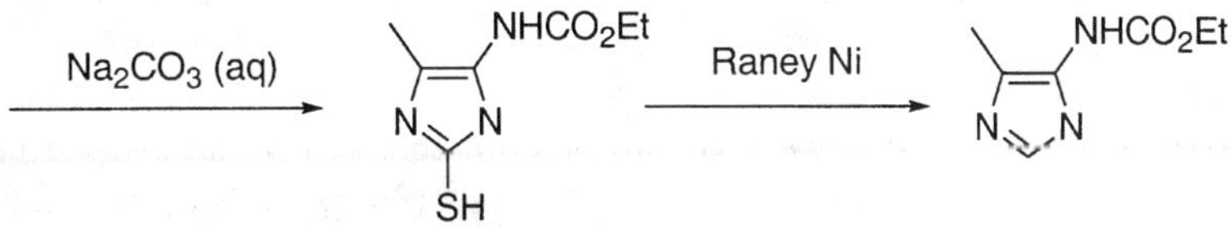

The series of reactions outlined above were found to differ depending on the nature of the nitrile group, particularly on the electronegative character of the group attached to the aminonitrile carbon atom. For example, thioureido compounds resulting from the reactions of carbethoxyisothiocyanate and amino aceto- or propionitriles isomerised rapidly into thiazoles but those from α-aminobenzylnitrile cyclized at a less favourable rate and using ethyl aminocyanoacetate did not yield any cyclic isomeride.[19]

To overcome the instability problems, the acyl group at the 2-position of the thiazole was replaced by a more electropositive group. Thus the reaction between α-aminonitriles and methyl isothiocyanate were explored.[19] Reaction of aminoacetonitrile with equimolar methyl isothiocyanate gave the thiazole 47. Hydrolysis with concentrated HCl gave 3-methyl-2-thiohydantoin 48. Reaction with dilute aqueous sodium carbonate gave 5-amino-2-mercapto-1-methylglyoxaline 49.

47

conc. HCl

48

Na$_2$CO$_3$ (aq)

49

7.3.6   Experimental

18

CS$_2$, EtOAc
0 °C

50

5-Amino-2-mercaptothiazole (50):[23]

A solution of sodium methoxide, prepared from sodium (23 g) and dry methanol (500 mL), was added drop-wise at 0 °C to a stirred suspension of aminoacetonitrile hydrochloride (18, 100 g, 1.08 mol) in dry methanol (100 mL). After stirring for 2 h at rt the precipitated sodium chloride was filtered off and the filtrate concentrated *in vacuo*. EtOAc (20 mL) was added and evaporated under reduced pressure to remove all traces of methanol. The oily residue was dissolved in dry EtOAc (100 mL) and anhydrous sodium sulfate added. After cooling, the precipitate was filtered off. The solution of crude aminoacetonitrile was used without further purification. This solution was added drop-wise during a period of 1 h to a vigorously stirred, ice-cooled solution of carbon disulphide (100 mL, 1.66 mol) in dry EtOAc (100 mL) under an N$_2$ atmosphere. Continued mechanical stirring and water-free conditions were essential. The mixture was stirred at 0 °C for 1 h. The resultant precipitate was filtered off, washed with Et$_2$O and dried, giving the product 50 as yellow crystals (99 g, 75 % on amount of sodium), m.p. 131 °C dec.; IR (KBr): *v* max 1630, 1500 cm^{-1}.

7.3.7   *References*

1.   Cook, A. H.; Heilbron, I.; Levy, A. L. *J. Chem. Soc.* **1947**, 1594.
2.   Cook, A. H.; Heilbron, I.; Levy, A. L. *J. Chem. Soc.* **1948**, 401.
3.   Freeman, F.; Kim, D. S. H. L. *J. Org. Chem.* **1991**, *56*, 4645.
4.   [R] (a) Hoffman, K. *Imidazole and Its Derivatives*; Interscience-Publishers: New York, **1953**; Part I, p 82.
        [R] (b) Hartman, G. D.; Sletzinger, M.; Weinstock, L. M. *J. Heterocycl. Chem.* **1975**, *12*, 1081.
5.   [R] Taylor, E. C.; McKillop, A. *The Chemistry of Cyclic Enaminonitriles and α-Aminonitriles*; Interscience: New York, **1970**.
6.   [R] Vernin, G. *Thiazole and Its Derivatives*; Metzger, J. V., Ed.; Wiley: New York, **1979**; Vol. 34, Part 1, p 289 and references therein.
7.   (a) Hennen, W. J.; Hinshaw, B. C.; Riley, T. A.; Wood, S. G.; Robins, R. K. *J. Org. Chem.* **1985**, *50*, 1741.
        (b) Pascual, A. *Helv. Chim. Acta* **1989**, *72*, 556; *Ibid.* **1991**, *74*, 531.
8.   El-Bayouki, K. A. M.; Basyouni, W. M. *Bull. Chem. Soc. Jpn.* **1988**, *61*, 3794.
9.   Ram, S.; Evans, W.; Wise, D. S., Jr.; Townsend, L. B.; McCall, J. W. *J. Heterocycl. Chem.* **1989**, *26*, 1053.
10.   Kanazawa, H.; Ichiba, M.; Tamura, Z.; Senga, K.; Kawai, K.; Otomasu, H. *Chem. Pharm. Bull.* **1987**, *35*, 35.
11.   Hurst, D. T.; Atcha, S.; Marshall, K. L. *Aust. J. Chem.* **1991**, *44*, 129.
12.   Cook, A. H.; Heilbron, I.; Levy, A. L. *J. Chem. Soc.* **1947**, 1598.
13.   Cook, A. H.; Fox, S. F. *J. Chem. Soc.* **1947**, 2337.
14.   Cook, A. H.; Heilbron, I.; Stern, E. S. *J. Chem. Soc.* **1948**, 2031.
15.   Sekiya, M.; Osaki, Y. *Chem. Pharm. Bull.* **1965**, *13*, 1319.
16.   Shaw, G.; Butler, D. N. *J. Chem. Soc.* **1959**, 4040.
17.   Cook, A. H.; Heilbron, I.; Smith, E. *J. Chem. Soc.* **1949**, 1440.
18.   Cook, A. H.; Heilbron, I.; Hunter, G. D. *J. Chem. Soc.* **1949**, 1443.
19.   Cook, A. H.; Downer, J. D.; Heilbron, I. *J. Chem. Soc.* **1948**, 2028.
20.   Sen, A. K.; Ray, S. *Indian J. Chem., Sect. B*, **1976**, *14*, 351.
21.   Cook, A. H.; Downer, J. D.; Heilbron, I. *J. Chem. Soc.* **1948**, 1262.
22.   Capp, C. W.; Cook, A. H.; Downer, J. D.; Heilbron, I. *J. Chem. Soc.* **1948**, 1340.
23.   Leysen, D. C.; Haemers, A.; Bollaert, W. *J. Heterocycl. Chem.* **1984**, *21*, 401.

Nadia M. Ahmad

7.4   Hurd–Mori 1,2,3-Thiadiazole Synthesis

7.4.1   Description

The Hurd–Mori 1,2,3-thiadiazole synthesis is the reaction of thionyl chloride with the N-acylated or tosylated hydrazone derivatives **1** to provide the 1,2,3-thiadiazole **4** in one simple step.[1-3]

In general, the reaction can be performed between 0–60°C with the majority of the reactions being run at room temperature. The reactivity of the hydrazones with either the acyl or tosyl leaving group with thionyl chloride depends on the substrate. However, the acylated hydrazones generally provide gaseous by-products where as the tosyl chloride reaction products have to be separated from the reaction mixture.

7.4.2   Historical Perspective

In 1955, Hurd and Mori first described the preparation of 1,2,3-thiadiazole as an unexpected product from the reaction of the hydrazone **5** and thionyl chloride.[4] The authors were attempting to prepare the six membered anhydride **7** in an analogous manner to the 5-membered anhydride **9**, prepared from **8** using thionyl chloride.[5] However, when the hydrazone **5** and thionyl chloride were mixed and heated at 60°C for 1 hour followed by cooling, the thiadiazole acid **6** precipitated out and was isolated by filtration. This serendipitous discovery led to a significant advance in the synthesis of thiadiazoles.

Following the discovery of this novel reaction, many publications have studied this reaction for the purpose of elucidating the mechanism, identifying substrates capable of undergoing the Hurd–Mori cyclization, and discovering novel uses for the thiadiazole heterocyclic systems. Many thiadiazole-containing analogs have been identified as anti-thrombotic agents,[6] antibacterial agents,[7-10] sedatives,[11] anti-inflammatory agents,[12] herbicides,[13] and, most recently, as plant activators or inducers of systemic acquired resistance (SAR) in plants.[14-16]

7.4.3 Mechanism

The mechanism of the Hurd–Mori reaction has been discussed extensively in the review by Stanetty.[3] The mechanism of the reaction was initially postulated by Hurd--Mori based on the isolation of intermediate **10**.[4] This intermediate was shown to transform into the desired thiadiazole upon heating in ethanol, either with or without acid. The reaction was thought to proceed via the four-membered intermediate **11**, which would release the volatile ethylformate as a by-product. In 1995, Kobori and co-workers were able to isolate and determine crystallographically a very similar intermediate structure to **10** in their mechanistic studies of the reaction.[17]

Hurd–Mori proposed mechanism:

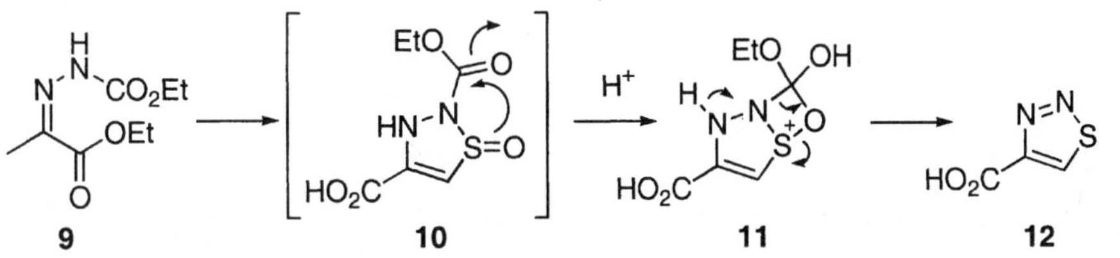

More elaborate mechanistic investigation on the reaction led to the discovery of a strong starting material substituent effect in the reaction pathway.[18] Depending on the substituent R, the interconversion of E and Z isomers can be quite facile,[19] which can result in the formation of reaction products from both isomers. The E-isomer **13** can cyclize to the intermediate **14** which would provide the thiadiazole directly or via intermediates **15** and **16**. The Z-isomer **13** cyclizes to the anhydride **18** which can ring open further and recyclize under the reaction conditions to provide the thiadiazole intermediate **20**, which then converts to **16** on the path to the product thiadiazole **17**.

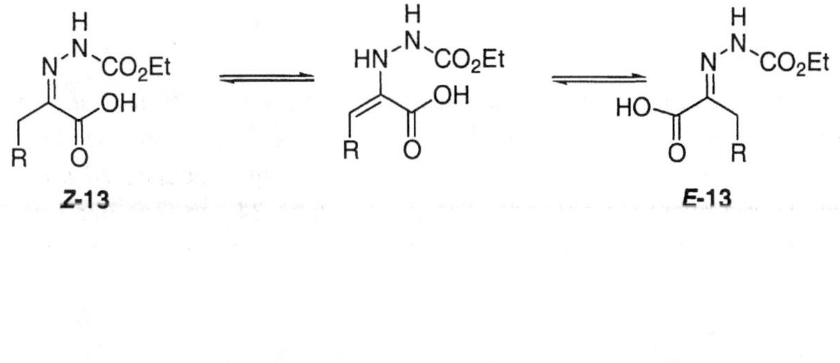

Proposed dual pathway mechanism:

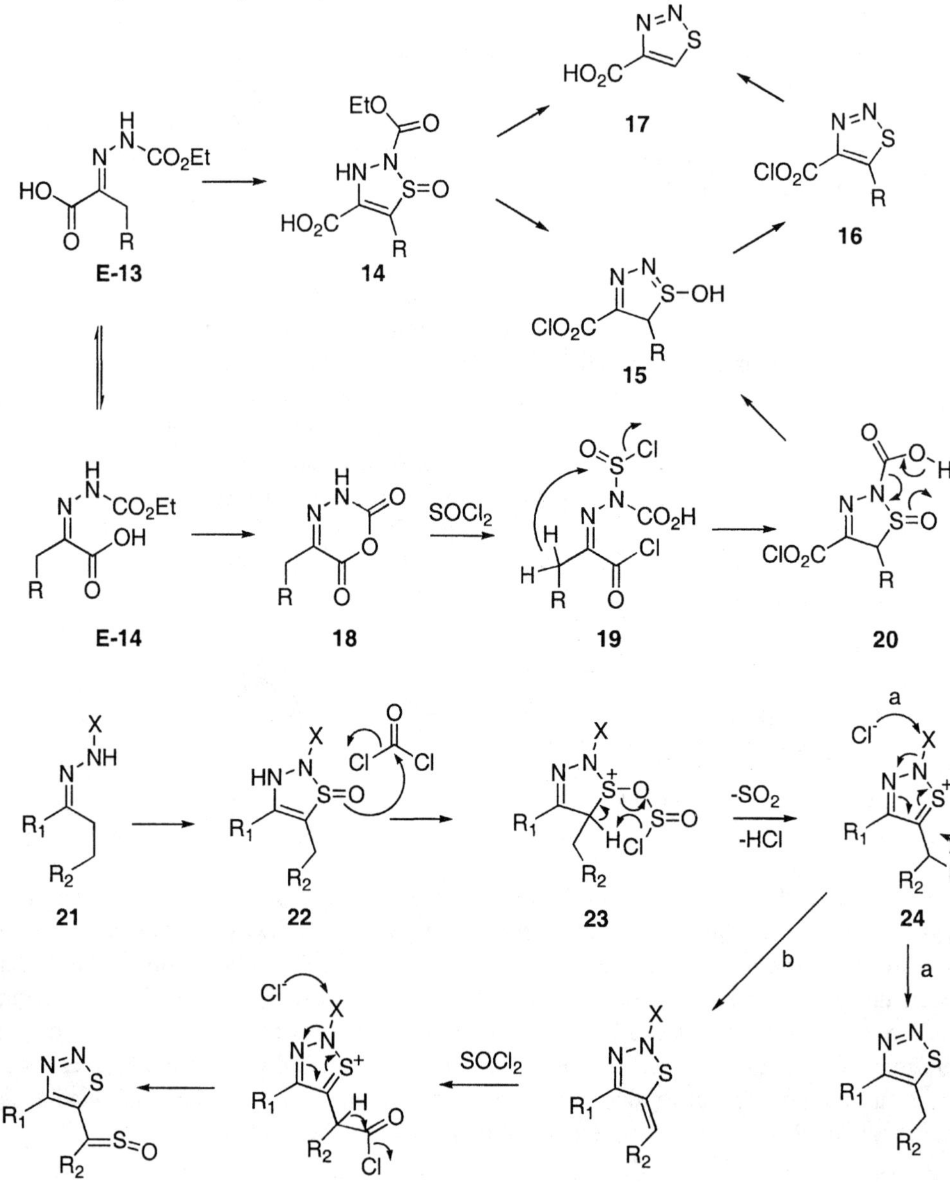

A third variation to the mechanism from the work of Britton *et al.* is explained in the above scheme.[20] The proposed mechanism is based on the Pummerer-type rearrangements of the intermediate **23** which gives the intermediate **24** after the loss of sulfur dioxide and hydrogen chloride. The intermediate **24** can then undergo loss of the leaving group **X** through route **a** from the attack of the chloride anion to give the thiadiazole **25**. Alternatively, loss of the proton from the methylene group via route **b** followed by further loss of the **X** leaving group can result in the intermediate **28**.

7.4.4 Variations and Improvements

The Hurd–Mori reaction is generally conducted at room temperature to 60°C with two to twenty-fold molar equivalents of thionyl chloride. The reaction can be run neat or in halogenated solvents. The commonly used leaving groups on the hydrazone are ethoxycarbonyl, aminocarbonyl or tosyl group. While both ethoxycarbonyl and aminoacarbonyl groups leave as gaseous products, simplifying workup and purification, the reactions with the tosyl group need to be purified by chromatography.

The thiadiazole ring cyclization occurs on the methylene moiety selectively when the hydrazone precursor has a choice between a methyl and a methylene group substituted with *n*-alkyl, CO_2Et, Ph, or chloride. However, depending on the substitution of the methylene group, selectivity can be shifted in favor of the methylene group or methyl group to give almost complete selectivity.[21] Subtituents that promote facile enolization of the methylene favors the cyclization on this group.

	33	**34**
R = SPh or SMe	>97	<3
R = t-Bu or N(CH₃)₂	0	100

7.4.5 Synthetic Utility

Preparation of thiadiazoles via the Hurd–Mori cyclization has led to the synthesis of a variety of biologically active and functionally useful compounds. Discussion of reactions prior to 1998 on the preparation of thiadiazoles have been compiled in a review by Stanetty *et al.*[3] Recent syntheses of thiadiazoles as intermediates for useful transformations to other heterocycles have appeared. For example, the thiadiazole intermediate **36** was prepared from the hydrazone **35** and converted to benzofuran upon treatment with base.[22] Similarly, the thiadiazole acid chloride **38** was converted to the hydrazine **39** which, upon base treatment, provided the pyrazolone, which can be sequentially alkylated *in situ* to provide the product **40**.[23]

Several thiadiazolo-triazoles **43** have been synthesized that show antifungal and cytotoxic properties.[24] Thiadiazoles **45** were prepared from hydrazones **44** by treating them neat with thionyl chloride at room temperatures. The thiadiazoles were formed regio-selectively on the methyl group of the hydrazone.[25]

A thiophene-annelated thiadiazole has been prepared from hydrazone **47**, which was obtained from the thiolactone **46**. Reaction of the hydrazone at room temperature with thionyl chloride resulted in an 8:1 mixture of **48** and **49**. Heating the reaction to 80°C in dichloroethane provided **48** exclusively.[26]

	48	49
CH$_2$Cl$_2$, 25°C	8	1
ClCH$_2$CH$_2$Cl, 80°C	1	0

The application of the Hurd–Mori reaction to the preparation of some potential fungicides from chiral hydrazones **50** and **52** shows that no racemization occurs under the reaction conditions.[27]

A general synthesis of phosphonyl thiadiazoles has been recently disclosed starting from the hydrazone **55**. The hydrazones were prepared from acyl phosphonates, which in turn were made from acid chlorides **54**. Thus treatment of the hydrazone **55** with thionyl chloride in the presence of DMF and sodium chloride provided the thiadiazoles in very high yields.[28] No explanation was provided for the use of DMF or sodium chloride.

High-speed synthesis of thiadiazoles has been recently completed on a solid support system using a "catch and release" technology to provide novel thiadiazoles. The solid-supported sulfonylhydrazine reacts with ketones to provide the solid phase hydrazones (catch) and formation of the thiadiazole with subsequent release of the thiadiazole upon cyclization provides the product without having to initiate cleavage of the product.[29]

7.4.6   Experimental

4-Methyl-5-[(1S,3R)-2,2-dimethyl-3cyanomethylcyclopropyl]-123-thiadiazole (51):[27]
A suspension of **50** (0.5 g) in dry dichloromethane (20 mL) was treated with thionyl chloride (20 equiv.) in one portion and the reaction mixture was stirred for 6 hr at room temperature. Water (20 mL) was added, the organic phase separated, dried with sodium sulfate and concentrated *in vacuo*. The oily residue was purified by chromatography on a silica gel column using hexane-ethyl acetate (1:2) as eluant to afford the thiadiazole **51** (0.23 g, 56%). Oil $[\alpha]^{30}_{578}$ +4.94 (c 0.0162, CHCl$_3$). ^{1}H NMR (CCl$_4$, acetone-d$_6$) δ 2.65 (s, 3H), 2.34 (dd, 1H), 2.20 (dd, 1H), 2.00 (d, 1H), 1.58 (ddd, 1H), 1.38 (s, 3H), 1.15 (s, 3H); MS, m/z (%) 208.09029 (M+H), 4.52). Calculated for C$_{10}$H$_{13}$N$_3$S+H. 208.090.

7.4.7   References

1   [R] Meier, H.; Hanold, N. In *Methoden der Organischen Chemie*: Houben-Weyl, Georg Thieme: Stuttgart – New York, **1994**; Vol. E8d, p.60.
2   [R] Thomas, E. W. In *Comprehensive Heterocyclic Chemistry;* Potts, K. T., Vol. Ed.; Katrizky, A. R.; Rees, C. W.; Series Eds.; Pergamon Press: London, **1984**; Vol. 6, Part 4B, p. 447. Thomas, E. W. In *Comprehensive Heterocyclic Chemistry II*; Storr, R. C., Vol. Ed.; Katrizky, A. R.; Rees, C. W.; Scriven, E. F.; Series Eds.; Pergamon Press: London, **1996**; Vol. 4, p. 289.
3   [R] Stanetty, P.; Turner, M.; Mihovilovic, M. D. In *Targets in Heterocyclic Systems* **1999**, *3*, 265-299.
4   Hurd, C. D.; Mori, R. I. *J. Am. Chem. Soc.* **1955**, *77*, 5359.
5   Leuchs, H. *Chem. Ber.* **1906**, *39*, 857.
6   Thomas, E. W.; Nishizawa, E. E.; Zimmermann, D. C.; Williams, D. J. *J. Med. Chem.* **1985**, *28*, 442.
7   Lewis, G. S.; Nelson, P. H. *J. Med. Chem.* **1979**, *22*, 1214.
8   Lalezari, I.; Shafiee, A.; Yazdany, S. *J. Pharm. Sci.* **1974**, *63*, 628.

9   Aoki, M.; Kumamoto, Y.; Hirose, T.; Okayama, S.; Sakai, S.; Enatu, A. *Chemotherapy (Tokyo)* **1986**, *34*, 394.

10   Maejima, T.; Inoue, M.; Mitsuhashi, S. *Antimicrob. Agents and Chemother.* **1991**, *35*, 104.

11   Ramsby, S. I.; Ogren, S. O.; Ross, S. B.; Stjernstrom, N. E. *Acta Pharm. Suecica* **1973**, *10*, 285.

12   Lau, C. K. In *U.S. 5677318 A.*; (Merck Frosst Canada, Inc., Can.). US, 1997, 13 pp.

13   Hanasaki, Y.; Tsukuda, K.; Watanabe, H.; Tsuzuki, K.; Murakami, M.; Niimi, N. (Tosoh Corp., Japan). *Eur. Pat. Appl.* EP 414511 A1 (1991),   64 pp..

14   Kunz, W.; Schurter, R. In *Eur. Pat. Appl.* EP 420803 A2.; (Ciba-Geigy A.-G., Switz.). Ep, 1991, 44 pp.

15   Kunz, W.; Jau, B. In *Eur. Pat. Appl.* EP 780372 A2; (Novartis Ag, Switz.). Ep, 1997, 28 pp.

16   Stanetty, P.; Kremslehner, M.; Jaksits, M. *Pest. Sci.* **1998**, *54*, 316-319.

17   Fujita, M.; Nimura, K.; Kobori, T.; Hiyama, T.; Kondo, K. *Heterocycles* **1995**, *41*, 2413.

18   Peet, N. P.; Sunder, S. *J. Heterocycl. Chem.* **1975**, *12*, 1191.

19   Zimmer, O.; Meier, H. *Chem. Ber.* **1981**, *114*, 2938.

20   Britton, T. C.; Lobl, T. J.; Chidester, C. G. *J. Org. Chem.* **1984**, *49*, 4773.

21   Fujita, M.; Kobori, T.; Hiyama, T.; Kondo, K. *Heterocycles* **1993**, *36*, 33.

22   D'Hooge, B.; Smeets, S.; Toppet, S.; Dehaen, W. *Chem. Commun. (Cambridge)* **1997**, 1753; Abramov, M. A.; Dehaen, W. *Synthesis* **2000**, 1529.

23   Hameurlaine, A.; Abramov, M. A.; Dehaen, W. *Tetrahedron Lett.* **2002**, *43*, 1015.

24   Shafiee, A. *J. Heterocycl. Chem.* **1976**, *13*, 301-4; Jalilian, A. R.; Sattari, S.; Bineshmarvasti, M.; Shafiee, A.; Daneshtalab, M. *Arch. der Pharma. (Weinheim, Germany)* **2000**, *333*, 347.

25   Attanasi, O. A.; De Crescentini, L.; Filippone, P.; Mantellini, F. *Synlett* **2001**, 557.

26   Stanetty, P.; Kremslehner, M.; Vollenkle, H. *J. Chem. Soc., Perkin Trans. 1* **1998**, 853-856; Stanetty, P.; Kremslehner, M.; Jaksits, M. *Pest. Sci.* **1998**, *54*, 316.

27   Morzherin, Y. Y.; Glukhareva, T. V.; Mokrushin, V. S.; Tkachev, A. V.; Bakulev, V. A. *Heterocycl. Commun.* **2001**, *7*, 173.

28   Chen, H.; Wang, W.-H.; Xue, M.; Cao, R.-Z.; Liu, L.-Z. *Heteroat. Chem.* **2000**, *11*, 413.

29   Hu, Y.; Baudart, S.; Porco, J. A., Jr. *J. Org. Chem.* **1999**, *64*, 1049.

Subas M. Sakya

7.5   Knorr Pyrazole Synthesis

7.5.1   Description

The Knorr pyrazole synthesis is the reaction of hydrazine or substituted hydrazines with 1,3-dicarbonyl compounds to provide the pyrazole or pyrazolone ring systems.[1–4] Unsubstituted hydrazine in addition to alkyl-, aryl-, heteroaryl-, or acyl-substituted hydrazines undergo reactions with the 1,3-dicarbonyl compounds to give the pyrazole ring system. While a symmetrical dicarbonyl system gives a single pyrazole isomer, unsymmetrically-substituted dicarbonyls can give one or both isomers (**3** and **4**). β-Ketoesters react with hydrazines to provide the pyrazolone ring system **6**.

R = H, Alkyl, Aryl, Het-aryl, Acyl, etc.

The reaction is generally performed between 0 and 100 °C with the majority of the reactions being run at reflux. Polar protic solvents such as methanol, ethanol, iso-propanol, and water are commonly used as solvents. Addition of acid or use of acetic acid as solvent generally helps push sluggish reactions. The use of β-ketoesters as the dicarbonyl partner occasionally requires added base for cyclization to occur to form the pyrazolone. When using alkyl hydrazine salts, base may be required to deprotonate the hydrazine for the reaction to take place.

7.5.2   Historical Perspective

Knorr reported the first pyrazole derivative in 1883.[5] The reaction of phenyl hydrazine and ethylacetoacetate resulted in a novel structure identified in 1887 as 1-phenyl-3-methyl-5-pyrazolone **9**. His interest in antipyretic compounds led him to test these derivatives for antipyretic activity which led to the discovery of antipyrine **10**.[6] He introduced the name pyrazole for these compounds to denote that the nucleus was derived from the pyrrole by replacement of a carbon with a nitrogen.[7] He subsequently prepared many pyrazole analogs, particularly compounds derived from the readily available phenyl hydrazine. The unsubstituted pyrazole wasn't prepared until 1889 by decarboxylation of 1H-pyrazole-3,4,5-tricarboxylic acid.[8]

Over the years, the pyrazole ring system has been studied in detail due to it's important properties in photography, dyes, and as pharmaceutical agents. The pyrazole ring system is a common core for many pharmaceutically active compounds. Several important drugs with the pyrazole ring system have been introduced to the market recently. Although there are many ways to prepare the pyrazole ring, the condensation of the 1,3-dicarbonyl and its variation remains the most common and facile way to assemble this ring system.

7.5.3 Mechanism

The mechanism of the Knorr reaction has been studied in detail by several labs using low temperature NMR methods.[9–12] Using low temperature flow NMR, a reaction of the 1,1-diacetyl cyclopropane with hydrazine was studied by Salivanov and coworkers for the mechanism of the cyclopropane ring opening with nucleophiles.[12] They observed an immediate formation of the intermediate **A** at –70°C within 10 seconds followed sequentially by formation of **B** and **C**. Because the signal intensity for intermediate **B** was small and constant ("quasi stationary") over the period of the reaction, it suggested a complex combination of consecutive reactions with at least the 3 intermediates **A-C** before forming the product. Similar intermediates **A-C** were observed for unsymmetrical diketones reacted with hydrazines or mono N-alkyl hydrazines. In addition, key intermediates 3,5-dihydroxypyrazolidine **14**, **15**, and the hydrazone **16** have been isolated.[14]

R = pentafluorophenyl

14   **15**   **16**

During studies of the reaction of heteroaryl hydrazine[15] and perfluoroaryl hydrazine[16] with 1,1,1-trifluoropentane-2,4-dione, the enolization of the dione **18** was studied with ^{19}F NMR and showed greater enolization of the trifluoromethyl ketone versus the methyl ketone. Thus the initial reaction of the free amine of the hydrazine takes place with the ketone resulting in the observed 10:1 selectivity of the reaction for **19** and **20**. Dehydration of the alcohol **19** to the pyrazole required strongly acidic conditions.[15–17]

17   **18**   **19**   10 : 1   **20**

The reaction conditions (neutral, acidic or basic) do have an affect on the regioselectivity of the reaction. Acidic reaction conditions have also been shown to preferentially provide one regioisomer over basic conditions for reactions of aryl hydrazines.[18–20] Extensive studies with 2-perfluoroacylcycloalkanones and mono-substituted hydrazines were studied to determine the selectivity of various alkyl-, aryl-, and heteroaryl-substituted hydrazines.[20] Reactions of the aryl hydrazine **21** with the trifluoromethyl-substituted cycloalkanone **22** under neutral conditions (methanol, reflux) gave a mixture of isomers **23** and **24**, whereas the reaction of the pyridyl hydrazine **25** was shown to give exclusively **26**.

A recent paper by Singh *et al.* summarized the mechanism of the pyrazole formation via the Knorr reaction between diketones and monosubstituted hydrazines.[17] The diketone is in equilibrium with its enolate forms **28a** and **28b** and NMR studies have shown the carbonyl group to react faster than its enolate forms.[10,16] Computational studies were done to show that the product distribution ratio depended on the rates of dehydration of the 3,5-dihydroxy pyrazolidine intermediates of the two isomeric pathways for an unsymmetrical diketone **28**. The affect of the hydrazine substituent R on the dehydration of the dihydroxy intermediates **19** and **22** was studied using semi-empirical calculations.[17]

7.5.4 Variations and Improvements

Because of the potential for the formation of two product isomers from the reactions of hydrazines with unsymmetrical dicarbonyl compounds and the resulting difficulty in separations, many efforts have been reported in the literature to enhance the selectivity of the product formation. The strategy of masking one of the carbonyl groups which can be released *in situ* to provide a selective product has been highlighted extensively in a recent review.[3] For example, the reaction of 4,4-dimethoxybutan-1-one **35** with methyl hydrazine followed by deprotection and cyclization gave the isomer **37**.[21]

The use of diphenyl hydrazone **33** has been used in the synthesis of pyrazoles under modified conditions where the hydrazine is released *in situ*. Some reversal of regiochemistry is seen in the reaction with unsymmetrical dicarbonyls. With aryl hydrazine and diphenyl hydrazone, the ratio of **41** to **42** is 22 : 1 and 5 : 1, respectively.

For hydrazone: 5:1
For hydrazine: 22:1

A solventless synthesis of pyrazoles, a green chemistry approach, has been described where an equimolar amount of the diketone and the hydrazine are mixed in a mortar with a drop of sulfuric acid and ground up. After an appropriate length of time (~ 1 h) the product is purified to provide clean products. Even acyl pyrazoles **42** were obtained under the solvent-less reaction conditions in good yields.

R = Me 85%
R = Ph 95%

7.5.5 Synthetic Utility

The Knorr pyrazole synthesis has been extensively utilized in the preparation of a number of pyrazoles as metal chelators, photographic dyes, herbicides, and biologically active agents.[1–4] Recent applications of the Knorr pyrazole synthesis for preparation of novel and useful pyrazole intermediates and pharmaceutical agents will be highlighted.

The discovery of the potent and selective COX-2 inhibitor Celecoxib **47**, prepared using the Knorr pyrazole synthesis, was reported by Pharmacia in 1997.[19] The reaction was run under acidic conditions to provide the desired product in good yield and with good regioselectivity. The commercial route to the preparation of Celecoxib utilized the reaction of the aryl hydrazine with the diketone sodium salt, which was used without isolation, under acidic conditions to provide the desired product **47**. This preparation of the final pyrazole product is selective for the desired regioisomer **47** but still had to be separated from some 5% of the isomer by-products (**48** and **49**) via crystallization. An improvement of this process to obtain at least 98% pure product has been reported in a recent patent filing.[22] An extensive analysis of the reaction conditions (e.g. concentration of the substrates, anhydrous conditions, type of acid addition, and choice of solvent, etc.) and changes in the final workup has improved the regioselectivity and purity of the product. The successful introduction of Celecoxib has provided an impetus for the synthesis of additional pyrazoles as COX-2 inhibitors.[23,24]

Sanofi-Synthélabo researchers discovered pyrazole **53** and analogs to have potent Cannabinoid receptor-1 (CB-1) antagonist/inverse agonist activity and have progressed **53** into development for treatment of obesity and alcohol dependence.[25, 26] The synthesis of **53** was accomplished by heating the diketone sodium salt **51** with the aryl hydrazine hydrochloride in acetic acid to provide the intermediate **52**, which was further derivatized to make **53**. Since the discovery of **53** (SR-141716), additional efforts to make other pyrazoles as CB-1 antagonists have appeared.[27]

Carpino *et al.* recently disclosed the synthesis of the fused pyrazolinone-piperidine dipeptide **56** with potent growth hormone secretagogue activity. The synthesis of the intermediate pyrazolone was accomplished by reacting the ketoester **54** with methyl hydrazine in refluxing ethanol.[28]

Application of the Knorr pyrazole synthesis has also been demonstrated on solid support.[29,30] To prepare trisubstituted pyrazoles, the diketone was linked to the solid support to make **57** using a linker with an amide bond. Alkylation of the diketone followed by condensation of the hydrazine with the resulting diketone gave the desired pyrazoles as mixtures of isomers. Subsequent cleavage of the amide bond linker then provided the pyrazole amides **59**.[29]

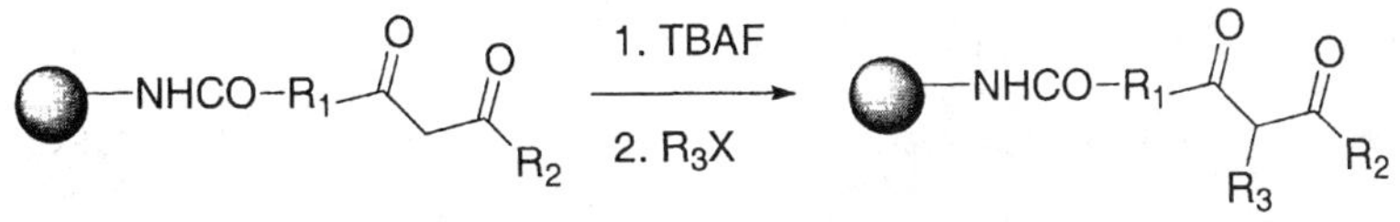

57 → 58

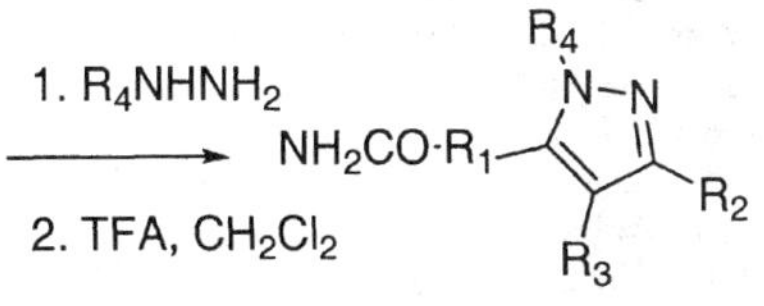

59

7.5.6 Experimental

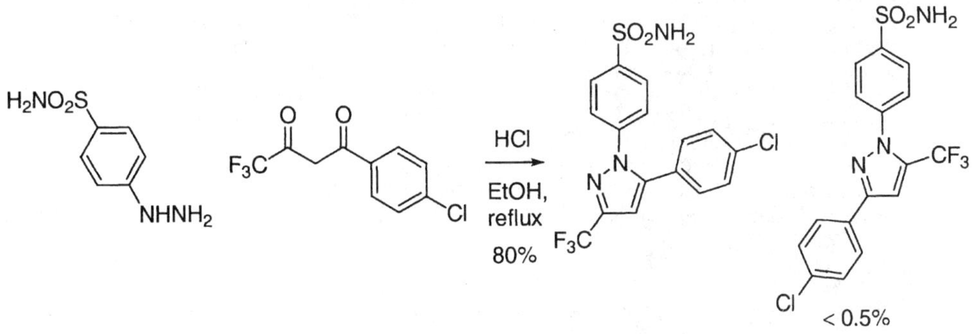

60   61   62   63

Preparation of 4-[5-(4-Chlorophenyl)-3-(trifluoromethyl)-1H-pyrazol-1-yl]benzenesulfonamide 62.[19]

(4-Sulfamoylphenyl)hydrazine hydrochloride (982 mg, 4.4 mmol) was added to a stirred solution of the dione 61 (1.00g, 4.0 mmol) in 50 mL of EtOH. The mixture was heated to reflux and stirred for 20 h. After cooling to room temperature, the reaction mixture was concentrated *in vacuo*. The residue was taken up in EtOAc, washed with water and brine, dried over $MgSO_4$, filtered, and concentrated in vacuo to give a light brown solid. Recrystallization from EtOAc and isooctane furnished pyrazole 62 (1.28 g, 80%): ^1H NMR ($CDCl_3/CD_3OD$) δ 5.2 (s, 2H), 6.8 (s, 1H), 7.16 (d, J = 8.5 Hz, 2H), 7.35 (d, J = 8.5 Hz, 2H), 7.44 (d, J = 8.7, 2H), 7.91 (d, J = 8.7, 2H); ^{13}C NMR ($CDCl_3/CD_3OD$) δ 106.42 (d, J = 0.03 Hz), 121.0 (q, J = 276 Hz), 125.5, 126.9, 127.3, 129.2, 130.1, 135.7, 141.5, 143.0, 143.9 (q, J = 37 Hz), 144.0; ^{19}F NMR ($CDCl_3/CD_3OD$) δ −62.9. HPLC analysis showed that the purified material contained ≤ 0.5% of the regioisomeric pyrazole 63 (4-[3-(4-chlorophenyl)-5-(trifluoromethyl)-1*H*-pyrazol-1-yl]benzenesulfonamide).

7.5.7 *References*

1.   [R] Jacobs, T. L., In *Heterocyclic Compounds*, Elderfield, R. C., Ed.; Wiley: New York, **1957**, *5*, 45.
2.   [R] Elguero, J., In *Comprehensive Heterocyclic Chemistry II*, Katrizky, A. R.; Rees, C. W.: Scriven, E. F. V., Eds; Elsevier: Oxford, **1996**, *3*, 1.
3.   [R] Stanovnik, E.; Svete, J. in *Science Of Synthesis*, **2002**, *12*, 15; Ed. by Neier,R.; Thieme Chemistry.
4.   [R] *Houben-Weyl*, **1967**, *10/2*, 539, 587, 589, 590.
5.   Knorr, L. *Ber Dtsch. Chem. Ges.* **1883**, *16*, 2597.
6.   Knorr, L. *Ber Dtsch. Chem. Ges.* **1884**, *17*, 546, 2032.
7.   Knorr, L. *Ber Dtsch. Chem. Ges.* **1885**, *18*, 311.
8.   Knorr, L., *Justus Liebigs Ann. Chem.* **1887**, *238*, 137.
9.   Selivanov, S.I.; Bogatkin, R.A.; Ershov. B.A. *Zh. Org. Khim.* **1981**, *17*, 886.
10.   Selivanov, S.I.; Bogatkin, R.A.; Ershov. B.A. *Zh. Org. Khim.* **1982**, *18*, 909.
11.   Selivanov, S.I.; Goldova, K.G.; Abbasov, Ya.A.; Ershov, B.A. *Zh. Org. Khim.* **1984, 20**, 1494.
12.   Selivanov, S.I.; Golodova, K.G.; Ershov. B.A. *Zh. Org.Khim.* **1986**, *22*, 2073.
13.   Zefirov, N. S.; Kozhushkov, S. I.; Kuznetsova, T. S. *Tetrahedron* **1996**, *42*, 709.
14.   Osadchii, S. A.; Barkhash, V. A. *Seriya Khimicheskaya* **1971**, *8*, 1825.
15.   Singh, S. P.; Kumar, D.; Jones, B. G.; Threadgill, M. D. *J. Fluorine Chem.* **1999**, *94*, 199.
16.   Song, L.; Zhu, S. *J. Fluorine Chem.* **2001**, *111*, 201.
17.   Singh, S. P.; Kumar, D.; Batra, H.; Naithani, R.; Rozas, I.; Elguero, J. *Can. J. Chem.* **2000**, *79*, 1109.
18.   Lyga, J. W.; Patera, R. M.; Plummer, M. J.; Halling, B. P.; Yuhas, D. A. *Pestic. Sci.* **1994**, *42*, 29.
19.   Penning, T. D.; Talley, J. J.; Bertenshaw, S. R.; Carter, Collins, P.W.; Docter, S.; Graneto, M.J.; Lee, L.F.; Malecha, J. W.; Miyashiro, J. M.; Rogers, R. S.; Rogier, D. J.; Yu, S. S.; Anderson, G. D.; Burton, E. G.; Cogburn, J. N.; Gregory, S. A.; Koboldt, C. M.; Perkins, W. E.; Seibert, K.; Veenhuizen, A. W.; Zhang, Y. Y.; and Isakson, P. C. *J. Med. Chem.* **1997**, *40*, 1347.
20.   Sevenard, D. V.; Khomutov, O. G.; Kodess, M. I.; Pashkevich, K. I.; Loop, I.; Lork, E.; and Röschenthaler G.-V. *Can. J. Chem.* **2001**, *79*, 183.
21.   Burness, D. M. *J. Org. Chem.* **1956**, *21*, 97.
22.   Letendre, L. J.; Mcghee, W. D.; Snoddy, C.; Klemm, G.; Gaud, H.T. *WO 03/099794 A1*, **2003**.
23.   Tsuji, K.; Nakamura, K.; Konishi, N. *Chem. Pharm. Bull.* **1997**, *45*, 987.
24.   De Leval, X.; Julemont, F.; Delarge, J.; Sanna, V.; Pirotte, B.; Dogne, J-M. *Expert Opin. Ther. Patents* **2002**, *12*, 969.
25.   Barth, F. *EP 0656354 A1*, **1995**.
26.   Barth, F.; Casellas, P.; Millan, J.; Oustric, D.; Rinaldi, M.; Sarran, M. *WO 97/21682*, **1997**.
27.   Stoit, A. R.; Lange, J. H. M.; Hartog, A. P.; Ronken, E.; Tipker, K.; Stuivenberg, H.H.; Dijksman, J. A. R.; Wals, H. C.; Kruse, C. G. *Chem. Pharm. Bull.* **2002**, *50*, 1109.
28.   Carpino, P. A.; Lefker, B. A.; Toler, S. M.; Pan, L. C.; Hadcock, J. R.; Cook, E. R.; DiBrino, J. N.; Campeta, A. M.; DeNinno, S. L.; Chidsey-Frink, K. L.; Hada, W. A.; Inthavongsay, J. ; Mangano, F.M.; Mullins, M. A.; Nickerson, D. F.; Ng, O. ; Pirie, C. M.; Ragan, J. A.; Rose, C. R.; Tess, D. A.; Wright, A. S.; Yu, L.; Zawistoski, M.P.; DaSilva-Jardine, P. A.; Wilson T. C.; Thompson, D. D. *Bioorg. Med. Chem.* **2003**,*11*, 581.
29.   Marzinzik, A. L.; Felder, E. R. *Tetrahedron Lett.* **1996**, *37*, 1003.
30.   Shen, D-M.; Shu, M.; Chapman, K. T. *Org. Lett.* **2000**, *2*, 2789.

Subas M. Sakya

PART 3 Six-Membered Heterocycles   301

Chapter 8. Pyridines   302

Pyridine is a polar, stable, relatively unreactive liquid (bp 115°C) with a characteristic strong penetrating odor that is unpleasant to most people. It is miscible with both water and organic solvents. Pyridine was first isolated, like pyrrole, from bone pyrolysates. Its name is derived from the Greek for fire (pyr) and the suffix "idine" used to designate aromatic bases. Pyridine is used as a solvent, in addition to many other uses including products such as pharmaceuticals, vitamins, food flavorings, paints, dyes, rubber products, adhesives, insecticides, and herbicides. Pyridine can also be formed from the breakdown of many natural materials in the environment.

Figure 8.1.1

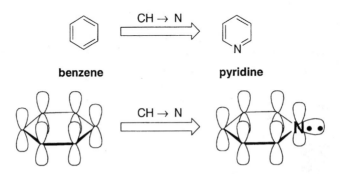

Pyridine is the prototypic electron-poor 6-membered ring heterocycle conceptually obtained by replacing one of the CH units of benzene with nitrogen (Figure 8.1.1). The aromaticity originally found in the benzene framework is maintained in

Figure 8.1.2

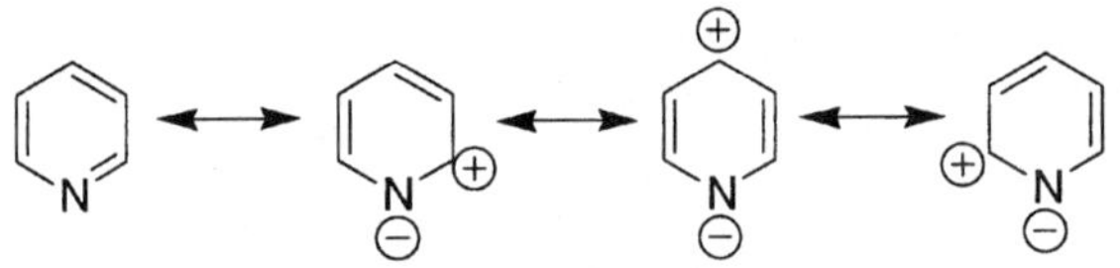

pyridine via overlap with the p orbital on the nitrogen atom that is parallel to the π-system. Compared to benzene, the resonance pictures (Figure 8.1.2), as well as, the natural atomic charges (calculated at RHF//6-31G**) of pyridine (Figure 8.1.3) predict its electron deficient nature and rationalize its net dipole.

Figure 8.1.3

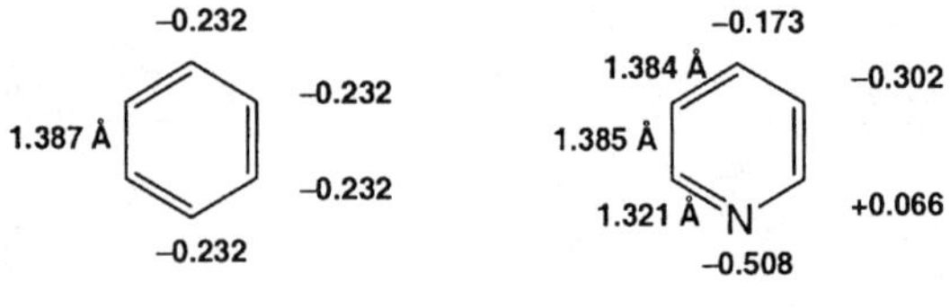

The consequences of this replacement gives pyridine a reduced susceptibility to electrophilic substitution compared to benzene, while being more susceptible to nucleophilic attack. An avenue of chemistry not possible with benzene, is the formation of pyridinium salts, by donation of the nitrogen lone pair electrons. The resultant salts are still aromatic, however much more polarized. This is reflected by the apparent acidity of the corresponding conjugate acid (pK_a = 5.2) compared to the acidity of the corresponding conjugate acid of piperidine (pK_a = 11.12). Synthetically, pyridine is prepared, commercially, by the gas phase, high temperature reaction of crotonaldehyde, formaldehyde, steam, air and ammonia over a silica-alumina catalyst in 60–70% yield. However, in the laboratory, the challenge is in the preparation of substituted pyridine derivatives in a fashion that allows one to control regioselectivity and chemoselectivity in the most efficient manner.

In taking a rational approach to this challenge, there are three possible retrosynthetic options (Figure 8.1.4). The classical approach towards heterocyclic systems is disconnection at the C–N bonds. This gives rise to ammonia and dicarbonyl systems, the reverse of a condensation reaction. A second strategy involves the disconnection through two C–C bonds giving rise to a 2-atom fragment and a 4-atom fragment, the reverse of a cycloaddition reaction. The final strategy is less obvious and could involve formation of the pyridine nucleus via some rearrangement of an alternative heterocyclic system or an appropriately substituted pyridine system.

Figure 8.1.4

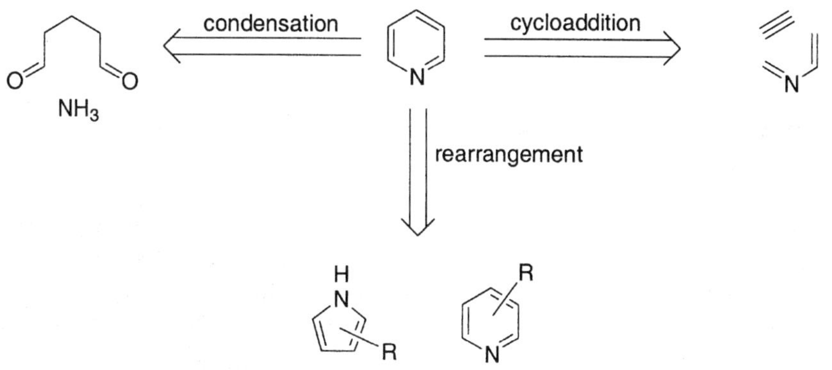

8.1. Preparation via Condensation Reactions

The various "name reactions" that have appeared in the literature over the years make use of a common retrosynthetic strategy, namely, disconnection at the C–N bond. Depending on how and at what time in the synthesis these bonds are formed dictates how the synthetic approach is classified, thus the "name" applied to the procedure (see Figure 8.1.5).

Figure 8.1.5

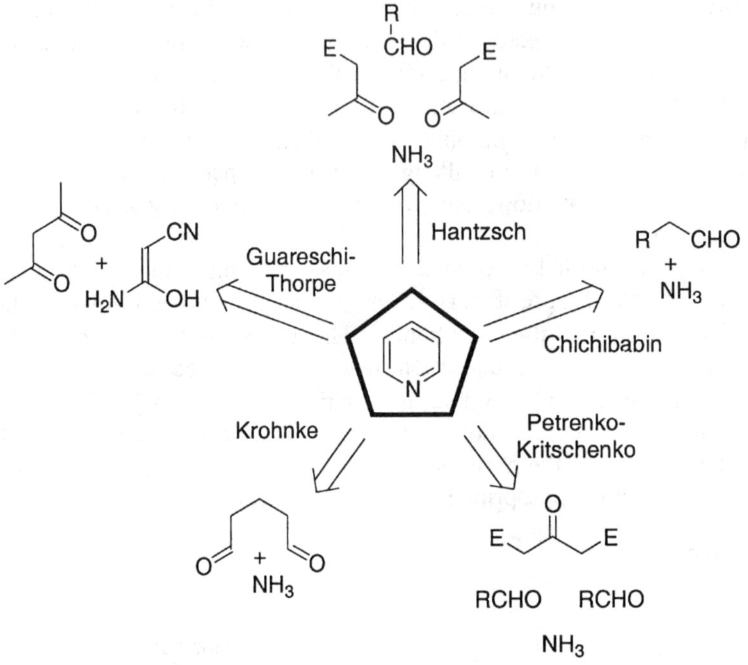

8.1.1 Hantzsch (Dihydro)-Pyridine Synthesis

8.1.1.1 Description

The Hantzsch pyridine synthesis[1] involves the condensation of two equivalents of a β-dicarbonyl compound, one equivalent of an aldehyde and one equivalent of ammonia. The immediate result from this three-component coupling, 1,4-dihydropyridine **1**, is easily oxidized to fully substituted pyridine **2**. Saponification and decarboxylation of the 3,5-ester substituents leads to 2,4,6-trisubstituted pyridine **3**.

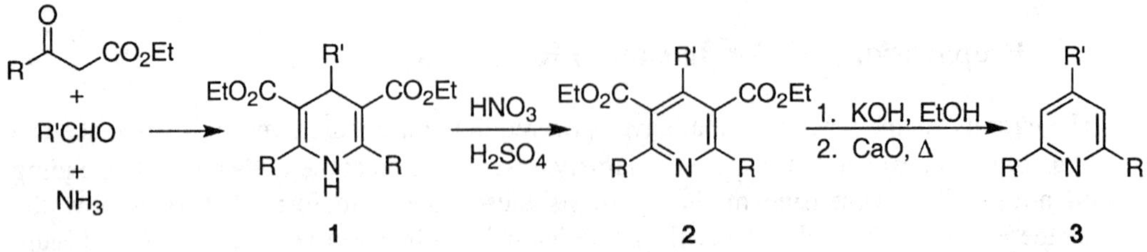

8.1.1.2 Historical perspective

While there are a number of related reactions that can assemble the pyridine nucleus, the oldest of these classical reactions is due to Arthur Hantzsch. In 1882[2] he reported the first synthesis of 1,4-dihydro-2,6-dimethylpyridine-3,5-dicarboxylates from a refluxing mixture of a β-keto ester, an aldehyde and ammonium hydroxide solution in ethanol. This chemistry later regained prominence, initially with the discovery in the 1930s of a "hydrogen-transferring" coenzyme. The structural determination of NADH, in the 1950s, indicated it contained a 1,4-dihydropyridine nucleus. The Hantzsch chemistry was used to construct model systems of NADH to understand the mechanistic details of this biological reducing agent. More recently, there is continued interest in the Hantzsch reaction as a result of the pharmacology of 1,4-dihydropyridines. In the 1970s, nifedipine **4** was introduced as a cardiovascular and antihypertensive agent. The calcium channel antagonistic pharmacology of these dihydropyridine derivatives led to a great interest in these compounds. Norvasc® (amlodipine) **5** is currently the most prescribed medicine for hypertension with expected sales in 2004 of over $4 billion and is currently the fifth largest selling drug worldwide. Throughout this timeframe, the role this reaction has played in the various approaches toward the synthesis of pyridine-containing natural products has been great.

4	**5**
nifedipine	**amlodipine (Norvasc®)**

8.1.1.3 Mechanism

While a number of pathways have been examined (see Scheme 8.1.1), consensus appears to have been reached from isolation of intermediates, as well as, ^{13}C- and ^{15}N-NMR spectroscopic studies.[3] It is now believed that one equivalent of β-dicarbonyl species **6** reacts with ammonia to form aminocrotonate **7** and the second equivalent of **6** undergoes an aldol reaction with the aldehyde to form chalcone **8**.[4] This was deduced from the following observations. A pathway in which ammonia ultimately reacts with both equivalents of **6** (either directly or via **6** + **7**) to form dienamine **12** was considered but rejected. Detailed studies[3a] indicated **12** was indeed being produced during the initial phase of the reaction but disappeared long before product appeared. It was concluded the formation of **12** was reversible and once formed was a metastable side-product that quickly reverted back to starting material. A more definitive conclusion[3a] was made utilizing preformed **7** and **8**, which when allowed to react together, resulted in the formation of **11**. Indeed, if **8** were exposed to aqueous ammonia, the formation of **11** was observed. This was rationalized by an ammonia catalyzed retro-aldol reaction to **6** followed by the typical forward reaction.

Scheme 8.1.1

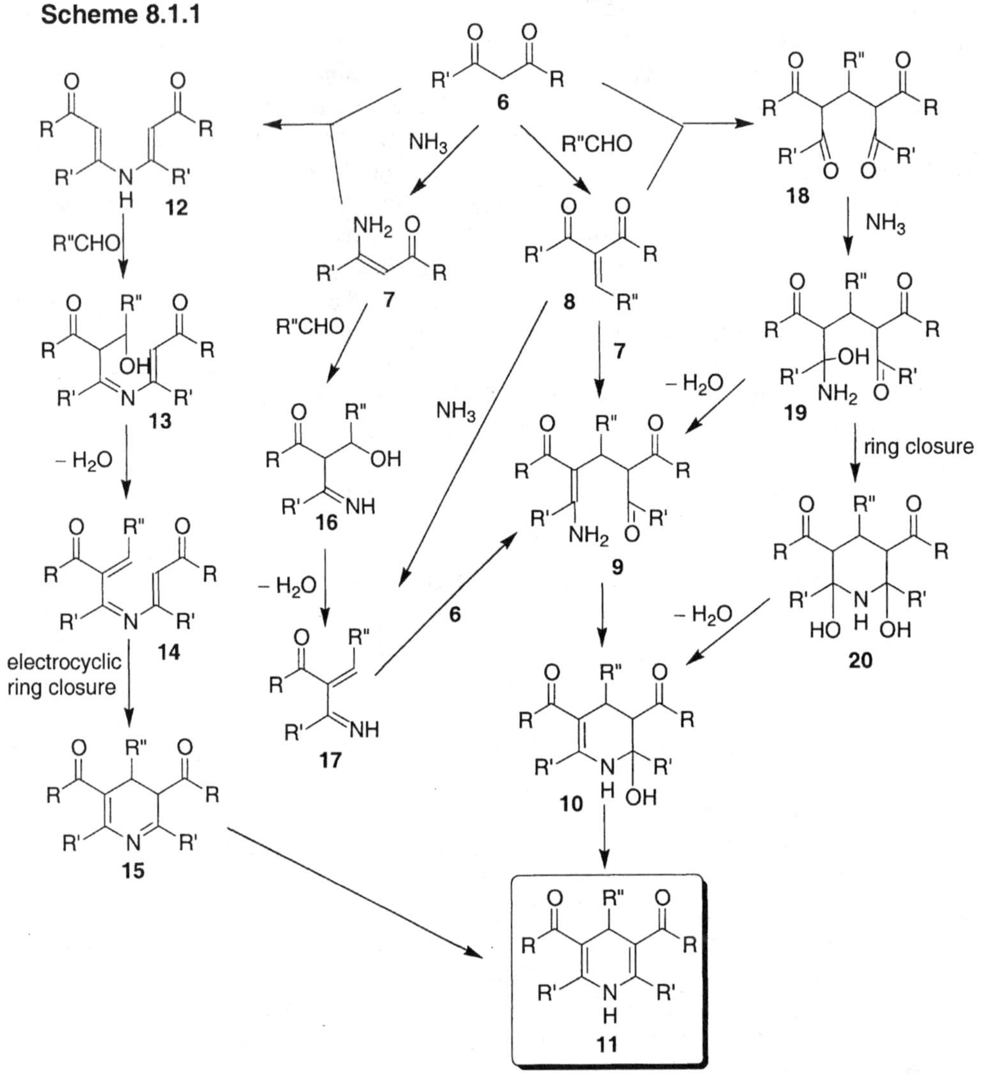

Once formed, **7** and **8** undergo a Michael reaction that gives rise to ketoenamine **9**. Ring closure, to form **10**, and loss of water then afforded 1,4-dihydropyridine **11**. The presence of **9** and **10** could not be detected; thus ring closure and dehydration were deduced to proceed faster than the Michael addition. This has the result of making the Michael addition the rate-determining step in this sequence. Conversely, if the reaction is run in the presence of a small amount of diethylamine, compounds related to **10** could be isolated.[4d] Diol **20** has been isolated in an unique case (R' = CF$_3$).[5] Attempts to dehydrate this compound under a variety of conditions were unsuccessful. Stereoelectronic effects related to the dehydration may be the cause. In related heterocyclic ring formations, it has been determined[6] that dehydration (**20** → **10**) is about 10^6 times slower than diol formation (**19** → **20**). Therefore, one would expect **20** to accumulate and this was not observed. Furthermore, **18** (R' = Me) was unstable and existed in the aldol initiated closed-form, that is, 3-hydroxycyclohexanone.[3b]

8.1.1.4 Variations

Subsequent to Hantzsch's communication for the construction of pyridine derivatives, a number of other groups have reported their efforts towards the synthesis of the pyridine heterocyclic framework. Initially, the protocol was modified by Beyer and later by Knoevenagel to allow preparation of unsymmetrical 1,4-dihydropyridines by condensation of an alkylidene or arylidene β-dicarbonyl compound with a β-amino-α,β-unsaturated carbonyl compound.[4a,b] Following these initial reports, additional modifications were communicated and since these other methods fall under the "condensation" approach, they will be presented as variations, although each of them has attained the status of "named reaction".

8.1.1.4.1 Guareschi–Thorpe pyridine synthesis

The Guareschi–Thorpe pyridine synthesis is closely related to the Hantzsch protocol.[7] The primary point of difference lies in the use of cyanoacetic esters. This modification assembles pyridine **23** by the condensation of acetoacetic esters **21** with cyanoacetic esters **22** in the presence of ammonia. A second variation[8] of this method involves reaction of cyanoacetic ester **22** with β-diketone **24** in the presence of ammonia to generate the 2-hydroxypyridine **25**.

Mechanistically,[9] one could envision a process initiated by ester/amide exchange brought about by the cyanoacetic ester with ammonia (**22 → 27**). Amide **27** could then undergo an aldol reaction with the β-diketone **24** to give **28** which then cyclizes to afford **26**.

Guareschi imides are useful synthetic intermediates. They are formed from a ketone reacting with two equivalents of the cyanoacetic esters and ammonia. This transformation is illustrated in the formation of 4,4-dimethylcyclopentenone **30**.[10] The synthesis was initiated with the Guareschi reaction of 3-pentanone **27** with **28** to generate imide **29**. This product was hydrolyzed to the diacid and esterified. Cyclization of the diester via acyloin condensation followed by hydrolysis and dehydration afforded the desired target **30**.

Guareschi imide

8.1.1.4.2 Chichibabin (Tschitschibabin) pyridine synthesis[11]

First described in 1905, the Chichibabin reaction was carried out by passing vapors of aliphatic aldehyde **31** and ammonia over alumina at 300–400°C to produce the corresponding pyridine derivative **32**. As a consequence, this method generates 2,3,5-trisubstituted pyridines.

From a mechanistic standpoint,[12] ammonia serves two functions: 1) it behaves as a base to catalyze an aldol reaction between 2 equivalents of **31** to generate the corresponding enal **33**, and 2) it is the source of nitrogen for the resultant pyridyl ring. This occurs through formation of enamine **34** with a third equivalent of **31**. The Michael addition of **34** to **33** followed by cyclization gives rise to **32**.

The close relationship of this reaction scheme with that for the Hantzsch protocol becomes obvious upon careful examination. The use of AcOH/NH$_4$OAc as the solvent/nitrogen source has dramatically improved on the experimental ease and yield of this reaction.[13]

8.1.1.4.3 Bohlmann–Rahtz pyridine synthesis

Bohlmann and Rahtz, in 1957,[14] reported the preparation of 2,3,6-trisubstituted pyridines. Their method employed the Michael addition of acetylenic ketones 35 with enamines 36. The δ-aminoketones 37 are typically isolated and subsequently heated at temperatures greater than 120°C to facilitate the cyclodehydration to afford 38. Again one can see the parallels in this mechanism with that for the Hantzsch protocol. However, in this case the pyridine is formed directly removing the need for the oxidation step in the Hantzsch procedure.

Recently the Bohlmann–Rahtz synthesis has received greater attention. Baldwin[15] has employed this method for the construction of heterocyclic substituted α-amino acids. Exposure of alkynyl ketone 39 to 3-aminocrotoyl ester 40 resulted in the Michael product 41. Thermolysis then gave rise to the desired pyridyl-β-alanines 42.

Moody's synthesis[16] of promothiocin **43** provided evidence that the Bohlmann–Rahtz method can be used for the rapid synthesis of complex pyridines. Oxazaole **44** was treated with alkynyl ketone **45** to afford **46** in 83% yield. The ester moiety of **46** was elaborated into a thiazole substituent providing entry into the northeast quadrant of **43**.

43

44   **45**   **46**

Moody and coworkers[17] have also applied this methodology to the preparation of a library of functionalized pyridine scaffold **49** with two points of diversity by coupling ynone **47** with enamine **48**.

47   **48**   **49**

Further improvements to this method have been reported by Bagley.[18] The requirement of harsh thermal conditions to facilitate the cyclodehydration can be minimized by simply adding acetic acid or Amberlyst 15.[18a] Alternatively, one could use Lewis acids[18b–d] such as $ZnBr_2$ or $Yb(OTf)_3$ in catalytic amounts.

8.1.1.4.4 Kröhnke pyridine synthesis[19]

Kröhnke observed that phenacylpyridinium betaines could be compared to β-diketones based on their structure and reactivity, in particular, their ability to undergo Michael additions. Since β-dicarbonyls are important components in the Hantzsch pyridine synthesis, application of these β-dicarbonyl surrogates in a synthetic route to pyridine was investigated. Kröhnke found that glacial acetic acid and ammonium acetate were the ideal conditions to promote the desired Michael addition. For example, N-phenacylpyridinium bromide **50** cleanly participates in a Michael addition with benzalacetophenone **51** to afford 2,4,6-triphenylpyridine **52** in 90% yield.

While this reaction to pyridine **56** occurs in a single pot, it is proposed to proceed via the 1,5-diketo derivative **55** obtained by a Michael addition of the pyridinium species **53** to enone **54**. Although one does not typically isolate this intermediate, it has been obtained in reactions of the isoquinolinium series.[19b]

A route to pyridines which involves an isolated 1,5-dicarbonyl compound, has been reported.[20] Aldol reaction of enone **57** with methylketone **58** generated 1,5-diketone **59**. When this was submitted to the reaction conditions for a Kröhnke reaction, thiopyridine **60** was isolated.

Newkome[21] employed this methodology in his synthesis of halogenated terpyridines. The requisite pyridinium ion **62** was prepared from the corresponding ketone **61** via a Ortoleva–King reaction.[22] Ketone **61** also served as the starting material for the enone component, in this case the Mannich base **63**. Exposure of **62** and **63** to acidic ammonium acetate generated terpyridine **64**.

Kelly[23] applied this chemistry to the synthesis of cyclosexipyridine **66**. This is an example of an intramolecular variation to this method. Masked enal **65** was prepared and treated with the standard reagents. The acidic medium liberated the aldehyde from its acetal protection. This *in situ* formation of the reactive species, similar to the above example, then undergoes cyclization to the expected pyridine derivative **66**.

A "green chemistry" variation[24] makes use of solventless conditions to minimize the waste stream from reactions of this type. To a mortar are added aldehyde **67**, ketone **68** and solid sodium hydroxide. The mixture is ground and within 5 minutes aldol product **69** is produced. Addition of the second ketone and further grinding affords the 1,5-diketone **70**, which can be isolated and cyclized to pyridine **71** with ammonium acetate. The authors report that this method can substantially reduce the solid waste (by over 29 times) and is about 600% more cost effective than previously published procedures.

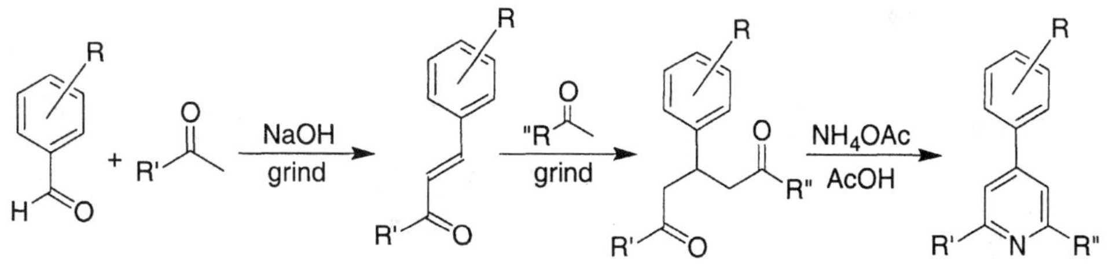

67 **68** **69** **70** **71**

8.1.1.4.5 Petrenko–Kritschenko piperidone synthesis[25]

While mechanistically this reaction is related to the Robinson-Schopf reaction for the generation of the tropinone skeleton, it also has similarities to the Hantzsch reaction. Here the heterocyclic ring **75** is assembled by the condensation of an equivalent of acetonedicarboxylic ester **72** with 2 equivalents of aldehyde **73** in the presence of ammonia or primary amine **74**.

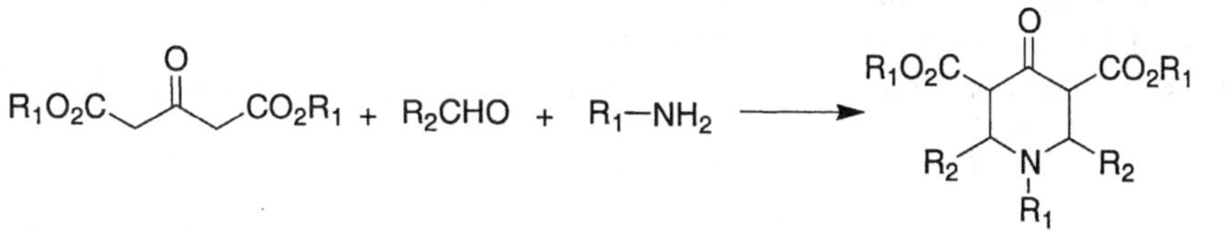

72 **73** **74** **75**

In an approach to opioid receptor ligands,[26] diazabicyclononanones were prepared in a double Petrenko–Kritschenko reaction. Diester **76**, in the presence of methylamine and aryl aldehydes, was converted to piperidone **77**. This was immediately resubmitted to the reaction conditions; however, in this iteration formaldehyde replaced the aryl aldehyde component. The outcome of this reaction produced **78** which was further investigated for its use in rheumatoid arthritis.

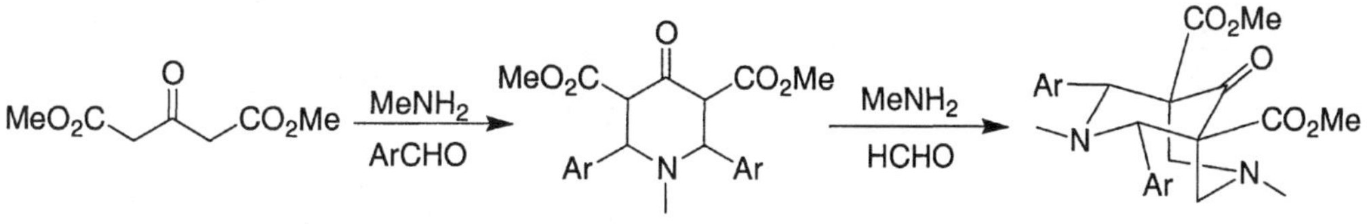

76 **77** **78**

Stevens[27] reported an interesting variation in which the amine and the aldehydes were linked by a tether. Combining bisacetalamine **79** with **76** under acidic conditions generated the single diastereomer **80**. This tricyclic species was then converted into the ladybug defense alkaloid precoccinelline **81**.

79   80   81

8.1.1.5 Improvements or modifications

The long history associated with the Hantzsch pyridine synthesis has produced numerous approaches and methods to the reaction protocol in order to control the various factors directing the course of the reaction.

It has been shown that TMSI is capable of mediating the reaction at room temperature.[28] The classical three component coupling was carried out using aldehyde **82** and ketoester **83** with ammonium acetate in acetonitrile at room temperature with *in situ* generated TMSI. This gave a 73–80% yield of 1,4-dihydropyridines **84** in 6–8 h. The best results were obtained with 1 equivalent of TMSCl and 1 equivalent of NaI.

82   83   84

The modified two component Hantzsch (Baeyer–Knoevenagel modification) was also examined. Shorter reactions times (2–3 h) were noted in this variation using **82** and **85** with slightly better yields (78–85%) being observed for the formation of **84**.

82   85   84

Indium trichloride has been found to catalyze the synthesis of tetrasubstituted pyridines.[29] Neat oxime **86** and acetoacetate **87** were exposed to indium trichloride at 150–160°C to produce pyridines **88** in 6–8 h with 55–80% yield. The indium catalyzes the Michael addition of **87** with **86**, as well as cyclization and reduction of the intermediate *N*-oxide.

86 **87** **88**

Combinatorial approaches have been applied to this chemistry.[30] In a method amenable to split and pool, PAL, or Rink resin, **89** is modified with an acetoacetate to generate the solid supported aminocrotonate **90**. Either a two- or three-component Hantzsch protocol is followed to produce **91**. Treatment with TFA carries out the cleavage from the resin and the cyclization to dihydropyridine **92**.

89 **90** **91** **92**

A limitation of this approach was the fact that the cyclization could not be accomplished on the resin. This would preclude further functionalization of the core. Therefore an alternate approach was to link the resin to the core via an aminoalcohol spacer as in **93**.[31] Furthermore, since linkage was conducted through the β-ketoester component rather than through the nitrogen atom, dihydropyridines **94** could now be formed on the solid support. When the 4-aryl substituent of **94** was nitro, on-resin reduction to the corresponding amine was possible. This allowed for further addition of diversity elements to the core scaffold before cleavage from the resin.

93 **94**

Microwave chemistry has been found to be a useful method for accelerating reactions or catalyzing reactions that are difficult to carry out by other methods. A modification of the Hantzsch method to directly obtain pyridines has been communicated.[32] A dry medium using ammonium nitrate / bentonitic clay system with microwave irradiation affords pyridines **96** in a single pot within 5 minutes. When the pyridine is not the major product (> 75% yield), the dealkylated pyridine **97** becomes an important side reaction. It was thought the microwave irradiation in conjunction with the acidic clay decomposes the ammonium nitrate into ammonia and nitric acid. These species then initiate the cyclization and *in situ* oxidation, respectively.

Variations on this theme have been reported.[33] One example utilized silica gel and urea with **95** and **82** under microwave irradiation to afford dihydropyridines **96** in 3–5 minutes and in 70–90% yield.

The previous methods used commercial microwave ovens. When a Smith Synthesizer was employed where one could control temperature and pressure, further improvements in time and yield were noted for the conversion of **95** and **82** into **96**.[34] Optimal conditions included the use of aqueous ammonium hydroxide as solvent and nitrogen source. The method was efficient enough to execute on a 4 × 6 array using the dicarbonyl and the aldehyde as points of diversity. The library of 24 compounds was obtained in 39–89% yields and 53–99% purity.

The Bohlmann–Rahtz reaction has also been improved by the use of microwave technology.[35] The optimum conditions involved irradiation of **97** with **98** in DMSO for 20 minutes at 170°C to generate pyridine **99** in 69% yield.

The immediate outcome of the Hantzsch synthesis is the dihydropyridine which requires a subsequent oxidation step to generate the pyridine core. Classically, this has been accomplished with nitric acid. Alternative reagents[1c] include oxygen, sodium nitrite, ferric nitrate/cupric nitrate, bromine/sodium acetate, chromium trioxide, sulfur, potassium permanganate, chloranil, DDQ, Pd/C and DBU. More recently, ceric ammonium nitrate (CAN) has been found to be an efficient reagent to carry out this transformation.[36] When **100** was treated with 2 equivalents of CAN in aqueous acetone, the reaction to **101** was complete in 10 minutes at room temperature and in excellent yield.

The oxidation has also been accomplished with Claycop (montmorillonite K-10 clay supported cupric nitrate).[37] The reaction of **96** to **102** was complete in 1.5–7 h with 81–93% yields. The time can be reduced to 5–10 minutes using ultrasound with minimal effect on yields. The major limitation of this protocol was the observation that only R = aryl gave product. Oxidation of 4-alkyl substituents was inert to these conditions with recovery of starting **96**.

An obvious outcome of the Hantzsch synthesis is the symmetrical nature of the dihydropyridines produced. A double protection strategy has been developed to address this issue.[38] The protected chalcone **103** was reacted with an orthogonally protected ketoester to generate dihydropyridine **104**. Selective deprotection of the ester at C3 could be accomplished and the resultant acid coupled with the appropriate amine. Iteration of this sequence with the C5 ester substituent ultimately gave rise to the unsymmetrical 1,4-dihydropyridine **105**.

103   **104**   **105**

Dihydropyridine Z0947 (**108**) has been identified as a potassium channel opener for use in urinary urge incontinence and an asymmetric synthesis was required for long-term manufacturing in support of clinical trials.[39] Unsaturated ketone **106** was added to a solution of **107** in acetonitrile containing TMSCl. The product was not isolated but treated with aqueous ammonia and ammonium chloride. After work-up, **108** was obtained in 95% *ee*.

106   **107**   **108**

Two- and three-component Hantzsch reactions using *C*-glycosylated reagents have been reported as an alternate method for conducting asymmetric syntheses of 1,4-dihydropyridines.[40] Reaction of **109**, **110** and **97** generate **111** with R_1 = sugar. Alternatively, **112** and **113** produce **111** with R_2 = sugar. While the yields were acceptable (60–90%), the diastereomeric ratio varied from 30–60%.

109   **97**   **111**   **112**   **113**

Pyridinium species comprise collagen cross-links that are formed during bone remodeling by osteoblasts. These cross-links are released into the serum and excreted in urine during bone resorption. Thus these compounds could be useful as clinical reference standards in the diagnosis of osteoporosis. A one-pot preparation of these pyridinium species was reported.[41] Alkylation of amine **115** with bromoketone **114** using potassium carbonate generates diketone **116** which subsequently undergoes an intramolecular aldol to close the ring and is rapidly aromatized with oxygen to afford the collagen cross-link **117**.

A unique method to generate the pyridine ring employed a transition metal-mediated 6-*endo-dig* cyclization of *N*-propargylamine derivative **120**.[42] The reaction proceeds in 5–12 h with yields of 22–74%. Gold (III) salts are required to catalyze the reaction, but copper salts are sufficient with reactive ketones. A proposed reaction mechanism involves activation of the alkyne by transition metal complexation. This lowers the activation energy for the enamine addition to the alkyne that generates **121**. The transition metal also behaves as a Lewis acid and facilitates formation of **120** from **118** and **119**. Subsequent aromatization of **121** affords pyridine **122**.

If one replaces one of the two equivalents of β-dicarbonyl with urea, such that the reaction is now carried out with one equivalent of aldehyde **123**, one equivalent of β-dicarbonyl **124** and an equivalent of urea **125** in acidic ethanol solution, then dihydropyrimidines **126** are formed. This class of reactions has been named Biginelli reactions[43] and are reviewed in section 10.6

8.1.1.6 Experimental
8.1.1.6.1 Three-component coupling[44]

To a solution of methyl 3-oxobutanoate **127** (580 mg, 5 mmol) and 1-methyl-2-methylthio-1*H*-imidazole-5-carboxaldehye **128** (390 mg, 2.5 mmol) in 5 mL of absolute methanol was added a solution of ammonium hydroxide (25%, 0.4 mL). The reaction was heated at reflux overnight before cooling to room temperature and removing the solvent. The crude product was purified by preparative TLC to afford 526 mg of dimethyl 1,4-dihydro-2,6-dimethyl-4-(1-methyl-2-methylthio-5-imidazolyl)-3,5-pyridine-dicarboxylate **129** (60%) as a solid, mp = 200–201 °C (MeOH).

8.1.1.6.2 Two-component coupling[45]

Equimolar amounts (0.5 mmol) of **130** and **95** in 5 mL of absolute methanol were heated at 100°C for 24 h. The reaction was cooled to room temperature and the solvent removed. The crude product was recrystallized from methanol to afford **131** in 98% yield (mp = 125–126°C).

Equimolar amounts (0.5 mmol) of **131** and chloranil in 2 mL of THF were heated at reflux for 4 h before cooling to room temperature and removing the solvent. Purification by preparative TLC gave **132** as an oil in 58% yield.

8.1.1.7 References

1. [R] (a) Eisner, U.; Kuthan, J. *Chem. Rev.* **1972**, *72*, 1–42
   [R] (b) Stout, D. M.; Meyers, A. I. *Chem. Rev.* **1982**, *82*, 223–243
   [R] (c) Sausins, A.; Duburs, G. *Heterocycles* **1988**, *27*, 291–314
   [R] (d) Lavilla, R. *J.C.S. Perkin Trans.* 1 **2002**, 1141–1156
2. Hantzsch, A. *Justus Liebieg's Ann. Chem.* **1882**, *215*, 1–83
3. (a) Katritzky, A. R.; Ostercamp, D. L.; Yousaf, T. I. *Tetrahedron* **1986**, *42*, 5729–5738
   (b) Katritzky, A. R.; Ostercamp, D. L.; Yousaf, T. I. *Tetrahedron* **1987**, *43*, 5171–5186
4. (a) Bottorff, E. M.; Jones, R. G.; Kornfeld, E. C.; Mann, M. J. *J. Am. Chem. Soc.* **1951**, *73*, 4380–4383
   (b) Berson, J. A.; Brown, E. *J. Am. Chem. Soc.* **1955**, *77*, 444–477
   (c) Marsi, K. L.; Torre, K. *J.Org. Chem.* **1964**, *29*, 3102–3103
   (d) Meyers, A. I.; Sircar, J. C.; Singh, S. *J. Heterocyclic Chem.* **1967**, *4*, 461–462
5. Singh, B.; Lesher, G. Y. *J. Heterocyclic Chem.* **1980**, *17*, 1109–1110
6. Cocivera, M.; Effio, A.; Chen, H. E.; Vaish, S. *J. Am. Chem. Soc.* **1976**, *98*, 7362–7366
7. (a) Baron, H.; Remfry, F. G. P.; Thorpe, J. F. *J. Chem. Soc.* **1904**, *85*, 1726–1761
   (b) Vogel, A. I. *J. Chem. Soc.* **1934**, 1758–1765
8. McElvain, S. M.; Lyle, R. E. Jr. *J. Am. Chem. Soc.* **1950**, *72*, 384–389
9. Collins, D. J.; Jones, A. M. *Aust. J. Chem.* **1989**, *42*, 215–221
10. Holder, R. W.; Daub, J. P.; Baker, W. E.; Gilbert, R. H. III; Graf, N. A. *J.Org. Chem.* **1982**, *47*, 1445–1451
11. Sprung, M. M. *Chem. Rev.* **1940**, *40*, 297–338
12. (a) Frank, R. L.; Seven, R. P. *J. Am. Chem. Soc.* **1949**, *71*, 2629–2635
    (b) Frank, R. L.; Riener, E. F. *J. Am. Chem. Soc.* **1950**, *72*, 4182–4183
    (c) Farley, C. P.; Eliel, E. L. *J. Am. Chem. Soc.* **1956**, *78*, 3477–3484
13. Weiss, M. *J. Am. Chem. Soc.* **1952**, *74*, 200–202
14. Bohlmann, F.; Rahtz, D. *Chem. Ber.* **1957**, *90*, 2265–2272
15. (a) Adlington, R. M.; Baldwin, J. E.; Catterick, D.; Pritchard, G. J.; Tang, L. T. *J. Chem. Soc. Perkin Trans.* 1 **2000**, 303–305
    (b) Adlington, R. M.; Baldwin, J. E.; Catterick, D.; Pritchard, G. J.; Tang, L. T. *J. Chem. Soc. Perkin Trans.* 1 **2000**, 2311–2316
16. Bagley, M. C.; Bashford, K. E.; Hesketh, C. L.; Moody, C. J. *J. Am. Chem. Soc.* **2000**, *122*, 3301–3313
17. Bashford, K. E.; Burton, M. B.; Cameron, S.; Cooper, A. L.; Hogg, R. D.; Dane, P. D.; McManus, D. A.; Matrunola, C. A.; Moody, C. J.; Robertson, A. A. B.; Warne, M. R. *Tetrahedron Lett.* **2003**, *44*, 1627–1629
18. (a) Bagley, M. C.; Dale, J. W.; Bower, J. *Synlett* **2001**, 1149–1151
    (b) Bagley, M. C.; Dale, J. W.; Hughes, D. D.; Ohnesorge, M.; Phillips, N. G.; Bower, J. *Synlett* **2001**, 1523–1526
    (c) Bagley, M. C.; Dale, J. W.; Bower, J. *Chem. Commun.* **2002**, 1682–1683
    (d) Bagley, M. C.; Dale, J. W.; Ohnesorge, M.; Xiong, X.; Bower, J. *J. Comb. Chem.* **2003**, *5*, 41–44
19. (a) Krohnke, F.; Zecher, W. *Angew. Chem. Intl. Ed.* **1962**, *1*, 626–632
    (b) Krohnke, F. *Synthesis* **1976**, 1–24
20. (a) Potts, K. T.; Cipullo, M. J.; Ralli, P.; Theodoridis, G. *J. Am. Chem. Soc.* **1981**, *103*, 3584–3585
    (b) Potts, K. T.; Cipullo, M. J.; Ralli, P.; Theodoridis, G. *J. Am. Chem. Soc.* **1981**, *103*, 3585–3586
21. Newkome, G. R.; Hager, D. C.; Kiefer, G. E. *J. Org. Chem.* **1986**, *51*, 850–853
22. King, C. *J. Am. Chem. Soc.* **1944**, *66*, 894–895
23 Kelly, T. R.; Lee, Y.–J.; Mears, R. J. *J. Org. Chem.* **1997**, *62*, 2774–2781
24. Cave, G. W. V.; Raston, C .L. *J. Chem. Soc. Perkin Trans.* 1 **2001**, 3258–3264
25. Petrenko–Kritschenko, P.; Zoneff, N. *Ber.* **1906**, *39*, 1358–1361
26. Siener, T.; Cambareri, A.; Kuhl, U.; Englberger, W.; Haurand, M.; Kogel, B.; Holzgrabe, U. *J. Med. Chem.* **2000**, *43*, 3746–3751
27. Stevens, R. V.; Lee, A. W. M. *J. Am. Chem. Soc.* **1979**, *101*, 7032–7035
28. Sabitha, G.; Reddy, G. S. K. K.; Reddy, Ch. S.; Yadav, J. S. *Tetrahedron Lett.* **2003**, *44*, 4129–4131
29. Saikai, P.; Prajapati, D.; Sandhu, J. S. *Tetrahedron Lett.* **2003**, *44*, 8725–8727
30. Gordeev, M. F.; Patel, D. V.; Gordon, E. M. *J. Org. Chem.* **1996**, *61*, 924–928
31. Breitenbucher, J. G.; Figliozzi, G. *Tetrahedron Lett.* **2000**, *41*, 4311–4315
32. Penieres, G.; Garcia, O.; Franco, K .; Hernandez, O.; Alvarez, C. *Heterecyclic Commun.* **1996**, *2*, 359–360
33. (a) Yadav, J. S.; Reddy, D. V. S.; Reddy, P. T. *Synthetic Commun.* **2001**, *31*, 425–430
    (b) Kidwai, M.; Saxena, S.; Mohan, R.; Venkataramanan, R. *J. Chem. Soc. Perkin Trans.* 1 **2002**, 1845–1846
34. Ohberg, L.; Westman, J. *Synlett* **2001**, 1296–1298

35.   Bagley, M. C.; Lunn, R.; Xiong, X. *Tetrahedron Lett.* **2002**, *43*, 8331–8334
36.   Pfister, J. R. *Synthesis* **1990**, 689–690
37.   Maquestiau, A.; Mayence, A.; Vanden Eynde, J.–J. *Tetrahedron Lett.* **1991**, *32*, 3839–3840
38.   Marzabadi, M. R.; Hong, X.; Gluchowski, C. *Tetrahedron Lett.* **1998**, *39*, 5293–5296
39.   Ashworth, I.; Hopes, P.; Levin, D.; Patel, I.; Salloo, R. *Tetrahedron Lett.* **2002**, *43*, 4931–4933
40.   (a) Dondoni, A.; Massi, A.; Minghini, E. *Synlett* **2001**, 89–92
      (b) Dondoni, A.; Massi, A.; Minghini, E.; Bertolasi, V. *Helv. Chim. Acta* **2002**, *85*, 33313348
41.   Adamczyk, M.; Johnson, D. D.; Reddy, R. E. *Tetrahedron: Asymmetry* **2000**, *11*, 2289–2298
42.   Abbiati, G.; Arcadi, A.; Bianchi, G.; DiGuuseppe, S.; Marinelli, F.; Rossi, E. *J. Org. Chem.* **2003**, *68*, 6959–6966
43.   Kappe, C. O. *Tetrahedron* **1993**, *49*, 6937–6963
44.   Foroumadi, A.; Analuie, N.; Rezvanipour, M.; Sepehri, G.; Najafipour, H.; Sepehri, H.; Javanmardi, K.; Esmaeeli, F. *Il Farmaco* **2002**, *57*, 195–199
45.   van Rhee, A. M.; Jiang, J.–L.; Melman, N.; Olah, M. E.; Stiles, G. L.; Jacobson, K. A. *J. Med. Chem.* **1996**, *39*, 2980–2989

Paul Galatsis

8.2 Preparation via Cycloaddition Reactions

A synthetically powerful method, an approach based on cycloaddition chemistry, allows one to assemble the pyridine ring in one step. Not only is this method efficient, "atom economy," but also its convergency allows for the preparation for highly substituted systems in which one can, in principle, control all five positions on the pyridine ring. A versatile example of this methodology is the Boger reaction. It has been applied to the synthesis of a very diverse set of targets.

8.2.1 Boger Reaction

8.2.1.1 Description[1]
The Boger pyridine synthesis involves the reaction of triazine 1 with activated alkene 2 in a hetero-Diels–Alder fashion. The intermediate bicyclic species 3 is unstable and a facile cycloreversion takes place due to the loss of nitrogen gas to afford the appropriately substituted pyridine derivative 4.

| 1 | 2 | 3 | 4 |

8.2.1.2 Historical perspective
Prior to the delineation of the concept of "conservation of orbital symmetry" by Woodward and Hoffmann,[2] Bachmann and Deno[3] reported that all Diels–Alder reactions

| 5 | 6 |

are not the same. The nature of the diene and dienophile could have an impact on the progression of the reaction. Later Carboni and Lindsey[4] published the reaction of tetrazines 5 with unsaturated compounds to generate 6. Since the diene in these types of Diels–Alder reactions is electron deficient, it is said to proceed with inverse electron demand. The inverse electron demand nature of these Diels–Alder reactions were not fully understood until the FMO analysis that accompanied the Woodward–Hoffmann rules.

Independently, Kondrat'eva reported that oxazoles[5] **7** would undergo reactions with alkenes to afford pyridine derivatives **8**.

An example of this methodology was its use in the synthesis of vitamin B_6, pyridoxine **12**.[6] Cycloaddition of oxazole **9**, prepared from ethyl *N*-acetylalanate and P_2O_5, with maleic anhydride initially gave **10**. Upon exposure to acidic ethanol, the oxabicyclooctane system fragments to afford pyridine **11**. Reduction of the ester substituents with $LiAlH_4$ generated the desired product **12**.

Modification of the oxazole, as in **13** allowed for the formation of vitamin B_6 analogs, such as **15** via intermediate pyridine **14**.

More recently, Neunhoeffer[7] showed that 1,3,5-triazines **16** could react with electron rich dienophiles, such as **17**, to produce pyrimidines **18**.

With this foundation, Boger communicated the use of 1,2,4-triazines as a dependable, azadiene equivalent for Diels–Alder approaches to substituted pyridines.[8] Electron rich olefin **19**, prepared from the corresponding ketone, was allowed to participate in an inverse electron demand Diels–Alder reaction with 1,2,4-triazine **1** to afford pyridines **20**. The reaction conditions were quickly improved with the *in situ* generation of **19**.

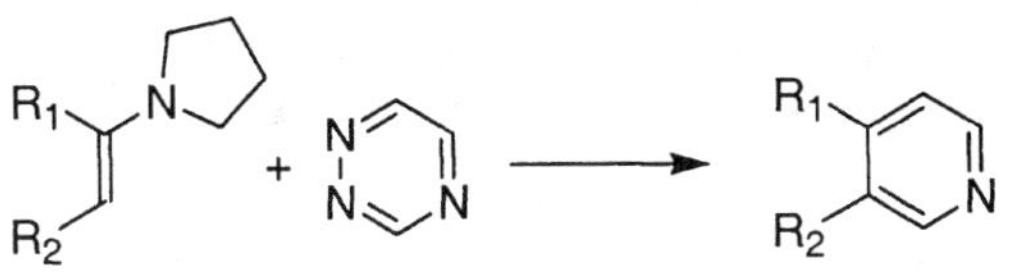

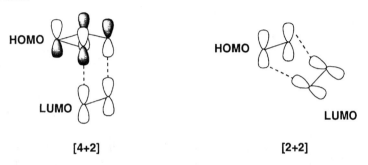

19   **1**   **20**

8.2.1.3 Mechanism[1]

The Diels–Alder reaction involves the cycloaddition of a diene with a dienophile to produce a cyclohexenyl system (see Figure 8.2.1). Since four π electrons of the diene react with 2 π electrons of the dienophile, these reactions are called [4 + 2]

Figure 8.2.1

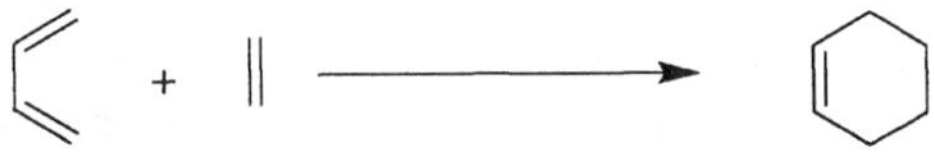

cycloadditions. To be more correct the terminology is $4\pi_S + 2\pi_S$ based on the Woodward and Hoffmann rules for the "conservation of orbital symmetry".[2] For any reaction to proceed, an electronically filled molecular orbital (MO) of one reacting partner (nucleophile) must be of similar energy and have the same orbital symmetry of the electronically empty MO of the other reacting partner (electrophile). The two MO's of greatest concern, the basis of Frontier Molecular Orbital theory (FMO), are the highest occupied MO (HOMO) and the lowest unoccupied MO (LUMO). For the Diels-Alder reaction, the appropriate HOMO–LUMO pairing is depicted in Figure 8.2.2. If the relative energy difference between these MO's is too great, then the reaction will not

Figure 8.2.2

HOMO

LUMO

HOMO

LUMO

[4+2]   [2+2]

proceed or at best be very sluggish and give poor yields. The "s" in the description denotes suprafacial, since both the HOMO and LUMO approach each other from the same face (syn addition). Other reactions are classified as antarafacial (a), when approach occurs on opposite faces of one partner as in the [2 + 2] cycloaddition of alkenes. If the two components have substituents, then the question of regio- and stereoselectivity becomes important. For example, consider the case were each component has only one substituent. The *ortho* isomer is typically the favored product over the *meta* isomer (Figure 8.2.3).

The theoretical rationalization described by FMO theory has to do with the value of the coefficients of the MO's on the atoms of the reactants. Atoms having p-orbitals of

Figure 8.2.3

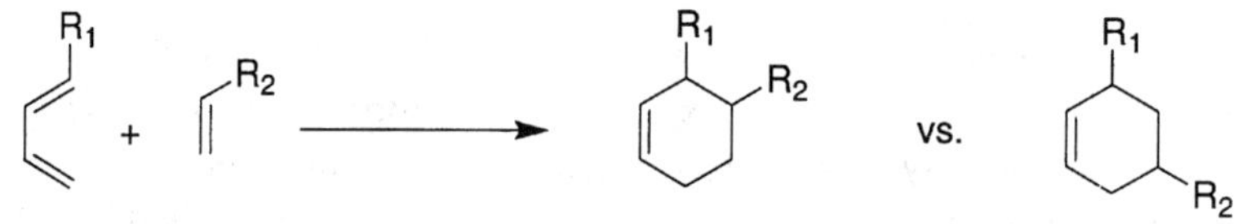

ortho   meta

similar coefficient magnitude pair in the transition state, thus giving rise to the *ortho* isomer. Stereochemistry is a result of the fact that the diene can approach the dienophile from one of its two faces, in addition to the two faces of the diene. With unsaturated dienophile substituents, the *endo* transition state is favored over the *exo* transition state as a result of secondary orbital overlap (the Alder endo rule). This would give rise to the relative stereochemistry illustrated in Figure 8.2.4.

Figure 8.2.4

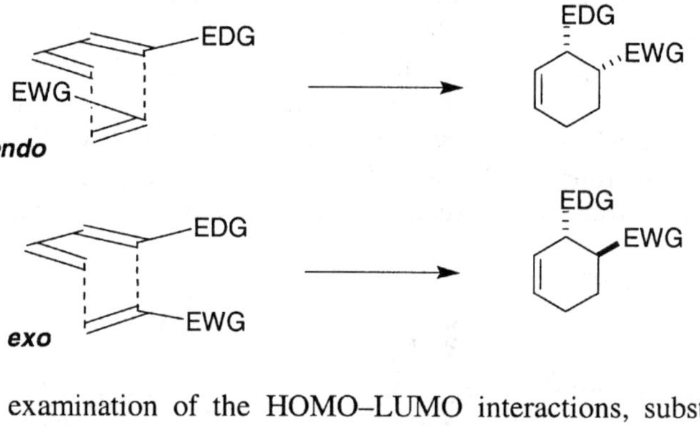

Upon closer examination of the HOMO–LUMO interactions, substituents can have a dramatic effect on the nature of this reaction. For the reference system, butadiene with ethylene, the HOMO–LUMO energies can be calculated and graphed as in Figure 8.2.5. Placing substituents on this core scaffold will change the relative energies for all the MO's but of greater concern is how this effects the HOMO–LUMO interactions. In the classical Diels–Alder reaction, an electron-donating group (EDG) is placed on the diene. This has the effect of raising the HOMO energy. When an electron-withdrawing group (EWG) is placed on the dienophile, its LUMO energy is lowered. This combination lowers the energy difference between the HOMO and LUMO, relative to the reference system, and has the net effect of rendering the reaction more facile. If one reverses these substitutions, EDG on the dienophile and EWG on the diene, the effect on the MO energies is similar. The HOMO energy of the dienophile is raised and the

LUMO energy of the diene is lowered. However, for the cycloaddition to occur, the dienophile is now the nucleophile and the diene is now the electrophile. Since the nature of the reacting partners is inverted relative to the classical case, it is called an inverse electron demand Diels–Alder reaction. Thus the Diels–Alder reaction can proceed, in practical terms, in one of two electronic modes: a) the normal mode which is $HOMO_{diene}$-controlled or b) the inverse electron demand or $LUMO_{diene}$-controlled process.

Figure 8.2.5

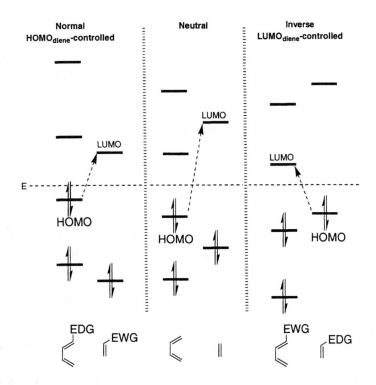

Application of this theory to the Boger pyridine synthesis reveals that it is a $LUMO_{diene}$-controlled Diels–Alder reaction. Since the 1,2,4-triazine is electron deficient as a result of the ring nitrogen atoms, one must pair this diene with electron rich dienophiles to allow optimal HOMO–LUMO pairing for this reaction to become facile.

With the energetics of the Diels–Alder adducts set correctly, the corresponding adduct gives rise to a bicyclic system. Under the reactions conditions, enamine **21** cycloadds with triazine **22**. The initially formed triazobicyclooctane **23** is unstable and a facile cycloreversion (retro-cycloaddition) occurs to liberate nitrogen gas. The concomitantly formed dihydropyridine **24** rapidly loses pyrrolidine to aromatize the ring thus producing the corresponding substituted pyridine **25**.

An alternative mechanism could be proposed that proceeded through an azete intermediate.[9] In the formation of **28** from **26** and **27**, initial rearrangement of **26** into **29**

could occur. Loss of nitrogen gas would generation **30**. Cycloaddition of **30** with **27** would afford **31** that could undergo an electrocyclic ring opening to produce **28**. However, the calculated activation energy of 115 kcal/mol for the conversion of **26** to **29** makes this step too unfavorable. As a result of this determination, this mechanism was rejected in favor of the Diels–Alder-based mechanism. Calculations also indicated that the Diels–Alder reaction is very asynchronous. In the transition state one of the two bonds is formed to a much larger extent than the other bond (Figure 8.2.6).

Figure 8.2.6

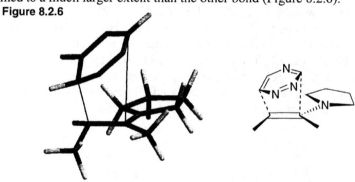

8.2.1.4 Variations and improvements

Kondrat'eva pyridine synthesis. This methodology to pyridine rings continues to be applied in total synthesis. An approach to the antitumor compound ellipticine **34**[10] makes use of a Diels–Alder reaction of acrylonitrile and oxazole **32** to form pyridiyl derivative **33**. Addition of methyllithium and hydrolysis transforms **33** into **34**.

An intramolecular variation of this reaction has been applied to the construction of the alkaloid eupolaurmine **37**.[11] Thermolysis of **35** generates **36**, an advanced intermediate for the synthesis of **37**.

The intramolecular Kondrat'eva reaction was also used in an approach to the antineoplastic alkaloid amphimedine **38**.[12] The advanced intermediate **39** was prepared and thermolysis afforded **40**. Attempts to complete the synthesis by closing the final ring failed.

A final example of this synthetic methodology is the chiral synthesis of the cyclopenta[c]pyridine ring system of (–)-plectrodorine **43**.[13] Oxazole **41** was heated in o-dichlorobenzene to produce **42**. This compound was then further elaborated to the monoterpene alkaloid **43**.

Boger developed his pyridine synthesis out of a need to construct a pentasubstituted pyridine in an approach to the formal total synthesis of antitumor antibiotic streptonigrin **44**.[14] The requisite triazine **48** was produced by using this methodology in an iterative sense with two different aza-heterocycles. Reaction of thioamidate **46** with tetrazine **47**, in a cycloaddition/cycloreversion sequence, afforded 1,2,4-triaziene **48**. Exposure of enamine **49** to **48** resulted in the formation of **50**. This compound was elaborated into a compound that intercepted Kende's synthesis of **44**.[14e]

Boger also used this chemistry in an approach to the fused-tricyclic half of the structurally related compound lavendamycin **45**.[15]

48 +

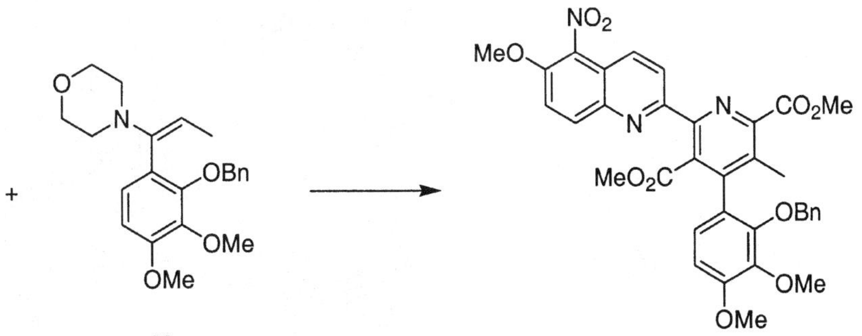

49

50

The synthesis of phomazarin **54** utilized the highly oxygenated alkene **52** paired with triazine **51** to produce **53**.[16] Further manipulations transformed this fully substituted pyridine into **54**.

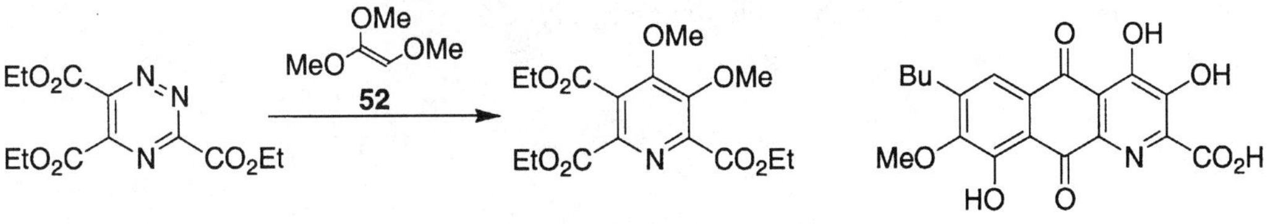

51

53

54

Taking a cue from Neunhoeffer, Boger[17] explored the use of the symmetrical triazine **57** with enamines. In contrast to 1,2,4-triazine, **57** generated the corresponding pyrimidine **60** rather than the pyridine derivative. This investigation led to the development of chemistry directed to the synthesis of desacetamidopyrimidoblamic acid **56** and ultimately deglyco-desacetamidobleomycin A_2 **55**. Triazine **57** could be treated with ynamine **58** followed by acid to produce pyrimidine **60**. A more direct method involved amidine **59** with **57** to form **60** in one step.

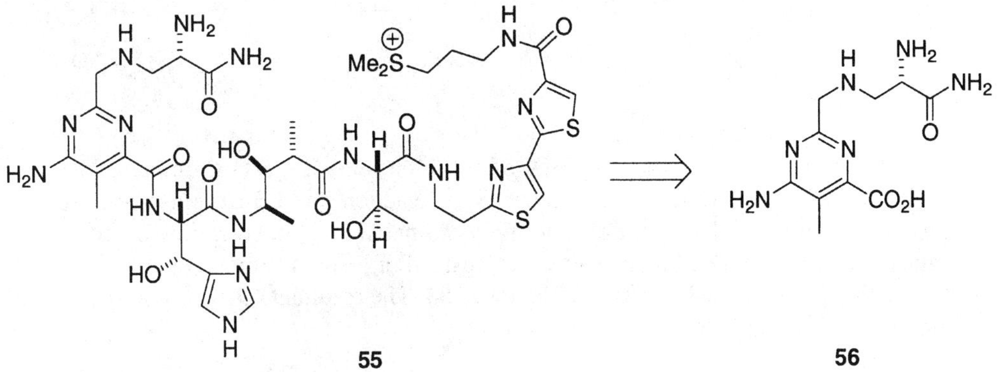

55

56

Further variations on this methodology were explored in chemistry directed to the synthesis of antitumor antibiotic CC-1065 **61**.[18] Intramolecular cycloaddition with concomitant loss of nitrogen transformed **62** into **63**. Further manipulation gave **64** which served as a building block in the assembly of **61**.

In an approach to the AB rings of rubrolone **65**, Boger[19] examined the use of oxazinones as a replacement for triazines. Reaction of 1,3-oxazin-6-one **66** with enamines **67** produced the corresponding pyridines **70**. The reaction proceeds in a manner analogous to the triazines; however, instead of losing nitrogen, these systems lose CO_2 via the intermediate bicyclo[2.2.2]octanes **68**. The resultant **69** then loses pyrrolidine as in the triazine example.

65

66   **67**   **68**   **69**   **70**

The total syntheses of fredericamycin **71**[20] and camptothecin **72**[21] made use of similar strategies. *N*-Sulfonyl-1-aza-1,3-butadienes in conjunction with electron rich dienophiles participated in the inverse electron demand Diels–Alder reaction to afford pyridines after treatment with base.

71   **72**

For **71**, ketene acetal **73** was paired with azabutadiene **74**. The cycloadduct was immediately treated with base to afford **75**. This compound was eventually converted into **71**.

73   **74**   **75**

Using a slightly different ketene acetal **77**, reaction with **76** and exposure to sodium ethoxide generated **78**. This served as a precursor to **72**.

Taylor has reported a number examples of intramolecular variations directed towards heterocyclic systems.[22]   The following two reactions are representative. Intramolecular addition of triazine **79**, after loss of nitrogen afforded **80**. Alternatively, triazine **81** generated bicyclic systems **82** which could be oxidized to **83**.

Snyder[23] has conducted similar chemistry but with the goal of generating carbon skeletons for the total synthesis of alkaloids.   Using indole **84** as a dienophile, the canthine alkaloid skeleton **85** was produced.   Access to aspidosperma alkaloids was obtained when **86** was transformed into **87**.

86   **87**

This methodology has been applied to the construction of non-natural compounds that are of interest for their physical properties.[24] Methano-1-aza[10]annulenes are of interest for study by ^{1}H-NMR. The methano-bridge resides in a highly shielded region of the molecule due to the ring current. Study of this type of compounds has led to our greater understanding of aromaticity. Reaction of benzocyclopropane **88** with triazines **89** afforded substituted annulenes **91**, through intermediate **90**.

88   **89**   **90**   **91**

The metal chelating 2,2´-bipyridines, useful in material science, can be obtained by this method.[25] For example, bis-triazines **92**, in the presence of two equivalents of enamine **67**, will produce bipyridines **93**.

92   **67**   **93**

Modern technology has influenced these reactions. Ultrasound assisted versions of these reactions has been reported.[26] Ultrasound irradiation facilitated the Diels-Alder reaction of dimethylhydrazone **94** with **95**. The resultant pyridine **96** are afforded in shorter reaction times and increased yields.

94   **95**   **96**

Microwave irradiation generates pyridine **98** from triazine **97** and enamine **67**. Again, the new technology reduces reaction time and the alternative conditions provide reaction manifolds not obtainable using traditional methods.[27]

97   **67**   **98**

Microwave technology has also been extended to the preparation of 1,2,4-triazines.[28]   The rapid synthesis of diverse triazines **101** can now be accomplished in excellent yield and purity from **99** and **100**.

99   **100**   **101**

There are other Diels–Alder approaches to pyridines but they do not proceed in one step. For example, Weinreb[29] reported the intramolecular Diels–Alder of oximino malonates tethered to a diene. Thermolysis of **102** produced **103** that upon treatment with base afforded pyridine **104**.

102   **103**   **104**

Alternative cycloaddition reactions have been reported that generate the pyridine ring system.[30]   Fully substituted pyridines **106** can be assembled using a novel organometallic method. Two different alkynes and a nitrile can be stitched together using a zirconium catalyst. The reaction was proposed to proceed via the intermediacy of azazirconacyclopentadiene **105**.

105 **106**

Reaction[31] of lithiated allene with methoxymethyl isothiocyanate afforded **107**, after trapping with methyl iodide. The newly formed **107** isomerizes under mild conditions to triene **108**. This compound is ideally setup to experience an electrocyclization to dihydropyridine **109**. Heating in the presence of acid facilitates aromatization of **109** to pyridines **110**.

107 **108** **109** **110**

8.2.1.5 Experimental
Preparation of Triazine **114**[32]

111 **112** **113** **114**

To a solution of **112** (2.0 g, 43.0 mmol) in 50 mL of dry THF at –65°C was added a solution of **111** (4.45 g, 34.0 mmol) in 100 mL of absolute ethanol containing 5 mL of acetic acid cooled to – 65°C in one portion. After stirring for 15 min., dry triethylamine (4.8 g, 510 mmol) was added. The reaction continued for 24 h with slow warming to room temperature before reducing the volume to ~10 mL. The crude **113** was brought to pH 10 with potassium carbonate. The aqueous solution was continuously extracted with chloroform, dried (K_2CO_3), evaporated onto neutral alumina, placed on a column of neutral alumina (50 g) and eluted with chloroform. The solvent was evaporated and the residue crystallized from ethanol to yield **113** (2.86 g 55%). The yellow solid had a mp = 72.5–73.8°C.

To a solution of **113** (0.50 g, 3.29 mmol) in 8 mL absolute ethanol was added a solution of KOH (0.30 g, 5.36 mmol) in 15 mL of ethanol. After 24 h, the precipitate was collected and recrystallized from 95% ethanol. The material (0.231 g, 1.42 mol) was acidified with 1.5 mL of 1.0N HCl and freeze dried. The solid was heated at 120°C in a sealed tube for 5 min. before diluting with 5 mL chloroform. The volume was reduced to 1 mL and then purified by preparative GC to afford **114** (46 mg, 40%).

Preparation of 2,3,5-tricarboethoxypyridine 117[33]

| 51 | 116 | 117 |

To a solution of **51** (1.49 g, 5.0 mmol) in 22.7 mL of chloroform was added **116** (2.22 g, 20 mmol). The reaction was heated to 60°C for 26 h before removing the solvent. The crude product was chromatographed on silica to afford **117** as a yellow oil (1.35 g, 92%).

8.2.1.5 References

1.   [R] (a) Boger, D. L. *Tetrahedron* **1983**, *39*, 2869–2939
     [R] (b) Boger, D. L. *Chem. Rev.* **1986**, *86*, 781–793
     [R] (c) Behforouz, M.; Ahmadian, M.*Tetrahedron* **2000**, *56*, 5259–5288
     [R] (d) Buonora, P.; Olsen, J.–C.; Oh, T. *Tetrahedron* **2001**, *57*, 6099–6138
     [R] (e) Jayakumar, S.; Ishar, M. P. S.; Mahajan, M. P. *Tetrahedron* **2002**, *58*, 379–471
2.   Woodward, R. B.; Hoffmann, R. "The Conservation of Orbital Symmetry" VCH Publishers **1970**
3.   Bachmann, W. E.; Deno, N. C. *J. Am. Chem. Soc.* **1949**, *71*, 3062–3072
4.   Carboni, R. A.; Lindsey, R. V. Jr. *J. Am. Chem. Soc.* **1959**, *81*, 4342–4346
5.   Turchi, I.. J.; Dewar, M. J. S. *Chem. Rev.* **1975**, *75*, 389–437
6.   (a) Firestone, R. A.; Harris, E. E.; Reuter, W. *Tetrahedron* **1967**, *23*, 943–955
     (b) Muhlradt, P. F.; Morino, Y.; Snell, E. E. *J. Med. Chem.* **1967**, *10*, 341–344
7.   Neunhoeffer, H.; Bachmann, M. *Chem. Ber.* **1975**, *108*, 3877–3882
8.   (a) Boger, D. L.; Panek, J. S. *J. Org. Chem.* **1981**, *46*, 2179–2182
     (b) Boger, D. L.; Panek, J. S.; Meier, M. M. *J. Org. Chem.* **1982**, *47*, 895–897
9.   (a) Yu, Z.–X.; Dang, Q.; Wu, Y.–D. *J. Org. Chem.* **2001**, *66*, 6029–6036
     (b) Prieto, P.; Cossio, F. P.; Carrillo, J. R.; Hoz, A.; de la Diaz–Ortiz, A.; Moreno, A. *J. Chem. Soc. Perkin Trans. 2* **2002**, 1257–1263
10. Kozikowski, A. P.; Hasan, N. M. *J. Org. Chem.* **1977**, *42*, 2039–2040
11. (a) Levin, J. I.; Weinreb, S. M. *J. Am. Chem. Soc.* **1983**, *105*, 1397–1398
     (b) Levin, J. I.; Weinreb, S. M. *J. Org. Chem.* **1984**, *49*, 4325–4332
12. Subramanyam, C.; Noguchi, M.; Weinreb, S. M. *J. Org. Chem.* **1989**, *54*, 5580–5585
13. Ohba, M.; Izuta, R.; Shimizu, E. *Tetrahedron Lett.* **2000**, *41*, 10251–10255
14. (a) Boger, D. L.; Panek, J. S. *J. Org. Chem.* **1982**, *47*, 3763–3765
     (b) Boger, D. L.; Panek, J. S.*J. Org. Chem.* **1983**, *48*, 621–623
     (c) Boger, D. L.; Panek, J. S. *Tetrahedron Lett.* **1983**, *24*, 4511–4514
     (d) Boger, D. L.; Panek, J. S. *J. Am. Chem. Soc.* **1985**, *107*, 5745–5754
     (e) Kende, A. S.; Lorah, D. P.; Boatman, R. J. *J. Am. Chem. Soc.* **1981**, *103*, 1271-1273.
15. Boger, D. L.; Panek, J. S. *Tetrahedron Lett.* **1984**, *25*, 3175–3178
16. Boger, D. L.; Hong, J.; Hikota, M.; Ishida, M. *J. Am. Chem. Soc.* **1999**, *121*, 2471–2477
17. (a) Boger, D. L.; Schumacher, J.; Mullican, M. D.; Patel, M.; Panek, J. S. *J. Org. Chem.* **1982**, *47*, 2673–2675
     (b) Boger, D. L.; Dang, Q. *Tetrahedron* **1988**, *44*, 3379–3390
     (c) Boger, D .L.; Dang, Q. *J. Org. Chem.* **1992**, *57*, 1631–1633
     (d) Boger, D. L.; Menezes, R. F.; Dang, Q. *J. Org. Chem.* **1992**, *57*, 4333–4336
     (e) Boger, D. L.; Kochanny, M. J. *J. Org. Chem.* **1994**, *59*, 4950–4955
18. (a) Boger, D. L.; Coleman, R. S. *J. Org. Chem.* **1984**, *49*, 2240–2245
     (b) Boger, D. L.; Coleman, R. S.; Invergo, B. J. *J. Org. Chem.* **1987**, *52*, 1521–1530
19. Boger, D. L.; Wysocki, R. J. Jr. *J. Org. Chem.* **1989**, *54*, 714–718
20. Boger, D. L. *J. Heterocyclic Chem.* **1996**, *33*, 1519–1531
21. Blagg, B. S. J.; Boger, D. L. *Tetrahedron* **2002**, *58*, 6343–6349
22. (a) Taylor, E. C.; French, L. G. *Tetrahedron Lett.* **1986**, *27*, 1967–1970

     (b) Taylor, E. C.; Macor, J. E.; Pont, J. L. *Tetrahedron* **1987**, *43*, 5145–5158

     (c) Taylor, E. C.; Pont, J. L. *Tetrahedron Lett.* **1987**, *28*, 379–382

     (d) Taylor, E. C.; Macor, J. E. *J. Org. Chem.* **1987**, *52*, 4280–4287

     (e) Taylor, E. C.; Macor, J. E. *J. Org. Chem.* **1989**, *54*, 4984–4989

23. (a) Benson, S. C.; Li, J.–H.; Snyder, J. K. *J. Org. Chem.* **1992**, *57*, 5285–8587

     (b) Li, J.–H.; Snyder, J. K. *J. Org. Chem.* **1993**, *58*, 519–519

     (c) Benson, S. C.; Lee, L.; Snyder, J. K. *Tetrahedron Lett.* **1996**, *37*, 5061–5064

     (d) Wan, Z.–K.; Snyder, J. K. *Tetrahedron Lett.* **1998**, *39*, 2487–2490

24. Martin, J. C.; Muchowski, J. M. *J. Org. Chem.* **1984**, *49*, 1040–1043

25. (a) Pabst, G. R.; Schmid, K.; Saur, J. *Tetrdhedron Lett.* **1998**, *39*, 6691–6694

     (b) Bushby, N.; Moody, C. J.; Riddick, D. A.; Waldron, I. R. *J. Chem. Soc. Perkin Trans. 1* **2001**, 2183–2193

     (c) Branowska, D.; Rykowski, A. *Synlett* **2002**, 1892–1894

     (d) Branowska, D. *Synthesis* **2003**, 2096–2100

     (e) Stanforth, S. P.; Tarbit, B.; Watson, M. D. *Tetrahedron Lett.* **2003**, *44*, 693–694

26. Villcampa, M.; Perez, J. M.; Avendano, C.; Menendez, J. C. *Tetrahedron* **1994**, *50*. 10047–10054

27. Diaz–Ortiz, A.; de la Hoz, A.; Prieto, P.; Carrillo, J. R.; Moreno, A.; Neunhoeffer, H. *Synlett* **2001**, 236–237

28. Zhao, Z.; Leister, W. H.; Stauss, K. A.; Wisnoski, D. D.; Lindsey, C. W. *Tetrahedron Lett.* **2003**, *44*, 1123–1127

29. Bland, D. C.; Raudenbush, B. C.; Weinreb, S. W. *Organic Lett.* **2000**, *2*, 4007–4009

30. Takahashi, T.; Tsai, F.–Y.; Kotora, M. *J. Am. Chem. Soc.* **2000**, *122*, 4994–4995

31. Nedolya, N. A.; Schlyakhtina, N. I.; Klyba, L. V.; Ushakov, I. A.; Fedorov, S. V.; Brandsma, L. *Tetrahedron Lett.* **2002**, *43*, 9679–9681

32. (a) Krass, D.; Paudler, W. W. *Synthesis* **1974**, 351–352

     (b) Paudler, W. W.; Barton, J. M. *J. Org. Chem.* **1966**, *31*, 1720–1722

33. Boger, D. L.; Panek, J. S.; Yasuda, M. *Organic Syn.* **1987**, *66*, 142–150

Paul Galatsis

8.3 Preparation via Rearrangement Reactions

In addition to the formation of the pyridine framework by *de novo* approaches (see section 8.1) or by the cycloaddition/cycloreversion sequence (see section 8.2), one can employ reactions that proceed through a rearrangement pathway. The Boekelheide reaction (see section 8.3.1) involves the rearrangement of an existing pyridine skeleton to a more functionalized scaffold, while the Ciamician-Dennstedt reaction (section 8.3.2) generates the pyridine nucleus by rearrangement of an alternative heterocycle.

8.3.1 Boekelheide Reaction

8.3.1.1 Description

The Boekelheide reaction and related reactions involves treating pyridine *N*-oxides[1] **1** with acylating agents to afford rearranged products **2**. Traditionally, the rearrangement occurs at the α-position but variations and/or side-products of this reaction afford γ-position modification.

8.3.1.2 Historical perspective

Katada,[2a,b] working in the labs of Ochiai,[2c] first described the reaction of *N*-oxide **3** with acetic anhydride. The resultant rearrangement produced α-pyridone **4**. Shortly thereafter, several groups[3-6] working independently reported modifications of this chemistry. Implementation of these modifications using 2-alkyl pyridines resulted in the

formation of 2-alkoxymethyl pyridines. One of these groups was that of Boekelheide and his name has now become linked with this established procedure for introducing oxygen functionality into alkyl groups alpha to the aromatic N atom.

8.3.1.3 Mechanism

Several groups have contributed pieces to the puzzle of how this reaction proceeds and understanding of the events that transpire has evolved with time. While on the surface, this reaction looks analogous to the Polonovsky reaction[7] (**5** → **6**) or the Pummerer

5

6

reaction.[8] A mechanism invoking an electrocyclic rearrangement is not consistent with the experimental observations discussed below. Okuda and Kobayashi[3] suggested a mechanism that involved an ionization and anionotropic rearrangement. *N*-oxide **7** is in equilibrium with its tautomeric form **8**, which is trapped by acetic anhydride to form anhydrobase **9**. Ionization of the N–O bond (**10**) followed by acetate attack at the primary carbocation produces the expected product **11**. Bullitt and Maynard,[5] following

7   **8**   **9**   **10**   **11**

the cue from Pachter,[9] invoked the intermediacy of anhydrobase **9** more directly through

7   **12**   **9**   **11**

acylation of starting material *N*-oxide **7**. These two mechanisms have some very subtle differences but Boekelheide[4] departed from an anionic mechanism and favored a free radical-based mechanism. His free radical mechanism was based on the observations that styrene polymerized only in the presence of intermediates present in this reaction. A sequence of events consistent with this begins with acylated N-oxide **12**. Homolysis of the N–O bond, generating the radical pair **13**, was followed by rearrangement of the pyridyl radical cation to **14** that was then rapidly trapped by the acetate radical to give rise to **11**. A free radical-based mechanism also rationalized other reaction attributes of an induction period followed by a strongly exothermic phase as part of the time course for the transformation.

12   **13**   **14**   **11**

Subsequent work by Traynellis and Martello[10] did not support the radical mechanism. When they conducted experiments similar to Boekelheide but with the radical scavengers benzoquinone or *m*-dinitrobenzene present, the yield of polystyrene dropped off dramatically but the yield of the desired product was unaffected.[11] Thus, this work combined with that of others who followed-up on these observations, effectively eliminated the free radical mechanism as an option for the Boekelheide reaction. The remaining options included an ion-pair or a rearrangement mechanism. The latter option was further eliminated from contention by ^{18}O-labeling studies,[12] initially reported by Oae and latter confirmed by Katrizky. The intramolecular nature of the reaction was confirmed but the two oxygen atoms of the acetoxy moiety become equivalent during the course of the reaction. This observation precludes an electrocyclic rearrangement (**9**→**15**) and was further confirmed when it was determined that scrambling of the oxygen atoms occurs during the reaction rather than before or after the rearrangement.

9   **15**   **11**

Additional work by Traynelis and Pacini[12c,13] employing spectroscopic and [2]H-labeling studies showed that the anhydrobase does not accumulate during the course of the reaction and that the formation of the anhydrobase is irreversible. In aggregate, this data indicates formation of the anhydrobase is the rate-determining step of the reaction and the subsequent rearrangement to products is fast. An ion-pair mechanism is now the generally accepted explanation of the events for this reaction. This was further confirmed by the observations that electron withdrawing groups favor the reaction and that 2-neopentylpyridine-1-oxide **16** gave rise to alkene **19**.[12b,d] The formation of alkene **19** can be rationalized as follows. Heterolysis of the N–O bond of acylated **16** would give ion pair **17**. The more stable **18** could be obtained from **17** by resonance. This neopentyl carbocation undergoes a [1,2]-Wagner-Meerwein methyl shift to give, after loss of a proton, **19**.

16 **17** **18** **19**

8.3.1.4 Variations and improvements

The Boekelheide reaction has found utility in a number of synthetic applications. A notable example of its application to natural product synthesis was described by Nicolaou[14] in his model system for an approach to the thiopeptide antibiotic thiostrepton, in particular, the elaboration of the quinaldic acid moiety. The tetrahydroquinoline **21** was converted to the N-oxide by m-CPBA oxidation. Subsequent treatment with TFAA, to carry out the Boekelheide reaction, was followed by hydrolysis of the resultant ester to produce **22** as a mixture of alcohols.

20

21   **22**

Building blocks, useful for supramolecular or material science,[15] have also been prepared using the Boekelheide reaction. Thus bipyridyl derivative **23** was subjected to the standard sequence of reactions (oxidation, rearrangement, and hydrolysis) to afford the diol **24**.

23   **24**

The Boekelheide reaction has been applied to the synthesis of non-natural products with the preparation of quaterpyridines serving as an example.[16] The sequence began with the 2,4-linked bipyridyl-*N*-oxide **25**. Execution under the typical reaction conditions produced the expected bis-pyridone **26**. Treatment with POCl₃ afforded the corresponding dichloride that was submitted to a palladium-catalyzed coupling with 2-stannyl pyridine to produce the desired quaterpyridine **27**.

25   **26**   **27**

The Boekelheide reaction has found utility in other synthetic methodology. An approach to 2,3-pyridynes[17] made use of this chemistry in the preparation of the key intermediate **30**. Treatment of **28** with acetic anhydride produced the desired pyridone **29**. Lithiation was followed by trapping with trimethylsilyl chloride and exposure to triflic anhydride gave the pyridyne precursor **30**. Fluoride initiated the cascade of reactions that resulted in the formation of 2,3-pyridyne **31** that could be trapped with appropriate dienes in Diels–Alder reactions.

28   **29**   **30**   **31**

Carrying out the Boekelheide reaction in tandem,[4a,18] can lead to the corresponding aldehyde rather than the typical alcohol product. Thus, acetic anhydride initiated rearrangement of **7** followed by *N*-oxide formation with hydrogen peroxide gave **32**. Repeating the exposure to acetic anhydride afforded the bis-acetoxy pyridine **33**. This diacetoxy acetal was directly hydrolyzed to afford the corresponding aldehyde **34**.

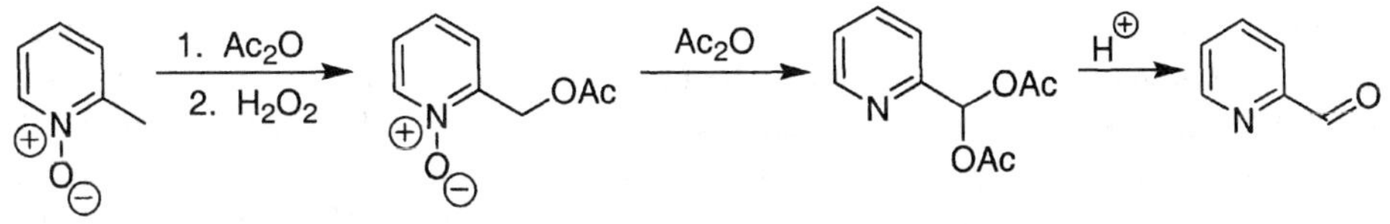

7 **32** **33** **34**

This chemoselectivity stands in contrast to that of 2,6-disubstituted pyridines. For example,[4a] 2,6-dimethylpyridine **35** was reacted with hydrogen peroxide and acetic anhydride to produce the expected acetoxy derivative **36**. A second iteration of the previous reaction conditions did not afford an aldehyde, as in the previous example, but 2,6-bis-acetoxy derivative **37**.

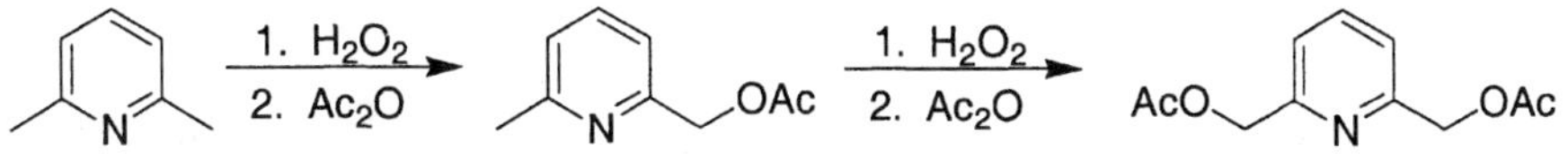

35 **36** **37**

An approach to the construction of Fe(II)-binding agent pyrimine **40**, isolated from *Pseudomonas* species, employed the bis-homophenylalanine **38**.[19] Initiation of the Boekelheide reaction with TFAA and hydrolysis gave the advanced intermediate **39** that provided access to the natural product.

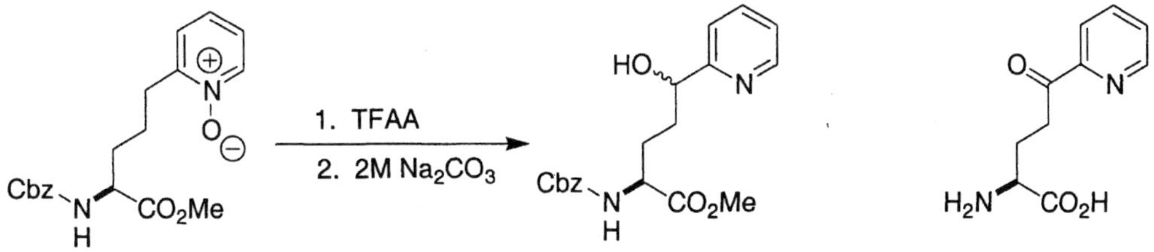

38 **39** **40**

The formation of a library of 2-substituted quinolines employed a variation on the Boekelheide reaction.[20] Treatment of N-oxide **41** with isobutylchloroformate did not result in the typical rearrangement. However, subsequent exposure to Grignard reagents resulted in loss of the carbonate with concomitant formation of the 2-substitute derivatives **42**.

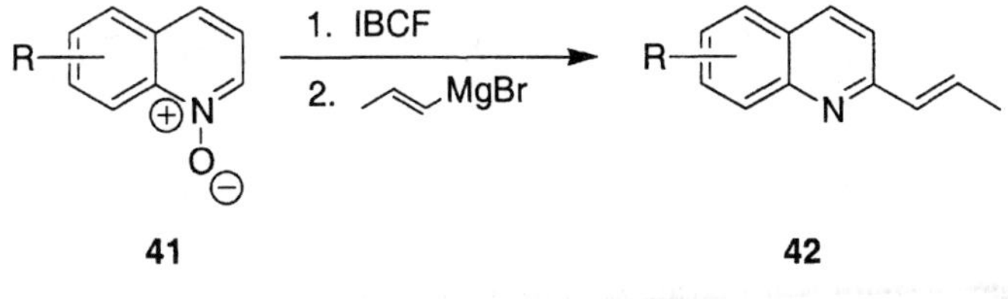

41 **42**

The Reissert–Henze reaction[21] offers a method to prepare cyanopyridines and extends the Reissert and Kaufmann reactions in the quinoline and isoquinoline series, respectively. Starting with an *N*-oxide, the reaction sequence is initiated by *O*-alkylation rather than *O*-acylation of the *N*-oxide. The resultant *N*-alkoxypyridyl compound is then subjected to cyanide to afford the 2-cyano-derivative with loss of the alkoxy group. This chemistry was used in the preparation of *S*-adenosylmethionine decarboxylase inhibitors of the type **43**.[22] Ethylation of *N*-oxide **44** with diethylsulfate afforded intermediate **45** which upon exposure to cyanide, produced the corresponding pyridine derivative **46**.

43

44   **45**   **46**

An improvement in this protocol makes use of trimethylsilylcyanide.[23]   This reagent carries out both steps in one pot (**47 → 48**). The trimethylsilyl group activates the N-oxide with concomitant displacement by cyanide

47   **48**

One can also use diethyl phosphorocyanidate (DEPC)[24] to effect a similar transformation. This was the reagent of choice in the generation of 1-substituted-4-oxygenated-β-carbolines (**49→50**).

49   **50**

One can extend the strategy employed in the Reissert–Henze reaction to nucleophiles other than cyanide. Phosphorus-based systems were employed in the generation of ligands for use as iron chelates.[25]   Alkylation of *N*-oxide **3** with dimethylsulfate gave **51** that was treated with lithium diethylphosphite to produce pyridine derivative **52**. Oxidation of the pyridyl nitrogen and repeating the initial two steps gave **53**. This could be hydrolyzed to the diphosphonic acid and used as the ligand.

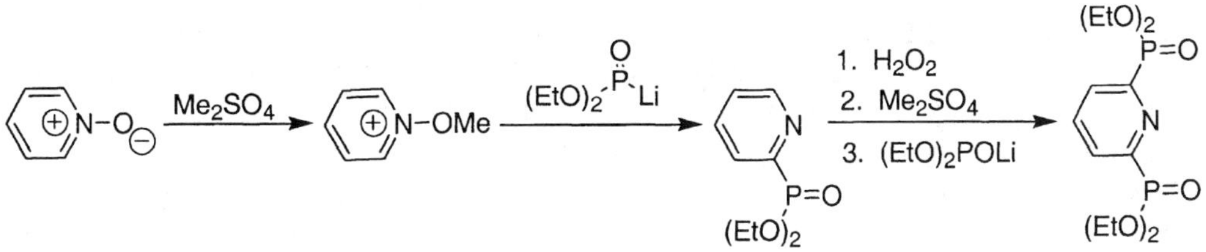

3   **51**   **52**   **53**

Minisci reactions have also been applied to these compounds.[26] *N*-oxide formation by exposure to *m*-CPBA and *O*-methylation with Meerwein's reagent converted **54** into **55**. Nucleophilic attack of the hydroxymethyl radical, generated with ammonium sulfate, provides an alternate route to 2-hydroxymethyl pyridines **56**.

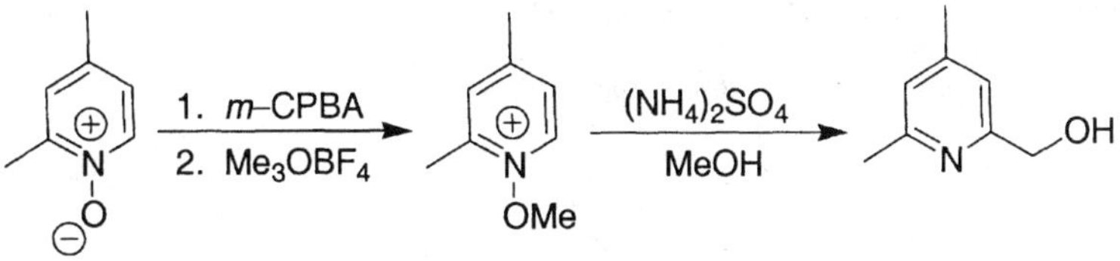

54   **55**   **56**

Finally, treating the *N*-oxide directly with Tebbe reagent provides a rapid method[27] of introducing the 2-methyl substituent (**57** → **58**).

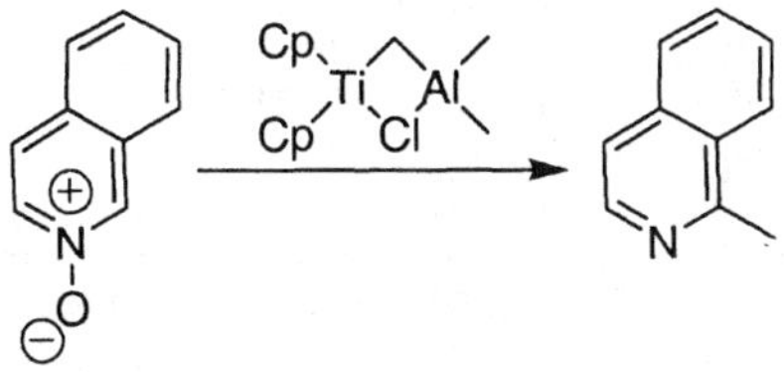

57   **58**

8.3.1.5 Experimental[28]

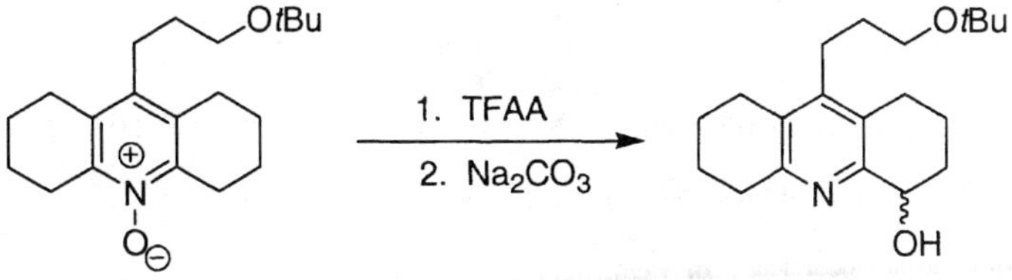

59   **60**

To *N*-oxide **59** (22.42 g, 70.4 mmol) in 400 mL of dry CH_2Cl_2 was slowly added TFAA (25 mL, 177 mmol). A slight rise in temperature of the reaction was observed. After 1 h at rt, the solution was concentrated and the residue dissolved in 100 mL of CH_2Cl_2 and saponified with 300 mL of 2 M Na_2CO_3 solution. The biphasic mixture was vigorously stirred for 3 h. The aqueous phase was extracted twice with CH_2Cl_2, the combined organic phases were washed with H_2O, brine and dried ($MgSO_4$). After concentration, 21.38 g of a pale yellow crystalline solid was obtained. Washing with hexanes afforded 19.85 g (89%) of **60** as white crystals (mp = 99°C) analytically pure.

8.3.1.6 References

1. [R] (a) Katritzky, A. R.; Lam, J. N. *Heterocycles* **1992**, *33*, 1011–1049
    (b) Katritzky, A. R.; Lagowski, J. M. "Chemistry of the Heterocyclic N-oxides" Academic Press, NY, 1971
2. (a) Katada, M. *J. Pharm. Soc. Jpn.* **1947**, *67*, 15–19
    (b) McKillop, A.; Bhagrath, M. K. *Heterocycles* **1985**, *23*, 1697–1701
    (c) Ochiai, E. *J. Org. Chem.* **1953**, *18*, 534–551
3. (a) Kobayashi, G.; Furukawa, S. *Pharm. Bull. Jpn.* **1953**, *1*, 347–349
    (b) Okuda, S. *Pharm. Bull. Jpn.* **1955**, *3*, 316–318
4. (a) Boekelheide, V.; Linn, W. J. *J. Am. Chem. Soc.* **1954**, *76*, 1286–1291
    (b) Boekelheide, V.; Harrington, D. L. *Chem. Ind.* **1955**, 1423–1424
    (c) Boekelheide, V.; Lehn, W. L. *J. Org. Chem.* **1961**, *26*, 428–430
5. Bullitt, O. H. Jr.; Maynard. J. T. *J. Am. Chem. Soc.* **1954**, *76*, 1370–1371
6. Berson, J. A.; Cohen, T. *J. Am. Chem. Soc.* **1955**, *77*, 1281–1283
7. Bell, S. C.; Childress, S. J. *J. Org. Chem.* **1962**, *27*, 1691–1695
8. [R] Padwa, A.; Gunn, D. E. Jr.; Osterhout, M. H. *Synthesis* **1997**, 1353–1377
9. Pachter, I. J. *J. Am. Chem. Soc.* **1953**, *75*, 3026–3027
10. (a) Traynelis, V. J.; Martello, R. F. *J. Am. Chem. Soc.* **1958**, *80*, 6590–6593
    (b) Traynelis, V. J. Sr.; Gallagher IHM, A. I.; Martello, R. F. *J. Org. Chem.* **1961**, *26*, 4365–4368
11. (a) Cohen, T.; Fager, J. H. *J. Am. Chem. Soc.* **1965**, *87*, 5701–5710
    (b) Koenig, T. *J. Am. Chem. Soc.* **1966**, *88*, 4045–4049
    (c) Koenig, T.; Wieczorek, J. S. *J. Org.Chem.* **1968**, *33*, 1530–1532
12 (a) Oae, S.; Kitao, T.; Kitaoka, Y. *J. Am. Chem. Soc.* **1962**, *84*, 3359–3362
    (b) Bodalski, R.; Katritzky, A. R. *Tetrahedron Lett.* **1968**, 257–260
    (c) Tamagaki, S.; Kozuka, S.; Oae, S. *Tetrahedron Lett.* **1968**, 4765–4768
    (d) Bodalski, R.; Katritzky, A. R. *J. Chem. Soc. (B)* **1968**, 831–838
    (e) Oae, S.; Kozuka, S. *Tetrahedron* **1964**, *20*, 2671–2676
13. (a) Traynelis, V. J.; Pacini, P. L. *J. Am. Chem. Soc.* **1964**, *86*, 4917–4922
    (b) Muth, C. W.; Dallak, R. S.; DeMatte, M. L.; Chovenec, G. F. *J.Org. Chem.* **1968**, *33*, 2762–2767
14. Nicolaou, K. C.; Safina, B. S.; Funke, C.; Zak, M.; Zecri, F. J. *Angew.Chem. Int. Ed.* **2002**, *41*, 1937–1940
15. Schubert, U. S.; Kersten, J. L.; Pemp, A. E.; Eisenbach, C. D.; Newkome, G. R. *Eur. J. Chem.* **1998**, 2573–2581
16. Zoltewicz, J. A.; Cruskie, M. P. Jr.; Dill, C. D. *Tetrahedron* **1996**, *52*, 4239–4244
17. Walters, M. A.; Shay, J. J. *Tetrahedron Lett.* **1995**, *36*, 7575–7578
18. Ginsburg, S.; Wilson, I. R. *J. Am. Chem. Soc.* **1957**, *79*, 481–485
19. Collier, P. N.; Campbell, A. D.; Patel, I.; Taylor, R. J. K. *Tetrahedron* **2002**, *58*, 6117–6125
20. Fakhfakh, M. A.; Franck, X.; Fournet, A.; Hocquemiller, R.; Figadere, B. *Tetrahedron Lett.* **2001**, *42*, 3847–3850
21. (a) Feely, W. E.; Beavers, E. M. *J. Am. Chem. Soc.* **1959**, *81*, 4004–4007
    (b) Fife, W. K.; Scriven, E. F. V. *Heterocycles* **1984**, *22*, 2375–2394
22. (a) Stanek, J.; Caravatti, G.; Frei, J.; Furet, P.; Mett, H.; Schneider, P.; Regenass, U. *J. Med. Chem.* **1993**, *36*, 2169–2171
    (b) Stanek, J.; Caravatti, G.; Capraro, H.-G.; Furet, P.; Mett, H.; Schneider, P.; Regenass, U. *J. Med. Chem.* **1993**, *36*, 46–54
23. (a) Fife, W. K. *J. Org. Chem.* **1983**, *48*, 1375–1377
    (b) Fife, W. K. *Heterocycles* **1984**, *22*, 1121–1124
24. (a) Harusawa, S.; Hamada, Y.; Shioiri, T. *Heterocycles* **1981**, *15*, 981–984

(b) Susuki, H.; Iwata, C.; Sajurai, K.; Tokumoto, K.; Takahashi, H.; Hanada, M.; Yokoyama, Y.; Murakami, Y. *Tetrahedron* **1997**, *53*, 1593–1606

25.   Chen, D.; Martell, A. E.; Motekaitis, R. J.; McManus, D. *Can. J. Chem.* **1998**, *76*, 445–451

26.   Biyouki, M. A. A.; Smith, R. A. J.; Bedford, J. J.; Leader, J. P. *Synth. Commun.* **1998**, *28*, 3817–3825

27.   Nicolaou, K. C.; Koumbis, A. E.; Snyder, S. A.; Simonsen, K. B. *Angew. Chem. Int. Ed.* **2000**, *39*, 2529–2533

28.   For improved experimental conditions see: Fontenas, C.; Bejan, E.; Haddou, H. A.; Balavoine, G. G. A. *Synth. Commun.* **1995**, *25*, 629–633

Paul Galatsis

8.3.2 Ciamician–Dennstedt Rearrangement

8.3.2.1 Description

The Ciamician–Dennstedt reaction involves the reaction of a pyrrole (**1**) with the carbene generated from chloroform and a base to provide a 3-chloropyridine (**2**, Scheme 8.3.1).

Scheme 8.3.1

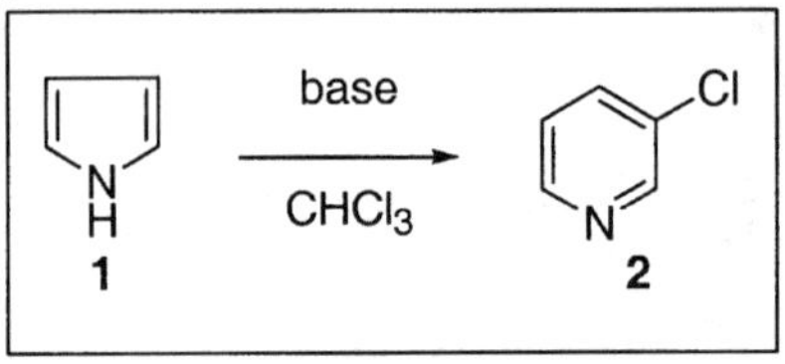

8.3.2.2 Historical Perspective

Ciamician and Dennstedt reacted the potassium salt of pyrrole with chloroform in ether and isolated, after much purification, 3-chloropyridine, which was confirmed by crystallization with platinum.[1] While the pyrrole salt can be used as the base, the chloroform carbene is typically formed with an alkali alcohol. Forty years later, Robinson and co-workers made 3-chloroquinolines from indoles using the Ciamician–Dennstedt reaction.[2]

8.3.2.3 Mechanism

The Ciamician–Dennstedt reaction can be thought of as the complement to the Reimer–Tiemann reaction (Scheme 8.3.2).[3] The first step of both reactions is cyclopropanation of one of the carbon-carbon double bonds of a pyrrole with a dichlorocarbene, resulting in intermediate **3**. The Ciamician–Dennstedt reaction results from cleavage of the internal C–C bond and elimination of chloride (path *a*), while the Reimer–Tiemann reaction results from cleavage of the exocyclic bond, and subsequent hydrolysis of the dichloromethyl moiety to furnish aldehyde **5** (path *b*).

Scheme 8.3.2

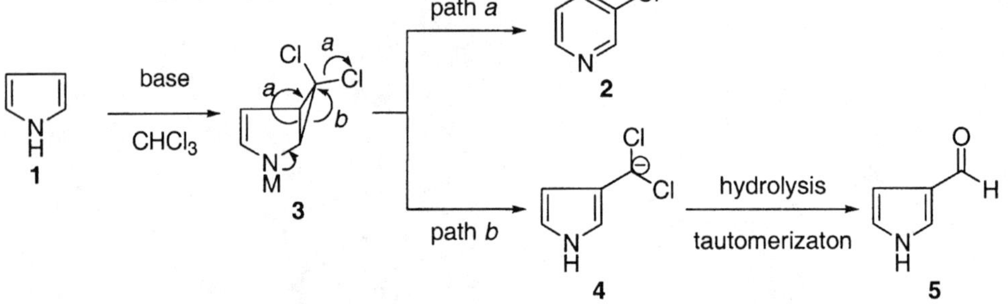

It has been suggested that the strain of the [3.1.0] ring system plays a role in facilitating the reaction. In a carbocyclic system, the [3.1.0] ring system opens 200 times faster than the analogous [4.1.0] ring system.[4]

8.3.2.4 Variations and improvements

The highest yields in the Ciamician–Dennstedt reaction have been achieved using phase transfer catalysts (Table 8.3.1).[5] In the reaction, the pyrrole or indole and a phase transfer catalyst (PTC, in this case benzyltriethylammonium chloride) are dissolved in chloroform and aqueous sodium hydroxide is added. Yields are typically in the 40s to 60s (rather than in the 20s for a typical Ciamician–Dennstedt reaction). More recently, yields as high as 80% have been reported using tetra-*n*-butylammonium hydrogen sulphate as the phase transfer catalyst.[6]

Table 8.3.1. Comparison of Phase Transfer Catalysts on Yield of Ciamician–Dennstedt reaction

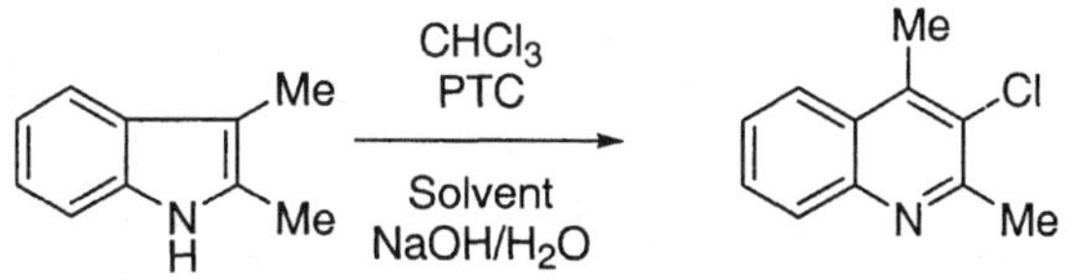

PTC	Temperature (°C)	% Yield 7	Solvent	Reference
BnEt$_3$NCl	rt	55	CHCl$_3$	5a
BnEt$_3$NCl	40	63	1:1 CHCl$_3$:benzene	5b
*n*Bu$_4$NHSO$_4$	rt	79	CHCl$_3$	6

The use of sodium tribromoacetate as the dibromocarbene precursor has been investigated and found to provide the Ciamician–Dennstedt product in higher yield than the traditional alkoxide/alcohol reaction conditions.[7] Deprotonation of bromoform with sodium ethoxide in ethanol and reaction of the resultant carbene with **6** provides quinoline **9** in 9% yield; thermolysis of sodium tribromoacetate in the presence of **6** furnishes **9** in 20% yield (Scheme 8.3.3).

Scheme 8.3.3

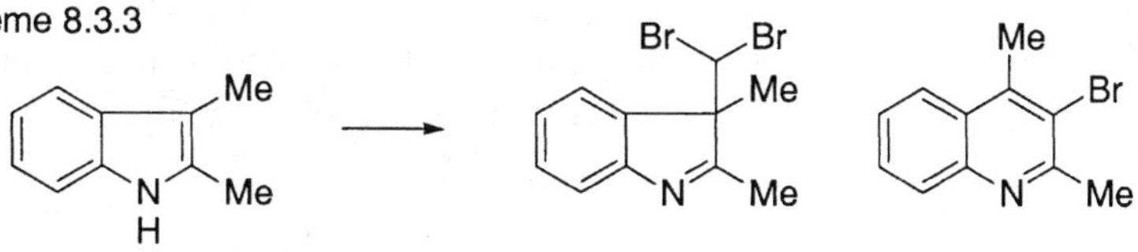

Conditions	Yield	Yield
	8	**9**
NaOEt, CHBr$_3$, EtOH, 50–55 °C	6.9%	8.8%
NaOCOBr$_3$, MeOCH$_2$CH$_2$OMe, Δ	3.2%	20.5%

Flash vapor pyrolysis of chloroform has been used to effect the Ciamician–Dennstedt reaction on pyrrole,[8] as well as pyrazoles,[9] indoles,[10] and imidazoles.[11] Phenyl(trichloromethyl)mercury has been used as a dichlorocarbene precursor in the Ciamician–Dennstedt reaction.[12] Dichlorocarbene generated electrochemically has been added to indoles to furnish the 3-chloroquinoline in yields comparable to traditional alkoxide/alcohol reaction conditions.[13]

8.3.2.5 Synthetic utility

The Ciamician–Dennstedt reaction has been used to prepare macrocycles. Reaction of 2,3-alkyl linked indole derivatives **10**, **11**, **13**, and **15** with phenyl(trichloromethyl)-mercury furnishes metaquinolophanes **12**, **14**, and **16** (Scheme 8.3.4).[14] Attempts to make the linker chain shorter than six carbons failed.

Scheme 8.3.4

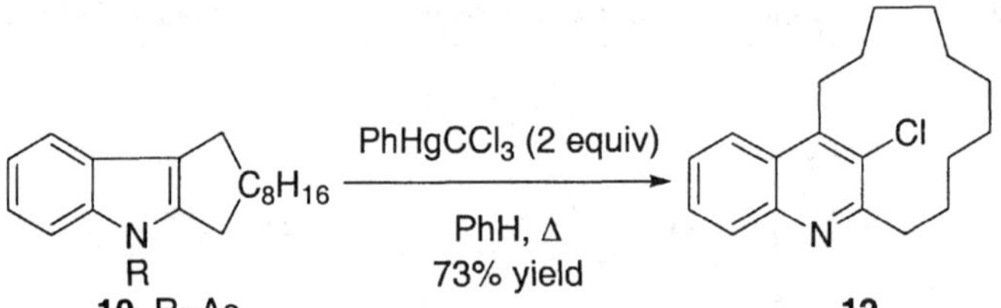

10, R=Ac
11, R=H

12
R=Ac, 76%
R=H, 69%

13 **14**

15 **16**

In a similar manner, metapyridophanes **18** can be prepared by reaction of sodium trichloroacetate with pyrrole **17** (Scheme 8.3.5). The transformation can also be achieved with phenyl(bromodichloromethyl)mercury, albeit in lower yield.

Scheme 8.3.5

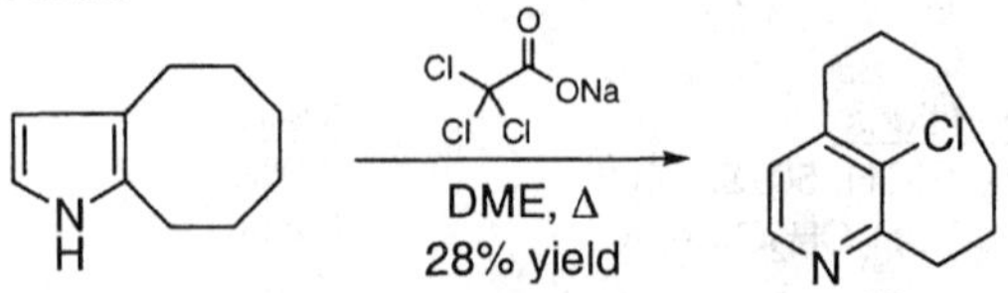

17 **18**

The Ciamician–Dennstedt reaction is currently the only way to date to make calix[4]pyridine **20**.(Scheme 8.3.6)[15] Four sequential treatments of calix[4]pyrrole **19** with sodium trichloroacetate results in all four possible geometric isomers of **20** (all four chlorines in the a position, three chlorines in the a position, two isomers with two chlorines in the a position) in 26% overall yield.

Scheme 8.3.6

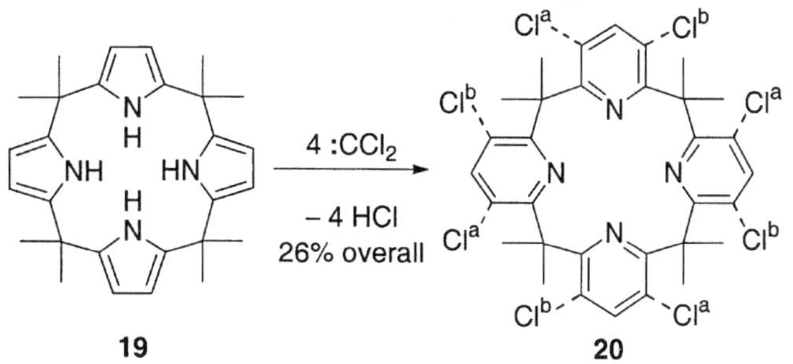

Though beyond the scope of this book, the carbocyclic analog of the Ciamician–Dennstedt reaction has been investigated.[16]

8.3.2.6 Experimental
3-Chloro-2,4-dimethyl-quinoline (7)[6]

To a stirred solution of 2,3-dimethyl indole (**6**, 1.45 g, 10 mmol, 1.0 equiv) and tetra-*n*-butylammonium sulfate (3.40g, 10 mmol, 1.0 equiv) in chloroform (150 mL) was added potassium hydroxide (50% aqueous solution, 20 mL) over 30 minutes. The stirring was continued for six hours, at which time the mixture was extracted with chloroform, the chloroform-water mixture was washed with water, and the organic layer concentrated. Silica gel chromatography provided 2,4-dimethyl-3-chloroquinoline (**7**, 1.52 g, 79% yield).

8.3.2.7 References

1   Ciamician, G. L. and Dennsted, M. *Chem. Ber.* **1881**, *14*, 1153.
2   Kermack, W. O.; Perkin, W. H.; Robinson, R. *J. Chem. Soc.* **1922**, *122*, 1896.
3   (a) [R] Wynberg, H. *Chem. Rev.* **1960**, *60*, 169. (b) [R] Wynberg, H. and Meijer, E. W. *Organic Reactions*, **1982**, *28*, 1.
4   Skell, P. S. and Sandler, S. R. *J. Am. Chem. Soc.* **1958**, *80*, 2024.
5   (a) Kwon, S.; Nishimura, Y.; Ikeda, M.; Tamura, Y. *Synthesis*, **1976**, 249. (b) De Angelis, F.; Gambacorta, A.; Nicoletti, R. *Synthesis*, **1976**, 798.
6   Joshi, K. C.; Renuka, J.; Seema, J. *J. Indian Chem. Soc.* **1993**, *70*, 567.
7   Rees, C. W. and Smithen, C. E. *J. Chem. Soc.* **1964**, 938.
8   (a) Rice, H. L. and Londergan, T. E. *J. Am. Chem. Soc.* **1955**, *77*, 4678. (b) Busby, R. E.; Iqbal, M.; Khan, M. A.; Parrick, J.; Shaw, C. J. G. *J. C. S. Perkin Trans. 1* **1979**, 1578.
9   Busby, R. E.; Iqbal, M.; Parrick, J.; Shaw, C. J. G. *J. C. S. Chem. Comm.* **1969**, 1344.
10   Busby, R. E.; Hussain, S. M.; Iqbal, M.; Khan, M. A.; Parrick, J.; Shaw, C. J. G. *J. C. S. Perkin Trans. 1* **1979**, 2782.
11   Busby, R. E.; Khan, M. A.; Parrick, J.; Shaw, C. J. G.; Iqbal, M. *J. C. S. Perkin Trans. I*, **1980**, 1427.
12   (a) Parham, W. E.; Rinehart, J. K. *J. Am. Chem. Soc.* **1967**, *89*, 5668. (b) Parham, W. E.; Davenport, R. W.; Biasotti, J. B. *Tetrahedron Lett.* **1969**, *7*, 557. (c) Gambacorta, A.; Nicoletti, R.; Cerrini, S.; Fedeli, W.; Gavuzzo, G. *Tetrahedron*, **1980**, *36*, 1367 and references cited therein. (d) Botta, M.; De Angelis, F.; Gambacorta, A. *Tetrahedron*, **1982**, *38*, 2315 and references cited therein.
13   De Angelis, F.; Inesi, A.; Feroci, M.; Nicoletti R. *J. Org. Chem.* **1995**, *60*, 445.
14   (a) ref 12(b). (b) Parham, W. E.; Davenport, R. W.; Biasotti, J. B. *J. Org. Chem.* **1970**, *35*, 3775.

15   Král, V.; Gale, P. A.; Anzenbacher, P. Jr.; K. Jursíková; Lynch, V.; Sessler, J. L. *J. Chem. Soc. Chem. Comm.* **1998**, 9.

16   (a) For the preparation of substituted naphthalenes, see Parham, W. E. and Wright, C. D. *J. Am. Chem. Soc.* **1957**, *79*, 1473 and references cited therein. (b) For the preparation of metacyclophanes, see ref 12(a). (c) For the preparation of substituted benzenes, see Dehmlow, E. V. and Bollmann, C. *Tetrahedron*, **1995**, *51*, 3755.

Derek A. Pflum

8.4 The Zincke Reaction

8.4.1 Description

The Zincke reaction is an overall amine exchange process that converts N-(2,4-dinitrophenyl)pyridinium salts (e.g., **1**), known as Zincke salts, to N-aryl or N-alkyl pyridiniums **2** upon treatment with the appropriate aniline or alkyl amine.[1] The Zincke salts are produced by reaction of pyridine or its derivatives with 2,4-dinitrochlorobenzene. This venerable reaction, first reported in 1904[2] and independently explored by König,[3] proceeds via nucleophilic addition, ring opening, amine exchange, and electrocyclic reclosure, a sequence that also requires a series of proton transfers. By

Scheme 8.4.1

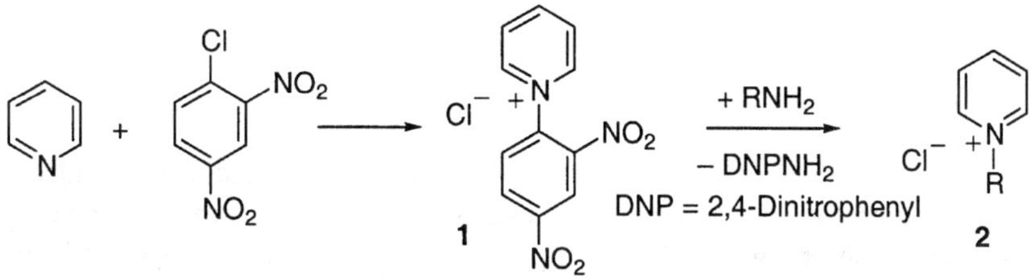

manipulating reaction conditions, including solvent and the Zincke salt counterion, the Zincke process has been applied to the preparation of a wide range of pyridinium salts, including library synthesis on solid support. The synthetic utility of the Zincke reaction derives from its efficacy in producing pyridinium salts unattainable by direct N-arylation or N-alkylation. For example, electron deficient, weakly nucleophilic pyridines can nevertheless be converted to the corresponding N-(2,4-dinitrophenyl)pyridinium salts and then to N-aryl or N-alkyl pyridiniums via the Zincke reaction. Additionally, while direct N-arylation of pyridines by nucleophilic aromatic substitution requires that the aryl halide electrophile be activated by electron withdrawing groups, the Zincke reaction tolerates a wide variety of substituted anilines as nucleophiles, resulting in a versatile route to N-aryl pyridinium salts. Pyridine N-alkylation, meanwhile, often fails with α-substituted electrophiles, and when these reactions do proceed, stereogenicity at the leaving group-bearing center is often compromised via the S_N1 process. With the Zincke reaction, however, α-chiral alkyl amines provide the corresponding N-alkyl pyridinium salts with retention of configuration. These attributes, along with the synthetic versatility of the pyridinium products, have made the Zincke reaction a key transformation in numerous preparations of alkaloids, NAD+/NADH analogs, cytotoxic metabolites, and other medicinally relevant substances.

8.4.2 Historical Perspective

In 1904, Zincke reported that treatment of N-(2,4-dinitrophenyl)pyridinium chloride (**1**) with aniline provided a deep red salt that subsequently transformed into N-phenyl pyridinium chloride **5** (Scheme 8.4.2).[2] Because the starting salt **1** was readily available from the nucleophilic aromatic substitution reaction of pyridine with 2,4-dinitrochlorobenzene, the Zincke reaction provided access to a pyridinium salt (**5**) that would otherwise require the unlikely substitution reaction between pyridine and chlorobenzene. The intermediate red salt proved to be the conjugated iminium species **3**, incorporating two aniline units. Independently, König characterized similar salts, now known as König salts, in the reactions of cyanogen bromide-activated pyridine with anilines, but assigned the ring closed structure (e.g., **4**).[3] Following these initial reports, Zincke and co-workers extended the reaction to a variety of substituted pyridines[4] as well as to the isoquinoline-derived system **6** (Scheme 8.4.3).[5]

Scheme 8.4.2

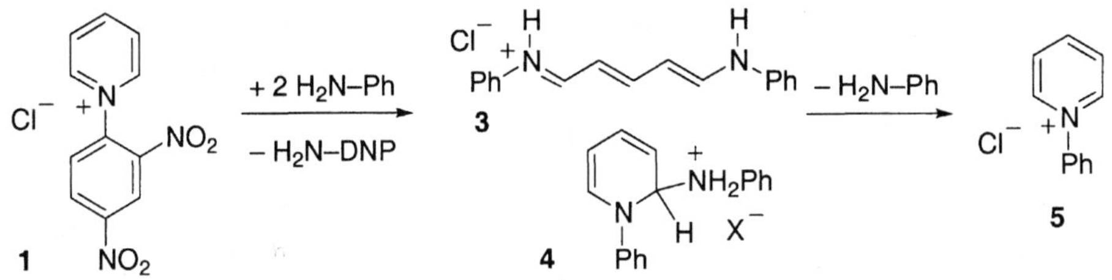

Later in the 20th century, Vompe[6] and Stepanov[7] delineated efficient procedures for the preparation of the so-called Zincke salts (e.g., **1**) from pyridines and 2,4-dinitrochlorobenzene, involving, for example, reflux in acetone. Vompe[8] and Lukes[9] also noted that electron-donating substituents on the pyridinium ring of the Zincke salt retarded reaction with amines at the 2-position of the pyridinium ring, sometimes leading instead to attack at the C-1′ position of the 2,4-dinitrobenzene ring, with displacement of the pyridine.

Scheme 8.4.3

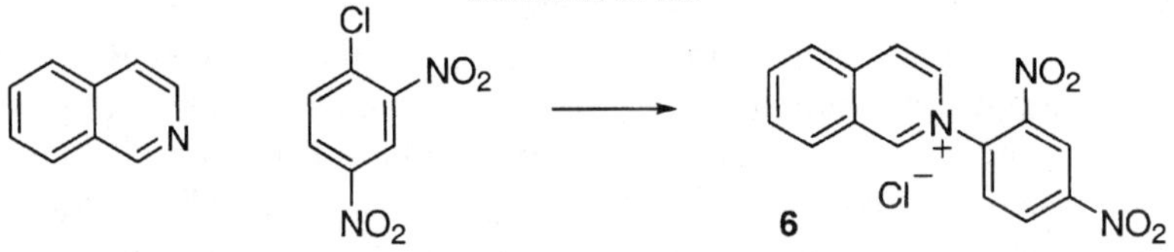

During the 1950s and 1960s Hafner used König salts, derived from the reaction of *N*-methyl aniline with Zincke salt **1**, for azulene synthesis.[10] The Zincke reaction also achieved prominence in cyanine dye synthesis[11] and as an analytical method for nicotinamide determination.[12]

Zincke salts have played an important role in the synthesis of NAD$^+$/NADH co-enzyme analogs since a 1937 report on the Zincke synthesis of dihydropyridine **7** for use in a redox titration study.[13] The widely utilized nicotinamide-derived Zincke salt **8**, first synthesized by Lettré[14] was also used by Shifrin in 1965 for the preparation and study of NAD$^+$/NADH analogs.[15] In 1972, Secrist reported using **8** for synthesis of simplified NAD$^+$ analogs such as **10** for use in spectroscopic studies (Scheme 8.4.4).[16] Subsequent utilization of **8** is discussed later in this article.

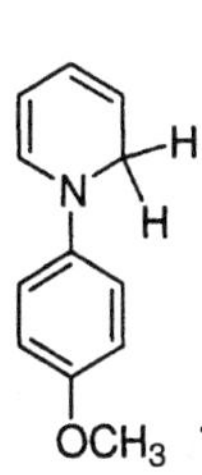

Scheme 8.4.4

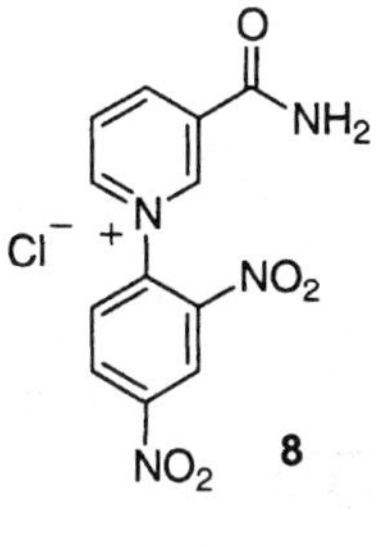

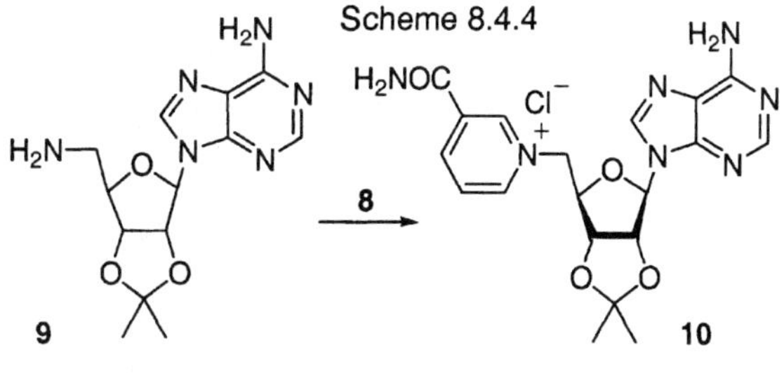

8.4.3 Mechanism

Early mechanistic studies on the Zincke reaction focused on the initial pyridinium ring opening upon addition of aniline, leading to the König salt **3** (Scheme 8.4.5). Van den Dunghen *et al.* proposed a rate-determining electrocyclic ring opening after finding that the process was first order in the Zincke salt **1**, first order in aniline, and first order in added hydroxide.[17] Oda and Mita also found the ring opening to be first order in **1**, but observed a second order dependence of ring opening rate on aniline concentration in the absence of added hydroxide, emphasizing the importance of proton transfer in the process.[18] Related studies have also been conducted on ring opening of an *N,N*-dimethyl carbamoyl pyridinium salt.[19]

Scheme 8.4.5

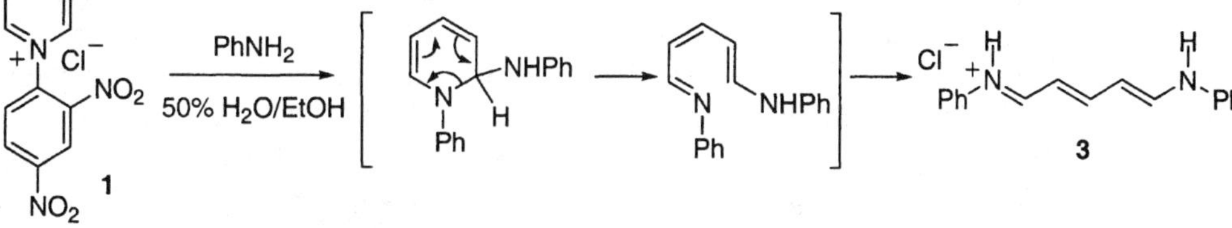

Subsequently, Marvell studied the ring-closing kinetics of the isolated König salt **3**, monitoring its conversion to phenyl pyridinium chloride by UV spectroscopy.[20,21] In the absence of any added base, the protonated material **11** was in equilibrium with a small amount of the free base **12** (Scheme 8.4.6) (K = 8.5×10^{-8} M at 42.6 °C), and cyclization via the trans→cis isomerized form **13** occurred with first order kinetics (k = 8.94×10^{-6} s^{-1} at 42.6 °C). Added base increased the rate up to 1 equiv but not thereafter, corresponding to full conversion of **11** to the deprotonated form **12**. Various bases, Et$_3$N, Bu$_3$N, and methoxide, all gave the same result. In the presence of excess base, the first-order rate constant was k' ~ 1.0×10^{-3} s^{-1} at 42.6 °C. On the basis of these results, ring closure was proposed to occur upon deprotonation and isomerization to the cis **13**. Cyclization and extrusion of aniline would then provide the *N*-phenyl pyridinium product **15**.

Scheme 8.4.6

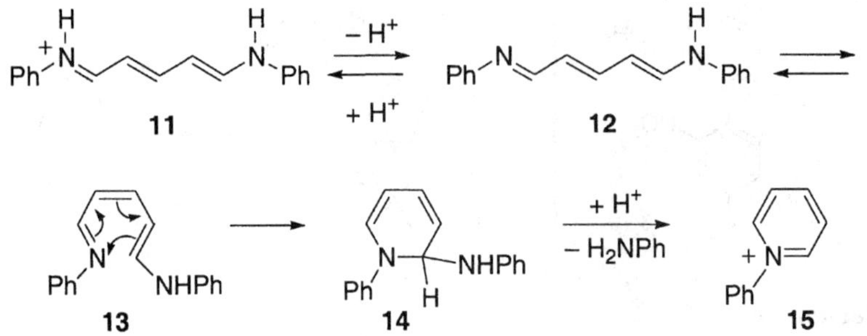

Scheme 8.4.7

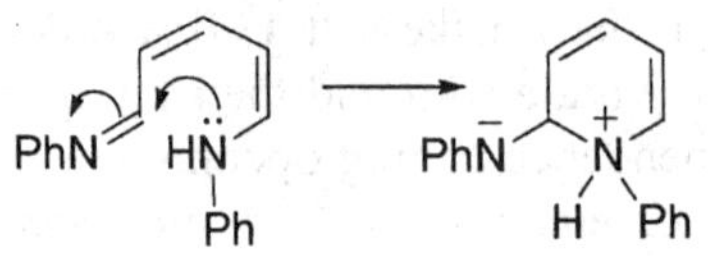

Several lines of evidence suggested that the ring closure occurred as an electrocyclization, rather than a polar nucleophilic addition (that is 16→17, Scheme 8.4.7).[21,22] The entropy of activation for the process 11→15 was close to zero, not consistent with buildup of charge in the transition state for the rate-determining cyclization step, which would entail organization of the solvent. The rate of reaction was not affected by added LiClO$_4$ and actually increased slightly with decreasing solvent polarity (MeOH → 1/1 MeOH/dioxane), again at odds with the polar ring closure mechanism. Finally, when various aryl-substituted König salts 18 were used in cyclization rate studies (Scheme 8.4.8), Marvell and Shahidi found only a minor dependence on electronic factors,[22] with the 4-nitro substitution giving cyclization just five times faster than the 4-dimethylamino substrate.

Scheme 8.4.8

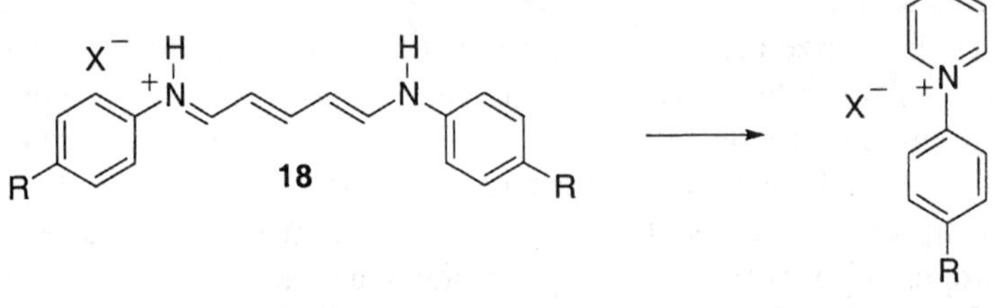

In all cases, entropies of activation were close to zero, and enthalpies of activation were similar across the series. The first-order dependence on concentration of König salt and the independence of rate on the type of added base were also maintained in the substituted systems. There was no good fit of rate data for the substrates 18 to the Hammett equation using either σ or σ^+, but using σ^+ as the better-fitting option gave $\rho = +0.35$, again indicating little dependence on electronic effects, more consistent with an electrocyclic ring closure than the polar nucleophilic addition alternative.

In terms of the final loss of aniline after ring closure, the fact that reactions using Et$_3$N and Bu$_3$N, (ammonium ion as proton source) occurred at the same rate as the reactions with methoxide base (MeOH as proton source) suggested a lack of general acid catalysis. Also, it was found that varying the amount of available acid did not change the rate of cyclization appreciably.[21]

In summary, these results were interpreted to support rate-determining electrocyclization for the ring closure, starting from the all-trans iminium 11, via the cis conformation 13 of the neutral form, followed by fast proton transfer and elimination of aniline (Scheme 8.4.6).

Scheme 8.4.9

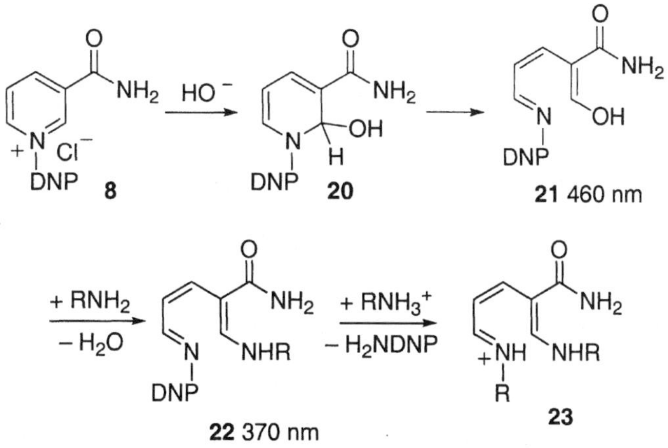

Ise and co-workers investigated the reactions of N-(2,4-dinitrophenyl)-3-carbamoyl pyridinium chloride (8) with amines, with hydroxide, and also with a combination of the two. They also measured the effect of added electrolytes as a mechanistic probe.[23] Their UV-spectroscopic studies suggested rate-determining addition of hydroxide to the pyridinium salt, followed by electrocyclic ring opening. Intermediate 21 (Scheme 8.4.9) formed even in the absence of added hydroxide in the solvent system used (EtOH/H$_2$O). Sequential exchange reactions with two equivalents of the amine and loss of 2,4-dinitroaniline would provide the König salt intermediate 23 (in equilibrium with the all-trans form) for the ring closure sequence delineated by Marvell. Intermediates 21 and 22 were identified spectroscopically, and it was found that formation of 21 was slowed by addition of electrolytes, indicating that it arose from the reaction of oppositely charged ions. Meanwhile, 22 diminished more rapidly in the presence of hydrophobic polymeric anions, suggesting a reaction with a cation, RNH$_3^+$. In related studies, Kaválek and Sterba also reported kinetic studies on the reactions of Zincke salts with anilines.[24]

From these various studies, the overall picture that emerged for the Zincke reaction mechanism is outlined in Scheme 8.4.10.[1a] When water is excluded, initial nucleophilic addition would be by the amine, leading to **26** after deprotonation and ring opening. Loss of 2,4-dinitroaniline occurs, yielding the unsaturated iminium species **27**. The conjugated system of intermediate **27** is in rapid equilibrium with the all-trans form **28** and with the neutral form **29**, even in the absence of additional base. Ring closure to **31** occurs from the cis isomer **30** via an electrocyclic process rather than a nucleophilic addition mechanism, and loss of an amine molecule follows protonation, leading to the final pyridinium salt product **32**.

Scheme 8.4.10

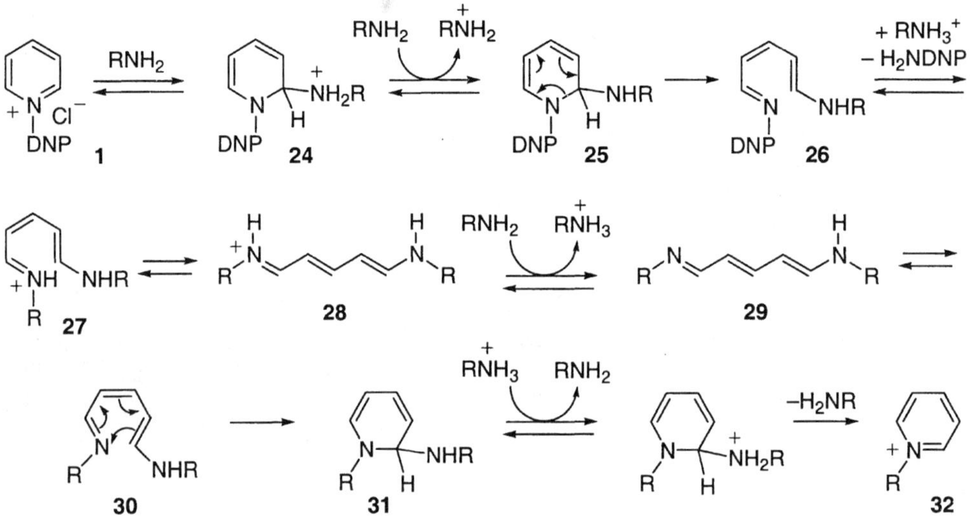

In reporting studies using polymer-bound amines in reactions with Zincke salts,[25] Kurth has pointed out that the mechanism outlined in Scheme 8.4.10 would require interactions among amino sites on the polymer for proton-transfer steps and for formation of an intermediate analogous to **29**. Furthermore, two-point attachment to the solid phase might restrict such an intermediate from adopting the appropriate cis conformation for ring closure. Kurth and Eda examined the effects of resin cross linking and amine loading levels on the efficiency of the solid-phase Zincke reaction.[25] There was little effect in going from 1% to 2% cross linking, but yields increased markedly at higher amine loadings, suggesting that site–site interactions were indeed required. Importantly, when Et$_3$N was added, the Zincke process occurred efficiently even at the lower amine loadings, suggesting that with Et$_3$N available to mediate proton-transfer steps an alternative pathway was operable (Scheme 8.4.11), involving only one polymer-bound amine site.

Scheme 8.4.11

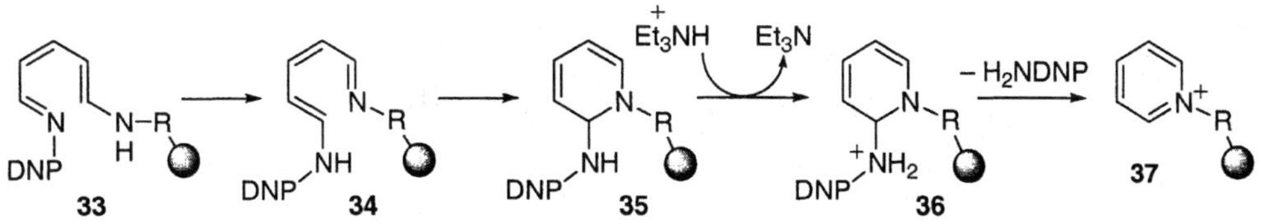

8.4.4 Variations and Improvements

In addition to their reactions with amines, Zincke salts also combine with other nitrogen nucleophiles, providing various N-substituted pyridine derivatives. Pyridine N-oxides result from the reaction with hydroxylamine, as exemplified for the conversion of Zincke salt **38** to the N-oxide **39**.[26,27] Reactions of Zincke salts with hydrazine, meanwhile, lead to N-aminopyridinium salts,[26,27,28] such as the 3-picolinium chloride derivative **41**, formed by reaction with Zincke salt **40** in water, followed by refluxing in dioxane.[26]

Scheme 8.4.12

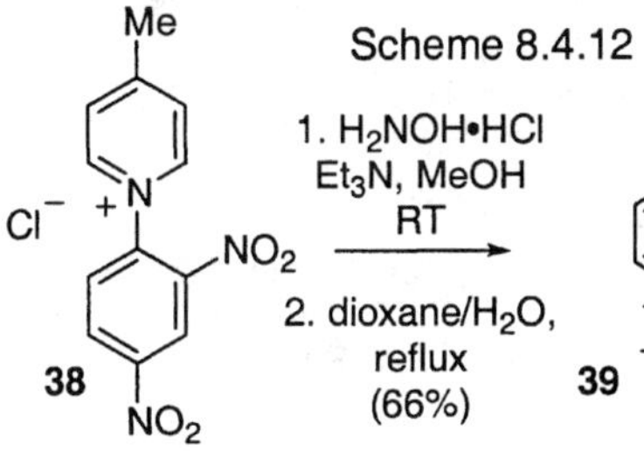

Scheme 8.4.13

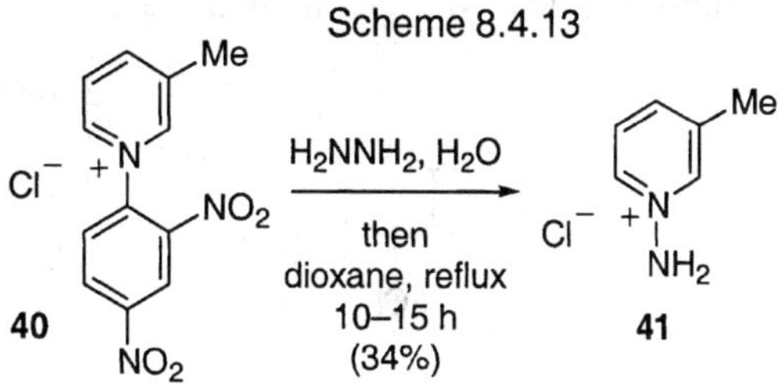

With N-acyl or N-sulfonyl hydrazines as nucleophiles, Zincke salts serve as sources of iminopyridinium ylides and ylide precursors.[27,28,29] Reaction of the nicotinamide-derived Zincke salt **8** with ethyl hydrazino urethane **42** provided salt **43**, while the tosyl hydrazine gave ylide **44** (Scheme 8.4.14).[30] Benzoyl hydrazines have also been used in reactions with Zincke salts under similar conditions.[31,32] N-amino-1,2,3,6-tetrahydropyridine derivatives such as **47** (Scheme 8.4.15), which showed anti-inflammatory activity, are also accessible via this route, with borohydride reduction of the initially formed ylide **46**.[29b]

Scheme 8.4.14

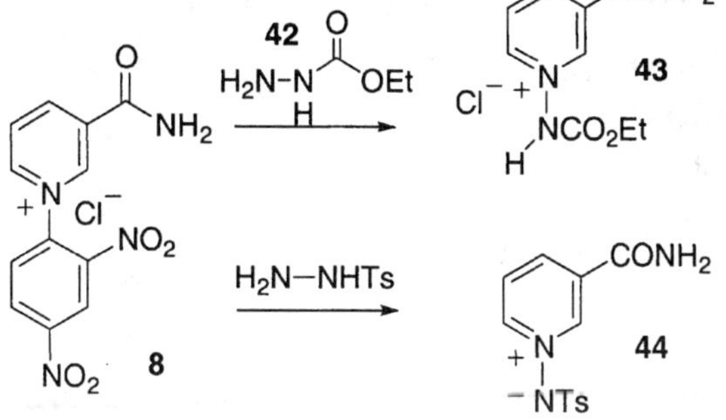

Scheme 8.4.15

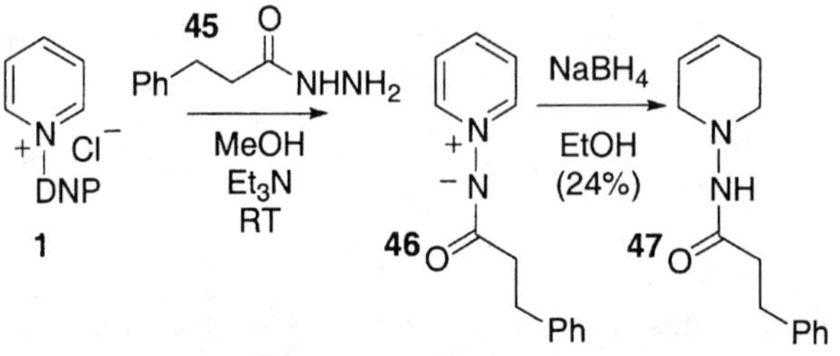

Zincke-type salts derived from other aromatic nitrogen heterocycles also undergo Zincke reactions. The isoquinolinium salt **6** (Scheme 8.4.16) permitted incorporation of a phenyl ethylamine chiral auxiliary, providing salt **48**.[33] In this context and others (*vide infra*), Marazano and co-workers found that refluxing *n*-butanol was a superior solvent system for the Zincke process. Additionally, the stereochemical integrity of the α-chiral amino fragment was reliably maintained.

Scheme 8.4.16

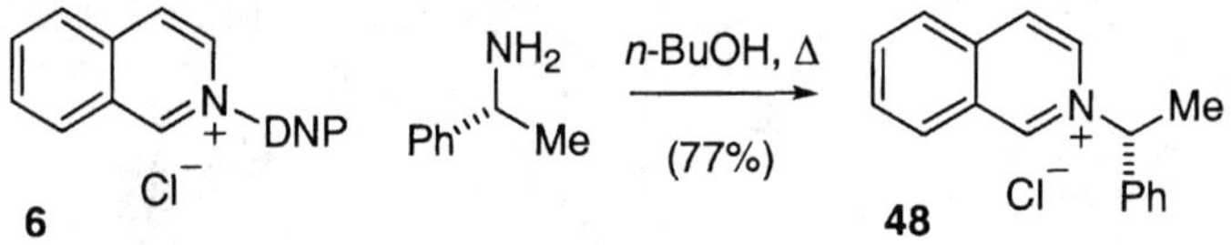

With a chiral phenylglycinol nucleophile (Scheme 8.4.17), use of the chloride Zincke salt **6** (*cf.* Scheme 8.4.16) gave decomposition of the salt back to isoquinoline and 2,4-dinitrochlorobenzene. The desired reaction was enabled by exchanging chloride for the weakly nucleophilic dodecyl sulfate anion. The resulting salt **49** also had improved solubility and smoothly underwent Zincke reaction to **50**.[33a] In the case of 5,5-dimethoxy isoquinoline-derived salts **51** and **52**, Zincke reactions worked best in CH_2Cl_2 solvent,[33a] and this solvent was broadly useful for Zincke reactions of electronically deactivated (i.e. electron-rich) pyridiniums as well (*vide infra*).

Scheme 8.4.17

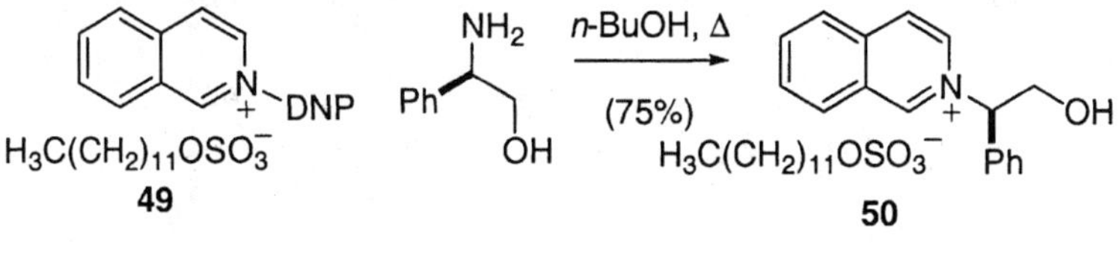

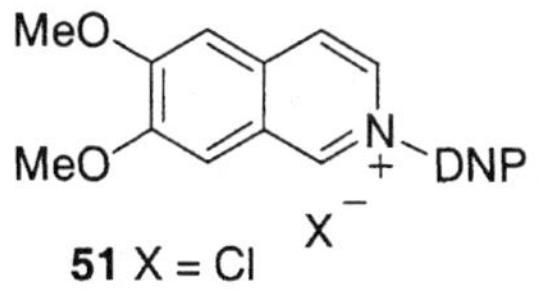

MeO, [structure]
MeO [structure with N+-DNP, X⁻]

51 X = Cl
52 X = $H_3C(CH_2)_{11}OSO_3$

The 2,7-naphthyridine system **53** (Scheme 8.4.18) was combined with 2,4-dinitrochlorobenzene and 2-amino glycerol for in situ reaction of the resulting Zincke salt. The resulting naphthyridinium **54** was trapped by Bradsher cycloaddition with (Z)-vinyl ether **55**, providing tetracycle **56** (X-ray) upon internal addition of one of the diastereotopic hydroxymethyl groups to the resulting iminium.[34] This approach was also extended to the use of chiral 2,7-naphthyridinium salts, prepared via the analogous Zincke process.[35]

Scheme 8.4.18

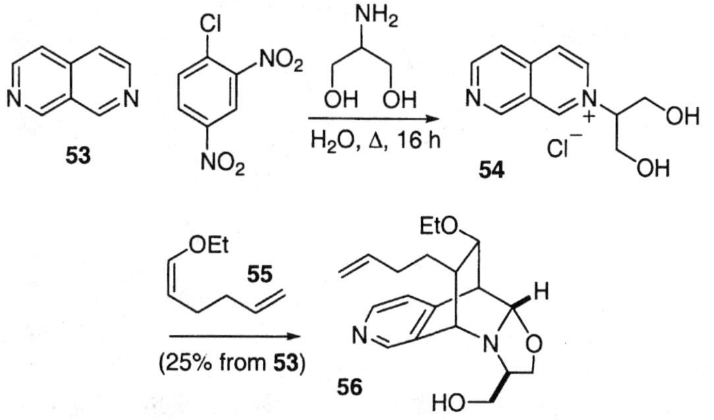

The Zincke reaction has also been adapted for the solid phase. Dupas *et al.* prepared NADH-model precursors **58**, immobilized on silica, by reaction of bound amino functions **57** with Zincke salt **8** (Scheme 8.4.19) for subsequent reduction to the 1,4-dihydropyridines with sodium dithionite.[36] Earlier, Ise and co-workers utilized the Zincke reaction to prepare catalytic polyelectrolytes, starting from poly(4-vinylpyridine). Formation of Zincke salts at pyridine positions within the polymer was achieved by reaction with 2,4-dinitrochlorobenzene, and these sites were then functionalized with various amines. The resulting polymers showed catalytic activity in ester hydrolysis.[37]

Scheme 8.4.19

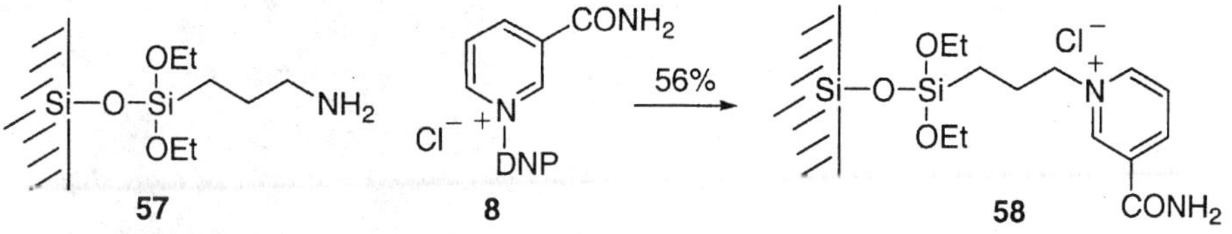

In search for activators of the cystic fibrosis transmembrane conductance regulator protein, Kurth and co-workers prepared benzo[c]quinolizinium-mimicking structures via solid-phase Zincke reaction.[38] For example, dodecyl sulfate Zincke salt **59** (Scheme 8.4.20) reacted with resin-bound phenylglycinol, providing the solid-supported Zincke product **60**. Toluene solvent provided advantageous swelling of the resin, and Et$_3$N was added to circumvent the need for two-site participation in the Zincke reaction[25] (see section 8.4.3 for the mechanism). Cleavage from the resin, ion exchange, and purification provided pyridinium chloride **61** in 81% overall isolated yield and 96% purity, as judged by HPLC analysis.

Scheme 8.4.20

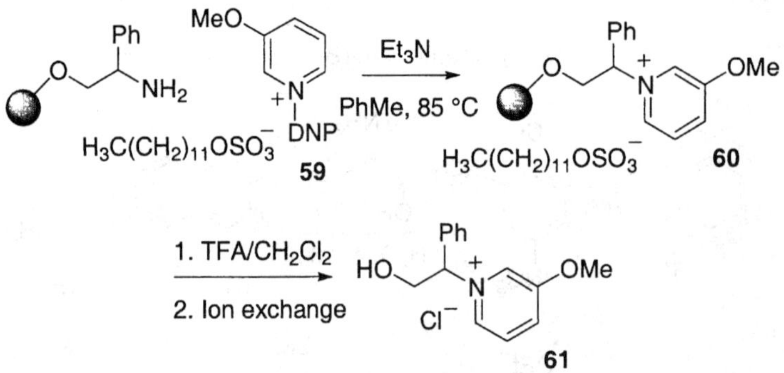

Eda and Kurth applied a similar solid-phase combinatorial strategy for synthesis of pyridinium, tetrahydropyridine, and piperidine frameworks as potential inhibitors of vesicular acetylcholine transporter.[39] One member of the small library produced was prepared from amino-functionalized trityl resin reacting with a 4-phenyl Zincke salt to give resin-bound product **62** (Scheme 8.4.21). After ion exchange and cleavage from the resin, pyridinium **63** was isolated. Alternatively, borohydride reduction of **62** led to the 1,2,3,6-tetrahydropyridine **64**, which could be hydrogenated to the corresponding piperidine **65**.

Scheme 8.4.21

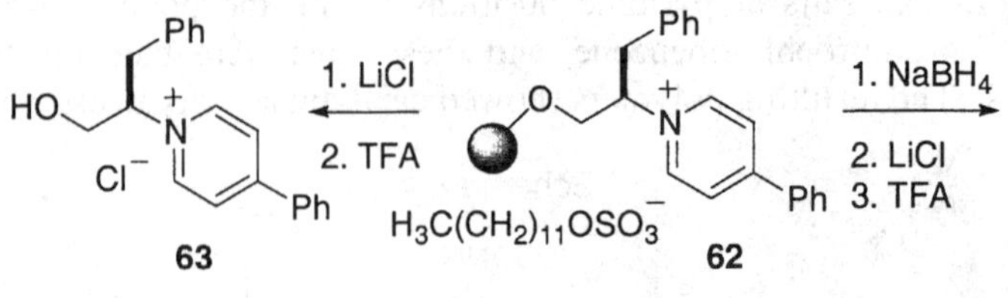

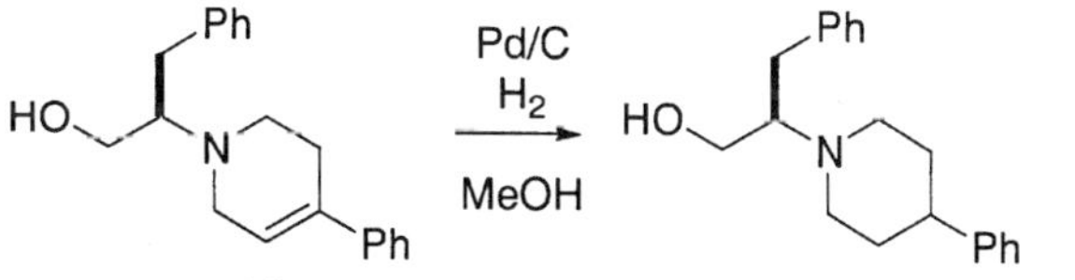

Prokai and co-workers have also used a solid-phase Zincke approach in preparing analogs of the tripeptide TRH (pGlu–His–Pro–NH$_2$), a hypothalamic releasing factor for regulation of pituitary function.[40]   The strategy was to deliver a prodrug having the central histidine replaced with a 1,4-dihydropyridine unit. The prodrug was expected to cross the hydrophobic blood-brain barrier, but to be trapped within the central nervous system upon oxidation to the hydrophilic pyridinium form. Diaminopropionic acid was incorporated into a resin-bound tripeptide, and Zincke reaction of amine **66** (Scheme 8.4.22) was carried out with nicotinamide-derived salt **8** in DMF with a catalytic amount of added pyridine. The pyridinium-containing tripeptide was cleaved from the resin and reduced to the dihydro form **68**, which showed in vivo activity, reducing sleeping time for mice drugged with pentobarbital.

Scheme 8.4.22

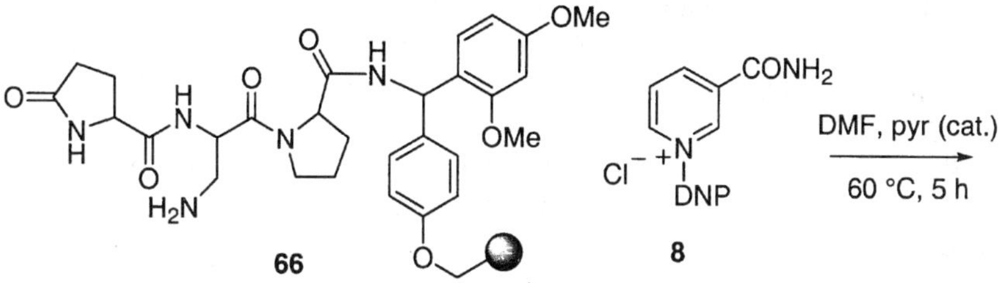

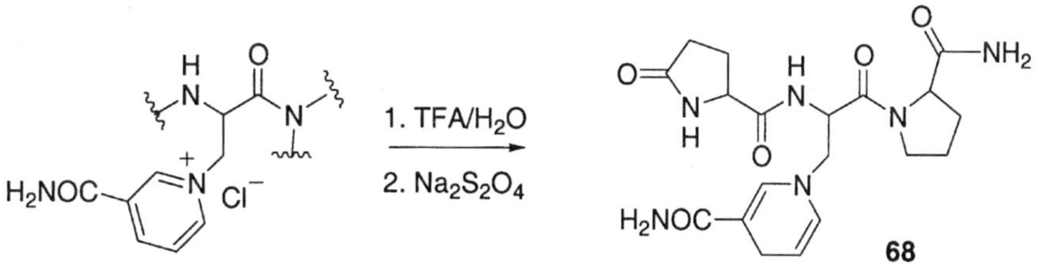

8.4.5 Synthetic Utility

The Zincke reaction has been of considerable use in the asymmetric synthesis of nitrogen heterocycles, because it permits incorporation of α-chiral amines with retention of configuration. Moreover, the resulting pyridinium salts are versatile building blocks for elaboration to a range of structures, including piperidine-containing natural products. In 1974, de Gee *et al.* demonstrated that a variety of α-chiral amines reacted with Zincke salt **71**, giving a series of *N*-(alkyl)-*p*-carbomethoxy pyridinium salts **72** (Scheme 8.4.23).[41]

Scheme 8.4.23

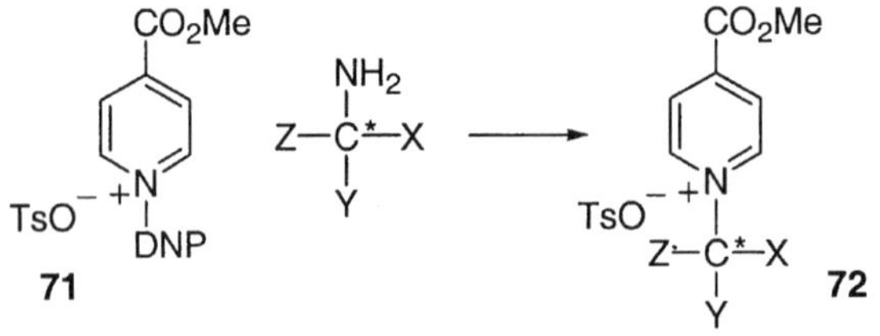

71   **72**

Scheme 8.4.24

8   **73 R =H**   **75 R =H**
                 74 R = Ac   **76 R = Ac**

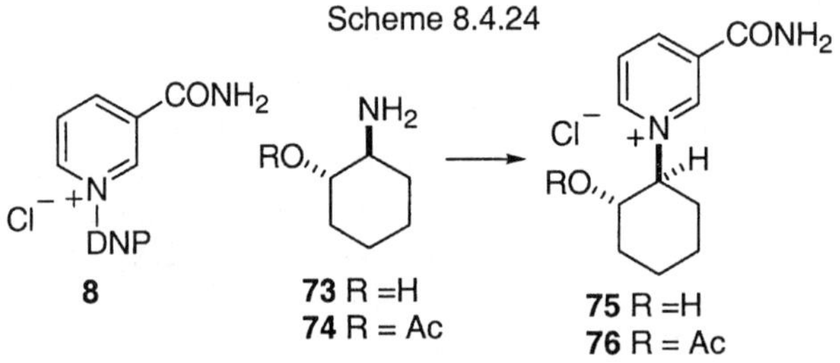

The configuration of the amine was retained, except in the case of amino acid derivatives, which racemized at the stage of the pyridinium salt product. Control experiments showed that, while the starting amino acid was configurationally stable under the reaction conditions, the pyridinium salt readily underwent deuterium exchange at the α-position in D_2O. In another early example, optically active amino alcohol **73** and amino acetate **74** provided chiral 1,4-dihydronicotinamide precursors **75** and **76**, respectively, upon reaction with Zincke salt **8** (Scheme 8.4.24).[42] The 1,4-dihydro forms of **75** and **76** were used in studies on the asymmetric reduction of α,β-unsaturated iminium salts.

Scheme 8.4.25

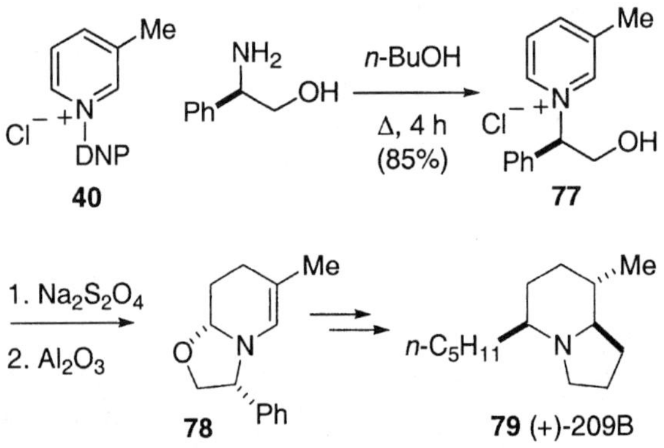

40   **77**

78   **79 (+)-209B**

Marazano and co-workers have used the Zincke reaction extensively to prepare chiral templates for elaboration to substituted piperidine and tetrahydropyridine natural products and medicinal agents. For example, 3-picoline was converted to Zincke salt **40** by reaction with 2,4-dinitrochlorobenzene in refluxing acetone, and treatment with R-(–)-phenylglycinol in refluxing n-butanol generated the chiral pyridinium **77**.[43] Reduction to the 1,4-dihydropyridine led to cyclization upon exposure to alumina. The resulting bicyclic oxazolidine **78** served as a framework for a sequence of moderately diastereoselective Grignard additions, providing the dart-poison alkaloid (+)-209B (the antipode of the natural material) after auxiliary cleavage and reductive amination.[43, 44, 45]

Scheme 8.4.26

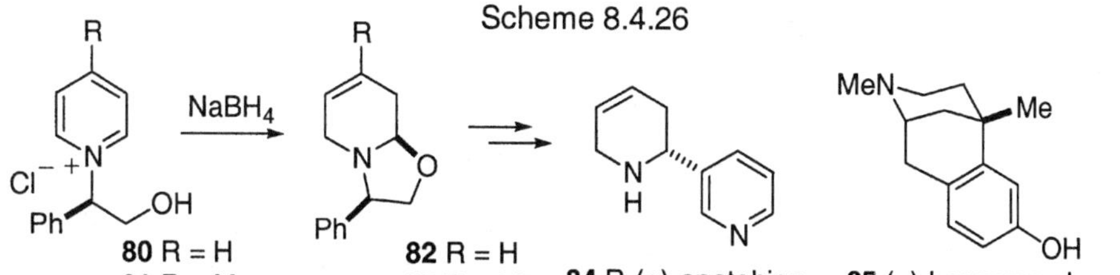

80 R = H
81 R = Me
82 R = H
83 R = Me
84 R-(+)-anatabine
85 (+)-benzomorphan

A similar oxazolidine-based strategy was employed in the syntheses of R-(+)-anatabine (**84**) and (+)-benzomorphan (**85**), utilizing Zincke-derived salts **80** and **81**, respectively (Scheme 8.4.26). In these cases, borohydride reduction, followed by cyclization, led to oxazolidines **82** and **83** as substrates for organometallic addition and subsequent elaboration to the natural products.[46,1a]

Diastereomeric oxazolidines **88** and **89** (Scheme 8.4.27) were available via 1,4-reduction and cyclization of Zincke product **87**, with **88** being the kinetic product under the reaction conditions and **89** (X-ray) the thermodynamically favored isomer in CDCl3 solution. Iterative reductive oxazolidine opening provided 3-aryl piperidine **91**, which was readily advanced to (–)-PPP (**92**), a selective dopaminergic receptor antagonist.[47]

Scheme 8.4.27

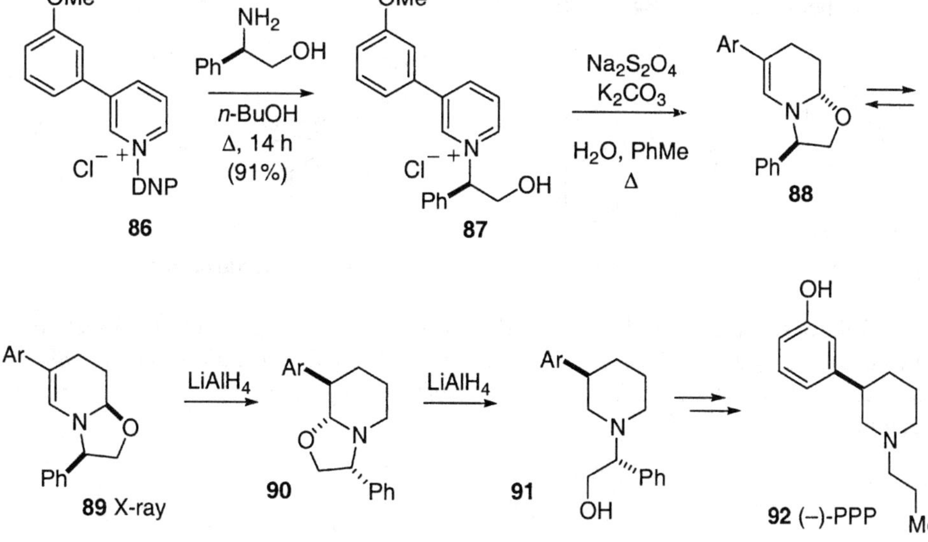

86
87
88

89 X-ray
90
91
92 (–)-PPP

Scheme 8.4.28

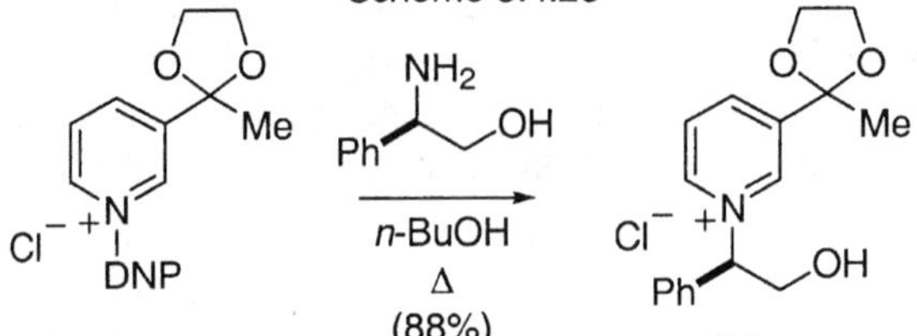

Scheme 8.4.29

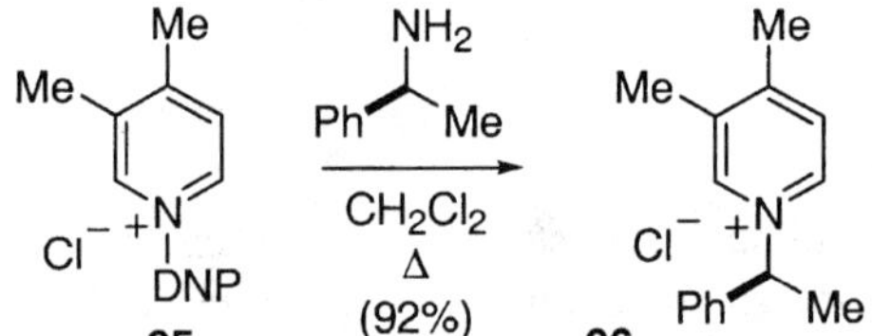

Marazano's conditions for the Zincke reaction were applicable to other variously substituted pyridiniums. For example, acetonide-protected salt **93** provided chiral product **94** in high yield (Scheme 8.4.28). The higher boiling alcohol solvent was required for good yields, and no amine racemization was observed. With electron-rich pyridinium salt **95** (Scheme 8.4.29), refluxing CH_2Cl_2 was the optimal solvent system, as use of *n*-BuOH gave only nucleophilic attack on the dinitrobenzene ring.[48] Piperidine natural products **98** and **99** were accessed via electron-rich Zincke salts *ent*-**96** and **97** (Scheme 8.4.30), prepared under the same CH_2Cl_2 conditions and advanced in a fashion similar to the route toward **84** and **85** (*cf.* Scheme 8.4.26).[49]

Scheme 8.4.30

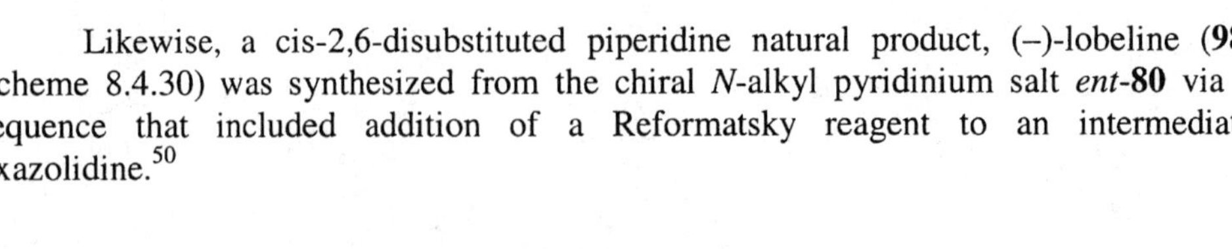

ent-**96**   **98** (+)-Normétazocine   **99** (+)- Nordextrorphan   **97**

Likewise, a cis-2,6-disubstituted piperidine natural product, (−)-lobeline (**98**, Scheme 8.4.30) was synthesized from the chiral *N*-alkyl pyridinium salt *ent*-**80** via a sequence that included addition of a Reformatsky reagent to an intermediate oxazolidine.[50]

Scheme 8.4.31

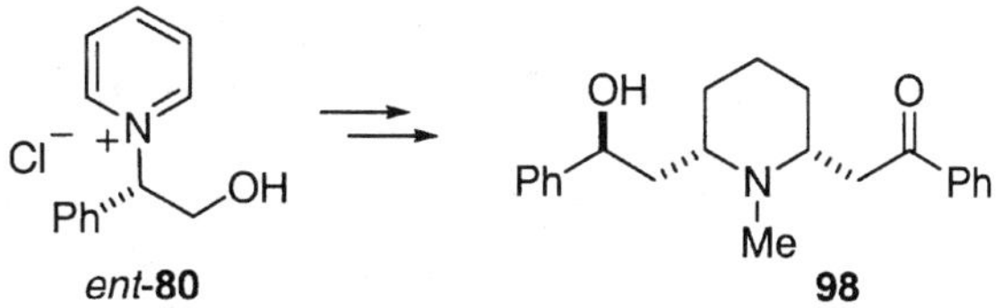

ent-**80**   **98**

Scheme 8.4.32

99 Ph   **100** Ph

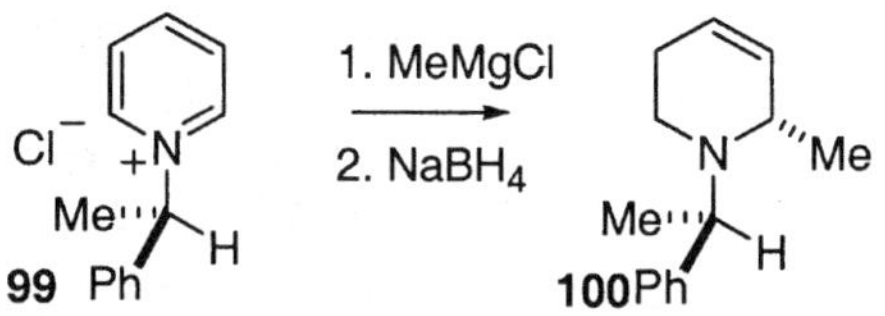

Direct addition of Grignard reagents to Zincke-derived chiral pyridinium salts such as **99**, meanwhile, allowed subsequent reduction to 1,2,3,6-tetrahydropyridines (e.g., **100**, Scheme 8.4.32). This strategy provided entry to asymmetric syntheses of (−)-lupetidin and (+)-solenopsin.[51] Tetrahydropyridines prepared by reduction of chiral pyridinium salts could also be epoxidized diastereoselectively,[52] and alkylated product **100**, for example, was readily advanced to the highly oxygenated piperidine **101** (Scheme 8.4.33).[53]

Scheme 8.4.33

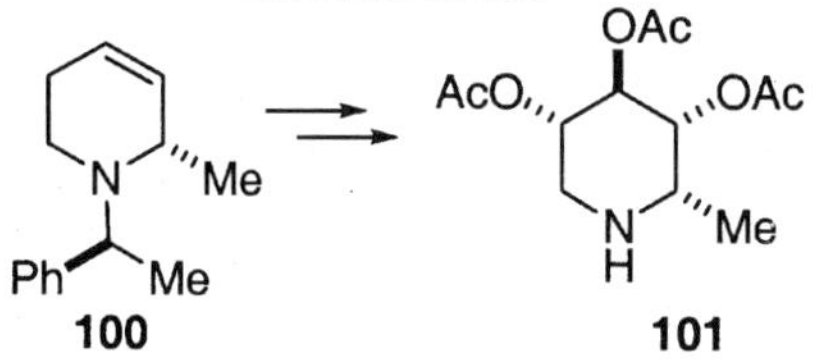

100   **101**

Scheme 8.4.34

102   **103**   **104** CO$_2$Me

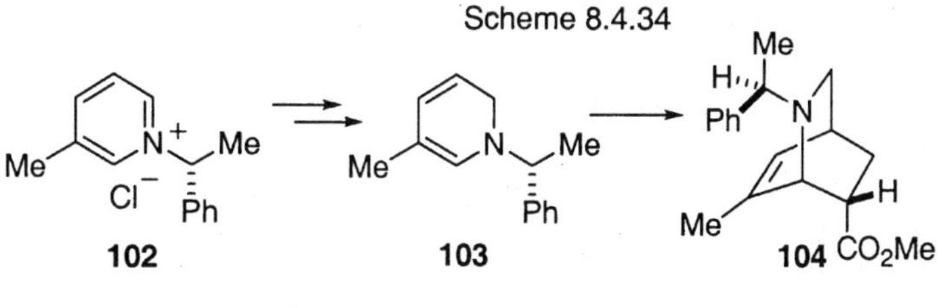

Chiral dihydropyridines such as **103** were also accessible from Zincke-derived N-alkyl pyridinium salt **102** (Scheme 8.4.34). The dihydropyridine underwent cycloaddition with methylacrylate, providing chiral isoquinuclidine derivative **104** as the major diastereomeric product.[54]

Scheme 8.4.35

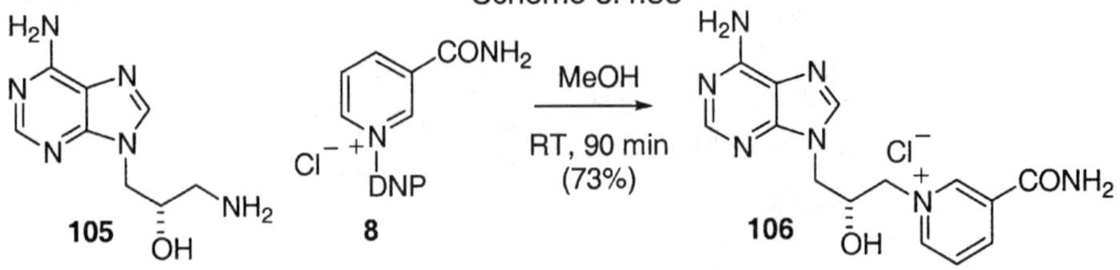

The utility of the Zincke reaction has been extended to the preparation of various NAD$^+$ and NADH analogs. Holy and co-workers synthesized a series of NAD$^+$ analogs containing nucleotide bases as a means to study through-space interaction between the pyridinium and base portions.[55] Nicotinamide-derived Zincke salt **8** was used to link with various adenine derivatives via tethers that contained hydroxyl (**105 → 106**, Scheme 8.4.35),[55a,b,d] phosphonate (**107→108**, Scheme 8.4.36),[55c] and carboxylate[55d] functionality, for example. For the Zincke reactions of an amino acid[55d] and amino phosphonate **107**,[55c] a base (DBU) was included in the reaction mixture to deprotonate the ammonium function of the zwitterionic starting material.

Scheme 8.4.36

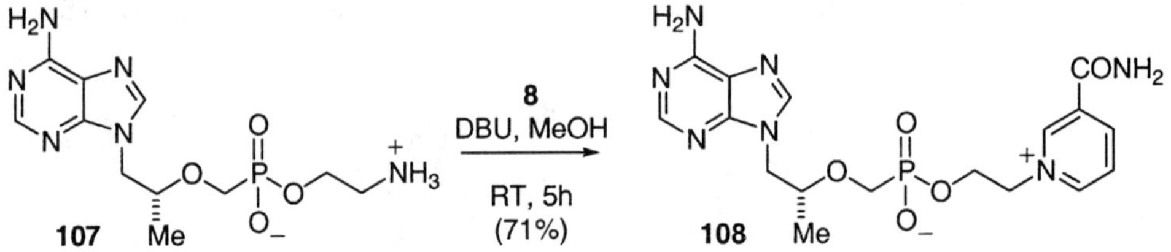

The Zincke reaction has also been used to prepare medicinal agents for chemical delivery via an *in situ* dihydropyridine/pyridinium redox strategy. One example was outlined in the solid-phase synthesis of tripeptide analog **70** (Scheme 8.4.22).[40] Woodard *et al.*, meanwhile, prepared a γ-aminobutyric acid (GABA) analog prodrug via Zincke reaction of pyridinium salt **8** with the γ-amino aldehyde acetal **109** (Scheme 8.4.37).[56,1a] The 1,4-dihydropyridine **110**, obtained after dithionite reduction, served as a precursor to GABA analog **111**. The neutral prodrug is sufficiently lipophilic for efficient delivery and is hydrolyzed and oxidized in vivo to the active form **111**. The same research group also used the nicotinamide-derived Zincke salt **8** in reactions with various anilines and primary amines to synthesize 1,4-dihydropyridines for a study on the effect of substitution on rate of reoxidation with K$_3$Fe(CN)$_6$.[57]

Scheme 8.4.37

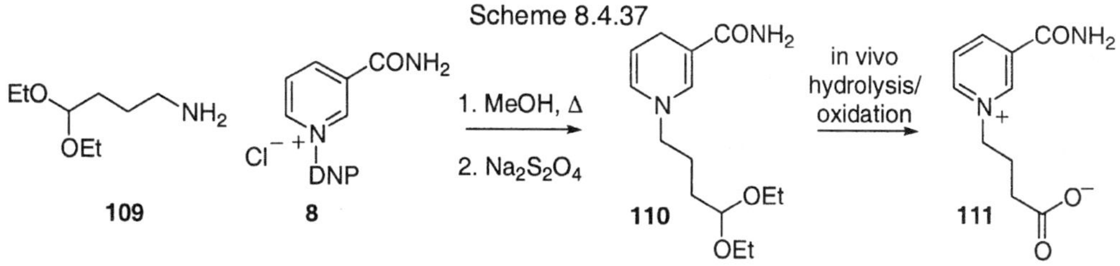

109 **8** → **110** in vivo hydrolysis/ oxidation **111**

1. MeOH, Δ

2. Na₂S₂O₄

Utilizing the Zincke reaction of salts such as **112** (Scheme 8.4.38), Binay *et al.* prepared 4-substituted-3-oxazolyl dihydropyridines as NADH models for use in asymmetric reductions.[58] They found that high purity of the Zincke salts was required for efficient reaction with *R*-(+)-1-phenylethyl amine, for example. As shown in that case (Scheme 8.4.38), chiral *N*-substituents could be introduced, and 1,4-reduction produced the NADH analogs (e.g. **114**).

Scheme 8.4.38

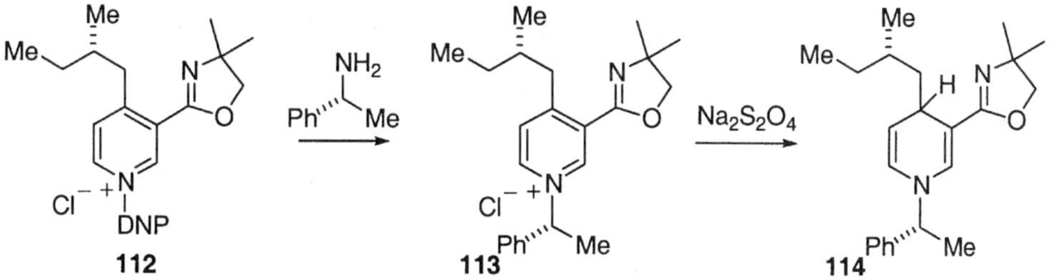

112 **113** **114**

Marazano and co-workers have also applied the reactions of tryptamine with various Zincke salts, including **115** (Scheme 8.4.39), in the synthesis of pyridinium salts such as **116**.[59] This type of product is useful for further conversion to dihydropyridine or 2-pyridone derivatives. For example, in a different study, Zincke-derived chiral pyridinium salts could be oxidized site-selectively with potassium ferricyanide under basic conditions as a means of chiral 2-pyridone synthesis (**117** → **118**, Scheme 8.4.40).[60]

Scheme 8.4.39

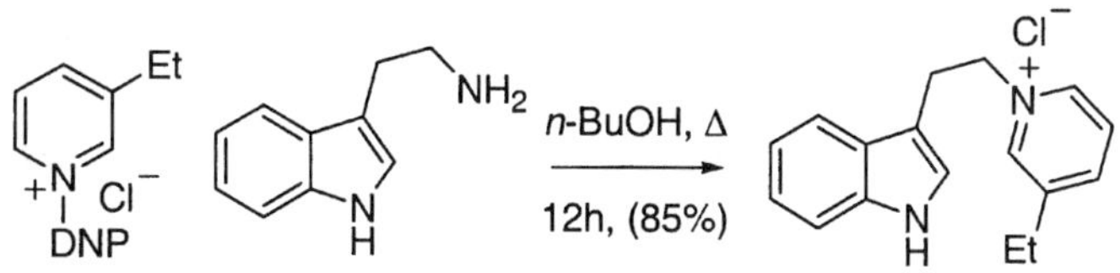

115 *n*-BuOH, Δ **116**

12h, (85%)

Scheme 8.4.40

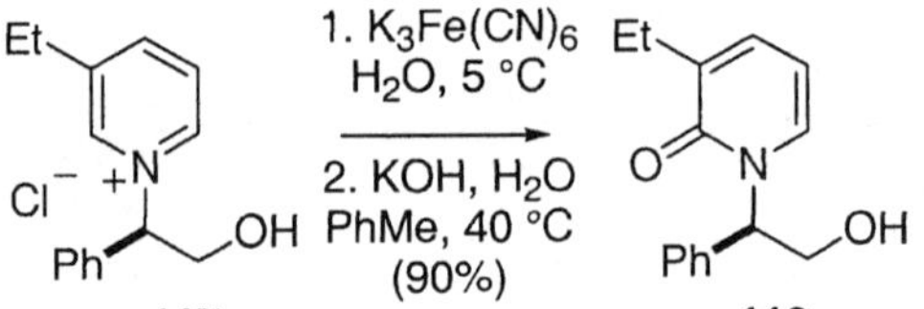

An intriguing application of Zincke processes occurred in Marazano's synthesis of dimeric, tetrameric, and even octameric pyridinium macrocycles, including cyclostellettamine B, a sponge-derived natural product.[61] The same strategy produced a synthesis of haliclamine A (121, Scheme 8.4.41), a cytotoxic sponge metabolite.[62] Intermediate 119, itself produced via a Zincke route, underwent an intramolecular Zincke reaction, providing macrocycle 120, which was reduced to the natural product.

Scheme 8.4.41

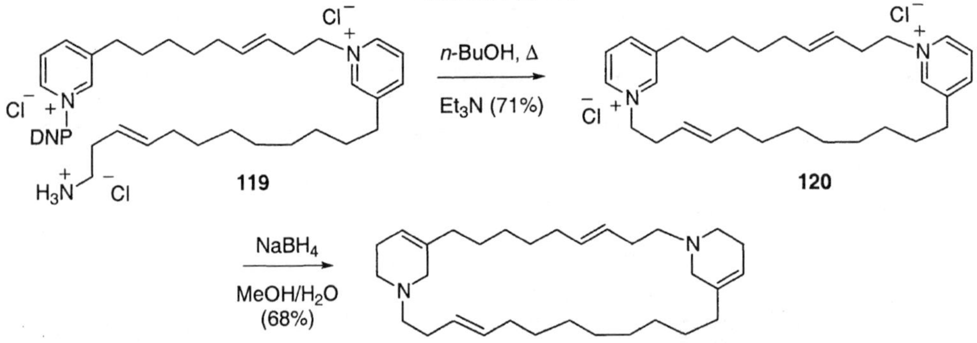

121 Haliclamine A

8.4.6   Experimental

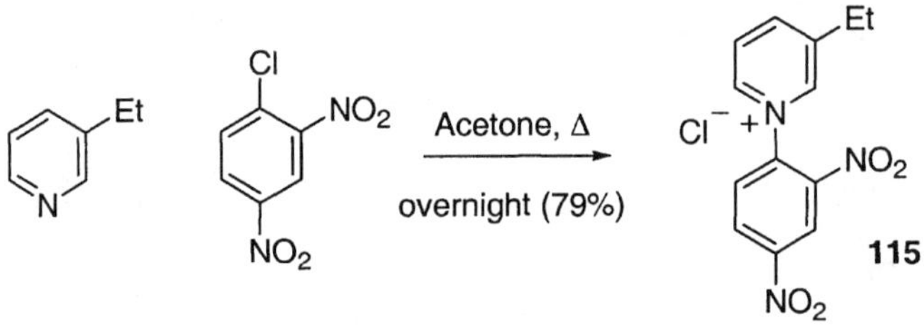

1-(2,4-Dinitrophenyl)-3-ethylpyridinium Chloride (115).[47]

A solution of 3-ethylpyridine (23 mL, 202.2 mmol) and 1-chloro-2,5-dinitrobenzene (40 g, 197.5 mmol) was refluxed in acetone (70 mL) overnight. The precipitate of salt 115 was collected by filtration as a white solid (48.38 g, 79%).

(–)-3-Ethyl-1-[(1R)-2-hydroxy-1-phenylethyl]pyridinium Chloride (122).[47]

R-(–)-Phenylglycinol (5 g, 36.5 mmol) was added to a solution of Zincke salt **115** (10.3 g, 33.3 mmol) in *n*-butanol (100 mL) at 20 °C. The resulting deep red solution was refluxed during 20 h. Removal of the solvent under reduced pressure left a residue that was treated with H$_2$O (70 mL). The precipitate (2,4-dinitroaniline hydrochloride) was eliminated by filtration, and the operation was repeated twice. The combined aqueous phase was basified with concentrated ammonia (5 mL) and washed twice with EtOAc (200 mL) in order to remove the remaining 2,4-dinitroaniline and the excess of *R*-(–)-phenylglycinol. Evaporation of water gave salt **122** (7.53 g, 86%) as a pale orange gum.

8.4.7　References

1.　　[R] (a) Cheng, W.-C.; Kurth, M. J. *Org. Prep. Proc. Int.* **2002**, *34*, 585. Related reviews: [R] (b) Becher, J. *Synthesis* **1980**, 589. [R] (c) Kost, A. N.; Gromov, S. P.; Sagitullin, R. S. *Tetrahedron* **1981**, *37*, 3423.

2.　　Zincke, Th. *Justus Liebigs Ann. Chem.* **1904**, *330*, 361.

3.　　König, W. *J. Prakt. Chem.* **1904**, *69*, 105.

4.　　(a) Zincke, Th.; Heuser, G.; Möller, W. *Justus Liebigs Ann. Chem.* **1904**, *333*, 296. (b) Zincke, Th.; Würker, W. *Justus Liebigs Ann. Chem.* **1905**, *338*, 107. (c) Zincke, Th.; Würker, W. *Justus Liebigs Ann. Chem.* **1905**, *341*, 365.

5.　　Zincke, Th.; Weisspfenning, G. *Justus Liebigs Ann. Chem.* **1913**, *396*, 103.

6.　　(a) Vompe, A. F.; Turitsyna, N. F.; Levkoev, I. I. *Doklady Akad. Nauk SSSR* **1949**, *65*, 839 [*Chem. Abstr.* **1949**, *43*, 6626h]. (b) Vompe, A. F.; Turitsyna, N. F. *Zhur. Obshchei Khim.* **1957**, *27*, 3282 [*Chem. Abstr.* **1958**, *52*, 9112d].

7.　　Stepanov, F. N.; Aldanova, N. A.; Yurchenko, A. G.; Dovgan, N. L. *Metody Polucheniya Khim. Reactivov i Preparatov, Gos. Kom. Sov. Min. SSSR po Khim.* **1962**, 86 [*Chem. Abstr.* **1964**, *60*, 15800e].

8.　　Vompe, A. F.; Turitsyna, N. F. *Doklady Akad. Nauk SSSR* **1949**, *64*, 341 [*Chem. Abstr.* **1949**, *43*, 4671a].

9.　　Lukes, R.; Jizba, J. *Chem. listy* **1957**, *51*, 2334 [*Chem. Abstr.* **1958**, *52*, 6348a].

10.　　(a) Hafner, K. *Justus Liebigs Ann. Chem.* **1957**, *606*, 79. (b) Hafner, K. *Angew. Chem.* **1958**, *70*, 419. (c) Hafner, K.; Asmus, K. D. *Justus Liebigs Ann. Chem.* **1964**, *671*, 31.

11.　　[R] Hamer, F. *Cyanine Dyes and Related Compounds*; Interscience: New York; 1964, pp 244–269.

12.　　(a) Vitler, S. P.; Spies, T. D.; Mathews, A. P. *J. Biol. Chem.* **1938**, *125*, 85. (b) Karrer, P.; Keller, H. *Helv. Chim. Acta* **1938**, *21*, 463.

13.　　Karrer, P.; Schwarzenbach, G.; Utzinger, G. E. *Helv. Chim. Acta* **1937**, *20*, 72.

14.　　Lettré, H.; Haede, W.; Ruhbaum, E. *Justus Liebigs Ann. Chem.* **1953**, *579*, 123.

15.　　Shifrin, S. *Biochim. Biophys. Acta* **1965**, *96*, 173.

16.　　Secrist, J. A. III; Leonard, N. J. *J. Am. Chem. Soc.* **1972**, *94*, 1702.

17.　　Van den Dunghen, E.; Nasielski, J.; Van Laer, P. *Bull. Soc. Chim. Belg.* **1957**, *66*, 661.

18.　　Oda, R.; Mita, S. *Bull. Chem. Soc. Jpn.* **1963**, *36*, 103.

19.　　Johnson, S. L.; Rumon, K. A. *Tetrahedron Lett.* **1966**, 1721.

20.　　Marvell, E. N.; Caple, G.; Shahidi, I. *Tetrahedron Lett.* **1967**, 277.

21.　　Marvell, E. N.; Caple, G.; Shahidi, I. *J. Am. Chem. Soc.* **1970**, *92*, 5641.

22.　　Marvell, E. N.; Shahidi, I. *J. Am. Chem. Soc.* **1970**, *92*, 5646.

23.　　Kunugi, S.; Okubo, T.; Ise, N. *J. Am. Chem. Soc.* **1976**, *98*, 2282.

24.　　(a) Kaválek, J.; Sterba, V. *Collect. Czech. Chem. Commun.* **1973**, *38*, 3506. (b) Kaválek, J.; Polansky, J.; Sterba, V. *Collect. Czech. Chem. Commun.* **1974**, *39*, 1049.

25.　　Eda, M.; Kurth, M. J. *Chem. Commun.* **2001**, 723.

26.　　Tamura, Y.; Tsujimoto, N.; Mano, M. *Chem. Pharm. Bull.* **1971**, *19*, 130.

27.　　Tamura, Y.; Tsujimoto, N. *Chem. Ind.* **1970**, 926.

28.   Ágai, B.; Lempert, K. *Tetrahedron* **1972**, *28*, 2069.
29.   (a) Knaus, E. E.; Redda, K. *J. Heterocyclic Chem.* **1976**, *13*, 1237. (b) Yeung, J. M.; Corleto, L. A.; Knaus, E. E. *J. Med. Chem.* **1982**, *25*, 191. (c) Redda, K. K.; Melles, H.; Rao, K. N. *J. Heterocyclic Chem.* **1990**, *27*, 1041.
30.   Epsztajn, J.; Lunt, E.; Katritzky, A. R. *Tetrahedron* **1970**, *26*, 1665.
31.   Tamura, Y.; Miki, Y.; Honda, T.; Lkeda, M. *J. Heterocyclic Chem.* **1972**, *9*, 865.
32.   Rao, K. N.; Redda, K. K. *J. Heterocyclic Chem.* **1995**, *32*, 307.
33.   (a) Barbier, D.; Marazano, C.; Das, B. C.; Potier, P. *J. Org. Chem.* **1996**, *61*, 9596. (b) Barbier, D.; Marazano, C.; Riche, C.; Das, B. C.; Potier, P. *J. Org. Chem.* **1998**, *63*, 1767.
34.   Magnier, E.; Langlois, Y. *Tetrahedron Lett.* **1998**, *39*, 837.
35.   Urban, D.; Duval, E.; Langlois, Y. *Tetrahedron Lett.* **2000**, *41*, 9251.
36.   Dupas, G.; Tintillier, P.; Tréfouel, T.; Cazin, J.; Losset, D.; Bourguignon, J.; Quéguiner, G. *New J. Chem.* **1989**, *13*, 255.
37.   Ise, N.; Okubo, T.; Kitano, H.; Kunugi, S. *J. Am. Chem. Soc.* **1975**, *97*, 2882.
38.   Eda, M.; Kurth, M. J.; Nantz, M. H. *J. Org. Chem.* **2000**, *65*, 5131.
39.   Eda, M.; Kurth, M. J. *Tetrahedron Lett.* **2001**, *42*, 2063.
40.   Prokai-Tatrai, K.; Perjési, P.; Zharikova, A. D.; Li, X.; Prokai, L. *Bioorg. Med. Chem. Lett.* **2002**, *12*, 2171.
41.   de Gee, A. J.; Sep, W. J.; Verhoeven, J. W.; de Boer, T. J. *J. Chem. Soc., Perkin Trans. I* **1974**, 676.
42.   Baba, N.; Makino, T.; Oda, J.; Inouye, Y. *Can. J. Chem.* **1980**, *58*, 387.
43.   Gnecco, D.; Marazano, C.; Das, B. C. *J. Chem. Soc., Chem. Commun.* **1991**, 625.
44.   Wong, Y.-S.; Gnecco, D.; Marazano, C.; Chiaroni, A.; Riche, C.; Billion, A.; Das, B. C. *Tetrahedron* **1998**, *54*, 9357.
45.   Mehmandoust, M.; Marazano, C.; Das, B. C. *J. Chem. Soc., Chem. Commun.* **1989**, 1185.
46.   Génisson, Y.; Mehmandoust, M.; Marazano, C.; Das, B. C. *Heterocycles* **1994**, *39*, 811.
47.   Wong, Y.-S.; Marazano, C.; Gnecco, D.; Génisson, Y.; Chiaroni, A.; Das, B. C. *J. Org. Chem.* **1997**, *62*, 729.
48.   Genisson, Y.; Marazano, C.; Mehmandoust, M.; Gnecco, D.; Das, B. C. *Synlett* **1992**, 431.
49.   Génisson, Y.; Marazano, C.; Das, B. C. *J. Org. Chem.* **1993**, *58*, 2052.
50.   Compère, D.; Marazano, C.; Das, B. C. *J. Org. Chem.* **1999**, *64*, 4528.
51.   Guilloteau-Bertin, B.; Compère, D.; Gil, L.; Marazano, C.; Das, B. C. *Eur. J. Org. Chem.* **2000**, 1391.
52.   Diez, A.; Vilaseca, L.; López, I.; Rubiralta, M.; Marazano, C.; Grierson, D. S.; Husson, H.-P. *Heterocycles* **1991**, *32*, 2139.
53.   Gil, L.; Compère, D.; Guillotequ-Bertin, B.; Chiaroni, A.; Marazano, C. *Synthesis* **2000**, 2117.
54.   (a) dos Santos, D. C.; de Freitas Gil, R. P.; Gil, L.; Marazano, C. *Tetrahedron Lett.* **2001**, *42*, 6109. (b) Mehmandoust, M.; Marazano, C.; Singh, R.; Gillet, B.; Césario, M.; Fourrey, J.-L.; Das, B. C. *Tetrahedron Lett.* **1988**, *29*, 4423.
55.   (a) Juricová, K.; Smrcková, S.; Holy, A. *Collect. Czech. Chem. Commun.* **1995**, *60*, 237. (b) Hocková, D.: Votavová, H.; Holy, A. *Tetrahedron: Asymmetry* **1995**, *6*, 2375. (c) Hocková, D.; Masojídková, M.; Holy, A *Collect. Czech. Chem. Commun.* **1996**, *61*, 1538. (d) Hocková, D.; Holy, A. *Collect. Czech. Chem. Commun.* **1997**, *62*, 948.
56.   Woodard, P. A.; Winwood, D.; Brewster, M. E.; Estes, K. S.; Bodor, N. *Drug Design and Delivery* **1990**, *6*, 15.
57.   Brewster, M. E.; Simay, A.; Czako, K.; Winwood, D.; Farag, H.; Bodor, N. *J. Org. Chem.* **1989**, *54*, 3721.
58.   Binay, P.; Dupas, G.; Bourguignon, J.; Queguiner, G. *Can. J. Chem.* **1987**, *65*, 648.
59.   Gnecco, D.; Juárez, J.; Galindo, A.; Marazano, C.; Enríquez, R. G. *Synth. Commun.* **1999**, *29*, 281.
60.   Gnecco, D.; Marazano, C.; Enríquez, R. G.; Terán, J. L.; del Rayo Sánchez S., M.; Galindo, A. *Tetrahedron: Asymmetry* **1998**, *9*, 2027.
61.   Kaiser, A.; Billot, X.; Gateau-Olesker, A.; Marazano, C.; Das, B. C. *J. Am. Chem. Soc.* **1998**, *120*, 8026.
62.   Michelliza, S.; Al-Mourabit, A.; Gateau-Olesker, A.; Marazano, C. *J. Org. Chem.* **2002**, *67*, 6474.

Christian M. Rojas

Chapter 9   Quinolines and Isoquinolines   **375**

9.1　Bischler–Napieralski Reaction

9.1.1　Description

The Bischler–Napieralski[1] reaction involves the cyclization of phenethyl amides **1** in the presence of dehydrating agents such as P_2O_5 or $POCl_3$ to afford 3,4-dihydroisoquinoline products **2**.[2-5] This reaction is one of the most commonly employed and versatile methods for the synthesis of the isoquinoline ring system, which is found in a large number of alkaloid natural products.[6] The Bischler–Napieralski reaction is also frequently used for the conversion of N-acyl tryptamine derivatives **3** into β-carbolines **4** (eq 2).

9.1.2　Historical Perspective

The synthesis of 3,4-dihydroisoquinolines via intramolecular reactions of phenethyl amides was first reported by August Bischler and Bernard Napieralski in 1893.[1] The authors described the conversion of N-acyl phenethylamide (**1**, R′ = Me) and N-benzoyl phenethylamide (**1**, R′ = Ph) to 1-methyl-3,4-dihydroisoquinoline (**2**, R′ = Me) and 1-phenyl-3,4-dihydroisoquinoline (**2**, R′ = Ph), respectively, in the presence of P_2O_5. This reaction has subsequently proven to be one of the most general methods ever developed for the synthesis of dihydroisoquinolines.

9.1.3　Mechanism

Despite the synthetic utility of this transformation, nearly eighty years elapsed between the discovery of the Bischler–Napieralski reaction and the first detailed studies of its mechanism.[7-9] Early mechanistic proposals regarding the Bischler–Napieralski reaction involved protonation of the amide oxygen by traces of acid present in P_2O_5 or $POCl_3$ followed by electrophilic aromatic substitution to provide intermediate **5**, which upon dehydration would afford the observed product **2**. However, this proposed mechanism fails to account for the formation of several side products that are observed under these conditions (*vide infra*), and is no longer favored.

Detailed mechanistic studies by Fodor demonstrated the intermediacy of both imidoyl chlorides (6) and nitrilium salts (7) in Bischler–Napieralski reactions promoted by a variety of reagents such as PCl_5, $POCl_3$, and $SOCl_2$).[7–9] For example, amide 1 reacts with $POCl_3$ to afford imidoyl chloride 6. Upon heating, intermediate 6 is converted to nitrilium salt 7, which undergoes intramolecular electrophilic aromatic substitution to afford the dihydroisoquinoline 2. Fodor's studies showed that the imidoyl chloride and nitrilium salt intermediates could be generated under mild conditions and characterized spectroscopically.[7–9] Fodor also found that the cyclization of the imidoyl chlorides is accelerated by the addition of Lewis acids ($SnCl_4$, $ZnCl_2$), which provides further evidence to support the intermediacy of nitrilium salts.[7–9]

Side reactions consistent with decomposition of intermediate nitrilium salt 7 have also been observed, including retro-Ritter reactions that afford alkenes (8), and VonBraun reactions that provide alkyl chlorides (9).[7–9]

In some instances the attack of the arene on the nitrilium salt occurs at the ipso carbon rather than the ortho carbon.[10] For example, the Bischler–Napieralski cyclization of phenethyl amide 10 affords a 2:1 mixture of regioisomeric products 11 and 12. The formation of 12 presumably results from attack of the ipso aromatic carbon on the nitrilium salt 13 followed by rearrangement of the spirocyclic carbocation 14 to afford 15, which upon loss of a proton yields product 12.[10]

Bischler–Napieralski reactions of N-acyl tryptamine derivatives **16** are believed to proceed via a related mechanism involving the initial formation of intermediate spiroindolenines (**17**) that rearrange to the observed 2-carboline products (**18**). The presence of these intermediates has been inferred by the observation of dimerized products that are presumably formed by the intermolecular trapping of the spiroindolenine by unreacted indole present in the reaction mixture.[11,12]

9.1.4 Variations and Improvements

In addition to P_2O_5 and $POCl_3$, which were originally utilized to effect the Bischler–Napieralski reaction, a number of other dehydrating reagents have been employed in these reactions. These reagents include PCl_5, $AlCl_3$, $SOCl_2$, $ZnCl_2$, Al_2O_3, $POBr_3$, and $SiCl_4$.[2] Mixtures of trifluoromethanesulfonic anhydride (Tf_2O) and DMAP have been employed as a very mild means of effecting Bischler–Napieralski cyclizations, and provide higher yields of products than $POCl_3$ in reactions of sensitive substrates.[13] A combination of $AlCl_3\cdot6H_2O/KI/H_2O/CH_3CN$ has been utilized to effect Bischler–Napieralski reactions in hydrated media.[14] Triphosgene has also been employed in Bischler–Napieralski reactions of 2-phenyl-1-(N-methylformamido)naphthalenes.[15]

A common modification of the Bischler–Napieralski reaction involves reduction of the dihydroisoquinoline product **2** to provide a tetrahydroisoquinoline derivative **19**.[2–5] A variety of different reducing agents have been employed, with $NaBH_4$ used with the greatest frequency.[2–5] In many cases the reduction is carried out on the crude product of the Bischler–Napieralski reaction; purification of the dihydroisoquinoline prior to reduction is usually not necessary.

One longstanding limitation of the Bischler–Napieralski reaction involves transformations of 1,2-diphenylethane derivatives, which frequently provide low yields of the desired 3-aryl isoquinoline derivatives due to a competing retro-Ritter reaction (e.g., **20**→**21** below).[2–4] The substitution pattern of the aromatic groups was shown to have an impact on the product distribution, with the highest yields of dihydroisoquinolines observed when one aryl ring is electron-rich and the other electron-deficient.[16] Dominguez demonstrated that the retro-Ritter process could be minimized by conducting the reaction in a nitrile solvent RCN, in which the R group of the solvent was the same as the R group on the amide portion of the substrate.[17] Although this variation of the reaction conditions led to substantially improved yields of 3-aryl isoquinolines **22**, this approach is somewhat impractical due to the limited numbers of nitriles that are sufficiently inexpensive and readily available for use as solvent.

A more practical solution to this problem was reported by Larson, in which the amide substrate **20** was treated with oxalyl chloride to afford a 2-chlorooxazolidine-4,5-dione **23**.[18] Reaction of this substrate with FeCl$_3$ affords a reactive N-acyl iminium ion intermediate **24**, which undergoes an intramolecular electrophilic aromatic substitution reaction to provide **25**. Deprotection of **25** with acidic methanol affords the desired dihydroisoquinoline products **22**. This strategy avoids the problematic nitrilium ion intermediate, and provides generally good yields of 3-aryl dihydroisoquinolines.

An interesting synthesis of quinolizidines was achieved using a vinylogous variation of the Bischler–Napieralski reaction. Angelastro and coworkers reported that treatment of amide **26** with PPSE (polyphosphoric acid trimethylsilyl ester) followed by reductive lactamization of product **28** afforded quinolizidine **29**.[19] This reaction is believed to proceed through nitrilium ion intermediate **27** in a manner analogous to the Bischler–Napieralski reaction.

Phenethyl carbamate derivatives **30** have also been employed in Bischler–Napieralski reactions;[2-5] cyclization of these substrates affords 3,4-dihydroisoquinolones **31**. These reactions have been conducted using a variety of different promoters including PPA,[20] POCl₃,[21] and Tf₂O.[13] Mixtures of P_2O_5 and $POCl_3$ appear to afford the best results in some cases.[22]

The adaptation of the Bischler–Napieralski reaction to solid-phase synthesis has been described independently by two different groups.[23,24] Meutermans reported the transformation of Merrifield resin-bound phenylalanine derivatives **32** to dihydroisoquinolines **33** in the presence of $POCl_3$. The products **34** were liberated from the support using mixtures of HF/p-cresol.[23] In contrast, Kunzer conducted solid-phase Bischler–Napieralski reactions on a 2-hydroxyethyl polystyrene support using the aromatic ring of the substrate **35** as a point of attachment to the resin.[24] The cyclized products **36** were cleaved from the support by reaction with i-butylamine or n-pentylamine to afford **37**.

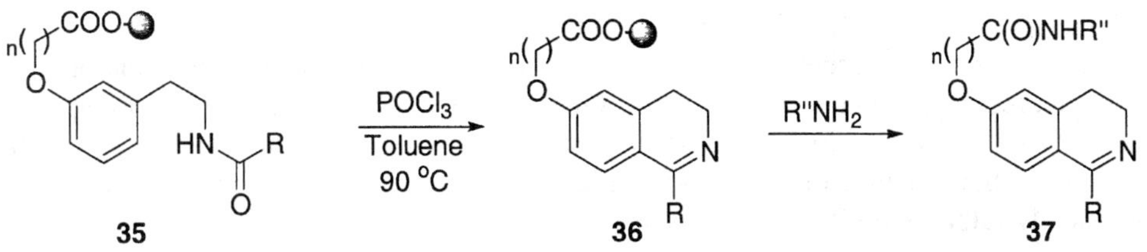

One important variation of the Bischler–Napieralski reaction is the Pictet–Gams modification, in which β-hydroxy or -alkoxy phenethylamides **38** are converted to isoquinolines **39**.[2-5] This transformation is covered in detail in section 9.12 of this text.

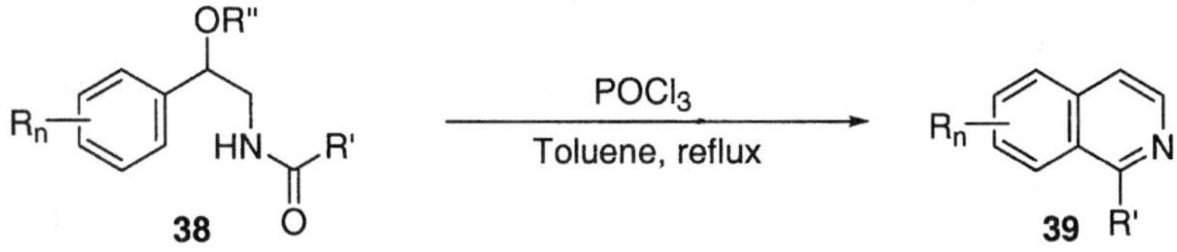

9.1.5 Synthetic Utility
9.1.5.1 Effects of substitution on the aromatic ring

As shown above in section 9.1.2, the mechanism of the Bischler–Napieralski reaction proceeds via an intramolecular electrophilic aromatic substitution reaction between an arene and a tethered nitrilium ion intermediate (generated *in situ* from an amide). Thus, the substituent effects observed in Bischler–Napieralski reactions parallel those observed in other electrophilic aromatic substitution processes.[2-5] Reactivity is typically increased by the presence of electron-donating groups on the aromatic ring, and decreased by the presence of electron-withdrawing groups. For example, reaction of dimethoxysubstituted 1,2-diarylethylamide **40** with EPP (ethyl polyphosphate) afforded an 89% yield of the desired dihydroisoquinoline **41**, whereas the reaction of substrate **42** that lacks the electron-donating substituents did not produce any dihydroisoquinoline product **43**, providing instead a 77% yield of *trans*-stilbene.[16] The difference in yield between these two reactions can be ascribed to the relative rates of electrophilic aromatic substitution and competing retro-Ritter reaction; the more electron-rich arene undergoes substitution much faster than the less electron-rich derivative.[2-5] These electronic effects also influence the regioselectivity of the Bischler–Napieralski reaction in a manner similar to electrophilic aromatic substitution reactions; the substitution typically occurs at the carbon bearing the greatest amount of electron-density and the least amount of steric hindrance.[2-5] In the above example a single regioisomer is obtained; the *o*-substituted product is not observed.

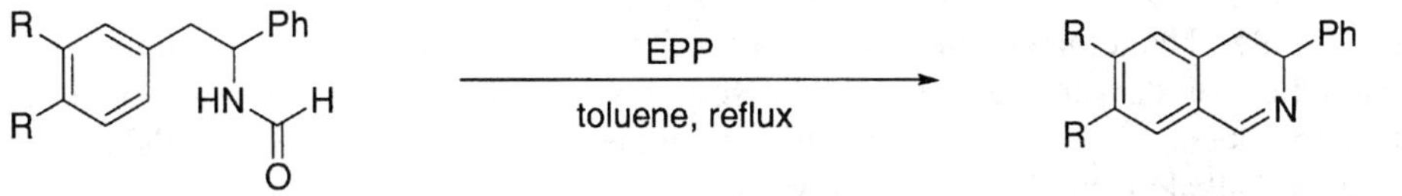

40 R – OMe

42 R = H

41 R = OMe   89% yield

43 R = H   0% yield

9.1.5.2 Effect of substituents on the phenethyl chain

Substitution on the phenethyl side chain of the substrate is usually well tolerated. For example, reaction of carbamate **44** with POCl3 afforded a 75% yield of the corresponding lactam **45**.[25] However, in some instances substituents on the chain lead to low yields, such as in the reaction of amide **46**, which provided only a 29% yield of the desired product **47** (albeit with 9:1 diastereoselectivity).[26]

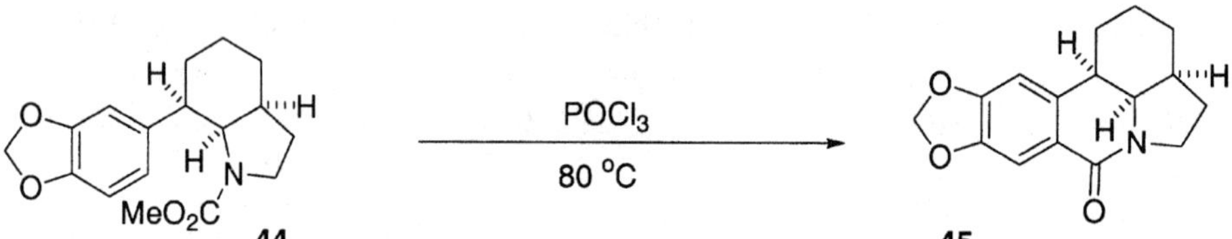

Racemization has been reported to occur in some Bischler–Napieralski reactions of 1-substituted phenethylamides.[27] However, this racemization can be suppressed by conducting the reactions at lower temperatures (0 °C–rt). For example, the product **49** obtained in reaction of **48** with P2O5 at 140 °C was found to be racemic, whereas the product obtained from a reaction conducted at room temperature retained optical activity.[27]

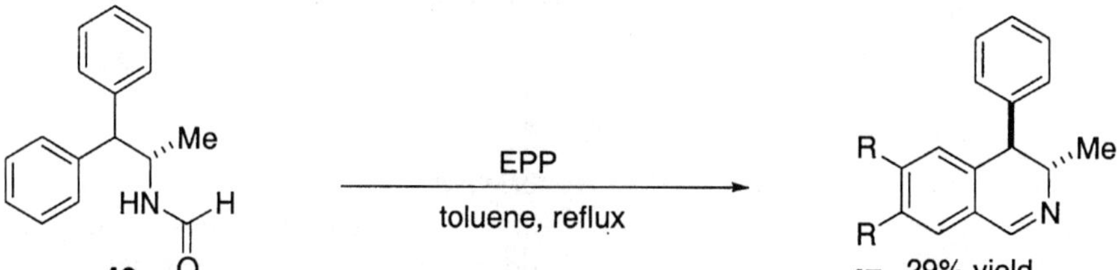

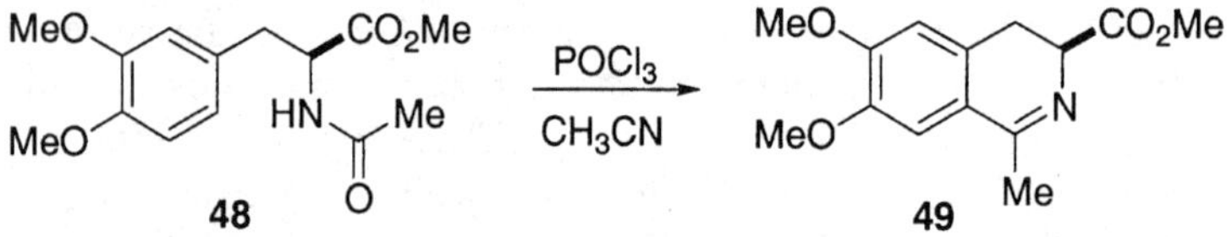

Temp.	α_D
140 °C	0°
rt	160°

In most cases the nature of the substituent on the amide does not have a large impact on the reactivity or selectivity in Bischler–Napieralski reactions.[2–5]

9.1.5.3 Applications in natural product synthesis

The Bischler–Napieralski reaction is one of the most widely used methods for the construction of dihydro- and tetrahydroisoquinoline units in the synthesis of alkaloid natural products.[2–5] A few representative examples of the Bischler–Napieralski reaction in complex alkaloid syntheses are shown below.

Wender[28] and Aube[29] have independently described the use of the Bischler–Napieralski reaction in the synthesis of Yohimban alkaloids. Aube's approach involved the cyclization of indole **50** followed by reduction of the resulting dihydroisoquinoline with NaBH$_4$ to afford **51** in 60% yield.[29] Wender's related Bischler–Napieralski reaction/NaBH$_4$ reduction of related substrate **52** provided an 86% yield of **53**.

A related route to the protoberberine skeleton was reported by Lete, in which substrate **54** underwent Bischler–Napieralski reaction followed by intramolecular alkylation of an *in situ* generated alkyl chloride to afford **55**.[30]

The Bischler–Napieralski reaction was employed by Bonjoch in the synthesis of melinonine-E and strychnoxanthine.[31] The preparation of polycyclic compound **57** was achieved in 53% yield by treating **56** with POCl$_3$ followed by reduction of the dihydroisoquinoline with NaBH$_4$.

Martin has achieved the synthesis of lycoramine (**59**) via a Bischler–Napieralski cyclization of **58** in the final step of the synthesis.[32] Treatment of **58** with POCl₃ followed by NaBH₄ provided the natural product **59** in 68% yield.

9.1.6   Experimental

N-Benzyl-6-methoxy-3,4-dihydroisoquinolone (61)[22]

To a solution of **60** (2.0 g, 6.7 mmol) in POCl₃ was added P₂O₅ (2.0 g, 13.5 mmol). The mixture was heated to reflux for 2 h, then cooled to rt. Excess POCl₃ was evaporated, and the residue was poured into ice water. The mixture was neutralized with Na₂CO₃ and extracted with ethyl acetate. The combined organic extracts were dried over anhydrous MgSO₄, filtered, and concentrated *in vacuo*. The crude product was purified by flash chromatography on silica gel (hexane/ether, 3:1) to afford 1.50 g (86%) of the title compound as a white solid, mp 103–105 °C. ^{1}H NMR (CDCl₃, 400 MHz) δ 8.14 (d, J = 8.5 Hz, 1 H), 7.38–7.30 (m, 5 H), 6.90 (dd, J = 2.5, 8.5 Hz, 1 H), 6.69 (d, J = 2.5 Hz, 1 H), 4.82 (s, 2 H), 3.88 (s, 3 H), 3.51 (t, J = 6.5 Hz, 2 H), 2.94 (t, J = 6.5 Hz, 2 H).

6-(3,4,5-Trimethoxyphenyl)-2,3,8,9-tetrahydro-[1,4]dioxino[2,3-g]isoquinoline (63)[33]

POCl₃ was added to a solution of amide **62** (0.075g, 0.02 mmol) in toluene (5 mL). The reaction mixture was heated to 110 °C with stirring for 5 h, then cooled to rt. A solution of 2 N NaOH (10 mL) was added and the mixture was extracted with ether (3 × 15 mL). The combined organic extracts were dried over anhydrous sodium sulfate, filtered, and concentrated *in vacuo*. The crude product was then purified by flash chromatography on silica gel (40% Ethyl acetate/hexanes) to afford the 68 mg (95%) of the title compound as a colorless oil. ^{1}H NMR (CDCl$_3$, 200 MHz) δ 6.88 (s, 1 H), 6.76 (s, 1 H), 6.73 (s, 2 H), 4.19 (m, 4 H), 3.80 (s, 9 H), 3.70 (m, 2 H), 2.61 (t, J = 6.0 Hz, 2 H).

9.1.7 References

1.   Bischler, A.; Napieralski, B. *Chem. Ber.* **1893**, *26*, 1903–1908.
2.   [R] Whaley, W. M.; Govindachari, T. R. *Org. React.* **1951**, *6*, 74–150.
3.   [R]Fowler, F. W. in *Comprehensive Heterocyclic Chemistry*, Katritzky, A. R.; Rees, C. W., Eds.; Pergamon: Oxford, 1984; Vol 2, pp 410–416.
4.   [R] Jones, G. in *Comprehensive Heterocyclic Chemistry II*, Katritzky, A. R.; Rees, C. W.; Scriven, D. F. V., Eds.; Elsevier: Oxford, 1996; Vol 5, pp 179–181.
5.   [R] Kametani, T.; Fukumoto, K. *Chem. Heterocycl. Cmpds.* **1981**, *38*, 139–274.
6.   [R] Bentley, K. W. *Nat. Prod. Rep.* **2003**, *20*, 342–365. See also earlier reviews in this series.
7.   Fodor, G.; Gal, J.; Phillips, B. A. *Angew. Chem. Int. Ed. Engl.* **1972**, *11*, 919–920.
8.   Nagubandi, S.; Fodor, G. *J. Heterocycl. Chem.* **1980**, *17*, 1457–1463.
9.   Fodor, G.; Nagubandi, S. *Tetrahedron* **1980**, *36*, 1279–1300.
10.   Doi, S.; Shirai, N.; Sato, Y. *J. Chem. Soc., Perkin Trans. 1* **1997**, 2217–2221.
11.   Frost, J. R.; Gaudilliere, B. R. P.; Wick, A. E. *J. Chem. Soc., Chem. Commun.* **1985**, 895–897.
12.   Frost, J. R.; Gaudilliere, B. R. P.; Kauffman, E.; Loyaux, D.; Normand, N.; Petry, G.; Poirier, P.; Wenkert, E.; Wick, A. E. *Heterocycles* **1989**, *28*, 175–182.
13.   Banwell, M. G.; Bissett, B. D.; Busato, S.; Cowden, C. J.; Hockless, D. C. R.; Holman, J. W.; Read, R. W.; Wu, A. W. *J. Chem. Soc., Chem. Commun.* **1995**, 2551–2553.
14.   Boruah, M.; Konwar, D. *J. Org. Chem.* **2002**, *67*, 7138–7139.
15.   Saito, T.; Yoshida, M.; Ishikawa, T. *Heterocycles* **2001**, *54*, 437–438.
16.   Aguirre, J. M.; Alesso, E. N.; Ibanez, A. F.; Tombari, D. G.; Iglesias, G. Y. M. *J. Heterocycl. Chem.* **1989**, *26*, 25–27.
17.   Dominguez, E.; Lete, E. *Heterocycles* **1983**, *20*, 1247–1252.
18.   Larsen, R. D.; Reamer, R. A.; Corley, E. G.; Davis, P.; Grabowski, E. J. J.; Reider, P. J.; Shinkai, I. *J. Org. Chem.* **1991**, *56*, 6034–6038.
19.   Marquart, A. L.; Podlogar, B. L.; Huber, E. W.; Demeter, D. A.; Peet, N. P.; Weintraub, H. J. R.; Angelastro, M. R. *J. Org. Chem.* **1994**, *59*, 2092–2100.
20.   Hendrickson, J. B.; Bogard, T. L.; Fisch, M. E.; Grossert, S.; Yoshimura, N. *J. Am. Chem. Soc.* **1974**, *96*, 7781–7789.
21.   Martin, S. F.; Tu, C. -y. *J. Org. Chem.* **1981**, *46*, 3763–3764.
22.   Wang, X. –j.; Tan, J.; Grozinger, K. *Tetrahedron Lett.* **1998**, *39*, 6609–6612.
23.   Meutermans, W. D. F.; Alewood, P. F. *Tetrahedron Lett.* **1995**, *36*, 7709–7712.
24.   Rolfing, K.; Thiel, M.; Kunzer, H. *Synlett* **1996**, 1036–1038.
25.   Banwell, M. G.; Harvey, J. E.; Hockless, D. C. R., Wu, A. W. *J. Org. Chem.* **2000**, *65*, 4241–4250.
26.   Nicolettti, M.; O'Hagen, D.; Slawin, A. M. Z. *J. Chem. Soc., Perkin Trans. 1.* **2002**, 116–121.
27.   Kametani, T.; Takagi, N.; Kanaya, N.; Honda, T. *Heterocycles* **1982**, *19*, 535–537.
28.   Wender, P. A.; Smith, T. E. *J. Org. Chem.* **1996**, *61*, 824–825.
29.   Aube, J.; Ghosh, S.; Tanol, M. *J. Am. Chem. Soc.* **1994**, *116*, 9009–9018.
30.   Sotomayor, N.; Dominguez, E.; Lete, E. *J. Org. Chem.* **1996**, *61*, 4062–4072.
31.   Quirante, J.; Escolano, C.; Merino, A.; Bonjoch, J. *J. Org. Chem.* **1998**, *63*, 968–976.
32.   Martin, S. F.; Garrison, P. J. *J. Org. Chem.* **1982**, *47*, 1513–1518.
33.   Capilla, A. S.; Romero, M.; Pujol, M. D.; Caignard, D. H.; Renard, P. *Tetrahedron* **2001**, *57*, 8297–8303.

John P. Wolfe

9.2   Camps Quinolinol Synthesis[1]

9.2.1   Description

The Camps quinoline synthesis entails the base catalyzed intramolecular condensation of a 2-acetamido acetophenone (1) to a 2-(and possibly 3)-substituted-quinolin-4-ol (2), a 4-(and possibly 3)-substituted-quinolin-2-ol (3), or a mixture.

9.2.2   Historical Perspective

Knorr showed that aniline could be condensed with ethylacetoacetate when heated to provide the acetoanilide (6).   Conrad and Limpach established that further heating

furnishes quinolin-2-ol 7.   Under different conditions, aniline is condensed with ethylacetoacetate to provide enamine 8, and further heating provides quinolin-4-ol 9. Camps isolated both 7 and 9 when he heated o-acetamidoacetophenone (10) to 104 °C in aqueous sodium hydroxide.[2]

9.2.3   Mechanism

Little work has been done to confirm the mechanism of the Camps reaction.   The presumed mechanism is shown.   Deprotonation α to the ketone portion 1 (11) followed by condensation with the amide furnishes **2**.[3] Alternatively, deprotonation α to the amide followed by condensation with the ketone results in the quinolin-2-ol (**3**).

9.2.4 Variations and Improvements

A general method for making Camps precursors has been developed.[4] Treatment of an anthranilic acid **15** with an acid anhydride or chloride in the usual way[5] results in the corresponding benzoxazinone (**16**). Subsequent treatment with the dianion of an *N*-substituted acetamide furnishes β-keto amide **17**. The reactions were run with crude **16**, yields typically 50–80% overall. The effect of substituents on the reaction has not been extensively investigated.

9.2.5 Synthetic Utility

As one might expect, most of the reported cases of Camps quinoline syntheses involve reactions in which only one of the carbonyl groups is enolizable, thus eliminating the regioselectivity problem.

The Camps reaction has been used to prepare a variety of anti-ulcer agents of the type **23**.[6] As can be seen from the yields, the reaction works equally well with electron donating and withdrawing aromatic rings.[7]

R	Yield
H	78
4-OMe	67
4-F	84
4-Cl	95
4-Me	72

An interesting use of the Camps quinoline synthesis is in the ring contraction of macrocycles.[8] Treatment of 9 member ring **24** with sodium hydroxide in water furnished quinolin-4-ol **25**, while **26** furnishes exclusively quinolin-2-ol **27** under the same reaction conditions (no yield was given for either reaction). The reaction does not work with smaller macrocycles. The authors rationalize the difference in reactivity based upon ground state conformation differences, but do not elaborate.

The isomeric ring contraction was applied to the synthesis of analogs of amsacrine.[9] As expected, ring contraction of 9-member ring **28** under the influence of sodium hydroxide provides cyclopentyl analog **29** in 90% yield.

9.2.6 Experimental[9]
2,4-Dimethyl-11H-1,5-diazabenzo[b]fluoren-10-ol (29)

A suspension of **28** (1g, 3.6 mmol) in 2N NaOH (53 mL) was stirred for 4 hours. The mixture was neutralized with con HCl (10.2 mL). The product was filtered, washed with water, and dried at room temperature to provide **29** as white crystals (849 mg, 90% yield).

9.2.7 References

1   For consistency, the products are drawn as the quinolin-2-ol or quinolin-4-ol, regardless of how they were drawn in the primary literature.

2   (a) Camps, R. *Chem. Ber.* **1899**, *32*, 3228. (b) Camps, R. *Arch. Pharm.* **1899**, *237*, 659.

3   β-Diketone **i** could be an intermediate in the pathway leading to **2**. No investigations into the intermediacy of this type of compound have been performed.

4   Clemence, F.; LeMartret, O.; Collard, J. *J. Heterocyclic Chem.* **1984**, *21*, 1345.

5   [R] Elderfield, R. C.; Todd, W. H.; Gerber, S. "Heterocyclic Compounds" Vol. 6, R. C. Elderfield, ed. J. Wiley and Sons, New York, *1957*, 576.

6   Hino, K.; Kawashima, K.; Oka, M.; Nagai, Y.; Uno, H.; Matsui, J. *Chem. Pharm. Bull.* **1989**, *37*, 110.

7   The authors also investigated R=3-CF₃ and 4-CF₃, but the yields were reported based upon the Grignard reagent (13% and 18% respectively) and are therefore excluded.

8   Witkop, B.; Patrick, J. B.; and Rosenblum, M. *J. Am. Chem. Soc.* **1951**, *73*, 2641.

9   Barret, R.; Ortillon, S.; Mulamba, M.; Laronze, J. Y.; Trentesaux, C.; Lévy, J. *J. Heterocyclic Chem.* **2000**, *37*, 241.

Derek A. Pflum

9.3   Combes Quinoline Synthesis

9.3.1   Description

The Combes reaction is a sequence of the following reactions: (a) condensation of an arylamine **1** with a 1,3-diketone, keto-aldehyde or dialdehyde **2** providing enamine **3**, and (b) cyclodehydration to provide quinoline **4**.

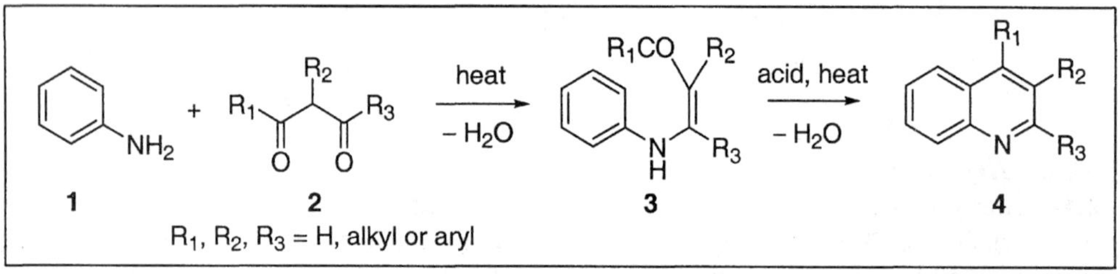

The first condensation has been conducted selectively on a variety of 1,3-diketones, 1,3-dialdehydes and β-keto-aldehydes. The first step works well on most simple anilines even when sterically congested and is mostly affected by basicity. The cyclodehydration step is affected by Friedel–Crafts type directing affects within the ring. Strong electron-withdrawing groups (EWG, i.e., nitro groups) attached to the aniline do not prohibit enamine formation, but do prohibit cyclodehydration. In cases in which keto-aldehydes are used, rearrangements have been reported to occur in the cyclization step.

9.3.2   Historical Perspective

In 1888, Combes described[1] condensation of 2,4-pentadione (acetylacetone) **5** with aniline **1** to provide enamine **6**. Subsequent warming in sulfuric acid provided quinoline **7**. An excellent study describing scope and limitations of the Combes reaction was published in 1928 by Roberts and Turner.[2] The authors noted that the ease of

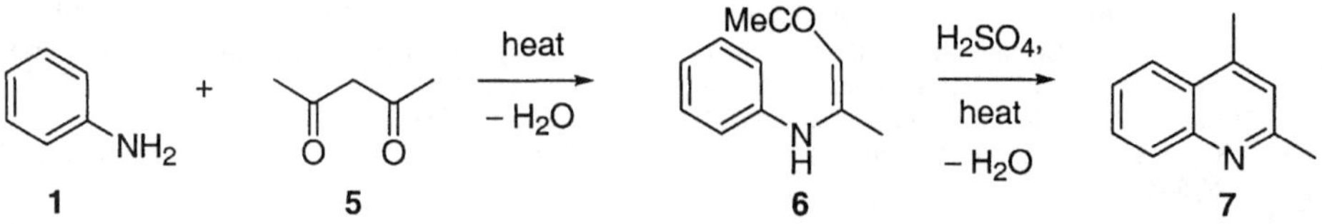

condensation was affected mostly by the basicity of the aniline and not sterics. It was also noted that strongly *ortho-*, *para-* directing groups located *meta-* to the nitrogen, facilitated cylclodehydration providing the quinoline. In these cases, the major product (if not exclusive) was reported to be the 7-substituted quinoline. In the absence of other groups, strongly *ortho-*, *para-* directing groups located *para-* to the nitrogen impeded the cyclodehydration step. The majority of the early reports used sulfuric acid to promote the cyclization. In 1927 Fawcett and Robinson[3] reported the use of ZnCl₂, HCl/AcOH, P₂O₅ or POCl$_3$ as well as H$_2$SO$_4$ to promote some difficult cyclizations. The cyclodehydration failed when 4-*N*-acetylanilide **12** was tried.

1:1 isomeric mixture

9.3.3 Mechanism

The rationale for the predominance of linear cyclization products versus angular cyclization products has been accepted as qualitative.[4] The mechanism of the Combes reaction has been argued.[5] It was initially proposed that cyclization to linear products was due to initial protonation of a more reactive site on the aromatic ring (1-position of **13** corrresponding to the 10-position of **15**) thus, blocking cyclization to angular products. Born[6] showed this not to be the case for the cyclization of 2-naphthyl amino-2-penten-4-one. No 10-deutero material was observed.

Condensation of an aniline with a dione with loss of water provides enamine **16**. Ketone protonation and cyclization forms **18** followed by loss of water provides quinoline **4**. Some have suggested the formation of dication **19** as a requirement to cyclization.[5] Cyclization of **19** to **20** and subsequent conversion to quinoline **4** requires loss of water and acid. Another rendering of the mechanism takes into account participation of an electron-donating group (EDG), which stabilizes intermediate **21**.

R_1, R_2, R_3 = H, alkyl or aryl

9.3.4   Variations and Improvements
9.3.4.1   Use of 3-keto-aldehydes

Petrow described[7] the formation of β-iminoketones from 3-keto-aldehydes and aniline. Cyclization in the presence of aniline hydrochloride and $ZnCl_2$ smoothly provides the desired quinoline **26**.   Bis-imine **24** is the proposed intermediate that undergoes cyclization. The aldimine is more reactive than the ketimine toward cyclization; thus, cyclization on the aldimine occurs. When the bis-imine is not formed, partial aniline migration can occur which results in mixtures of cyclized products.

Tilak and co-workers[8] subsequently reported the application of this methodology to prepare similar compounds. In this work, Tilak also described an extension of this method using lactic acid to release the aniline comprising the aldimine, which led to **28**. Enamine **28** would then undergo cyclization to give rearranged product **29a** (not from direct cyclization of **27** forming **29b**). When the traditional Petrow conditions were employed, mixtures were reported.

Conditions were also arrived at whereby one could obtain *without rearrangement* the cyclization product **31** from enamino-ketone **30**.[9]

9.3.4.2 Use of 1,3-dialdehydes

Uhle and Jacobs were first to utilize 2-cyano- and 2-nitromalonaldehyde (**33**)[10] as 3-carbon components in the Combes reaction and elegantly applied this to the preparation of compounds relating to ergot alkaloids. Morley and Simpson[11] performed a short study on the cyclization using **33** as a 3-carbon building block for the preparation of 3-nitroquinolines. In this work, the importance of having aniline hydrochloride in the reaction media was demonstrated. The condensation of *p*-toluidine with **33** provides **34**. Cyclization using AcOH at reflux provided 3-nitroquinoline **36** in good yield. Cyclization failed when nitro substituents were attached to the anil.

9.3.4.3 Conditions for enamine formation and various acids used in the cyclodehydration step

The first step of the Combes reaction has been reported to occur by merely mixing an aniline and diketone in an alcoholic solvent, or neat, with slight warming. Dilute aqueous acid (2 M HCl), AcOH, ZnCl$_2$ as well as CaCl$_2$,[12] or other types of drying agents have been used to promote the first step.

Many acids other than sulfuric acid have been used for the challenging cyclodehydration step. It is important to note that when the concentration of sulfuric acid is below 70%, hydrolysis of the imine or enamine occurs.[13] As previously mentioned, HCl/AcOH, ZnCl$_2$, PPA, POCl$_3$, and lactic acid have been successfully applied to promote the cyclization. Chloroacetic acid was found to perform similar to lactic acid.[9] Concentrated HCl,[14] *p*-TsOH,[15] and HF[16,17] have proven beneficial in generating linear products. Johnson and co-workers contrasted cyclization reactions of 2-naphthyl-amino-2-penten-4-one **37** promoted by ZnCl$_2$ and HF. They found that ZnCl$_2$ provided angular product **39** while HF gave rise to linear product **40**. Sulfuric acid was also used and gave sulfate salts of the linear product.

9.3.4.4 Substitution effects on the selectivity of cyclization

Previously mentioned was the importance of directing substituents of the anil in expediting cyclization and that when powerful EWGs are present, cyclization might be inhibited entirely. A study was conducted to consider the steric influence of methyl substituents on the anil and their effect on the cyclization.[18] It was proposed that a methyl group on the anil had a profound effect on the site of cyclization. For example, while **42** was isolated in only 22% yield, the other isomer **43** was not observed. This work also confirmed the work by Roberts and Turner[2] who claimed that reaction of *m*-Cl-aniline with acetylacetone followed by H_2SO_4 promoted cyclization of **44** provided **only** 7-chloro-2,4-dimethyl quinoline **45**.

These effects are not entirely understood because cyclizations forming 5,8-dimethyl quinolines are not problematic as shown below by the conversion of **46** into **47**.

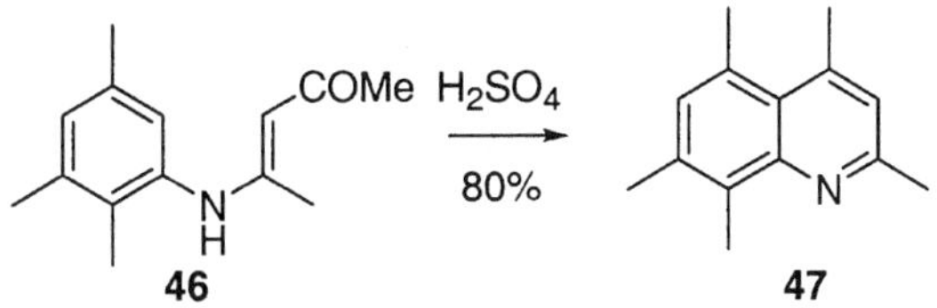

9 3.4.5 Selectivity of cyclizations of unsymmetrical diones

We have previously discussed that keto-aldehydes react with anilines first at the aldehyde carbon to form the aldimine. Subsequent condensation with another aniline formed a bis-imine or enamino-imine. The aniline of the ketimine normally cyclizes on the aldimine (**24** → **26**). Conversely, cyclization of the aldimine could be forced with minimal aniline migration to the ketimine using PPA (**30** → **31**). The use of unsymmetrical ketones has not been thoroughly explored; a few examples are cited below. One-pot enamine formation and cyclization occurred when aniline **48** was reacted with dione **49** in the presence of catalytic *p*-TsOH and heat. Imine formation occurred at the less-hindered ketone, and cyclization with attack on the reactive carbonyl was preferred.[15]

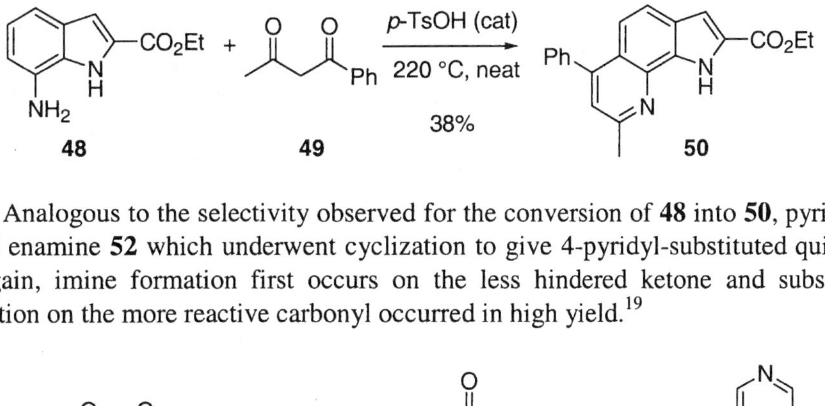

Analogous to the selectivity observed for the conversion of **48** into **50**, pyridyl **51** formed enamine **52** which underwent cyclization to give 4-pyridyl-substituted quinoline **53**. Again, imine formation first occurs on the less hindered ketone and subsequent cyclization on the more reactive carbonyl occurred in high yield.[19]

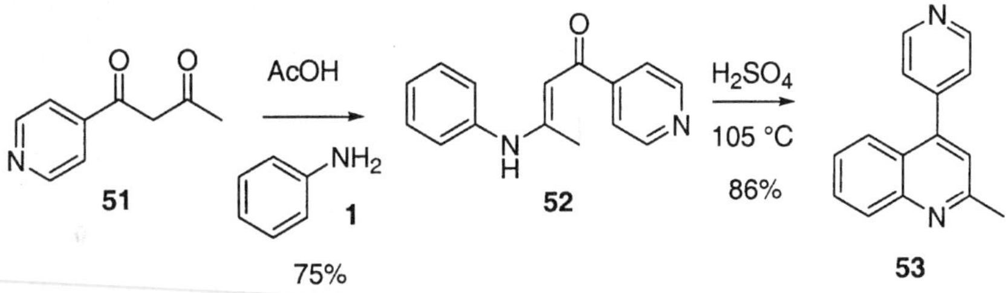

Imine preparation from a tricarbonyl system followed by PPA promoted cyclization was reported to provide quinoline **55** in excellent yield.[20] This cyclization failed when the corresponding aldimine was used.

54   **55**

9.3.4.6 Selectivity of cyclizations with some heterocycles

The Combes reaction has been applied to the preparation of carbazoles related to ellipticine. In that case, the imine was not formed separately nor purified yet the desired product **57** was isolated in 35%.[21]

56   **5**   **57**

35%

Imine formation by reaction of aniline **58** and dione **49** under thermal conditions gave a mixture of imines. Cyclodehydration using PPA gave nearly a 1:1 mixture of isomers **59** and **60**. These authors attempted thermal cyclization conditions (similar to Gould-Jacobs type conditions) to affect cyclization of this mixture and failed. Also, these authors reported difficulty in the clean formation of the imine. They observed large amounts of the *N*-acetyl compound presumably coming from fragmentation of the imine at the reported temperature.[22]

58   **59, 48%**   **60, 41%**

9.3.5   *Experimental*[23]

1   **61**   **62**   **63**

Preparation of 2,4-Diethylquinoline (63)

A mixture of 3,5-heptanedione (61, 14.1 g, 0.11 mol) and freshly distilled aniline (1, 9.3 g, 0.1 mol) was boiled under gentle reflux for 1.5 hrs. and allowed to cool. Distilled water (50 mL) was added and the product was extracted with benzene and dried over anhydrous magnesium sulphate. Removal of the solvent left anil 62, 16 g, 81% yield, as a yellow oil which was used directly in the next stage without further purification.

Sulfuric acid (conc., 90 mL) was cooled to below 5 °C and anil (16 g) was added in 1 mL portions over a period of 15 min. The solution was then heated on the water bath for an additional 30 min. After cooling, the reaction mixture was made alkaline with solid NaOH and the product was extracted with ether and dried over anhydrous $MgSO_4$. Removal of the solvent left 2,4-diethylquinoline (63), 8.7 g, 47% yield as a pale yellow liquid, bp. 286–287 °C/760 mmHg, lit. bp. 282.8–284.8 °C. Anal. Calcd for $C_{13}H_{15}N$: C, 84.3; H, 8.2; N, 7.6. M: 185.3. Found: C, 84.1; H, 8.3; N, 7.8.

9.3.6   References

1.   Combes, A. *Bull. Soc. Chim. France* **1888,** *49,* 89.
2.   (a) Roberts, E. and Turner, E. *J. Chem Soc.* **1927,** 1832.
3.   Fawcett, R. C. and Robinson, R. *J. Chem. Soc.* **1927,** 2254.
4.   [R] Yamashkin, S. A.; Yudin, L. G. and Kost, A. N. *Chemistry of Heterocyclic Compounds,* **1993,** 845.
5.   (a) [R] Jones, G. In *Heterocyclic Compounds,* John Wiley & Sons, Inc., New York, **1977,** Quinolines Vol 32, pps.119–125. (b) [R] Elderfield, R. C. In *Heterocyclic Compounds,* Elderfield, R. C., John Wiley & Sons, Inc., New York, **1952,** vol. 4, pps.36–38.
6.   Born, J. L. *J. Org. Chem.* **1972,** *37,* 3952.
7.   Petrow, V. A. *J. Chem. Soc.* **1942,** 693.
8.   Tilak, B. D.; Berde, H.; Gogte, V. N. and Ravindranathan, T. *Indian J. of Chem.* **1970,** *8,* 1.
9.   Berde, H. V.; Gogte, V. N.; Namjoshi, A. G. and Tilak, B. D. *Indian J. of Chem.* **1972,** *10,* 9.
10.   Ulhe, F. C. and Jacobs, W. A. *J. Org. Chem.* **1945,** *10,* 76.
11.   Morley, J. S. and Simpson, J. C. E. *J. Chem. Soc.* **1948,** 2024.
12.   Lempert, H. and Robinson, R. *J. Chem. Soc.* **1934,** 1419.
13.   Bonner, T. G. and Barnard, M. *J. Chem. Soc.* **1958,** 4176.
14.   El Ouar, M.; Knouzi, N., El Kihel, A.; Essassi, E. M., Benchidmi, M.; Hamelin, J.; Carrie, R. and Danion–Bougot, R. *Synth. Commun.* **1995,** *25,* 1601.
15.   El Ouar, M.; Knouzi, N. and Hamelin, J. *J. Chem. Research (S)* **1998,** 92.
16.   Johnson, W. S.; Woroch, E. and Mathews, F. J. *J. Am. Chem. Soc.* **1947,** *69,* 566.
17.   Johnson, W. S and Mathews, F. J. *J. Am. Chem. Soc.* **1944,** *66,* 210.
18.   Osborne, A. G. *Tetrahedron* **1983,** *39,* 2831.
19.   Hey, D. H. and Williams, J. M. *J. Chem. Soc.* **1950,** 1678.
20.   [R] Popp, F. D. and McEwen, W. E. *Chem. Rev.* **1958,** *58,* 321.
21.   Alunni–Bistocchi, G.; Orvietani, P., Bittoun, P., Ricci, A. and Lescot, E. *Pharmazie* **1993,** *48,* 817.
22.   Sanna, P.; Carta, A. and Paglietti, G. *Heterocycles,* **1999,** *51,* 2171.
23.   Claret, P. A. and Osborne, A. G. *Org. Prep. Proceed. Int.* **1970,** *2,* 305.

Timothy T. Curran

9.4   Conrad–Limpach Reaction

9.4.1   Description

The Conrad–Limpach reaction is a sequence of the following reactions: (a) condensation of an arylamine 1 with the ketone or aldehyde of a β-ketoester or α-formylester 2 providing enamine 3, and (b) cyclization with loss of alcohol to yield 4-hydroxy-quinoline 4.

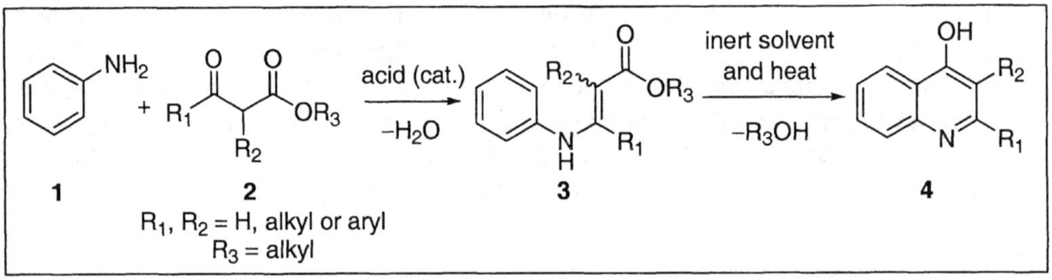

The first condensation is conducted selectively on a variety of β-ketoesters and α-formylesters. The first step works well on most simple anilines even when sterically congested and is mostly affected by basicity. Formation of intermediate 3 is problematic when strong electron-withdrawing groups (EWG) are attached to the aniline (e.g., nitro). The cyclization step is promoted thermally in inert solvents as well as using acidic solvents at elevated temperature. When there exists an opportunity to form isomers on cyclization (e.g., *m*-substituted anilines) a mixture of the 5- and 7-substituted quinolines usually results.

9.4.2   Historical Perspective

In 1887, Conrad and Limpach described[1] the condensation of ethyl acetoacetate 5 with aniline 1 to provide enamine 6. Subsequent warming of the mixture provided quinoline 7. Limpach reported several years later that the yield of the cyclization step was improved when an inert solvent (e.g., mineral oil) was employed. While the cyclization step was normally quite facile at 240–280 °C, the physical properties and the methods described for the preparation of enamino-esters were inconsistent.

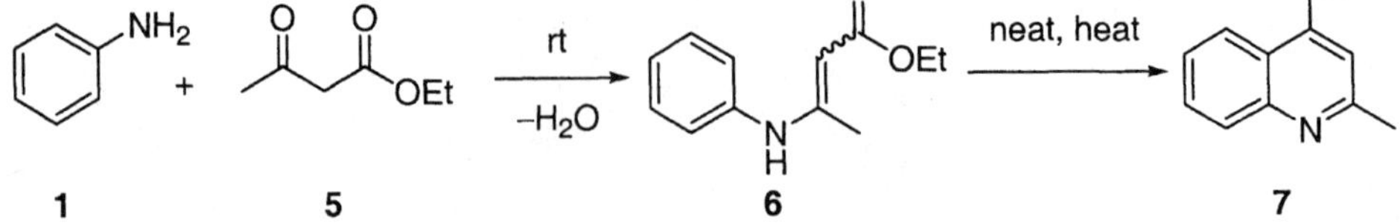

These early contradictions were eventually resolved and led to the correction by Knorr[2] of his initially proposed structure. While Conrad and Limpach described the reaction of aniline 1 with ethyl acetoacetate 5 which ultimately yielded 4-hydroxy-2-methylquinoline (7) via initial reaction of the amine with the ketone, Knorr described reaction of the same components, yet the first condensation occurred at the ester. Subsequent acid-promoted cyclization (e.g., Combes) provided 4-methyl-2-quinolone (9).[2,3,4] Aiding the confusion was the observation that anilide **8** and enamino-ester **6** could interconvert under appropriate reaction conditions. A study was conducted by Hauser and Reynolds[5] determining important factors for formation of either **6** or **8**. Heating **1** and **5** at high temperature (130–140 °C) usually provided the anilide (ie., **8**) unless the alcohol ejected had low volatility (e.g., *n*-amyl alcohol). Alternatively, reaction of **1** and **5** with catalytic acid in an alcohol in the presence or absence of a drying agent (e.g., CaSO$_4$) at room temperature or with warming were favorable conditions to form the desired enamino-ester for use in the Conrad–Limpach reaction.

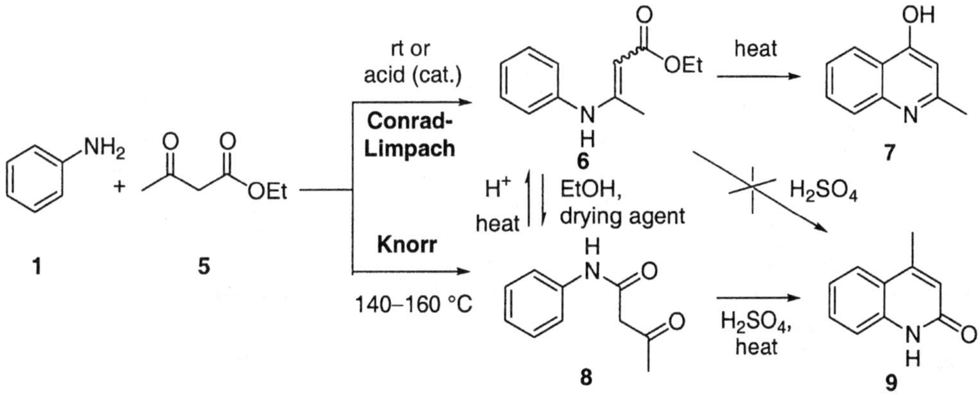

9.4.3 Mechanism

The proposed mechanism for the Conrad–Limpach reaction is shown below. Condensation of an aniline with a β-keto-ester (i.e., ethyl acetoacetate **5**) with loss of water provides enamino-ester **6**. Enolization furnishes **10** which undergoes thermal cyclization, analogous to the Gould–Jacobs reaction, via 6π electrocyclization to yield intermediate **11**. Compound **11** suffers loss of alcohol followed by tautomerization to give 4-hydroxy-2-methylquinoline **7**. An alternative to the proposed formation of **10** is ejection of alcohol from **6** furnishing ketene **13**, which then undergoes 6π electrocyclization to provide **12**.

9.4.4 Variations and Improvements
9.4.4.1 General: Conditions for each step and selectivity of m-substituted anilines

As previously mentioned, Hauser and Reynolds[5] reported on factors governing the first step of the Conrad–Limpach reaction but they were by no means exhaustive. Other than the conditions reported above for the first step, HCl/MeOH,[6] CHCl$_3$ or CHCl$_2$ (neat or with acid catalyst),[7] PhMe or PhH with removal of water with or without acid catalyst,[8, 9] or EtOH/AcOH/CaSO$_4$[10] were reported to provide the desired enamino-ester from an aryl amine and β-keto-ester. Hauser and Reynolds also noted that o-nitroaniline and o-nitro-p-methoxyaniline failed to form the desired enamino-ester under conditions which they reported.

The cyclization step has been reported to work well in triglyme,[11] mineral oil, paraffin, Dowtherm A™, Ph$_2$O or polyphosphoic acid (PPA). PPA has been used to promote the entire reaction in a single process (*vide infra*).

The selectivity of the cyclization using enamino-esters **18-20** derived from *m*-halogenated anilines **14-16**, provided mixtures of 5- and 7-substituted quinolines.[12, 13] In all of these cases, the cyclization gave either equal amounts of the 5- and 7- isomers or in the case of *m*-iodoaniline, about a 1:2 ratio was observed. During the time of these publications, it was the desire of the authors to obtain the 7-substituted quinolines, which were potential drugs for the treatment of malaria.

14, X = Cl
15, X = Br
16, X = I

17

18, X = Cl, 78-88%
19, X = Br, 78-88%
20, X = I, 78-90%

21, X = Cl, 47%
22, X = Br, 45%
23, X = I, 32%

24, X = Cl, 50%
25, X = Br, 53%
26, X = I, 60%

Lisk and Stacy reported[14] a dependence of concentration on the formation of the isomers. Under concentrated conditions (2 parts Dowtherm A to 1 part enamino-ester) the 5-isomer was almost formed exclusively, while under dilute conditions (30:1), the 7-isomer predominated.

Another improvement was reported by Leonard *et al.* in their preparation of a promising antimalarial, Endochin. The improvement was the alkylation of intermediate enamino-ester **28** by reaction with NaOEt followed by alkylation with an alkyl bromide, rather than forming **29** by reaction of **27** and a suitable β-keto-ester. This provided the important intermediate **29** required for cyclization to Endochin (**30**). Endochin was first reported by German scientists but was not publicly disclosed until the Department of Commerce made this information available after World War II.[15] Leonard was able to improve upon the chemistry reported by Andersag and Salzer in 1940 and isolated Endochin in 40% overall yield from *m*-anisidine (**27**).

27

5

28

29

Endochin, **30**

9.4.4.2 Use of acetylene dicarboxylate and α-formyl ester

Aryl amine **31** was found to react readily with acetylene dicarboxylate **32** to yield fumarate **33**.[16] Several similar reactions were reported and found to be general. The authors reported that a single olefinic isomer was observed by ^{1}H NMR for the enamino-esters. Cyclization of **33** then provided the desired quinolone **34** in good yield. As previously observed, use of 3- or 3,4-disubstituted anilines, gave mixtures of products.

More importantly, Peet and coworkers[17] reported the reaction of *o*-nitroaniline **35** with acetylene dicarboxylate **32** to provide fumarate **36**. Subsequent cyclization proved difficult under thermal conditions and only a 35% yield of quinolone **37** was isolated. Use of PPA for the cyclization improved the yield of **37** significantly. Using this modification allowed enamino-ester formation with a nitro-group attached to the arylamine.

Reactions using 2-formyl esters have also been reported. The anilines reacted[18] smoothly with **40** to give **41** which was cyclized without purification to give the desired quinolines **42** and **43** in 50 and 37% overall yield, respectively.

9.4.4.3 3-Heteroatom substituted keto-esters

Brassard[19] applied the Conrad–Limpach reaction as an approach to the A-B ring system of Phomazarin. While the overall yield was only 23%, he showed that a methoxy group was an acceptable substituent on enamino-ester **46** and for the subsequent cyclization to quinolone **47**.

Kemp[6] utilized a very similar oxo-diester **49**. Not only was this ketone reactive enough to form an enamine using *p*-nitroaniline **48**, but **49** underwent bis-addition of aniline **48**, providing **50** in 64% yield. Intermediate **50** smoothly cyclized to provide quinolone **51** which was utilized to prepare a diacylaminoepindolidione, a template to study the folding of β-pleated sheets.

9.4.4.4 Reactions with heterocyclic compounds and formation of heterocycles

Various heterocyclic compounds have been used as substrates for the Conrad–Limpach reaction. Amino-isoquinoline **52** was converted into **54** in 36% overall yield.[20]

The Conrad–Limpach reaction has been applied as a key step in the formation of pyrido[4,3-b]quinoline. Condensation of 3 different anilines **55** (R = H, Br, OMe) with keto-ester **56** provided the enamino-esters **57** in acceptable yields. Cyclization gave the desired quinolones **58** in good to moderate yield.[21]

Reaction of 2-aminopyridine **59** with β-keto-ester **60** in PPA provided pyrido-pyrimidine **61** in poor yield. Interestingly, upon heating isolated **61**, rearrangement occurred to provide napthyridone **62** in good yield.[22]

N-acetyl groups[20] attached to the aniline have been shown to withstand the Conrad–Limpach reaction. Phenols and alcohols also survived unless in proximity to a reactive center. Jaroszewski[23] reported the formation of **64** by reaction of aniline **63** with ethyl acetoacetate (**5**). Cyclization under thermal conditions in paraffin gave a mixture of quinolone **65** and quinoline **66**.

Although phenols have not participated in the Conrad–Limpach reaction under certain conditions thiophenols were not as innocent. Lee and coworkers[24] reported mixtures of thiochromenones and quinolones from reactions of amino-thiophenols with ethyl benzoyl acetate. Amino-thiophenol **67** reacted with ethyl benzoylacetate **68** in PPA to give a mixture of thiochromenone **70** and quinolone **69** in which the quinolone predominated.

9.4.5 Experimental[25]

Preparation of 2-Methyl-4-Hydroxyquinoline (7)

In a 1-L rbf attached to a Dean-Stark trap, equipped with a reflux condenser is placed distilled aniline (**1**, 46.5 g, 45.5 mL, 0.5 mol), commercially available ethyl acetoacetate (**5**, 65 g, 63.5 mL, 0.5 mol), benzene (100 mL) and glacial AcOH (1 mL). The flask is heated at about 125 °C, and the water which distills out of the mixture with the refluxing benzene is removed at intervals. Refluxing is continued until no more water separates (9 mL collects in about 3 hrs) and then for an additional 30 min. The benzene is then distilled under reduced pressure, and the residue is transferred to a 125 mL modified Claisen flask with an insulated column. The flask is heated in an oil or metal bath maintained at a temperature not higher than 120 °C while the forerun of **1** and **5** is removed and at 140–160 °C the product distills giving 78–82 g, 76–80% yield of **6**.

Dowtherm A™ (150 mL) was placed in a 500 mL 3-necked, rbf equipped with a dropping funnel, a sealed mechanical stirrer, and an air condenser. The solvent was stirred and heated at reflux while enamino-ester **6** (65 g, 0.32 mol) was added rapidly through the dropping funnel. Stirring and refluxing continued for 10–15 min after the addition was completed. The ethanol formed in the condensation reaction may be allowed to escape from the condenser through a tube leading to a drain, or it may be collected by attaching a water-cooled condenser. The mixture is allowed to cool to rt, at which stage a yellow solid separates. Approximately 200 mL of petroleum ether (60–70 °C) is added; the solid is collected on a filter and washed with petroleum ether (100 mL). The crude product was recrystallized from boiling water with Darco or Norit (10 g) to give 43–46 g, 85–90% yield of **7** as white needles.

9.4.6 References

1.   [R] Reitsema, R. H., *Chem. Rev.* **1948**, *47*, pps. 47–51.
2.   [R] Jones, G. In *Heterocyclic Compounds*, John Wiley & Sons, Inc., New York, **1977**, Quinolines, Vol 32, 137–151.
3.   [R] Manske, R. H., *Chem Rev.* **1942**, *30*, 113–114.
4.   [R] Elderfield, R. C. In *Heterocyclic Compounds*, Elderfield, R. C., John Wiley & Sons, Inc., New York, **1952**, vol. 4, 31–36.
5.   Hauser, C. R.; Reynolds, G. A. *J. Am. Chem. Soc.* **1948**, *70*, 2402.
6.   Kemp, D. S.; Bowen, B. R. *Tetrahedron Lett.* **1988**, *29*, 5077.
7.   Steck, E. A.; Hallock, L. L.; Holland, A. J. *J. Am. Chem. Soc.* **1946**, *68*, 132.
8.   Moon, S.–S., Kang, P. M.; Park, K. S.; Kim, C. H. *Phytochemistry* **1996**, *42*, 365.
9.   Somanathan, R.; Smith, K. M. *J. Heterocycl. Chem.* **1981**, *18*, 1077.
10. Raban, M.; Martin, V. A.; Craine, L. *J. Org. Chem.* **1990**, *55*, 4311.
11. El–Desoky, S. I.; Kandeel, E. M.; Abd el–Rahman, A. H.; Shmidt, R. R. *Z. Naturforsch. B* **1998**, *53*, 1216.
12. Steck, E. A.; Hallock, L. L.; Holland, A. J. *J. Am. Chem. Soc.* **1946**, *68*, 380.
13. Steck, E. A.; Hallock, L. L.; Holland, A. J. *J. Am. Chem. Soc.* **1946**, *68*, 1241.
14. Lisk, G. F.; Stacy, G. W. *J. Am. Chem. Soc.* **1946**, *68*, 2686.
15. Leonard, N. J.; Hebrandson, H. F.; Van Heyningen, E. M. *J. Am. Chem. Soc.* **1946**, *68*, 1279.
16. Heindel, N. D.; Bechara, I. S.; Kennewell, P. D.; Molnar, J.; Ohnmacht, C. J.; Lemke, S. M.; Lemke, T. F. *J. Med. Chem.* **1968**, *11*, 1218.
17. Peet, N. P.; Baugh, L. E.; Sunder, S.; Lewis, J. E. *J. Med. Chem.* **1985**, *28*, 298.
18. Elderfield, R. C.; Wright J. B. *J. Am. Chem. Soc.* **1946**, *68*, 1276.
19. Guay, V.; Brassard, P. *J. Heterocycl. Chem.* **1987**, *24*, 1649.
20. Misani, F.; Bogert, M. T. *J. Org. Chem.* **1945**, *10*, 347.
21. Coscia, A. T.; Dickerman, S. C. *J. Am. Chem. Soc.* **1959**, *81*, 3098.
22. Deady, L. W.; Werden, D. M. *Synth. Commun.* **1987**, *17*, 319.
23. Jaroszewski, J. W. *J. Heterocycl. Chem.* **1990**, *27*, 1227.
24. Wang, H.–K.; Bastow, K. F.; Cosentino, L. M.; Lee, K.–H. *J. Med. Chem.* **1996**, *39*, 1975.
25. (a) Reynolds, G. A.; Hauser, C. R. *Org. Synth. Coll. Vol. III*, **1955**, 374 and (b) *ibid*, 593.

Timothy T. Curran

9.6   Doebner Quinoline Synthesis[1]

9.5.1   Description

The Doebner reaction is a three component coupling of an aniline (**1**), pyruvic acid (**2**), and an aldehyde (**3**) to provide a 4-carboxyl quinoline (**4**). That product can be decarboxylated to furnish quinoline **5**.

9.5.2   Historical Perspective

In 1883, Böttinger described the reaction of aniline and pyruvic acid to yield a methylquinolinecarboxylic acid.[2] He found that the compound decarboxylated and resulted in a methylquinoline, but made no effort to determine the position of either the carboxylic acid or methyl group. Four years later, Doebner established the first product as 2-methylquinoline-4-carboxylic acid (**8**) and the second product as 2- methylquinoline (**9**).[3]   Under the reaction conditions (refluxing ethanol), pyruvic acid partially decarboxylates to provide the required acetaldehyde *in situ*. By adding other aldehydes at the beginning of the reaction, Doebner found he was able to synthesize a variety of 2-substituted quinolines. While the Doebner reaction is most commonly associated with the preparation of 2-aryl quinolines, in this primary communication Doebner reported the successful use of several alkyl aldehydes in the quinoline synthesis.

Doebner also found that under certain conditions (specifically cold ether), the result of the three-component coupling was not the quinoline but diketopyrrolidine **10**.[4]

9.5.3   Mechanism

A possible mechanism is shown. Condensation of **6** and **3** provides **11**, which can be attacked by the enol of **2** (**12**) to provide **13**. Intramolecular condensation yields **16**, which is then oxidized to **17**. A common variant of the Doebner reaction involves first reaction of the aniline and aldehyde and subsequent addition of the pyruvic acid.[5] Keto acid **13** is also a potential intermediate in the formation of **10**.

Alternatively, the pyruvic acid can first condense with the aldehyde. Addition of the aniline to the β-position of **18** provides the same intermediate (**13**), as above. The mechanism could be substrate dependent.

The mechanism of oxidation (**16** → **17**) is not well understood. Both tetrahydroquinoline (**19**)[6] and *N*-alkylaniline (**20**)[7] have been isolated from reaction

mixtures, implying hydrogen transfer. Addition of oxidants such as nitrobenzene has not resulted in improved yields.[8] Quinoline (**23**) has reportedly been isolated in 95% yield, implying an oxidant other than an intermediate.[9] Yields of the Doebner reaction are typically < 50%, however.

9.5.4 Variations and Improvements

The electronic properties of the aniline are important in the Doebner reaction. The reaction works best with electronic donating groups. Anilines substituted with a chlorine at the meta position consistently give low yields,[5] but fluorine at the meta position seems to provide the quinoline smoothly.[10] The reaction can be regioselective; reaction of aniline **24** provides quinoline **26**.[11]

Both aromatic and aliphatic acids work, though aromatic aldehydes are far more common. Increasing electron density of the aromatic aldehyde lowers the yield.[12] Formaldehyde can also be used.[13]

Few examples with α-keto acids other than pyruvic acid have been reported. Typically, the yields with alternate acids are quite poor. When using pyruvic acid, the reaction gives higher yields of the desired product when freshly distilled pyruvic acid is used.[5]

9.5.5 Synthetic Utility

The Doebner reaction can provide the 4-carboxyl quinoline when the Pfitzinger reaction does not.[14] Pfitzinger reaction of pinnacolone with isatin did not provide the desired quinoline. Doebner reaction of aniline with acetaldehyde and pyruvic acid did furnish the quinoline, albeit in only 8% yield.

9.5.6 Experimental[15]

31   2   25   32

Preparation of 8-Methyl-2-phenyl-quinoline-4-carboxylic acid (32)

A solution of *o*-toluidine (28g, 260 mmol) in ethanol (50 mL) was added to a solution of pyruvic acid (33g, 380 mmol) and benzaldehyde (28g, 260 mmol) in ethanol (100 mL). The mixture was heated to reflux for 3 h and allowed to cool overnight. The resulting solid was collected by filtration, washed with ethanol, washed with benzene, and dried to give **32** (13.4g, 20% yield).

9.5.7 References

1   (a) [R] Elderfield, R. C. *Heterocyclic Compounds*; Elderfield, R. C., Ed.; John Wiley & Sons, Inc.: New York, **1952**, Volume 4, Quinoline, Isoquinoline and Their Benzo Derivatives, pp. 25-29. (b) [R] Jones, G. in *Heterocyclic Compounds*, John Wiley & Sons, Inc.: New York, **1977**, Vol. 32; Quinolines, pp. 125-131.

2   Böttinger, C. *Chem. Ber.* **1883**, *16*, 2357.

3   Doebner, O. *Chem. Ber.* **1887**, *20*, 277.

4   Doebner, O. *Annalen*, **1887**, *242*, 265.

5   Lutz, R. E.; Bailey, P. S.; Clark, M. T.; Codington, J. F.; Deinet, A. J.; Freek, J. A.; Harnest, G. H.; Leake, N. H.; Martin, T. A.; Rowlett, R. J.; Salsbury, J. M.; Shearer, N. H.; Smith, J. D.; Wilson, J. W. *J. Am. Chem. Soc.* **1946**, *68*, 1813.

6   Musajo, L. *Gazz. Chim. Ital.* **1930**, *60*, 673.

7   Simon, L. J.; Mauguin, C. *Compt. Rend.* **1906**, *143*, 466.

8   Carrara, G. *Gazz. Chim. Ital.* **1930**, *60*, 623.

9   Mathur, F. C.; Robinson, R. *J. Chem. Soc.* **1934**, 1520. The yield is reported as 95%, but the mass recovered from the reaction is not reported.

10   Aboul-Enein, H. Y.; Ibrahim, S. E. *J. Fluorine Chem.* **1992**, *59*, 233.

11   Rapport, M. M.; Senear, A. E.; Mead, J. J.; Koepfli, *J. Am. Chem. Soc.* **1946**, *68*, 2697. Regioisomer **27** was independently synthesized via the Pfitzinger reaction and was undetected in the Doebner reaction mixture.

12   Merchant, J. R.; Shah, R. J.; Bhandarkar, R. M. *Rec. Trav. Chim. Pas Bas.* **1962**, *81*, 131.

13   (a) Borsche, W. *Chem. Ber.* **1909**, *42*, 4072. (b) Borsche, W.; Wagner-Roemmich, M. *Annalen*, **1940**, *544*, 280.

14   see, for example, Mead, J. F.; Senear, A. E.; Koepfli, J. B. *J. Am. Chem. Soc.* **1946**, *68*, 2708.

15   Atwell, G. J.; Baguley, B. C.; Denny, W. A. *J. Med. Chem.* **1989**, *32*, 396.

Derek A. Pflum

9.6 Friedländer Quinoline Synthesis[1]

9.6.1 Description

The Friedländer quinoline synthesis combines an α-amino aldehyde or ketone (1) with another aldehyde or ketone with at least one methylene α to the carbonyl (2) to furnish a substituted quinoline. The reaction can be promoted by acid, base, or heat.

9.6.2 Historical Perspective

Friedländer reported the condensation of 2-aminobenzaldehyde (4) with acetaldehyde (5) to provide quinoline (6) in 1882.[2]

Interestingly, the first report of a reaction of this type was reported by Fischer and Rudolph earlier the same year Friedländer reported his quinoline synthesis.[3a] Heating of acetylaniline (7) in the presence of zinc chloride promotes acyl migration to furnish 2-aminoacetophenone (8) and 4-aminoacetophenone (9). These two molecules combine to furnish flavanilin (10).[3]

By 1922, the Friedländer reaction had been well enough established that it was being used to prepare derivatives for structure elucidation.[4]

9.6.3 Mechanism

Two possible mechanisms exist for the Friedländer reaction. The first involves initial imine formation followed by intramolecular Claisen condensation, while the second reverses the order of the steps. Evidence for both mechanisms has been found,[5] both intermediates have been observed, and the mechanism may change for the same two partners based upon reaction conditions.

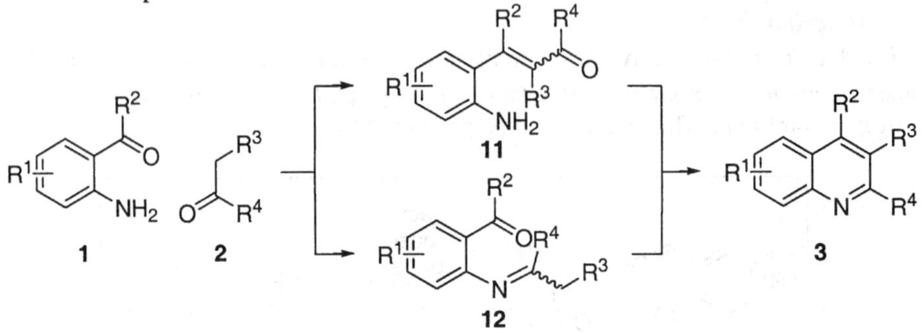

9.6.4   Variations and Improvements

The Friedländer reaction is quite versatile.   The primary limitation on the *o*-aminobenzaldehyde component is preparation of the starting material; as one might expect, these compounds are prone to self-condensation. Both electron rich[6] and electron poor[7] *o*-aminobenzocarbonyl compounds undergo the Friedländer reaction. When ketone partner **2** has only one available reactive methyl or methylene or is symmetrical, only one product is obtained. Even when two products can be formed, it is possible to choose reaction conditions such that only one product is isolated (*vide infra*). The reaction can be promoted by acid catalysis, sometimes with improved results.

Several variations of the Friedländer reaction exist and are well known enough to have their own names:

The Niementowski reaction involves condensation of an *o*-aminobenzoic acid (**13**) with **2** resulting in a quinolinol (**14**).[8]

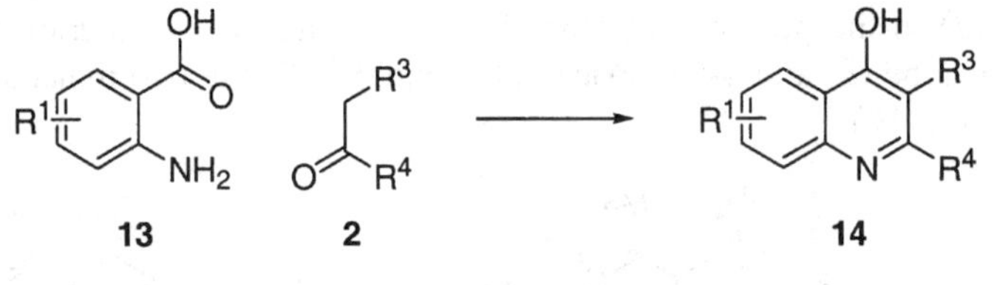

The Pfitzinger reaction describes the condensation of an *o*-aminophenylglyoxylic acid **16** (which can be generated *in situ* from an isatin **15**) with **2** results in a quinoline-4-carboxylic acid (**17**), which is the subject of its own chapter in this book.[9]

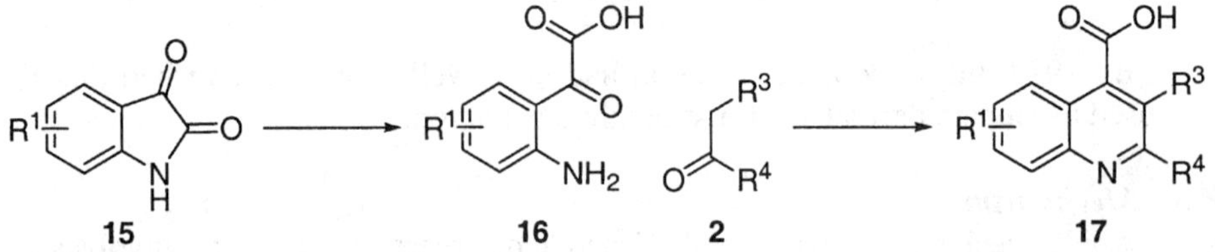

The Kepmter modification is when the hydrochloride salt of **1** is used as the starting material.[10]

The Fehnel modification describes the Friedländer reaction in acetic acid with a catalytic amount of sulfuric acid.[11]   In some cases, this modification provides the expected product in much improved yields.

conditions	yield
alkali alcohol	<5%
thermolysis/cat HCl	56%
AcOH/H$_2$SO$_4$	64%

The Borsche modification describes the protection of the aldehyde as an imine; frequently the starting material is the imine-protected o-nitrobenzaldehyde.[12] The nitro group is reduced with zinc and then treated with a ketone in the usual Friedländer way. Alternatively, reduction of the free o-nitrobenzaldehyde (21) with SnCl$_2$ in the presence of a ketone (e.g. 22) and ZnCl$_2$ results in good yields of the expected quinoline.[13]

The Henegar modification of the Friedländer reaction has been recently reported.[14]   The N-Boc protected derivative of o-aminobenzaldehyde (25, in this case prepared via directed ortho metallation of 24) is a stable, crystalline compound that can be stored for extended periods (in contrast with 4, which typically is freshly prepared). Treatment of 25 with ketone 26 in acetic acid results in deprotection of the aniline *in situ* and subsequent formation of 27, an intermediate in the synthesis of mappicine.

The Friedländer reaction originally was performed in ethanolic alkoxide.[2] Amine bases, such as piperidine, have been used.[15] Anion exchange resins have also been used.[16]

Alternatively, the Friedländer reaction can be promoted by thermolysis to 150-200 °C in the absence of solvents.[17]

Merck chemists have done a detailed investigation on the effect of reaction conditions on the yield and selectivity of the Friedländer reaction.[18]   Initially, the condensation of 2-aminonicotinaldehyde (**28**) with methyl ketone **29** under the influence of potassium hydroxide in ethanol resulted in an approximately 1:2 mixture of **30:31**. Pyrrolidines, and especially **32**, were found to be effective and selective catalysts for the preparation of the desired 2-monosubstituted-8-naphthyridine (>99% conversion, 87:13 **30:31**). Further improvements – increasing the reaction temperature to 65-70 °C and slow addition of the ketone to the mixture of aminoaldehyde and **32** – resulted in >99% conversion to a 96:4 mixture of **30:31**.

Conditions	Conversion	Ratio **30:31**
NaOH, rt	>99%	37:63
pyrrolidine, 5% H$_2$SO$_4$, rt	97%	86:14
32, 5% H$_2$SO$_4$, rt	>99%	87:13
32, 5% H$_2$SO$_4$, slow addition of **29**, 65 °C	>99%	94:6

1,3,3-trimethyl-6-azabicyclo[3.2.1]octane
(TBAO, **32**)

9.6.5　Synthetic Utility

The Friedländer reaction makes available a wide variety of 1,10-phenanthrolines and other macrocyclic chelators (e.g. **33**, **34** and **35**).[19] These polyaza-aromatic rings can bind a variety of metals or organic substrates.

9.6.6　Experimental[20]

2-Benzyl-1,2,3,4-tetrahydrobenzo[b][1,6]naphthyridine
A solution of **4** (16 g, 132 mmol), **36** (27.5 g, 145 mmol), and NaOMe (7.8 g, 145 mmol) in dry EtOH (300 mL) was refluxed for 3 h. The reaction mixture was evaporated *in*

vacuo to give a residue, which was dissolved in water and toluene. The toluene layer was evaporated *in vacuo* to give a residue, which was recrystallized from IPA/CH$_2$Cl$_2$ to give **37** (32.4 g, 90% yield) as colorless prisms (mp 125–126 °C).

9.6.7   References

1   The Friedlander has been the subject of excellent reviews. a) [R] Cheng, C.-C.; Yan, S.-J. *Org. Reactions*, **1982**, *28*, 37–201. b) [R] Jones, G. in *Heterocyclic Compounds*, Quinolines, vol. 32, **1977**; John Wiley & Sons, Inc.: New York, 181–191. c) [R] Elderfield, R. C. in *Heterocyclic Compounds*; Elderfield, R. C., Ed.; John Wiley & Sons, Inc.: New York, 1952; Volume 4, Quinoline, Isoquinoline and Their Benzo Derivatives, 45–47.

2   Friedlander, P. *Chem. Ber.* **1882**, *15*, 2572.

3   a) Fischer, O.; Rudolph, C. *Chem. Ber.* **1882**, *15*, 1500. b) Besthorn, E.; Fischer, O. *Chem. Ber.* **1883**, *16*, 68.

4   Armit, J. W.; Robinson, R. *J. Chem. Soc.* **1922**, *121*, 827.

5   a) For Claisen condensation followed by imine formation, see, for example, reference 4. b) For imine formation followed by Claisen condensation, see, for example, Fehnel, E. A.; Deyrup, J. A.; Davidson, M. B. *J. Org. Chem.* **1958**, *23*, 1996.

6   see, for example, Fernandez, M.; Lopez, F.; Tapia, R.; Valderrama, J. A. *Synth. Commun.* **1989**, *19*, 3087

7   see, for example, Bu, X.; Deady, L. W. *Synth. Commun.* **1999**, *29*, 4223.

8   von Niementowski, S. *Chem. Ber.*, **1894**, *27*, 1394.

9   Pfitzinger, W. *J. Prakt. Chem.*, **1886**, *33*, 100.

10   Kempter, G.; Andratschke, P.; Heilmann, D.; Krausmann, H.; Meitasch, M. *Chem. Ber.* **1964**, *97*, 16.

11   Fehnel, E. A. *J. Org. Chem.* **1966**, *31*, 2899

12   a) Borsche, W.; Ried, W. *Liebigs Ann. Chem.* **1943**, *554*, 269. b) Borsche, W.; Reid, W. *Chem. Ber.* **1943**, *76*, 1011.

13   McNaughton, B. R.; Miller, B. L. *Org. Letters*, **2003**, *5*, 4257.

14   Henegar, K. E.; Baughman, T. A. *J. Heterocycl. Chem.* **2003**, *40*, 601.

15   Stark, O. *Chem. Ber.* **1907**, *40*, 3425.

16   Yumada, S.; Chibata, I. *Chem. Pharm. Bull.* **1955** *3*, 21.

17   Borsche, W.; Sinn, F. *Liebigs Ann. Chem.*, **1939**, *538*, 283.

18   Dormer, P. G.; Eng, K. K.; Farr, R. N.; Humphrey, G. R.; McWilliams, J. C.; Reider, P. J.; Sager, J. W.; Volante, R. P. *J. Org. Chem.* **2003**, *68*, 467. This mirrors similar selectivity seen in the 4-aminonicotinaldehyde Friedlander reaction; see Hawes, E. M.; Gorecki, D. K. J.; Johnson, D. D. *J. Med Chem.* **1973**, *16*, 849 and Hawes, E. M.; Gorecki, D. K. J. *J. Heterocycl. Chem.* **1974**, *11*, 151.

19   see, for example a) Hoste, J. *Anal. Chim. Acta*, **1950**, *4*, 23. b) [R] Thummel, R. P. *Synlett*, **1992**, *1992*, 1. c) Reisgo, E. C.; Jin, X.; Thummel, R. P. *J. Org. Chem.* **1996**, *61*, 3017. d) Bartsch, R. A.; Eley, M. D. *Tetrahedron*, **1996**, *52*, 8979. e) Gladiali, S.; Chelucci, G.; Mudadu, M. S.; Gastaut, M.-A.; Thummel, R. P. *J. Org. Chem.* **2001**, *66*, 400.

20   Shiozawa, A.; Ichikawa, Y.-I.; Komuro, C.; Kurashige, S.; Miyazaki, H.; Yamanaka, H.; Sakamoto, T. *Chem. Pharm. Bull.* **1984**, *32*, 2522.

Derek A. Pflum

9.7 Gabriel–Colman Rearrangement

9.7.1 Description

The Gabriel–Colman rearrangement entails reaction of the enolate of a maleimidyl acetate (2) to provide isoquinoline 1,4-diol 3.[1]

1, G = CO, SO$_2$ 2   3

9.7.2 Historical Perspective

In 1900, Gabriel and Colman reported the preparation of phthalimidoyl acetate 4.[2] They had anticipated saponifying 4 with sodium ethoxide and were surprised to find, rather than hydrolysis, rearrangement to 5. The identity of the product was confirmed by hydrolysis of the newly formed ester and concomitant decarboxylation to provide 6, which was hydrogenated to the known isocarbostyril (7).

The Gabriel–Colman reaction has been used to prepare 3-alkyl isoquinoline 1,4-diols.[2] Phthalimides 8 and 9 rearrange as expected when treated with alkoxides. Further treatment with sodium ethoxide results in decarboxylation and the expected isoquinolinone 1,4-diols 12 and 13.

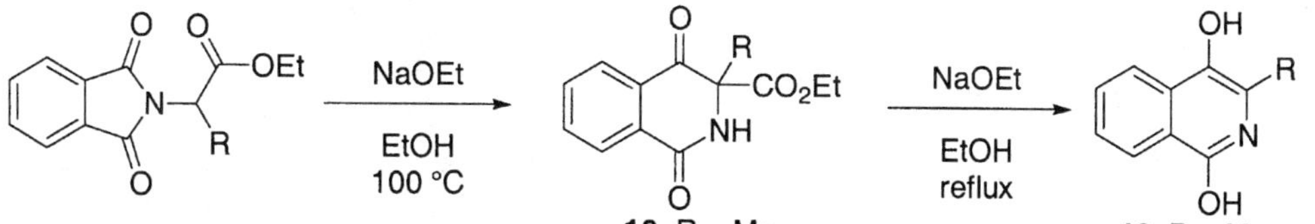

8, R = Me
9, R = Et

10, R = Me
11, R = Et (not observed)

12, R = Me
13, R = Et

Use of α-phthalimidoyl ketones has been described.[3] Treatment of either **14** or **15** with sodium methoxide in refluxing methanol affords the expected isoquinoline **16** or **17**, respectively.

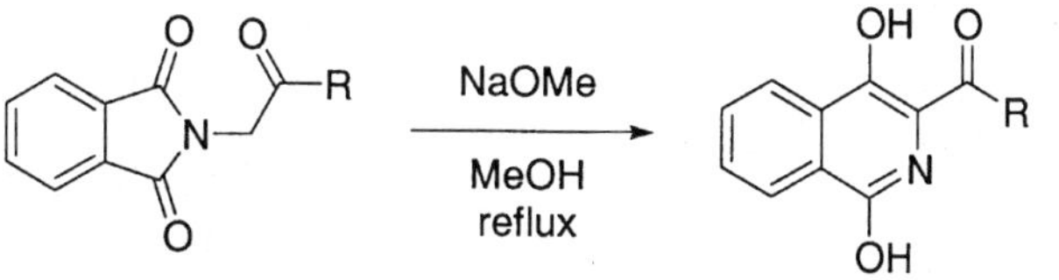

14, R = Me
15, R = Ph

16, R = Me
17, R = Ph

9.7.3 Mechanism

Two mechanisms have been proposed for the Gabriel–Colman rearrangement. The first involves initial formation of the ester enolate **19** from **18** by the alkoxide.[4] Cyclopropanation onto one of the phthalimide carbonyls provides intermediate **20**, which

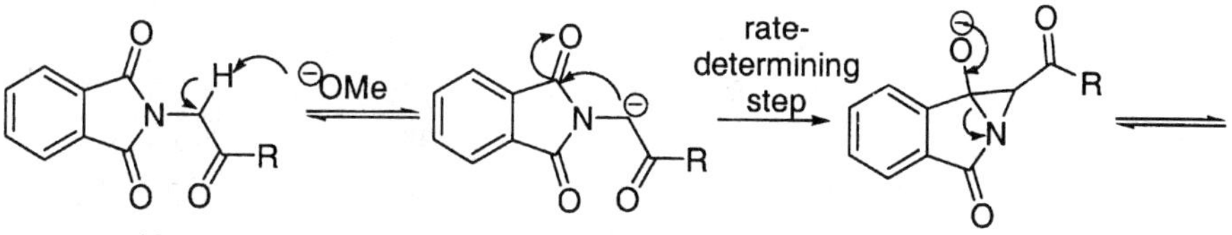

18

19

rate-determining step

20

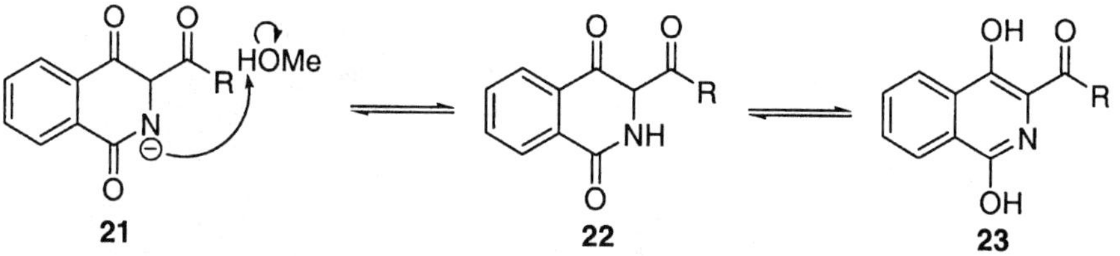

21

22

23

decomposes to **21** and subsequently aromatizes to afford isoquinoline **23**. This mechanism was proposed based upon rearrangements of benzyl ethers to carbinols to explain the GC reaction; no experimental evidence for this mechanism was provided.

24   **25**

The second proposed mechanism involves initial ring opening of the phthalimide.[5] Alkoxide attack on one of the imide carbonyls furnishes amide anion **26**. Proton transfer affords enolate **27**, which undergoes Diekmann type condensation followed by aromatization to afford the requisite isoquinoline **23**.

18   **26**   **27**

28   **23**

Although kinetic data allows both mechanisms (if steady-state approximations are made and assuming three-member ring formation is rate determining in the first mechanism and ring opening or closing is rate determining in the second), experimental evidence on the rearrangement of N-phenacylphthalimides **29** points to the ring

29   R = Me, **30**
                                   R = H, **31**

opening/ring closing mechanism.[6] The rate of ring closure of phthalamate **30** with methoxide is the same as the rate of rearrangement of **29**, indicating that **30** is a competent intermediate along the reaction pathway. Furthermore, if t-butoxide in t-butanol is substituted for methoxide in methanol, the observed rate of rearrangement of **30** is substantially slower. If mechanism 1 were operative, the steric bulk of base used would not be expected to impact the rate of reaction; however if mechanism 2 were operative, the steric bulk of the t-butyl ester would be expected to slow the reaction.

Finally, reaction of hydroxide with **29** furnishes **31** approximately 20 times faster than methoxide promoted rearrangement of **29** to **23** (R = Ph), suggesting ring closure is the rate-determining step.

9.7.4 Variations and Improvements

The Gabriel–Colman reaction can be used to prepare isoquinoline-1,4-diols regioselectively by the use of unsymmetrically substituted phthalimides. Reaction of phthalimide **32** with sodium ethoxide in ethanol provides a 1:7 mixture of **33:34**.[7] It was rationalized that attack at carbon *b* is preferred because of its greater steric accessibility and diminished electron density compared to carbon *a*. In spite of the reasonable regioselectivity observed in this reaction, the Gabriel–Colman reaction has not been substantially investigated in the preparation of non-symmetrically substituted isoquinolines.

32

33, 3%
results from attack at *a*

34, 21%
results from attack at *b*

The rearrangement has been found to be substrate specific. In some cases, the reaction proceeds as described above, *i.e.* using alkoxide in alcoholic solvent. In other cases, these conditions do not work well, or the reaction has been found to work better under pressure at elevated temperature in alcoholic solvents,[8] in DMSO,[9] DMF,[10] or toluene.[11] Rigorous exclusion of moisture and carbon dioxide is necessary.[6]

Calcium alkoxides do not promote the rearrangement and instead afford only ring opened product.[12] Otherwise, the effect of metal counterion on the reaction has not been investigated.

9.7.5 Synthetic Utility
9.7.5.1 Naphthyridine synthesis[13]

Gabriel and Colman extended the rearrangement to nitrogen containing heterocycles.[14] Cinchomeronylglycine ester **35** was treated with sodium methoxide in methanol to provide 2,7-naphthyridine **37**. The regiochemistry of the product was established by chemical degradation. That the product is the 2,7-naphthyridine and not the 2,6 isomer **39** can be rationalized by the reactivities of the different carbonyls. One carbonyl in the starting material can be viewed as a vinylogous α-imino amide and therefore relatively more electrophilic than the other carbonyl, which can be considered a vinylogous urea.

Shortly after Gabriel and Colman reported the rearrangement of cinchomeronylacetic ester, Fels reported a similar rearrangement of α-quinolinimidoacetic ester (40) to provide a 1,6-naphthyridine (41).[15] The structure of the isolated compound was not unambiguously determined for more than 30 years.[16] More recently, the reaction has been shown to produce both 41 and 42 in a 3:1 ratio.[17]

9.7.5.2 Benzothiazine synthesis

The most widely used variant of the Gabriel–Colman is the conversion of saccharine derivatives to benzothiazine derivatives. The reaction has been extensively studied as benzothiazines are important pharmacophores, particularly in the oxicam class of anti-inflammatories. The first reported instance of this transformation was in 1956 where 43 was treated with sodium methoxide to provide 44.[18] The rearrangement also works with esters[19] and some amides[20] in addition to ketones.

The mechanism of this variant of the Gabriel–Colman reaction has been investigated. Treatment of saccharine derivatives **45–48** with 1–2 equivalents of sodium alkoxide at room temperature provides esters **49–52** in good yields; treatment of **45–48** with sodium alkoxide at reflux provides the expected benzothiazines **53–56**.[21] Increased concentration leads to higher yields.

45, R = Me
46, R = Et
47, R = *i*-Pr
48, R = *t*-Bu

49, R = Me, 92%
50, R = Et, 88%
51, R = *i*-Pr, 90%
52, R = *t*-Bu, 86%

53, R = Me, 70% (one step)
54, R = Et, 62% (one step)
55, R = *i*-Pr, 85% (one step)
56, R = *t*-Bu, 64% (one step)

The benzothiazine equivalent of a 1,7 naphthyridine (**58**) has also been prepared.[22] The reaction did not work in alcoholic solvents, but when DMF was used **57** rearranged to provide the desired product in moderate yield.

9.7.6   Experimental[21]

4-Hydroxy-1,1-dioxo-1,2-dihydro-1λ^6-benzo [*e*] [1,2] thiazine-3-carboxylic acid isopropyl ester (55)

A solution of sodium isopropoxide was prepared from sodium (0.92 g, 40 mmol, 4 equiv.) in isopropanol (16 mL). The solution was refluxed in an oil bath and **47** (2.8 g, 10 mmol) was added all at once as a powder. After a few minutes, the orange slurry was poured into ice-cold concentrated hydrochloric acid. The solid was filtered, washed with water and recrystallized from ethanol to afford **55** (2.4 g, 85% yield): mp 170 °C; ^{1}H NMR (CDCl$_3$) δ 11.5 (s, 1H), 7.70–8.32 (m, 4H), 6.65 (S, 1H), 5.42 (m, 1H), 1.55 (d, 6H); IR (KBr, cm^{-1}) 3200, 1660, 1340, 1190.

9.7.7 *References*

1   For consistency, structures are drawn as the isoquinoline-1,4-diol throughout regardless of tautomer drawn in the original reference. The actual tautomeric structure is dependent on the pH of the solution. See Caswell, L. R.;Campbell, R. D. *J. Org. Chem.* **1961**, *26*, 4175.

2   Gabriel, S.; Colman, J. *Chem. Ber.* **1900**, *33*, 980.

3   Gabriel, S.; Colman, J. *Chem. Ber.* **1900**, *33*, 2630.

4   Hauser, C.R.; Kantor, S. W. *J. Am. Chem. Soc.* **1951**, *73*, 1437.

5   [R] Gensler, W. J. *Heterocyclic Compounds*, Vol. 4, R. C. Elderfield, Ed., John Wiley and Sons, Inc., New York, N.Y. **1952**, 378.

6   Hill, J. H. M. *J. Org. Chem.* **1965**, *30*, 620.

7   Koelsch, C. F.; Lindquist, R. M. *J. Org. Chem.* **1956**, *21*, 657.

8   Caswell, L. R.; Atkinson, P. C. *J. Heterocycl. Chem.* **1966**, *3*, 328.

9   Lombardino, J. G.; Wiseman, E. H.; McLamore, W. M. *J. Med. Chem.* **1971**, *14*, 1171.

10   Genzer, J. D.; Fontsere, F. C. U.S. Patent 3 960 856, 1976.

11   Lazer, E. S.; Miao, C. K.; Cywin, C. L.; Sorcek, R.; Wong. H.–C.; Meng, Z.; Potocki, I.; Hoermann, M. A.; Snow, R. ;J.; Tschantz, M. A.; Kelly, T. A.; McNiel, D. W.; Coutts, S. J.; Churchill, L.; Graham, A. G.; David, E.; Grob, P. M.; Engel, D. W.; Meier, H.; Trummlitz, G. *J. Med. Chem.* **1997**, *40*, 980.

12   Svoboda, J.; Palecek, J.; Dedek, V. *Collect. Czech. Chem. Commun.* **1986**, *51*, 1133.

13   For a review of the preparation of naphthyridines, including use of the Gabriel–Colman reaction, see [R] Allen, C. F. H. *Chem. Rev.* **1950**, *47*, 275. For the use of the Gabriel–Colman reaction in the preparation of 1,6-naphthyridines see p. 284, and for the use of the Gabriel–Colman reaction in the preparation of 2,7-naphthyridines see p. 287.

14   Gabriel, S.; Colman, J. *Chem. Ber.* **1902**, *35*, 1358.

15   (a) Fels, B. *Chem. Ber.* **1904**, *37*, 2129.   (b) *ibid*, 2137.

16   Ochai, E.; Arai, I., *J. Pharm. Soc. Japan* **1939**, *59*, 458.

17   Albert, A.; Hampton, A. *J. Chem. Soc.* **1952**, 4985.

18   (a) Abe, K.; Yamamoto, S.; Matsui, K. *J. Pharm. Soc. Japan* **1956**, *86*, 1058.   (b) Subsequently, a complementary rearrangement of i to ii was reported. Zinnes, H.; Comes, R. A.; Shavel, J. Jr. *J. Org. Chem.* **1964**, *29*, 2068; this is the only report of the benzothiazin-4-one being formed in the reaction.

19   Lombardino, J. G.; Wiseman, E. H.; McLamore, W. M. *J. Med. Chem.* **1971**, *14*, 1171.

20   Schapira, C. B.; Abasolo, M. I.; Perillo, I. A. *J. Heterocycl. Chem.* **1985**, *22*, 577.

21   Schapira, C. B.; Perillo, I. A.; Lamdan, S. *J. Hererocycl. Chem.* **1980**, *17*, 1281.

22   Zawisza, T.; Malinka, W. *Farmaco Ed. Sci.* **1986**, *41*, 892.

Derek A. Pflum

9.8 Gould–Jacobs Reaction

9.8.1 Description

The Gould–Jacobs reaction is a sequence of the following reactions: (1) condensation of an arylamine **1** with either alkoxy methylenemalonic ester or acyl malonic ester **2** providing the anilidomethylenemalonic ester **3**; (2) cyclization of **3** to the 4-hydroxy-3-carboalkoxyquinoline **4**; (3) saponification to form acid **5**, and (4) decarboxylation to give the 4-hydroxyquinoline **6**. All steps of this process will be described herein with emphasis on the formation of intermediates like **3** and **4**.

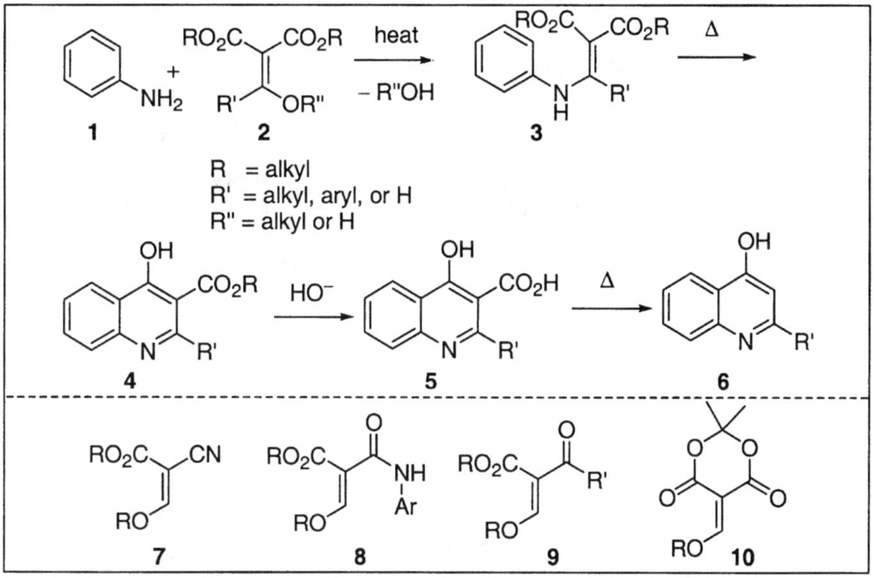

The first reaction can be conducted using various derivatives of methylenemalonic ester, such as malononitriles **7**, malonamides **8**, β-keto-esters **9** or Meldrum's acid **10**. Substitutions of the aryl ring (including fused rings) and within the aryl ring are well tolerated for this reaction.

9.8.2 Historical Perspective

In 1885, Just reported[1] the reaction of sodium diethylmalonate with *N*-phenylbenzimino chloride to provide **11**. Thermal cyclization provided **12**. This work was virtually untouched for several decades, but laid the groundwork for further development.

In 1939, Gould and Jacobs reported[1] the condensation of aniline with acetyl malonic ester (AME) and ethoxymethylenemalonate (EMME), respectively, followed by cyclization of anilinomethylenemalonic ester **13** and **14** in mineral oil to afford the esters **15** and **16**. Saponification of the esters gave the known acids **17** and **18**, respectively.

R = Me, R' = H (AME)
R = H, R' = Et (EMME)

30%, R = Me, **13**
70%, R = H, **14**

R = Me, **15**
60%, R = H, **16**

R = Me, **17**
R = H, **18**

Improvements to this report by Gould and Jacobs did not come for nearly seven years when a plethora of activities in the synthesis of quinolines was reported due to the need for antimalarials. Dowtherm™ (a mixture of diphenyl ether and biphenyl) or diphenyl ether[2] could be used for the cyclization and for the decarboxylation. Initially, these steps had used mineral oil and a Wood's metal bath, respectively. Much was learned about the selectivity, or lack of selectivity, in the cyclization step.[3] Normally, *m*-substituted anilines give the 7-substituted quinoline isomer. Whereas *m*-chloro aniline was initially reported to cleanly provide the 7-chloroquinoline, subsequent reports (with scale-up) claimed up to 15% of the 5-chloro isomer. Likewise, *m*-fluoroaniline was reported to give mixtures of the 7- and 5-fluoroquinolines in which the the 5-isomer was reported as being "detectable." A recent report on this substrate claimed not observing the 5-fluoro isomer.[4] *m*-Cyano-aniline was reported to exclusively provide the 5-cyanoquinoline. Formation of the silver salts was found to facilitate decarboxylation in cases in which powerful EWGs (electron-withdrawing groups, *i.e.*, NO_2) were attached to the arylamine.[5]

In order to expand the utility of the reaction, modification of the route to anilidomethylene malonic ester equivalents was developed. Simple condensation of triethyl orthoformate with cyanoacetic ester, acetoacetic ester, or malonic ester in the presence of an *m*-chloroaniline gave the desired anilinoacrylate.[2a] A higher yield of **23** was reported by reaction of a bisarylformamidine with malonic ester to give the acrylanilide **23** in 70% yield. Cyclization gave quinolines **24–26** bearing a nitrile, ketone, or amide at the 3-position. Note that cyclization occurred on the ester carbon rather than on the nitrile, ketone, or amide carbon.

Condensation of *N*-aryliminochlorides with malonic ester followed by thermal cyclization, as initially reported by Just, was found to be a general method for the preparation of 2, 3, 4-substituted quinolines.[2a] Various substituents on the aryl ring of the iminochloride proved uneventful, even though the conditions required to generate the iminochloride utilized PCl_5.

9.8.3   Mechanism

The mechanism of this reaction has not been thoroughly explored. Some work has been done in analysis of potential intermediates for the reaction, although these intermediates were generated using flash vacuum pyrolysis (FVP).[6] Materials in this experiment were trapped and IR spectrum suggested the formation of a ketene prior to cyclization.

This information coupled with the proposed mechanism of the Conrad–Limpach reaction, reasonably lead to the below proposed mechanisms.[1] Conjugate addition of aniline and elimination of alcohol provides the β-anilinoacrylate 14, which upon heating to 180–320 °C gives species, like 34a,b, which undergo 6π-electrocyclization to 35 or 36, respectively. Loss of ethanol from 36 gives 35 and tautomerization provides 4-

hydroxy-3-ethoxycarbonylquinoline **16**.   Saponification of ester **16** to the acid and decarboxylation (via β-keto ester tautomerization) gives the 4-hydroxyquinoline.   Under acid-promoted cyclization conditions (*i.e.*, PPA, POCl$_3$, H$_2$SO$_4$), Friedel–Crafts type cyclization could also be operable.

9.8.4   Variations and Improvements
9.8.4.1 Preparation and application of alternative methylenemalonate equivalents
Due to the commercial availability of EMME in good purity, there has not been a need to develop new methods to prepare β-anilino-acrylates; therefore, only a few alternatives have been reported.   One approach described thermal carbene generation from **37**, and rearrangement to form **38**.[7]   Cyclization in refluxing 1,2-dichlorobenzene (1,2-DCB) provided the 2, 3, 4-trisubstituted quinolines.   An electron-withdrawing group (EWG) on the carbene carbon was required for this reaction, and therefore led to the EWG substitution in the 2-position of the quinoline.

X = CO$_2$Me, **37**	41%	X = CO$_2$Me, **38**	64%	X = CO$_2$Me, **39**
X = COPh, **40**	47–52%	X = COPh, **41**	48%	X = COPh, **42**
X = COMe, **43**	62%	X = COMe, **44**	54%	X = COMe, **45**

The preparation and use of derivatized Meldrum's acid has led to an alternative preparation of 2-substituted quinolines (**49** and **50**)[8] and the preparation of pyridopyrimidines (**52**).[9]   When Meldrum's acid derivatives are used (as shown in this example) decarboxylation occurred under the cyclization conditions.   Three component coupling has been used to readily assemble the desired β-anilino-acrylate from reaction of Meldrum's acid, (EtO)$_3$CH and an aniline (e.g. **54** or **55**).[6]

R = m-CF₃-Ph, **46**   **48**   64%   R = m-CF₃-Ph, **49**
R = Me, **47**   46%   R = Me, **50**

51   **52**

86%, Ar = p-Me-Ph, **54**
21%, Ar = p-OH-Ph, **55**

53

9.8.4.2 Reaction media for cyclization and decarboxylation

Several high-boiling, inert solvents have been reported in various steps of the Gould–Jacobs reaction. Dowtherm,™ Ph₂O, cumene, tetraglyme,[10a] diphenyl methane,[10b] paraffin,[10c] and 1,2-DCB have been successfully employed for the cyclization and decarboxylation steps. Copper chromide/quinoline,[11a] vacuum sublimation,[11b] paraffin,[11c] quinaldine,[11d] or dibutyl phthalate[11e] proved adequate to promote decarboxylation. Cyclization[12] via continuous flow reactor (380 °C, 45 s loop) and decarboxylation have been described neat. The change in selectivity of cyclization is notable and will be addressed later (compare conversion of **51–52** with **56–57**).

75%, 94% purity
8% N-isomer observed

56   **57**

Alternatively, cyclization has been accomplished using various acids which dramatically altered the selectivity in the cyclization, a cyclization that failed under thermal conditions as in the preparation of **59** and **60**.[13] Often, the yield of the acid-promoted reaction was lower than the thermally-promoted cyclization. Of course, the temperature in which acid-promoted reactions were conducted was much lower than the normal 250 °C required for thermal cyclization. The selectivity was altered when an acid was used as in **58** → **59** and **60**, suggesting a change in mechanism.

PPE	76%, 4:3
Ac$_2$O/H$_2$SO$_4$	20%, 3:2
PPA	15%, 13:7
POCl$_3$	67%, 15:11*
Dowtherm A (thermal)	82%, 1:9

*The reaction using POCl$_3$ gave the 4-Cl quinolines.

N-cyclopropylquinolone **63** and *N*-fused pyrrolidinyl **66** and piperidinyl **69** quinolones were reported as intermediates for the preparation of potent antibacterial agents. Use of acids as reaction solvent allow for the cyclization of these *N*-substituted anilino-acrylates.[14] The reports of thermal cyclization of *N*-substituted-anilino-acrylates are rare.[15]

Cyclizations, which failed to occur under thermal conditions, have been forced by using strong acids as solvent. Such cyclizations required careful temperature control in order to cyclize while maintaining the 3-carboxyl substituent.[16]

Typical acids employed to promote the Gould–Jacobs reaction are polyphosphoric acid (PPA), ethyl polyphosphate (PPE), Ac_2O/H_2SO_4, $POCl_3$, or BF_3OEt_2[17] at 60–170 °C. Phosphorous oxychloride provides the 4-chloroquinoline directly. Overall, if a reaction can be promoted thermally, that is the method of choice due to the ease of running the reaction, the higher yield or recovery, and the potential to change selectivity under acid-promoted reaction conditions.

9.8.4.3 Alternative techniques applied to cyclization

Alternative techniques, such as flash vacuum pyrolysis (*vide infra*; **30** → **33**), have been applied to the Gould–Jacobs reaction. Use of microwave has in some cases provided an excellent yield of the desired 3-alkoxycarbonyl-4-hydroxy quinoline[18]. The reactions were conducted on a 10 mmol scale with superior overall purity to classical thermal conditions.

9.8.4.4 Substitution and its effect on selectivity in the cyclization

Nearly every substitution of the aromatic ring has been tolerated for the cyclization step using thermal conditions, while acid-promoted conditions limited the functionality utilized. Substituents included halogens, esters, nitriles, nitro, thio-ethers, tertiary amines, alkyl, ethers, acetates, ketals, and amides.[19] Primary and secondary amines are not well tolerated and poor yield resulted in the cyclization containing a free phenol.[11e] The Gould–Jacobs reaction has been applied to heterocycles attached and fused to the aniline.

3-Benzoxazol-4-fluoro-aniline **77** and benzthiazol-aniline **80** both provided the "linear" product in very good overall yield. This was considered to be the normal or predominate mode of cyclization.[20] Alternatively, fused amino benzoxazole **83** gave exclusively the angular product **85**, while substituted amino-benzoxazoles **86** and **89** gave 2:1 mixtures of "linear" (**86** and **89**) and "angular" products (**88** and **91**).[21]

R = H, **83**	94%, <5:>95	R = H, **84**
R = Me, **86**	65%, 2:1	R = Me, **86**
R = Ph, **89**	68%, 2:1	R = Ph, **90**

85
88
91

Cyclization of quinoxalines **92** and **94** gave angular products **93** and **95**.[22]

R = CO$_2$Et, **92**	70%	R = CO$_2$Et, **93**
R = CN, **94**	76%	R = CN, **95**

Both steric and electronic factors have been claimed to control the selectivity in the cyclization step. Not only the control of the selectivity on the ring closure but also the lack of activity toward cyclization was observed. In one example of this, methyl substituted aminoindole **96** provided cyclization product **99** while attempted cyclization of methyl ether **98** led to decomposition.[23]

R = Me, **96**	76%	R = Me, **99**
R = OMe, **98**	trace; turned tarry after 30 min	R = OMe, **100**

Reactions demonstrating steric control have been reported. Cyclization of the dimethyl acetal **103** led to a 9:1 ratio of **104:105** instead of a 1:1 ratio using unsubstituted dioxolane **101**.[24] Yet others reported sterics did not control the selectivity of cyclization.

R = H, **101**   not isolated, 1:1   R = H, **102**   **103**
R = Me, **103**   62% desired, 9:1   R = Me, **104**   **106**

Thermal cyclization of the 3-nitro anilide **107** gave a 1 : 1 mixture of **108** and **109** in excellent yield, though one would expect the nitro group to be more sterically demanding than a hydrogen.[25]   An   interesting   difference   in   the   reported

cyclization of **110** and **113** suggested that electronic effects predominantly controlled the observed mode of cyclization. Thus, a methylene substituent attached *para*- to the center

of cyclization, gave a 10:1 ratio of **111:112** (analogous to a methyl group being *para*-, **60:59**). An ether *para*- to the center of cyclization provided only one isomer.[26]

Although regio-isomers have been reported in some cases as being problematic, due to the wide range of substituents that withstand the cyclization, the Gould–Jacobs reaction is one of the more general methods used to prepare quinolines.

9.8.4.5 Selectivity of cyclization for some heterocycles

As previously described, the Gould–Jacobs reaction has been applied to heterocycles fused to anilines, and to some amino-substituted heterocycles. Selectivity of *N*- and *C*-cyclization of 2-aminopyridino-methylene malonates has been mentioned (**51** and **56**). The normal mode of cyclization of 2-aminopyridino-methylene malonates is on the nitrogen to form a pyridopyrimidine. If an electron-donating group (EDG) is in the 6-

position of a 2-aminopyridine (alkyl, OR, or Cl), the 1,8-napthyridine is obtained; otherwise, the pyrido-pyrimidine is obtained.[27] There are of course exceptions to this generality. Presumably, the formation of the 1,8-naphthyridine is due to the steric requirements for cyclization on nitrogen being much different than on carbon. Another example of the peculiarity of the cyclization of 2-aminopyridines.[28] Whereas pyridofuran **115** gave predominantly *N*-alkylated **116**, the straight-chain substituted **118** provided the

C-alkylated material **119**. The cyclization of a 3-aminopyridino-methylene malonate has been shown. The formation of amino-acrylate **121** was nearly quantitative and subsequent cyclization afforded a 50% yield of the 1,5-napthyridine. Saponification and decarboxylation gave the desired 1,5-napthyridine **124** in good yield.[29]

Formation of the amino-acrylate of aminopyrimidine intermediates was reported. In the absence of base, alkylation occurred on the carbon and not the nitrogen, followed by cyclization to give **125**. In the presence of base (EtONa), condensation occurred on the nitrogen. Cyclization under thermal conditions afforded **128**.[30]

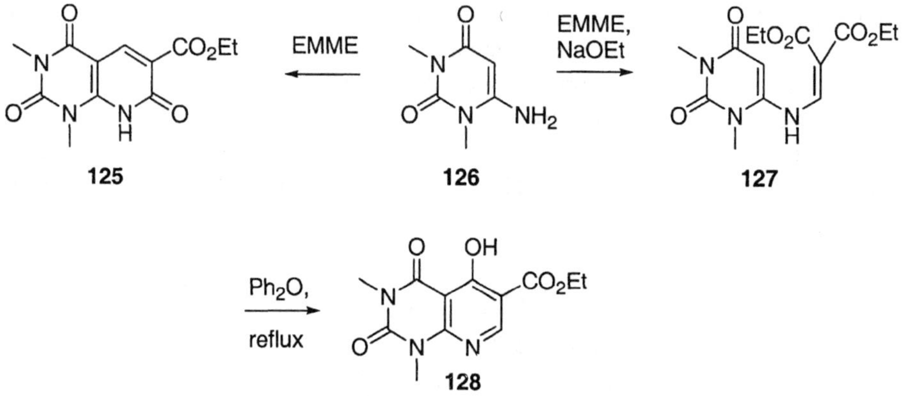

125   **126**   **127**

128

With pyrimidine systems, another anomaly occurred with which group underwent cyclization in the case of α-cyano-β-(pyrimidino)amino-acrylate. For example, cyclization of **129** occurred on the CN group providing the 5-amino-pyridopyrimidine **130**.

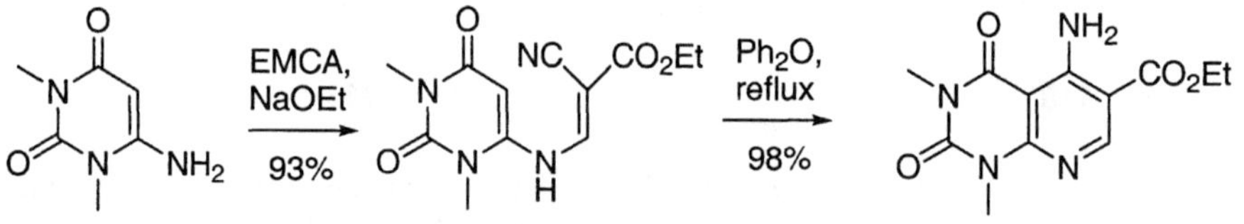

126   **129**   **130**

9.8.5   Experimental[4]

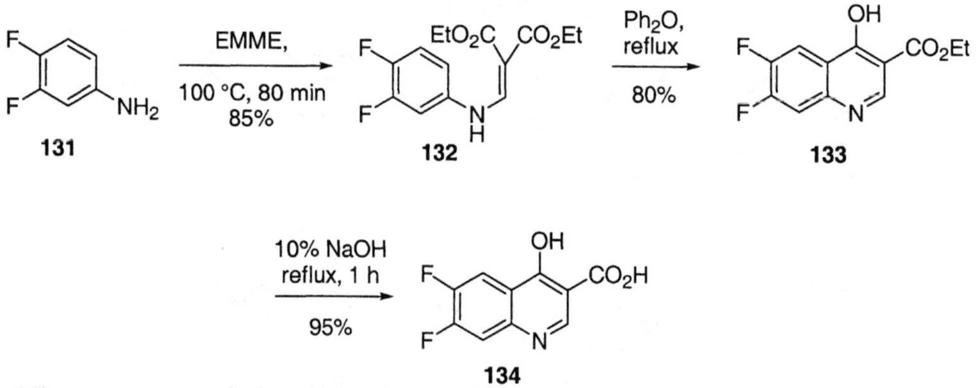

131   **132**   **133**

134

6,7-Difluoro-4-hydroxyquinoline-3-carboxylic acid (134)

A mixture of aniline **131** (20 mmols) and EMME (22 mmols) was warmed to 100 °C for 80 min with a light N_2 flow to remove the ethanol which was formed. Hexanes was added and the reaction mixture warmed and allowed to cool to rt to give malonate **132** as a slightly pink solid in 85% yield. For malonate **132**: mp = 76–77 °C; IR (KBr) 1739 (ester), 1688 (ester) cm^{-1}; ^{1}H NMR (CD$_3$COCD$_3$) 1.2 (3H, t, CH$_3$), 1.26 (3H, t, CH$_3$), 4.18 (2H, q, CH$_2$), 4.24 (2H, q, CH$_2$), 7.26 (1H, m, aromatic H), 7.4 (1H, m, aromatic H), 7.44 (1H, m, aromatic H), 8.4 (1H, d, vinyl H), 10.8 (1H, d, NH); MS (EI, 70 ev) m/z: 299 (100%).

A round-bottomed flask containing diphenyl ether (14 mL) was placed in an oil bath which was heated to 250 °C and malonate **132** (12 mmol) was added slowly. The mixture was kept under reflux for 1 h. During this time vapors evolved and a white solid formed. The solid was filtered and washed with hexane to remove excess Ph$_2$O. This provided quinoline ethyl ester **133** as a white solid in 80% yield. For quinoline **133**: mp = 238–239 °C; IR (KBr) 1698 (ester) cm^{-1}.

Ethyl ester **133** from above was treated with 10% NaOH and warmed to reflux for 1 h. The mixture was allowed to cool to room temperature and was acidified with a solution of 10% HCl to give acid **134** as a white solid in 95% yield. For acid **134**: mp = 276–278 °C; IR (KBr) 1712 (acid) cm^{-1}; ^{1}H NMR (DMSO-d$_6$) 7.88 (1H, m, J = 11, 6, 3 Hz, aromatic H), 8.17 1H, m, J = 10, 2 Hz, aromatic H), 8.9 (1H, s, aromatic H), 13.7 (H, broad s, CO$_2$H); ^{19}F NMR (TFA standard; DMSO-d$_6$) –48.4 (1F, m), –58.8 (1F, m); MS (EI, 70 ev) m/z: 225 (36%), 207 (100%); exact mass for C$_{10}$H$_5$NO$_3$F$_2$, 225.0237. Found: 225.0237.

9.8.6 References

1. [R] Jones, G. In *Heterocyclic Compounds, Quinolines* Vol. 32, chapter 2, pps. 146–150, 158, 159.
2. [R] Reitsema, R. H. *Chem Revs.* **1948**, *53*, 43.
3. (a) [R] Elderfield, R. C. In *Heterocyclic Compounds*, Elderfield, R. C., John Wiley & Sons, Inc., New York, **1952**, vol. 4, pps. 38–41. (b) Price, C. C.; Snyder, H. R.; Bullitt, O. H.; Kovacic, P. *J. Am. Chem. Soc.* **1947**, *69*, 374.
4. Leyva, E.; Monreal, E.; Hernandez, A. *J. Fluorine Chem.* **1999**, *94*, 7.
5. Baker, R. H.; Lappin, G. R.; Albisetti, C. J.; Riegel, B. *J. Am. Chem. Soc.* **1946**, *68*, 1267.
6. Briehl, H.; Lukosch, A.; Wentrup, C. *J. Org. Chem.* **1984**, *49*, 2772.
7. Ouali, M. S.; Vaultier, M.; Carrie, R. *Tetrahedron* **1980**, *36*, 1821.
8. Venugopalan, B.; de Souza, E. P.; Sathe, K. M.; Chatterhee, D. K.; Iyer, N. *Indian J. Chem.* **1995**, *34B*, 778.
9. Horvath, G.; Hermecz, I.; Gorvath, A.; Pongor–Csakvari, M.; Pusztay, L. *J. Heterocycl. Chem.* **1985**, *22*, 481.
10. (a) Adaway, T. J.; Budd, J. T.; King, I. R.; Krumel, K. L.; Kershner, L. D.; Maurer, J. L.; Olmstead, T. A.; Roth, G. A.; Tai, J. J.; Hadd, M. A. U. S. Patent 5 973 153, 1999. (b) Hermans, B. K. F.; Janssen, M. A. C.; Verhoeven, H. L. E.; Knaeps, A. G.; Van Offenwert, T. T. J. M.; Mostmans, J. H.; Willems, J. J. M.; Maes, B.; Vanparijs, O. *J. Med. Chem.* **1973**, *16*, 1047. (c) Duffin, G. F.; Kendall, J. D. *J. Chem Soc.* **1948**, 893.
11. (a) Carbonio, S.; DaSettimo, A.; Tonetti, I. *J. Heterocycl. Chem.* **1970**, *7*, 875. (b) Markees, D. G. *Helv. Chim. Acta* **1983**, *66*, 620. (c) Duffin, G. F.; Kendall, J. D. *J. Chem. Soc.* **1948**, 893. (d) Spencer, C. F.; Snyder, Jr., H. R.; Alaimo, R. J. *J. Heterocycl. Chem.* **1975**, *12*, 1319. (e) Sivasankaran, K.; Sardesai, K. S.; Sunthankar, S. V. *J. Sci. Industr. Res.* **1959** *18b*, 164.
12. Cablewski, T.; Gurr, P. A.; Pajalic, P. J.; Strauss, C. R. *Green Chemistry* **2000**, February, 25.
13. Agui, H.; Komatsu, T.; Nakagome, T. *J. Heterocycl. Chem* **1975**, *12*, 557.
14. (a) Tsuji, K.; Tsubouchi, H.; Ishikawa, H. *Chem. Pharm. Bull.* **1995**, *43*, 1678. (b) Hashimoto, K.; Okaichi, Y.; Nomi, K.; Miyamoto, H.; Bando, M.; Kido, M.; Fujimura, T.; Furuta, T.; Minamikawa, J. *Chem. Pharm. Bull.* **1996**, *44*, 642.
15. Zhang, M. Q.; Haemers, A.; Berghe, V. D.; Pattyn, S. R.; Bollaert, W.; Levshin, I. *J. Heterocycl. Chem.* **1991**, *28*, 673.
16. Wang, C. G.; Langer, T.; Kiamath, P. G.; Gu, Z.Q.; Skolnick, P.; Fryer, R. I. *J. Med. Chem.* **1995**, *38*, 950.
17. Agui, H.; Mitani, T.; Nakashita, M.; Nakagome, T. *J. Heterocycl. Chem.* **1971**, 357.
18. (a) Dave, C.; Joshipura, H. M. *Ind. J. Chem.* **2002**, *41B*, 650. (b) Dave, C. G.; Shah, R. D. *Heterocycles* **1999**, *51*, 1819.

19.   (a) Riegel, B.; Lappin, G. R.; Adelson, B. H.; Jackson, R. I.; Albisetti, Jr., C. J.; Dodson, R. M.; Baker, R. H. *J. Am. Chem Soc.* **1946**, *68*, 1264. (b) Koga, H.; Itoh, A.; Murayama, S.; Suzue, S.; Irkura, T. *J. Med. Chem.* **1980**, *23*, 1358.

20.   Richardson, T. O.; Shanbhag, V. P.; Adair, K.; Smith, S. *J. Heterocycl. Chem.* **1998**, *35*, 1301.

21.   Ilavsky, D.; Heleyova, K.; Nadaska, J.; Bobosik, V. *Collect. Czech. Chem. Commun.* **1996**, *61*, 268.

22.   Sabnis, R. W.; Rangnekar, D. W. *J. Heterocycl. Chem.* **1991**, *28*, 1105.

23.   Yamashkin, S. A.; Yurovskaya, M. A. *Chem. Heterocyclic Compds.* (Engl. Transl.) **1997**, *33*, 1284.

24.   (a) Cooper, C. S.; Klock, P. L.; Chu, D. T. W.; Fernandes, P. B. *J. Med. Chem.* **1990**, *33*, 1246. (b) Zhang, M. Q.; Haemers, A.; Berghe, D. V.; Pattyn, S. R.; Bollaert, W.; Levshin, I. *J. Heterocycl. Chem.* **1991**, *28*, 673.

25.   Suzuki, N.; Tanaka, Y.; Dohmori, R. *Chem. Pharm. Bull.* **1979**, *27*, 1.

26.   Cruickshank, P. A.; Lee, F. T.; Lupichuk, A. *J. Med. Chem.* **1970**, *13*, 1110.

27.   (a) Lappin, G. R. *J. Am. Chem. Soc.* **1949**, *70*, 3348. (b) Hirose, T.; Mishio, S.; Matsumoto, J.–I.; Minami, S. *Chem. Pharm. Bull.* **1982**, *30*, 2399.

28.   Hayakawa, I.; Suzuki, N.; Suzuki, K.; Tanaka, Y. *Chem. Pharm. Buli.* **1984**, *32*, 4914.

29.   Barlin, G. B.; Jiravinyu, C. *Aust. J. Chem.* **1990**, *43*, 1175.

30.   Heber, D.; Ravens, U.; Shulze, T. *Pharmazie* **1993**, *48*, 509.

Timothy T. Curran

9.9   Knorr Quinoline Synthesis

9.9.1   Description

The Knorr quinoline synthesis refers to the formation of α-hydroxyquinolines **4** from β-ketoesters **2** and aryl amines **1**.[1] The reaction usually requires heating well above 100°C. However, some cases do exist when the cyclization takes place in the presence of a catalytic amount of mineral acid at temperatures as low as −10 °C.[2] The intermediate anilide **3** undergoes cyclization by dehydration with concentrated sulfuric acid. The reaction is conceptually close to the Doebner–Miller[1] and Gould–Jacobs reactions.[3]

9.9.2   Historical Perspective

The first cyclization of acetanilide was carried out by Knorr in 1883 who subsequently demonstrated that the reaction was also applicable to acetoacylated aryl amines that have a vacant ortho position.[2]

9.9.3   Mechanism

In the first step of what is considered to be a fairly straightforward mechanism, the anilinic nitrogen reacts with the ester group of the β-ketoester **5** to provide the anilide **3**.[4,5] The latter can either be isolated or carried on directly. Upon warming in the presence of acid, the acetanilide cyclizes with subsequent loss of water to yield the quinolone product **9**.[1]

3   **7**   **8**

9   **4**

9.9.4   *Variations and Improvements*

β-Ketoacylation of amines has been performed by means of a variety of reagents. The use of β-keto esters requires the application of high temperatures and long reaction times[6] in order to achieve chemoselective reaction on the ester group under thermodynamic conditions. The use of mixed tin(II) amides as nucleophiles[7] in that reaction has also been reported although the generality of that protocol as a β-ketoacylation method has not been established. Diketene is a good acetoacetylating reagent,[8] but its volatility, low stability, and high toxicity, as well as the fact that its derivatives are not easily available, have promoted the search for alternatives.[9] More recently, it has been shown that β-keto thioesters react with amines in the presence of silver trifluoroacetate under exceptionally mild conditions.[10] This modified protocol has been applied in a synthesis of β-ketoanilides under relatively mild conditions.[11] The products are then cyclized to the respective quinoline targets under the standard acidic conditions.

In their general synthesis of quinoline-2,5,8(1*H*)-triones, Avendaño and co-workers use this modified protocol to access the key 2,5-dimethoxyanilide systems **11** required in their synthetic plan.[11] For more examples, see reference 11 and references cited within.

R= H (69%)
R= CH$_3$ (93%)
R= CH$_2$CH$_3$ (87%)
R= PhCH$_2$ (85%)
R= CH$_2$=CHCH$_2$ (69%)

10   **11**

9.9.5 *Synthetic Utility*

The Knorr synthesis is a venerable process that has found a number of useful synthetic applications.[12] Most recently, the Knorr protocol has found application in the synthesis of 2,5-dimethoxyquinolines of type **13**.[11] They are penultimate intermediates in the synthesis of 2,5,8-triquinones which are otherwise inaccessible via the standard Friedländer[13] and Vilsmeier–Haack protocols.[14]

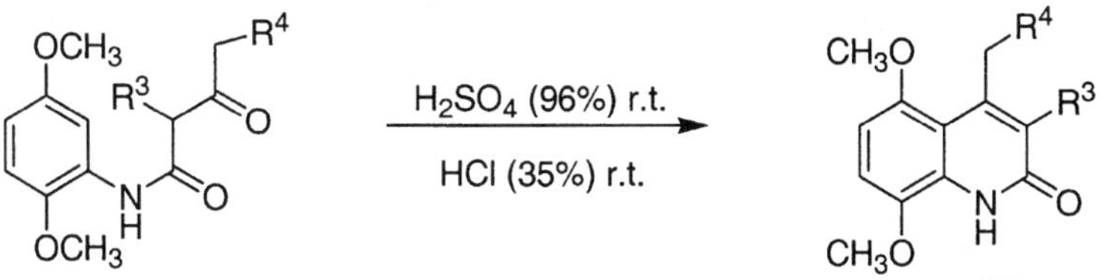

R[3]	R[4]	Yield (%)
CH_3	H	100
PhCH_2	H	96
CH_2=CHCH_2	H	37
H	CH_3	96
H	PhCH_2	97
H	CH_2=CHCH_2	26

The Knorr quinoline synthesis has been nicely extended by Hodgkinson and Staskun to include β-ketoesters that do not have protons at the 2 position of the starting keto-ester.[15] 2,2′-dichloroanilides of type **14** can cyclize to provide quinolines such as **15** and **16** in good respective yields.[15]

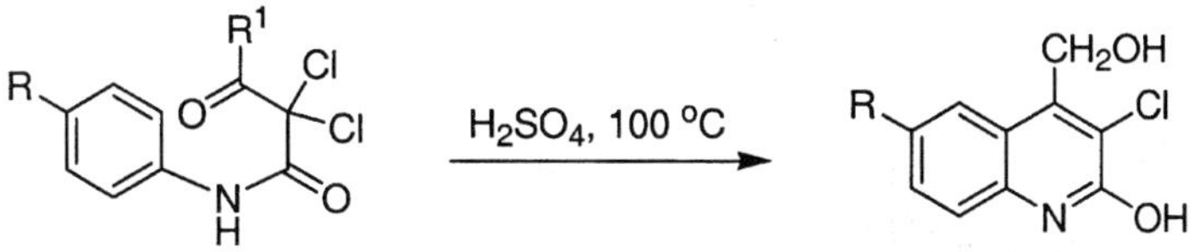

R = H (32%)   **15**

R = Cl (75%)   **16**

A variety of aryl systems have been explored as substrates in the Knorr quinoline synthesis. Most notable examples are included in the work of Knorr himself who has demonstrated the high compatibility of substituted anilines as nucleophilic participants in that reaction.[12] In the case of heteroaromatic substrates however, the ease of cyclization is dependent on the nature and relative position of the substituents on the aromatic ring.[4,5] For example, 3-aminopyridines do not participate in ring closure after forming the anilide presumably as a result of the unfavorable electronic effects of the nitrogen present in the pyridine ring which reduces the nucleophilicity of the ring via inductive destabilization of the required transition state.[4,5] The same is true for 3-quinolineamines such as **19**, which do not react at all whereas the 4-quinolineamines **17** provide the cyclized quinolines in a gratifying 56% yield.[4,5]

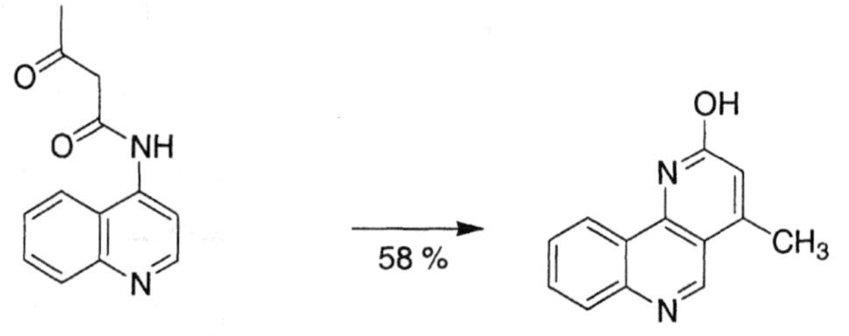

17   58 %   **18**

19   0 %   **20**

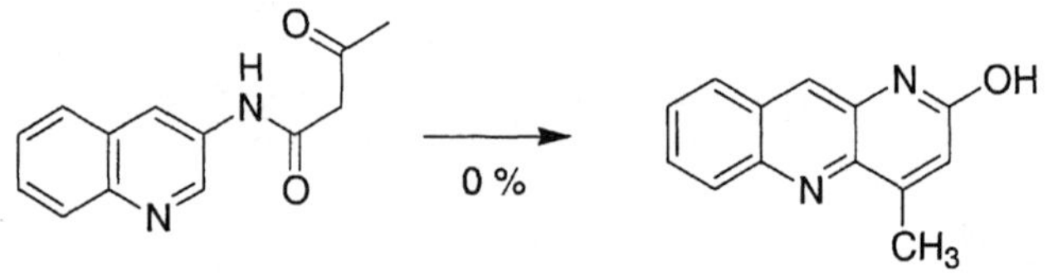

A facile way of promoting the cyclization is to increase the nucleophilicity of the aryl system when possible. In the total synthesis of diazadiquinomycins A and B, for example, the authors were able to effect a double Knorr cyclization with concomitant in situ oxidation to the internal diquinone **23** by deprotecting the hydroquinone thereby lowering the activation barrier for the desired transformation. If the hydroquinone is left protected as the di-MOM ether, the reaction does not take place.[16]

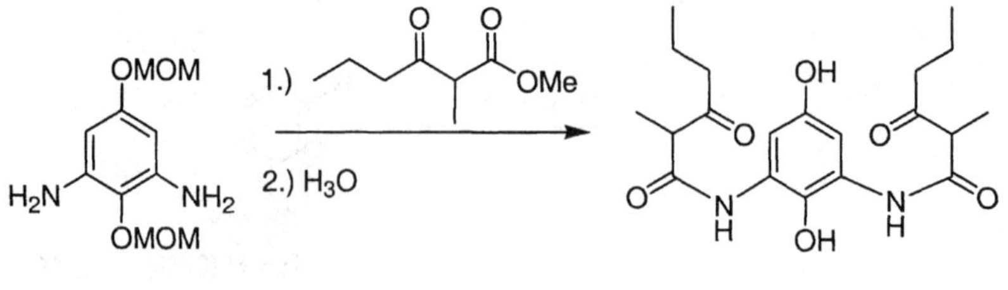

21   1.) ... 2.) H₃O   **22**

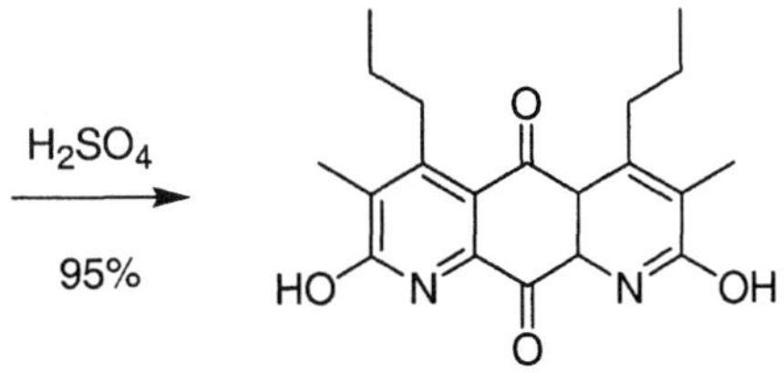

23

9.9.6 Experimental

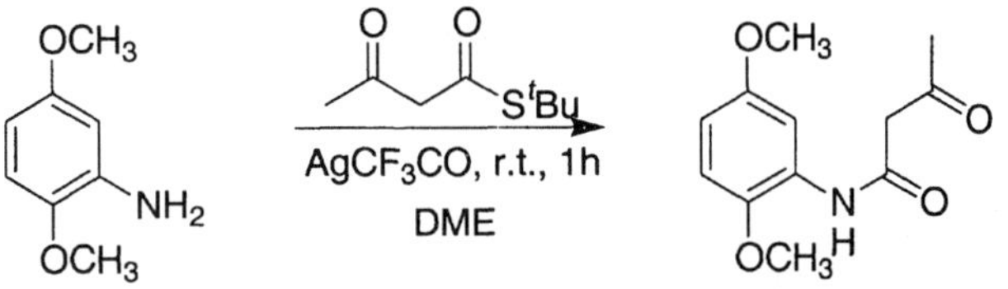

10   **24**

2,5-Dimethoxy-3-ketoanilides: General Procedure:[11]

To a solution of the β-keto thioester (0.750–0.300 g, 1.60–3.71 mmol) and 2,5-dimethoxyaniline (0.256–0.625 g, 1.68–4.08 mmol, 1.05–1.1 equiv) in DME (5–15 mL) was added freshly prepared silver trifluoroacetate (0.282–0.656 g, 1.28–2.97 mmol). After 10 min at r.t., more silver salt (0.141–0.328 g, 0.64–1.49 mmol, total 1.2 equiv) was added, and the suspension was stirred overnight at r.t. The dark-brown suspension was decanted from the precipitate of silver salts, which was washed with petroleum ether (3 × 20–50 mL). The combined organic phases were concentrated under reduced pressure to give a black oil. Chromatography (CH$_2$Cl$_2$ to Et$_2$O) yielded the corresponding 2,5-dimethoxy-3-oxoanilides as white solids or syrups.

5,8-Dimethoxyquinolines: General Procedures:[11]

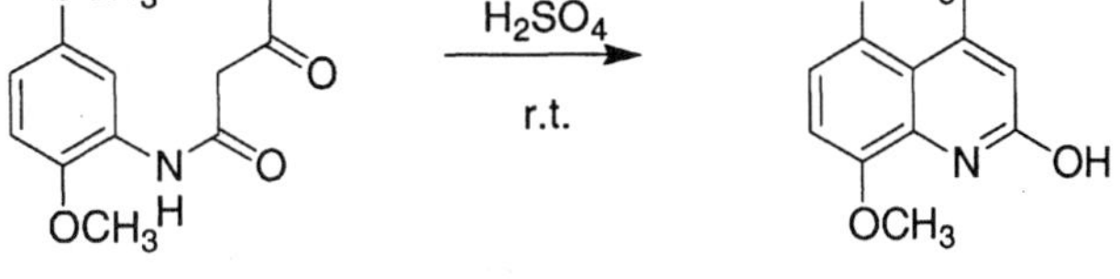

24   **25**

Method A: Cyclization with H$_2$SO$_4$:

A solution of *N*-(2,5-dimethoxyphenyl)-3-ketobutanamide (0.30–0.15 g, 1.20–0.54 mmol) in 96% H$_2$SO$_4$ (2–3 mL) was stirred at r.t. for 50 min–3 h, and was then quenched with ice, made basic with 25% aq NH$_4$OH, and extracted with CHCl$_3$ (3 times 25–50 mL). The combined organic layers were washed with brine (25–50 mL), dried (Na$_2$SO$_4$), and concentrated under reduced pressure to give a pale white solid. Chromatography (50% Et$_2$O/EtOAc to EtOAc) afforded the corresponding 4-mono or 3,4-disubstituted 5,8-dimethoxyquinolines as white solids.

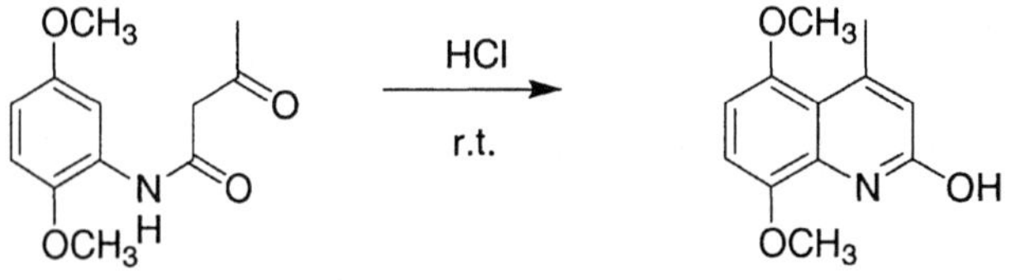

24   25

Method B: Cyclization with HCl:

A suspension of the amide (0.30 g, 0.92 mmol) in 35% HCl (25 mL) was vigorously stirred at r.t. for 4–7 d. The product was isolated following Method A and the chromatography was performed using Et_2O as the mobile phase.

9.9.7   References

1.   [R] Bergstrom, F. W. *Chem. Rev.* **1944**, *35*, 77.
2.   [R] Manske, R. H. F. *Chem. Rev.* **1942**, *30*, 113.
3.   [R] Reitsema, R. H. *Chem. Rev.* **1948**, *43*, 43.
4.   Hauser, C. R.; Reynolds, G. A. *J. Org. Chem.* **1950**, *15*, 1224.
5.   Hauser, C. R.; Reynolds, G. A. *J. Am. Chem. Soc.* **1948**, *70*, 2402.
6.   [R] Benetti, S.; Romagnoli, R.; De Risi, C.; Spalluto, G.; Zanirato, V. *Chem. Rev.* **1995**, *95*, 1065.
7.   Wang, W. B.; Roskamp, E. J. *J. Org. Chem.* **1992**, *57*, 6101.
8.   [R] Clemens, R. J. *Chem. Rev.* **1986**, *86*, 241.
9.   Clemens, R. J.; Hyatt, J. A. *J. Org. Chem.* **1985**, *50*, 2431.
10.   Ley, S. V.; Woodward, P. R. *Tetrahedron Lett.* **1987**, *28*, 3019.
11.   Lopez-Alvarado, P.; Avendano, C.; Menendez, J. C. *Synthesis* **1998**, 186.
12.   [R] Sumpter, W. C. *Chem. Rev.* **1944**, *34*, 393.
13.   Blanco, M. d. M.; Avendano, C.; Cabezas, N.; Menendez, J. C. *Heterocycles* **1993**, *36*, 1387.
14.   Alonso, M. A.; del Mar Blanco, M.; Avendano, C.; Menendez, J. C. *Heterocycles* **1993**, *36*, 2315.
15.   Hodgkinson, A. J.; Staskun, B. *J. Org. Chem.* **1969**, *34*, 1709.
16.   Kelly, T. R.; Field, J. A.; Li, Q. *Tetrahedron Lett.* **1988**, *29*, 3545.

Peter A. Orahovats

9.10   Meth-Cohn Quinoline Synthesis

9.12.1 Description

The Meth-Cohn quinoline synthesis involves the conversion of acylanilides **1** into 2-chloro-3-substituted quinolines **2** by the action of Vilsmeier's reagent in warmed phosphorus oxychloride (POCl$_3$) as solvent.[1]

9.10.2 Historical Perspective

The classical Vilsmeier–Haack reaction is one of the most useful general synthetic methods employed for the formylation of various electron rich aromatic, aliphatic and heteroaromatic substrates.[2]   However, the scope of the reaction is not restricted to aromatic formylation and the use of the Vilsmeier–Haack reagent provides a facile entry into a large number of heterocyclic systems.[3]   In 1978, the group of Meth-Cohn demonstrated a practically simple procedure in which acetanilide **3** (R = H) was efficiently converted into 2-chloro-3-quinolinecarboxaldehyde **4** (R = H) in 68% yield.[4] This type of quinoline synthesis was termed the *'Vilsmeier Approach'* by Meth-Cohn.[5]

Typically, an acetanilide (1 mol. equiv.) was treated with the Vilsmeier reagent generated from POCl$_3$ (7 mol. equiv.) and *N,N*-dimethylformamide (DMF, 2.5 mol. equiv.) at ~75 °C for 4 – 20 h. The reaction products were readily obtained by filtration after pouring the reaction mixture onto ice-water; minor reaction products were isolated after basification of the filtrate.   A variety of acetanilides were studied under these optimised reaction conditions and some significant observations were noted.[6] Activated acetanilides **3** [*e.g.* R = 4-Me (70%), 4-OMe (56%)] reacted faster and in better yield to give quinolines **4** than other strongly deactivated systems **3** [*e.g.* R = 4-Br (23%), 4-Cl (2%), 4-NO$_2$ (0%)] — in these cases, formamidines **5** and acrylamides **6** were the major reaction products.

5   **6**

Moreover, a +M substituent in the *meta*-position showed remarkable selectivity invariably yielding only 7-substituted quinolines, cyclisation occurring *para* to the substituent (*e.g.* **3**, R = 3-SMe yielded **4**, R = 7-SMe, 92%). The reaction was successfully extended to higher anilides **7** (R ≠ H) to yield 2-chloro-3-substituted quinolines **8**. In general, shorter reaction times were required for efficient reaction, reflecting the greater ease of formylation of the higher acylanilides **7** (R ≠ H) compared to acetanilide. The reaction tolerates a wide variety of functionality with acylanilides **7** [R = alkyl, aryl, $(CH_2)_nCl$, pyridyl and thienyl] having been shown to cyclise to 2-chloro-3-substituted quinolines **8** in good yield under the same reaction conditions.[7]

7   **8**

When *N*-substituted acylanilides **9** are treated under the same reaction conditions, the corresponding *N*-substituted-2-quinolones **10** are isolated in high yields.[8] This reaction was initially misinterpreted,[9] but it has since been demonstrated to follow a similar mechanistic pathway to the Meth-Cohn quinoline synthesis.[8]

9   **10**

R^1 = Me, Ph; R^2 = H , Me, $(CH_2)_nCl$

9.10.3 Mechanism[6]

In the Meth-Cohn quinoline synthesis, the acetanilide becomes a nucleophile and provides the framework of the quinoline (nitrogen and the 2,3-carbons) and the 4-carbon is derived from the Vilsmeier reagent. The reaction mechanism involves the initial conversion of an acylanilide **1** into an α-iminochloride **11** by the action of $POCl_3$. The α-chloroenamine tautomer **12** is subsequently *C*-formylated by the Vilsmeier reagent **13** derived from $POCl_3$ and DMF. In examples where acetanilides **1** (R^2 = H) are employed, a second *C*-formylation of **14** occurs to afford **15**; subsequent cyclisation and aromatisation by loss of dimethylamine finally affords the 2-chloro-3 quinolinecarboxaldehyde product **17**. Support for a mechanism involving *C*-formylation of **14** (R^2 = H) derives from the isolation of **6**. These *N*-aryl-3-dimethylamino-2-formacrylamides **6** have been successfully isolated from reaction mixtures in cases where the quinoline formation is slow, notably when the acetanilides bear electron-withdrawing groups on the aromatic ring (*e.g.* **1**, R^1 = 4-Cl), and they are clearly derived by hydrolysis of the iminium salt **15** on reaction work- up. Compounds **6** may themselves be efficiently converted to quinolines *via* hydrolysis to malondialdehydes **18** with aqueous ethanolic alkali, cyclisation to quinolones **19** with polyphosphoric acid (PPA), and subsequent conversion to quinolines **17** by POCl$_3$. In the case of anilides **1** (R^2 ≠ H), a second *C*-formylation is not possible and cyclisation of intermediate **14** occurs; aromatisation and loss of dimethylamine affords 2-chloro-3-substituted quinolines **2**.

R^1 = 6-Cl, 7-Cl, 8-Cl, 6-Br

9.10.4   Variations and Improvements

The Vilsmeier cyclisation of acetanilides by the conventional methods described above often requires long reaction times and elevated temperatures. Moreover, only activated acetanilides react efficiently to afford 2-chloro-3-substituted-quinolines; strongly deactivated systems afford mainly amidine **5** or acrylamide **6**.[6]

Rajanna *et al.*[10] have recently demonstrated that acetanilides, particularly deactivated ones **20** (*e.g.* R = Br, Cl, NO$_2$), undergo rapid cyclisation in micellar media to afford 2-chloro-3-quinolinecarboxaldehydes **21**. Cyclisation in the presence of, for example, 10mol% cetyl trimethylammonium bromide (CTAB) under Vilsmeier-Haack conditions afforded 2-chloro-3-quinolinecarboxaldehydes in good yield in 45–90 minutes.

R = Me (90%), OMe (86%),
NO$_2$ (82%), Br (90%), Cl (84%)

Rajanna *et al.* also demonstrated dramatic rate enhancements when ultrasonically irradiated Meth-Cohn quinoline syntheses were performed;[11] again, deactivated acetanilides **20** were found to undergo efficient cyclisation in good yield.

R = H (84%), Me (85%), OMe (82%),
Br (68%), Cl (65%)

Gupta *et al.* reported that the Vilsmeier–Haack cyclisation of acetanilides **20** using supported reagents and microwave-irradiation in solvent-free conditions is rapid and efficient.[12] Reaction yields are good, although only a few activated derivatives have been investigated.

R = H (79%), Me (58%), OMe (65%)

It has been shown[13] that acetamidothiophenes **22** can be converted to either chlorothieno[2,3-*b*]pyridines **23** or chlorothieno[2,3-*b*]pyridinecarboxaldehydes **24** using POCl$_3$ and DMF by appropriate choice of reaction conditions. However, unlike the acetanilides, initial ring formylation rather than side-chain formylation is believed to lead to the formation of the pyridine ring. These reactions have been extended to the synthesis of the isomeric thieno[3,2-*b*]- and thieno[3,4-*b*]pyridines, **25** and **26**, from 3-acetamidothiophene and 3-acetamido-2,5-dimethylthiophene, respectively.

In a useful extension to the Meth-Cohn quinoline synthesis, pyridoquinolin-2-ones **27** are readily prepared in a one-pot procedure by sequential treatment of an acetanilide **3**, firstly with the Vilsmeier reagent from DMF and POCl$_3$ to afford the intermediate **16**, which is then further reacted *in* situ with another secondary amide.[14]

3   **16**   **27** 22 - 59%

R^1 = H, 3-Me   R^2 = Et, Ph, PhCH$_2$, 4-MeO-C$_6$H$_4$-, 4-Cl-C$_6$H$_4$-

By replacement of DMF with its aza-analogue, *N*-nitrosodimethylamine, a synthetic route to quinoxalines **28** from acylanilides **3** is available. This method is, however, of limited synthetic value as yields are low.[15]

3   **28**

9.10.5 Synthetic Utility

The Meth-Cohn quinoline synthesis provides a versatile and reliable entry into 3-substituted-2-chloroquinolines with a wide variety of substrates having been demonstrated to be applicable. The reaction products are themselves key intermediates for further [*b*]-annelation of various ring systems[1,16] and diverse functional group interconversions, as the aldehyde, chloro and substituent groups are very versatile 'reactive handles'.[1,17] Two examples demonstrating these reactions can be found in recent applications in medicinal chemistry, in which the Meth-Cohn quinoline synthesis has been a pivotal starting reaction. E-ring modified derivatives of Camptothecin **29**, compounds having potential anti-cancer activity, were prepared using the Meth-Cohn quinoline synthesis as the key entry point.[18]

3   **4**   **29**

Similarly, *N*-methyl-D-aspartate (NMDA) antagonists **32** with analgesic activity were prepared, again using the Meth-Cohn quinoline synthesis as the key entry reaction, subsequent functional group manipulation giving the desired target compound.[19]

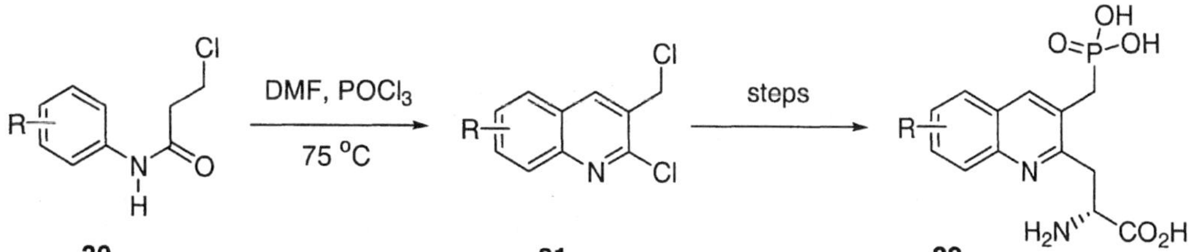

9.12.6 Experimental

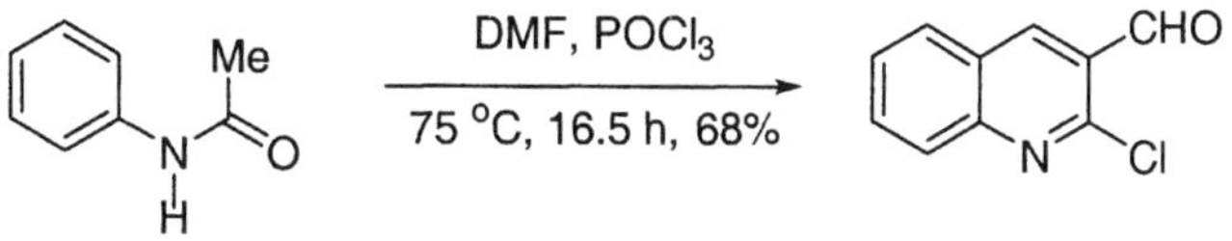

2-chloro-3-quinolinecarboxaldehyde (34)[6]:

To an ice-cooled solution of N,N-dimethylformamide (10.95 g, 11.6 mL, 0.15 mol) was added dropwise with stirring phosphoryl chloride (53.7 g, 32.3 mL, 0.35 mol). Acetanilide 33 (6.75 g, 0.05 mol) was then added and the reaction raised to 75 °C and stirred for a further 16.5 h. The reaction mixture was poured into ice-water (300 mL) and stirred for 0.5 h at < 10 °C. The precipitated solid was collected by filtration and washed well with water (100 mL), air dried and recrystallised from ethyl acetate to afford the product 34 (6.5 g, 68%) as a white solid, mp 148 – 149 °C; ^{1}H NMR (d$_6$-DMSO) δ 7.60 – 8.30 (m, 4H), 8.83 (s, 1H), 10.35 (s, 1H); IR (nujol, cm^{-1}) 1690.

References

1   [R] Meth-Cohn, O. *Heterocycles* **1993**, *35*, 539.

2   [R] Masson, C.M. *Tetrahedron* **1992**, *48*, 3659.

3   [R] Meth-Cohn, O.; Tarnowski, B. *Adv. Het. Chem.* **1982**, *31*, 207.

4   (a) Meth-Cohn, O.; Narine, B. *Tetrahedron Lett.* **1978**, 2045; (b) Meth-Cohn, O.; Narine, B.; Tarnowski, B. *Tetrahedron Lett.* **1979**, 3111.

5   The *'Reverse Vilsmeier Approach'* to the synthesis of quinolinium salts is discussed in: Meth-Cohn, O.; Taylor, D.L. *Tetrahedron* **1995**, *51*, 12869.

6   Meth-Cohn, O; Narine, B.; Tarnowski, B. *J. Chem. Soc., Perkin Trans. 1* **1981**, 1520.

7   For examples see: (a) Meth-Cohn, O.; Rhouati, S.; Tarnowski, B. *Tetrahedron Lett.* **1979**, 4885. (b) Blackburn, T. P.; Cox, B.; Guildford, A. J.; Le Count, D.J.; Middlemiss, D. N.; Pearce, R. J.; Thornber, C. W. *J. Med. Chem.* **1987**, *30*, 2252. (c) Alonso, M. A.; del Mar Blanco, M.; Avendaño, C.; Menéndez, J. C. *Heterocycles* **1993**, *36*, 2315.

8   Hayes, R.; Meth-Cohn, O; Tarnowski, B. *J. Chem. Research* **1980**, 414.

9   Schulte, K.E.; Bergenthal, D. *Arch. Pharm.* **1979**, *313*, 265.

10   Ali, M. M.; Tasneem; Rajanna, K. C.; Sai Prakash, P. K. *Synlett* **2001**, 251.

11   Ali, M. M.; Sana, S.; Tasneem; Rajanna, K. C.; Saiprakash, P. K. *Synth. Comm.* **2002**, *32*, 1351.

12   Paul, S.; Gupta, M.; Gupta, R *Synlett.* **2000**, 1115.

13   Meth-Cohn, O; Narine, B.; Tarnowski, B. *J. Chem. Soc., Perkin Trans. 1* **1981**, 1531.

14   Meth-Cohn, O.; Tarnowski, B. *Tetrahedron Lett.* **1980**, *21*, 3721.

15   Meth-Cohn, O; Rhouati, S.; Tarnowski, B.; Robinson, A. *J. Chem. Soc., Perkin Trans. 1* **1981**, 1537.

16   For examples see: (a) Hayes, R; Smalley, R. K. *J. Chem Res. (S)* **1988**, 14. (b) Kidwai, M.; Negi, N, *Monatsch. Chem.* **1997**, *128*, 85. (c) Hayes, R.; Meth-Cohn, O. *Tetrahedron Lett.* **1982**, *23*, 1613.

17   For examples see: (a) Meth-Cohn, O; Narine, B.; Tarnowski, B.; Hayes, R.; Keyzad, A.; Rhouati, S.; Robinson, A. *J. Chem. Soc., Perkin Trans. 1* **1981**, 2509. (b) Bhat, N. B.; Bhaduri, A. P. *J. Heterocycl. Chem.* **1984**, 1469. (c) Hayes, R; Meth-Cohn, O. *Tetrahedron Lett.* **1982**, *23*, 1613.

18   (a) Toyota, M.; Komori, C.; Ihara, M. *J. Org. Chem.* **2000**, *65*, 7110. (b) Mekouar, K.; Génisson, Y.; Leue, S.; Greene, A. E. *J. Org. Chem.* **2000**, *65*, 5212. (c) Lavergne, O.; Lesueur-Ginot, L.; Pla Rodas, F.; Kasprzyk, P. G.; Pommier, J.; Demarquay, D.; Prévost, G.; Ulibarri, G.; Rolland, A.; Schiano-Liberatore, A.-M.; Harnett, J.; Pons, D.; Camara, J.; Bigg, D. C. H. *J. Med. Chem.* **1988**, *41*, 5410.

19   (a) Swahn, B.-M.; Claesson, A.; Pelcman, B.; Besidski, Y.; Molin, H.; Sandberg, M. P.; Berge, O.-G. *Bioorg. Med. Chem. Lett.* **1996**, *6*, 1635. (b) Swahn, B.-M.; Andersson, F.; Pelcman, B.; Soderberg, J.; Claesson, A. *J. Labelled. Compounds Radio.* **1997**, *39*, 259.

Adrian J. Moore

9.11   Pfitzinger Quinoline Synthesis

9.11.1 Description

The Pfitzinger reaction entails the synthesis of quinoline-4-carboxylic acids **2** via condensation of isatic acids formed from isatins **1** and α-methylene carbonyl compounds in the presence of strong aqueous bases. Subsequent decarboxylation can afford the corresponding quinolines.[1-3]

9.11.2 Historical Perspective

In 1886 Pfitzinger reported a formal extension of the known Friedländer protocol for the synthesis of quinolic acids.[2] This new protocol relied on the use of isatin which is much more stable than the *ortho*-aminoaryl intermediates that are required in the Friedländer quinoline synthesis.[2] In this early paper, Pfitzinger reports that upon heating of isatin **3** in the presence of aqueous sodium hydroxide, the former is hydrolyzed to the isatic acid **4** which then in the presence of acetone reacts to give aniluvitonic acid **6**.[1]

However the earliest example in the literature of a reaction that proceeds via the putative intermediates invoked in the Pfitzinger reaction can be traced to the formation of quindoline-5-carboxylic acid **10**, a product of over-reduction of indigo.[2]

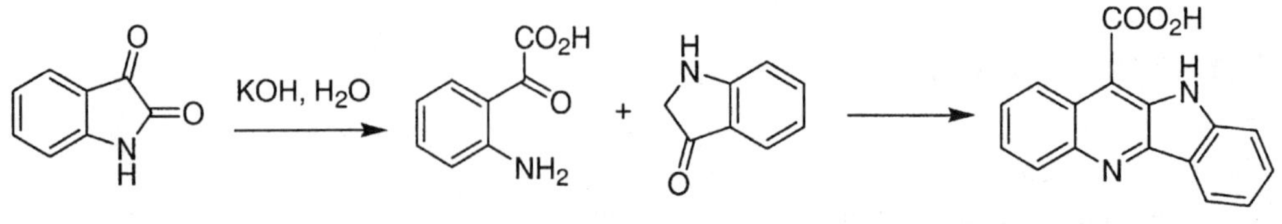

7   **8**   **9**   **10**

In 1903 Walther and co-workers recognized that imino-nitriles show parallel reactivity to that for the corresponding ketones in the presence of isatin under Pfitzinger conditions.[2]

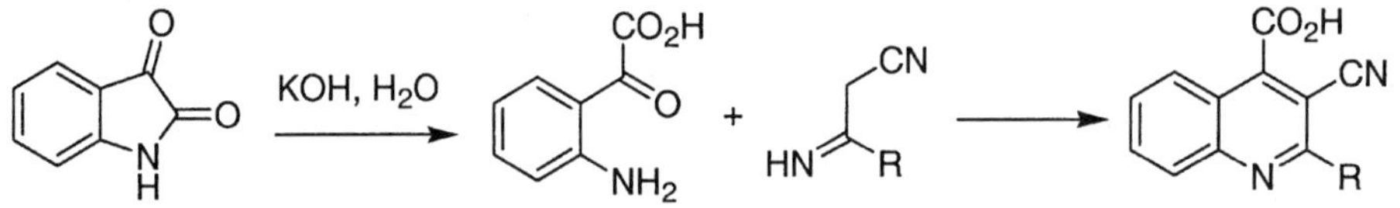

11   **12**   **13**   **14**

9.11.3 Mechanism

The mechanism is postulated to involve the initial formation of a Schiff base **17** from the condensation of the anilinic amine **16** with the carbonyl-containing substrate. This is followed by a Claisen condensation between the benzylic carbonyl and the activated α-methylene of the imine.[4,5]

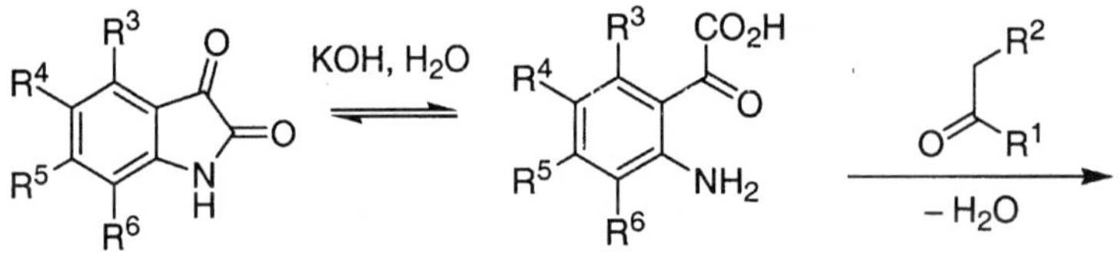

15   **16**

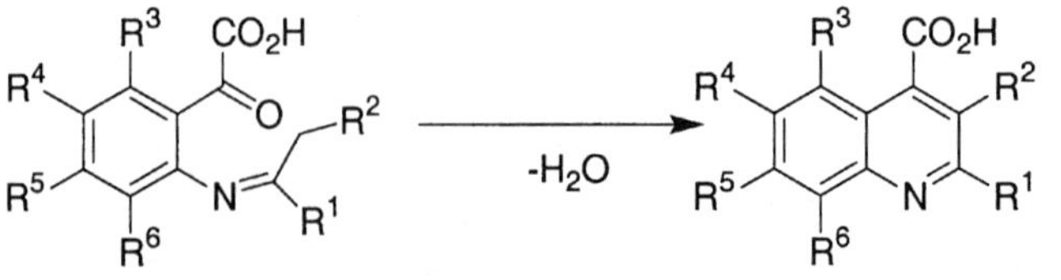

17   **18**

9.11.4 Variations and Improvements

Faced with the inapplicability of the standard basic conditions required for the Pfitzinger condensation in the context of their study, Lackey and Sternbach developed a modified protocol which allows for the formation of quinolinic acids under acidic conditions.[6]

They report that in the reaction between 5-chlorisatin **19** and 5,6-dimethoxindanone **20** under basic conditions at reflux for 16 hours, the desired quinolinic acid **22** is obtained in 38% yield with an unavoidable competing amount of the aldol product **21**. However, if the same reaction is carried out using aqueous acid conditions, the quinolinic acid is obtained in a reproducible 86% yield.[6]

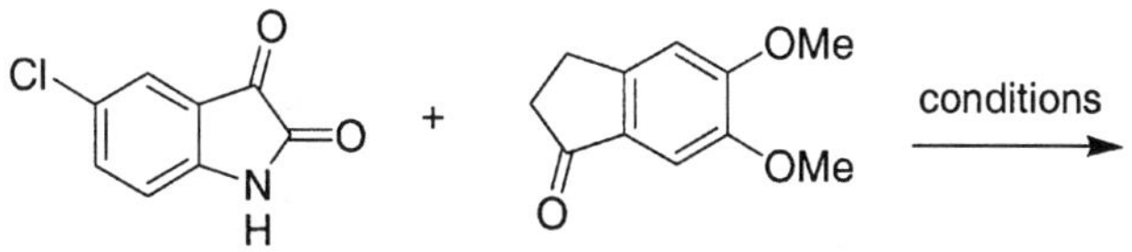

Conditions

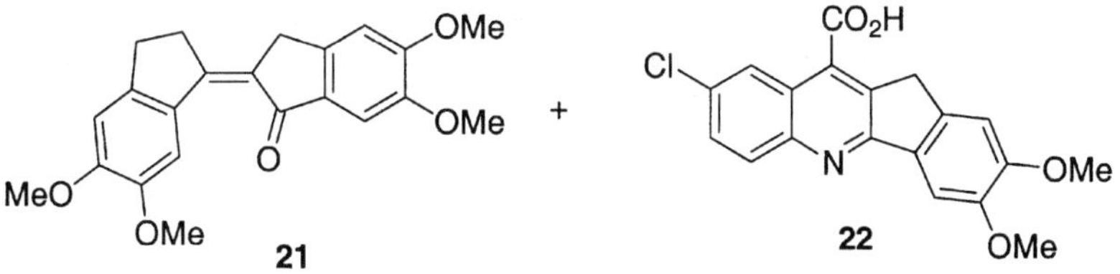

KOHaq., EtOH, reflux

AcOH, HCl, 75 °C

Yield (%)	Yield (%)
23	38
0	86

A rate dependency was observed in the case of acid-promoted Pfitzinger condensation: isatins substituted with electron-withdrawing groups reacted faster than the corresponding isatins with less electron-withdrawing substituents.[6]

9.11.5 Synthetic Utility

Buu–Hoi[4,7-11] and Mueller[11] have described detailed studies on the effect of steric hindrance in the Pfitzinger reaction. In a systematic approach Buu–Hoi has found that although 2,4-dimethylacetophenone and 2,6-dimethyl-4-*t*-butylacetophenone both condense readily with isatin, when the α-methyl group is more heavily substituted as in the corresponding benzyl aryl ketones, the yields drop drastically. For example 4-methyldesoxybenzoin gave 62% of the cinchoninic acid **26**, 2,4-dimethyldesoxybenzoin a low yield of **27**, and 2,4,6-trimethyldesoxybenzoin failed to react.[4,7,8]

23   **24**

25 $R^1 = R^2 = R^3 = H$ (72%)

26 $R^1 = R^3 = H$, $R^2 = CH_3$ (62%)

27 $R^1 = R^2 = CH3$, $R^3 = H$ (12%)

28 $R^1 = R^2 = R^3 = CH_3$ (0%)

Similarly, Mueller has shown that transposition of the methyl groups to the other ring of the ketones also reduces their reactivity in that reaction. For example, **29** condensed to yield 37% of the cinchoninic acid **31**, **32** was obtained in only 12%, and **33** was not produced at all.[11,12]

29   **30**

31 $R^1 = R^3 = H$, $R^2 = CH_3$ (37%)

32 $R^1 = R^2 = CH_3$, $R^3 = H$ (10%)

33 $R^1 = R^2 = R^3 = CH_3$ (0%)

Besides the possibility for nefarious effects due to steric hindrance, it should also be pointed out that due to the involvement of strong bases or acids (as in the modified version), the Pfitzinger is limited to substrates with tolerant functionalities.

9.11.5.1   *With ketones*

Buu–Hoi has shown that *n*-alkyl methyl ketones excluding ethyl methyl ketone, yield primarily 2-monosubstituted cinchoninic acids. It has been demonstrated that the products of the condensation of isatin with aryloxyketones are the corresponding 3-aryloxy-4-quinoline carboxylic acids rather than the isomeric 2-aryloxymethylcinchoninic acids.[13] In the case of simple α-alkoxyketones such as 1-alkoxyethyl methylketones, the preferred products are the 2-alkoxyalkylcinchoninic acids.[14] However, in the more highly substituted homologues, the 2-alkyl-3-alkoxy cinchoninic acids are the preferred products of the condensation.[15]

9.11.5.2 *With diversely substituted indanones*

Deady and co-workers have employed the Pfitzinger reaction extensively as a means to access diverse indenoquinolinecarboxylates as a part of a study on the latter's cytotoxicity.[16-18]

For more examples, see reference 18 and papers cited within.

9.11.5.3 *With acids*

Carboxylic acids with labile α-methylene protons react with isatin in the presence of strong aqueous base. In the total synthesis of methoxatin, the coenzyme of methanol dehydrogenase and glucose dehydrogenase, Weinreb employs a Pfitzinger condensation of an isatin **37** and pyruvic acid as a key step to provide the 4-quinolinic acid **38** in 50% yield under the standard basic conditions.[19]

9.11.6 Experimental

Method A:

Condensation of Isatin 3 with Ethyl Methyl Ketone 39.[20]

A mixture of 60 g (0.408 mol) of isatin, 200 mL of 34% potassium hydroxide in diluted alcohol solution, 88 g (1.22 mol) of ethyl methyl ketone and 375 mL of water were stirred and heated under reflux for 72 hours. About 125 mL of liquid was removed by distillation; the residue was made slightly acidic and filtered. The filtrate was made strongly acidic to precipitate the reaction product, which was collected by filtration, washed, dried, weighed 70 g (85% yield).

19   **92%**   **40**

Method B:

6-Chloro-3-methyl-2-phenylquinoline-4-carboxylic acid 40:[6]

5-Chlorisatin (200 mg, 1.10 mmol) was slurried in glacial AcOH (3 mL) at r.t. Propiophenone (148 mg, 1.10 mmol) was added and the reaction mixture was placed in a preheated oil bath set to 75 °C. The reaction was stirred for 5 min. before the addition of concentrated HCl (1.0 mL). The reaction was heated to 105 °C and stirred for 16 h. The reaction was then cooled to r.t. and water was added. The cream colored solid was collected by filtration, and was washed with EtOH (3 mL) and Et$_2$O (6 mL). The solid was dried under high vacuum to afford the product (302 mg, 92 %).

9.11.7 References

1.   Bergstrom, F. W. *Chem. Rev.* **1944**, *35*, 77.
2.   Manske, R. H. F. *Chem. Rev.* **1942**, *30*, 113.
3.   Sumpter, W. C. *Chem. Rev.* **1944**, *34*, 393.
4.   Buu, H.; Royer, R. *Bull. soc. chim.* **1946**, 374.
5.   BuuHoi, N. P.; Roussel, O.; Jacquignon, P. *Bull. Soc. Chim. Fr.* **1963**, 1125.
6.   Lackey, K.; Sternbach, D. D. *Synthesis* **1993**, 993.
7.   Buu, H.; Cagniant, P. *Bull. Soc. Chim. Fr.* **1946**, 134.
8.   Buu, H.; Cagniant, P. *Bull. Soc. Chim. Fr.* **1946**, 123.
9.   Buu, H.; Royer, R. *Rec. Trav. Chim. Pays-Bas Belg.* **1947**, *66*, 305.
10.   Buu-Hoi, N. P.; Sy, M.; Riche, J. *Bull. Soc. Chim. Fr.* **1960**, 1493.
11.   Mueller, G. P.; Stobaugh, R. E. *J. Am. Chem. Soc.* **1950**, *72*, 1598.
12.   Palmer, M. H.; McIntyre, P. S. *J. Chem. Soc. B: Physical Organic* **1969**, 539.
13.   Dowell, A. M., Jr.; McCullough, H. S.; Calaway, P. K. *J. Am. Chem. Soc.* **1948**, *70*, 226.
14.   Buu-Hoi, N. P.; Jacquignon, P. *Bull. Soc. Chim. Fr.* **1958**, 1567.
15.   Buu-Hoi, N. P.; Royer, R.; Xuong, N. D.; Jacquignon, P. *J. Org. Chem.* **1953**, *18*, 1209.
16.   Deady, L. W.; Kaye, A. J.; Finlay, G. J.; Baguley, B. C.; Denny, W. A. *J. Med. Chem.* **1997**, *40*, 2040.
17.   Deady, L. W.; Desneves, J.; Kaye, A. J.; Thompson, M.; Finlay, G. J.; Baguley, B. C.; Denny, W. A. *Bioorg. Med. Chem.* **1999**, *7*, 2801.
18.   Deady, L. W.; Desneves, J.; Kaye, A. J.; Finlay, G. J.; Baguley, B. C.; Denny, W. A. *Bioorg. Med. Chem.* **2001**, *9*, 445.
19.   Gainor, J. A.; Weinreb, S. M. *J. Org. Chem.* **1982**, *47*, 2833.
20.   Henze, H. R.; Carroll, D. W. *J. Am. Chem. Soc.* **1954**, *76*, 4580.

Peter A. Orahovats

9.12 Pictet–Gams Reaction

9.12.1 Description

The isoquinoline framwork is derived from the corresponding acyl derivatives of β-hydroxy-β-phenylethylamines. Upon exposure to a dehydrating agent such as phosphorous pentaoxide, or phosphorous oxychloride, under reflux conditions and in an inert solvent such as decalin, isoquinoline frameworks are formed.

9.12.2 Historical Perspective and Mechanism

The isoquinoline nucleus is found in many natural products[1] and is a useful template in medicinal chemistry.[2] The construction of the isoquinoline framework was first described by Pictet and Gams in 1909.[3] Their synthetic approach was used to construct the isoquinoline alkaloid, Berberine — a synthesis considered "classic" even though the final product has been identified as erroneous due to the authors unknowingly reporting the air oxidized side-product of Berberine.[3d] From 1909 until 1977 the reaction was considered capricious and its popularity as a useful synthetic tool quickly diminished. This was due to the fact that many expected products were not isolated. Indeed, a mixture of regio-isomers, or rearrangement products were commonly obtained. In fact, the Bischler–Napieralski reaction was the method of choice rather than employing the Pictet–Gams conditions for constructing the isoquinoline framework.

To understand the unpredictable nature of the Pictet–Gams reaction, Hartwig and Whaley conducted the first mechanistic studies in 1949.[4] Their work focused on substituent effects when directly attached to the ethylamine side chain. They also investigated a variety of dehydration agents in order to identify optimal reaction conditions. It was determined that formation of the isoquinoline structure was virtually impossible when alkyl or phenyl substituents were placed in the 4-position of the ethylamine side chain.

R_1 = Ethyl
R_2 = Phenyl

R_1 = Ethyl, 10% yield
R_2 = Phenyl, 0% yield

Hartwig and Whaley suggested that when substituents are placed in the 4-position of the ethylamine side chain, an oxazoline intermediate (**7**) is formed; a "side reaction" that was first mentioned by Krabbe[5] in 1940. However, Hartwig and Whaley *did not isolate* the putative intermediate.

5   **6**   **7**

Hartwig and Whaley also demonstrated that increasing the size of the alkyl side chain off the 3-position, resulted in low yields of expected product. In fact, when R_1 is larger than a propyl group, the expected products were not isolated (*n*-butyl gave a 1% of the corresponding isoquinoline!). Furthermore, isoquinolines were formed more

8   **9**

R_1	% Yield
H	91
Me	50
Et	26
n-Propyl	20
n-Butyl	1

readily when a phenyl group was placed in the 1-position, as compared to alkyl groups of comparable size (results not shown).

In 1968[6], the aforementioned reaction was repeated and found to produce a "rearranged" product (**11**). This was the first report of aryl migration with the Pictet–Gams conditions. The expected product was 1-methyl-3-phenyl isoquinoline, but only 1-methyl-4-isoquinoline (**11**) was observed. Interestingly, the authors did not suggest a mechanism for the formation of the isolated product.

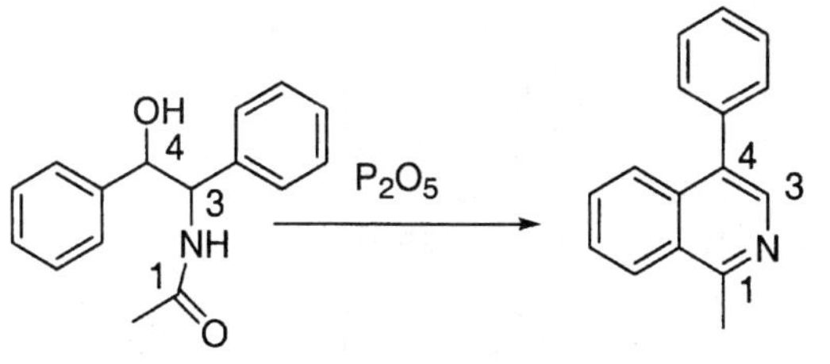

10 → **11**

In 1977 Fitton[7] *et al.* conducted the first detailed study that shed light on the true mechanism of the Pictet–Gams reaction. It was postulated that all Pictet–Gams reactions create an oxazoline intermediate (**15**) by the mechanism shown below:

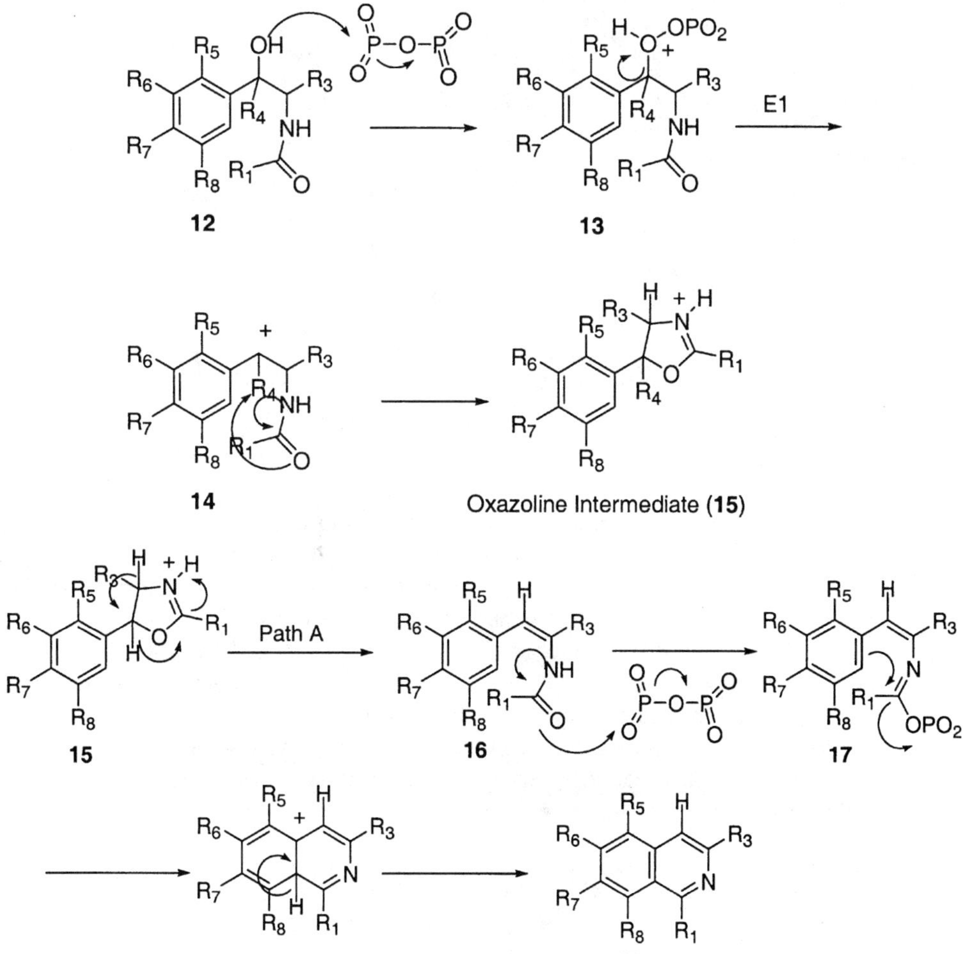

12 **13**

14 Oxazoline Intermediate (**15**)

15 Path A **16** **17**

18 **19**

    Once the oxazoline intermediate was formed, the reaction could undergo one of
two paths *or both* - depending upon the type of substituent present in the 4-position of the
phenylethylamine side chain.   It was observed that when R_3 was methyl or ethyl, the
products obtained were from path A:

    This was rationalized by the oxazoline intermediate (**15**), upon exposure to acid at
high temperatures (boiling decalin (b.p. = 190 °C)), collapsing to the vinyl amide
intermediate (**16**) that subsequently undergoes standard acid catalyzed ring closure to the
corresponding isoquinoline (**19**). When R_3 was *n*-propyl, a mixture of regioisomers were
isolated (relative amounts were not listed). Therefore, if the migratory aptitude for
substituents in the 3-position are low, then path **A** will prevail and the "expected" regioisomer isoquinoline should be observed.

    When R_3 was *n*-butyl, phenyl, benzyl or para-methoxybenzyl, the products
obtained were from path B:

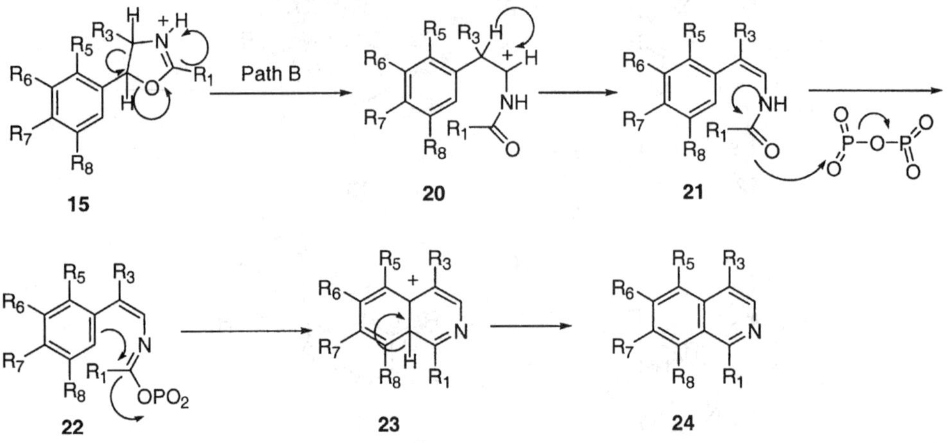

    These results are rationalized by the R3 group migrating (via E1 mechanism) to
the benzylic position of the oxazoline intermediate (**20**), causing collapse of the
intermediate, followed by standard mechanistic protocol for isoquinoline formation (**24**).
Therefore, if the migratory aptitude for substituents in the 3-position are high, then path **B**
will prevail and the "opposite" regio-isomer isoquinoline should be observed.   These
mechanistic findings provided the confidence necessary for organic chemists to once
again begin employing the Pictet–Gams reaction for organic synthesis.

    In 2000, Simig[9] *et al.* began to conduct structure activity relationships on **25** by
employing the Pictet–Gams reaction. Compound **25** had been identified as an anxiolytic
agent that does not show sedative side-effects.[8]

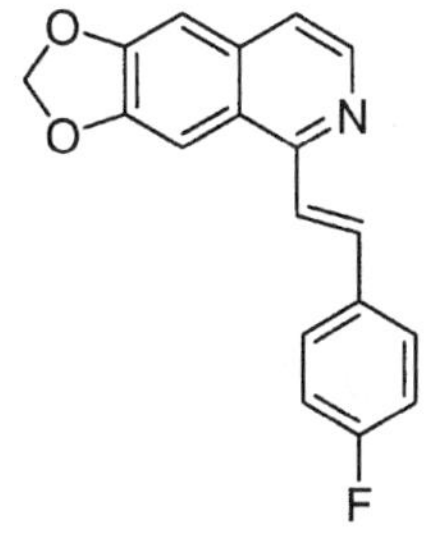

25

However, when carrying out the synthetic route, as shown below, isoquinoline production was not realized. Rather, the oxazolidine intermediate (**27**) was isolated. Fortunately, an X-ray of the intermediate was obtained that proved unambiguously, that 2-oxazolidines were an intermediate in the Pictet–Gams reaction.[10]

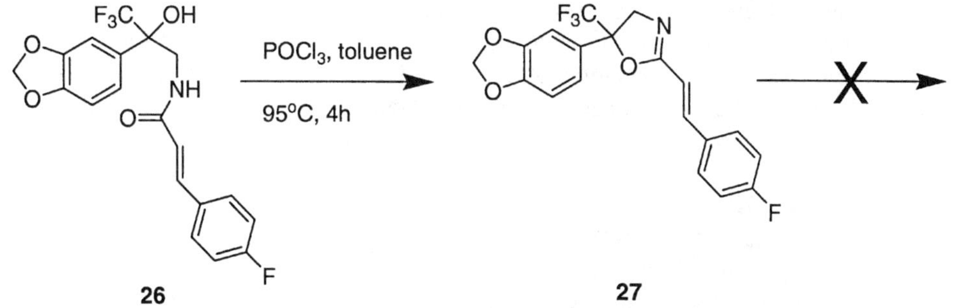

26   **27**

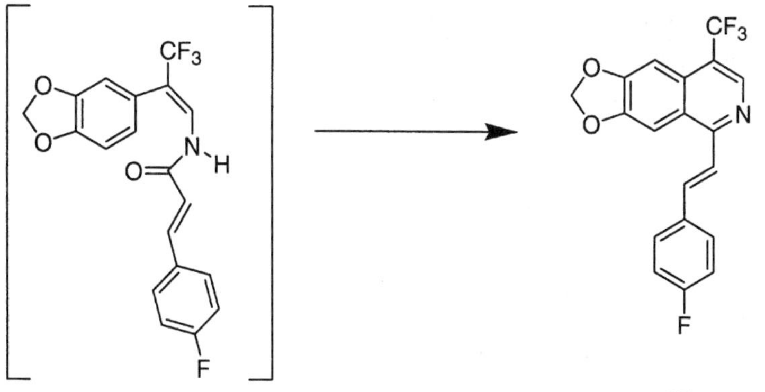

28   **29**

The oxazolidine intermediate (**27**) was expected to collapse via heterolysis of the carbon-oxygen bond through an E1 mechanism, and follow the "path A" mechanism. However, the electronegative trifluoromethyl group destabilizes the incipient positive charge in the benzylic position and consequently, transformation of the oxazolidine (**27**) to the isoquinoline (**29**) does not occur. Simig *et al.* circumvented this issue by replacing the hydroxyl group geminal to the trifluoro group with a methoxy moiety (**30**), so that dehydration cannot not occur. Next, the application of Bischler–Napieralski reaction conditions to **30**, produced the desired 6-member ring intermediate **31**. The lack of methanol elimination under

Bischler–Napieralski reaction conditions can be attributed, again, to the destabilizing ability of the trifluoromethyl group to the cationic transition state of the acid catalyzed elimination.[11] Formation of compound **29** was ultimately accomplished by base catalyzed methanol elimination-conditions; conditions that are quite unusual for isoquinoline formation.

9.12.3 Synthetic Utility

Even though the Pictet–Gams reaction requires strong acid and high temperatures to form the desired isoquinoline framework, it remains the method of choice when acid labile substituents are not present in the molecule. This is particularly true now that the mechanism has also been elucidated and the reaction more predictable.

Isoquinolines have been shown to useful as artificial receptors for resorcinol.[12] The Pictet–Gams reaction offers a short and efficient route to the complex tri-isoquinoline (**33**) artificial receptor.

In addition, isoquinolines have been used for receptor-mediated imaging and demarcation agents for certain types of cancers.[13] When the isoquinoline framework is attached to a cyclen-based fluorophore and then chelated with lanthanides, luminescence occurs, providing a good MRI contrast for receptor-mediated imaging.

43
cyclen-based fluorophore

9.12.4 Experimental[13]

44   **45**

A mixture of 2-chloro-*N*-(2-hydroxyl-1-methyl-2-phenylethyl)benzamide (**44**) (9.5g, 24.9 mmol) and P_2O_5 in *o*-chlorobenzene (150 mL) was refluxed overnight. Upon completion, the reaction was cooled to room temperature and then chilled to 0 °C. To the crude reaction mixture, 300 mL of water was *cautiously* added. The resulting dark solution was washed with toluene (2 × 50 mL). The aqueous layer was cooled to 0 °C and 50% NaOH added to final pH of 11. The resulting mixture was extracted with toluene (4 × 50 mL). The toluene fractions were combined, dried, filtered and concentrated *in vacuo*. The residue was crystallized from benzene to afford 1-(2-chlorophenyl)-3-methylisoquinoline (**45**) as a white solid (6.68g, 80%). M.P. = 107–108 °C; ^{1}H NMR (CDCl$_3$) δ 8.45 (s, 1H), 8.11 (d, 1H), 7.85 (dt, 1H), 7.41–7.68 (bm, 6H), 2.51 (s, 3H).

9.12.5 References

1.   Bentley, K. W. *Nat. Prod. Rep.*, **2003**, *20*, 342.
2.   (a) Brossi, A. *Heterocycles* **1978**, *11*, 521.; (b) Kumar, A., Katiyar, S. B., Agarwal, A., Chauhan, P. M. S. *Curr. Med. Chem.*, **2003**, *10*, 1137.
3.   (a) Pictet, A.; Gams, A. *Ber.*, **1909**, *42*, 2943; (b) Pictet, A.; Gams, A. *Ber.*, **1910**, *43*, 2384; (c) Pictet, A.; Gams, A. *Ber*, **1911**, *44*, 2480; (d) Buck, J. S.; Davis, R. M., *J. Am.Chem. Soc.*, **1930**, *52*, 660.
4.   Whaley, W. M.; Hartwig, W. H. *J. Org. Chem.*, **1949**, 650.
5.   Krabbe, W.; Polzin, G.; Culemeyer, K. *Ber.*, **1940**, *73b*, 652.
6.   Bindra, A. A.; Wadia, M. S.; Dutta, N. C. *Tetrahedron Lett.*, **1968**, 2677.
7.   (a) Fritton, A. O.; Frost, J. R.; Zakaria, M. M.; Andrew, G. *J. Chem. Soc., Chem. Commun.*, **1973**, 889; (b) Fritton, A. O.; Ardabilchi, N.; Frost, J. R.; Oppong–Boachire, F. *Tetrahedron Lett.*, **1977**, 4107.
8.   Balogh, G.; Doman, I.; Blasko, G.; Simig, G.; Kovacs, E.; Egyed, I.; Gacsalyi, I.; Bilkei–Gorzo, K.; Pallagi, K.; Szemeredi, K.; Kazo, K. (EGIS Pharmaceuticals Ltd), EP 680953, **1995**.
9.   Simig, G.; Poszavacz, L. *J. Het. Chem.*, **2000**, *37*, 343.
10.   (a) Kopczynski, T. *Polish J. Chem.*, **1994**, *68*, 73.; (b) Kopczynski, T. *Polish J. Chem.*, **1985**, *59*, 375. (c) Kopczynski, T.; Goszczynski, S. *Polish J. Chem.*, **1981**, *55*, 393.
11.   Simig, G.; Poszavacz, L. *Tetrahedron*, **2001**, *57*, 8573.
12.   Dyker, G.; Gabler, M.; Nouroozian, M.; Schulz, P. *Tetrahedron Lett.*, **1994**, *35*, 9697.
13.   Manning, H. C.; Goebel, T.; Marx, J. N.; Bornhop, D. *J. Org. Lett.*, **2002**, *4*, 1075.

Daniel D. Holsworth

9.13   Pictet–Hubert Reaction

9.13.1 Description

The Pictet–Hubert reaction describes the construction of the phenanthridine nucleus (2) by dehydration of acyl-*o*-xenylamines (1).   The application of zinc chloride at high temperature facilitates the dehydration.[1]   This reaction is also referred as the Morgan–Walls Reaction.

9.13.2 Historical Perspective / Improvements

Phenanthridine (R₁ and R = H) (2) was discovered in 1889 by Pictet and Ankersmit[2] from pyrolysis of benzylideneaniline at "bright-red heat". Small amounts of phenanthridine were isolated by this method. In 1896, Pictet and Hubert described the first synthetic protocol for the construction of phenanthridine (unstated yield) (*see Description Section 9.13.1*). The reaction conditions listed by Pictet and Hubert were not amenable to most substituents on the acyl-*o*-xenylamine framework and the reaction was not successful in many cases. In 1931 Morgan and Walls[3] recognized the potential therapeutic utility of the phenanthridine nucleus (compounds were found to exhibit trypanocidal activity),[4] and developed reaction conditions for the Pictet–Hubert reaction that were amenable to a larger number of substituents. The new conditions also improved the yield of the reaction.   The modifications were two fold: (1) Zinc chloride was replaced with the strong acid, phosphorus oxychloride; and (2) Nitrobenzene (b.p. 210° C) was added as solvent, due to its high boiling and good ionizing properties. When these two modifications were utilized, phenanthridine was isolated in 42% yield. Notice that these conditions are extremely similar to the Bischler–Napieralski reaction conditions.

Substituents	Product
R = methyl, R₁ = H	70%
R = ethyl, R₁ = H	80%
R = phenyl, R₁ = H	75%

There have also been reports of adding zinc chloride or tin chloride to the Morgan–Walls conditions to catalyze the reaction (*see experimental section*).[5]

9.13.3 Mechanism

The mechanism of the Pictet–Walls reaction has been investigated by Ritchie.[6]   Upon exposure of phosphorus oxychloride to the acyl-*o*-xenylamine (**5**), the iminophophorus oxychloride is formed (**6**). Upon high temperature, the imino cation (**7**) is envisioned to be produced (the counter-ion OPCl$_2$ or Cl is likely attached), followed by electrophilic addition of the positively charged carbon atom to the α-carbon of ring A (**7**). Elimination of the β-proton (**8**) regenerates aromaticity and the phenanthridine nucleus, **9**, is formed. The efficeincy of the cyclization, in which compound **8** can be formed, is dependant on at least three factors: (1) the formation of the carbonium ion; (2) the stability of the carbonium ion and; (3) the electron density at the heteronuclear ortho carbon atoms.

A heteroatom (in ring A) will control the electron density at the position of ring closure by its inductive and resonance effects. If the substituent increases the electron density of the ring, then ring closure will be facilitated by the heteroatom. If the heteroatom decreases electron density, then the ring closure will be hindered.

The acyl residue controls the formation and stability of the carbonium ion. If the carbonium ion is destabilized (by electron withdrawing groups), then cyclization to the phenanthridine nucleus will be sluggish. The slower the rate of cyclization, the greater the chance of side reactions with the cyclization reagent. Therefore, the yield of the phenanthridine will depend on the relative rates of cyclization and side reactions, which is controlled by the stability of the carbonium ion.

9.13.4 Synthetic Utility

The synthetic utility of the Pictet–Hubert reaction is limited. This is due to the requirement of having a substituent in the para position of the "A" ring (10) or no substituent on the "A" ring. Substituents present in the meta position of the "A" ring (12)

can cause loss of regioselectivity during ring closure since cyclization can occur ortho (14) or para (13) relative to the meta substituent. Both regioisomers are usually obtained in a 1:1 ratio.

The formation of the ortho regioisomer can be enhanced if electrostatic attraction between the carbonium ion and the meta substituent on the "A" ring can be accommodated.[7]

The Pictet–Hubert reaction has found utility in the production of phenanthridine molecules that act as DNA-intercalator antitumor and antiviral agents (17).[8, 9]

The phosphorus oxychloride / nitrobenzene conditions were sometimes replaced with polyphosphoric acid at 150°C. R groups used in the study were: H, Ph, o-aza, o-Cl, m-aza, m-Cl, p-aza, p-F, p-Cl, p-Br, p-I, p-OMe, p-OH, p-NO₂, p-NH₂, p-NHCOMe, p-NHSO₂Me. The yield of the cyclization step was not specified.

9.13.5 Experimental[9]

1-(3-cyanobenzamido)-3,8-dinitrobiphenyl (**18**) (6.63 g, 16.20 mmol) and phosphorus oxychloride (2.08 mL, 22.68 mmol) in nitrobenzene (40 mL) were refluxed for 1h; then tin chloride (0.38 mL, 3.24 mmol) was added. After 2h, the solution was cooled and triturated with boiling anhydrous ethanol (120 mL) for 10 min. After filtration, the solid product was washed with hot anhydrous ethanol to yield 9-(3-cyanophenyl)-2,7-dinitrophenanthridine (**19**) in 98% yield. M. P. = 302–304°C; ^{1}H NMR (d$_6$-DMSO) δ 9.38 (1H, d, $J = 8.9$), 9.29 (1H, d, $J = 9.1$), 8.98 (1H, d, $J = 2.3$), 8.95 (1H, dd, $J = 8.9$, 2.2), 8.80 (1H, d, $J = 2.2$), 8.64 (1H, dd, $J = 9.1$, 2.3), 8.40 (1H, bs), 8.26 (1H, d, $J = 7.7$), 8.23 (1H, d, $J = 7.7$), 7.97 (1H, t, $J = 7.7$).

9.13.6 References

1.   Pictet, A.; Hubert, A. Ber., **1896**, 29, 1182.
2.   Pictet, A.; Ankersmit, H. J. Ber., **1889**, 22, 3339.
3.   Morgan, G. T.; Walls, L. P. J. Chem. Soc., **1931**, 2447.
4.   Browning, C. H.; Morgan, G. T.; Robb, J. V. M.; Walls, L. P. J. Path. Bact. **1938**, 46, 203.
5.   Fodor, G.; Nagubandi, S. Tetrahedron **1980**, 1279.
6.   Ritchie, E. J. Proc. Roy. Soc. N. S. Wales **1945**, 78, 147.
7.   Caldwell, A. G.; Walls, L. P. J. Chem. Soc., **1948**, 188.
8.   Atwell, G. J.; Baguley, B. C.; Denny, W. A. J. Med. Chem., **1988**, 31, 774.
9.   Peytou, V.; Condom, R.; Patino, N.; Guedj, R.; Aubertin, A.–M.; Gelus, N.; Bailly, C.; Terreux, R.; Cabrol–Bass, D. J. Med. Chem., **1999**, 42, 4042.

Daniel D. Holsworth

9.14 Pictet–Spengler Isoquinoline Synthesis

9.14.1 Description

The Pictet–Spengler reaction is one of the key methods for construction of the isoquinoline skeleton, an important heterocyclic motif found in numerous bioactive natural products. This reaction involves the condensation of a β-arylethyl amine **1** with an aldehyde, ketone, or 1,2-dicarbonyl compound **2** to give the corresponding tetrahydroisoquinoline **3**.[1] These reactions are generally catalyzed by protic or Lewis acids, although numerous thermally-mediated examples are found in the literature. Aromatic compounds containing electron-donating substituents are the most reactive substrates for this reaction.

R = H, hydroxy, alkoxy, alkyl

R_1, R_2 = H, alkyl, Ar, carbonyl

This reaction is also a key method for the formation of tetrahydro-β-carbolines **5** from indole bases **4** and aldehydes, ketones, or 1,2-dicarbonyl compounds **2**.[2-5] These reactions are similarly acid-catalyzed or thermally-induced and have been utilized in the synthesis of numerous indole alkaloids.

R = H, hydroxy, alkoxy, alkyl

R_1, R_2 = H, alkyl, Ar, carbonyl

9.14.2 Historical Perspective

In 1911, Amé Pictet and Theodor Spengler reported that β-arylethyl amines condensed with aldehydes in the presence of acid to give tetrahydroisoquinolines.[6] Phenethylamine **6** was combined with dimethoxymethane **7** and HCl at elevated temperatures to give tetrahydroisoquinoline **8**. Soon after, the Pictet–Spengler reaction became the standard method for the formation of tetrahydroisoquinolines.

A few years later, Tatsui developed this process for use with indole bases and prepared 1-methyl-1,2,3,4-tetrahydro-β-carboline **11** from tryptamine **9** and acetaldehyde **10** under acid catalysis.[7]

9.14.3 Mechanism

The Pictet–Spengler reaction is an acid-catalyzed intramolecular cyclization of an intermediate imine of 2-arylethylamine, formed by condensation with a carbonyl compound, to give 1,2,3,4-tetrahydroisoquinoline derivatives. This condensation reaction has been studied under acid-catalyzed and superacid-catalyzed conditions, and a linear correlation had been found between the rate of the reaction and the acidity of the reaction medium.[8] Substrates with electron-donating substituents on the aromatic ring cyclize faster than the corresponding unsubstituted compounds, supporting the idea that the cyclization process is involved in the rate-determining step of the reaction.

Under acidic conditions, imine **12** is protonated to give the iminium ion **13** which undergoes an electrophilic aromatic substitution reaction to form the new carbon-carbon bond. Rapid loss of a proton and concomitant re-aromatization gives the tetrahydroisoquinoline **14**.

The Pictet–Spengler condensation of indole bases and carbonyl compounds to form β-carbolines involves a slightly different mechanism than the isoquinoline synthesis. Under acidic conditions, imine **15** is protonated to give the iminium ion **16** which then can cyclize to give two different cationic intermediates. A 5-endo-trig cyclization onto C(3) of the indole ring gives the spiroindolenine **17**. Alternatively, a 6-endo-trig cyclization onto C(2) gives the β-carboline carbonium ion **18**. Deuterium labeling studies of a closely related system provided direct evidence that the spiro intermediate is involved in the Pictet–Spengler condensation reaction, and that its formation is fast and reversible.[5,9,10] It is unclear whether carbonium ion **18** is formed from a direct attack at the indole 2-position by the iminium cation **16** or if this intermediate results from a rearrangement of spiroindolenine **17**.[11] Loss of a proton results in the formation of β-carboline **19**.

9.14.4 Variations and Improvements

The Pictet–Spengler reaction has been carried out on various solid support materials[12-17] and with microwave irradiation activation.[18-20] Diverse structural analogues of (−)-Saframycin A have been prepared by carrying out the Pictet–Spengler isoquinoline synthesis on substrates attached to a polystyrene support.[21] Amine **20** was condensed with aldehyde **21** followed by cyclization to give predominantly the *cis* isomer tetrahydroisoquinoline **22** which was further elaborated to (−)-Saframycin A analogues.

22 → (-)-Saframycin A Analogues

One complication of the Pictet–Spengler condensation of benzylisoquinolines **24** is regiochemical control in the closure of ring C when activating substituents are present on the D ring. Experimentally, the ring-closure reaction yields predominantly the 10,11-disubstituted product **23** rather than the 9,10-disubstituted product **25**.[22]

10,11-disubstitution   9,10-disubstitution

This problem has been circumvented by utilizing the *ipso*-directing capability of a silicon substituent at the 2′ position of the D ring to facilitate ring closure at this position.[22,23] Benzylisoquinoline **26** condensed with formaldehyde to give predominantly product **28** when R = H. However, when R = SiMe₃, **27** was the only product observed and was isolated in excellent yield.[23] Several members of the protoberberine family of natural products have been synthesized using this strategy.

	27	**28**
R = H	6%	31%
R = TMS	98%	0%

Thioorthoesters,[24] α-chloro-α-phenylthioketones,[25] and perhydro-1,3-hetero-cycles[26] have found utility as synthetic equivalents of aldehydes in the Pictet–Spengler condensation reaction. Oxazinane **29** was combined with *N*-benzyl-L-tryptophan methyl ester **30** in the presence of trifluoroacetic acid to give the tetrahydro-β-carboline **31** in good yield as predominantly the *trans* diastereomer.[26]

93:7
trans : cis

The oxa–Pictet–Spengler reaction has been used with success to prepare dihydrofurano[2,3-*c*]pyrans[27] and isochromans[28] from 1-(3-furyl)alkan-2-ols and 2-(3',4'-dihydroxy)phenylethanol, respectively. Furanyl alcohol **32** reacted with isobutyraldehyde **33** in the presence of *p*-toluenesulfonic acid to give the corresponding *cis*-5,7-diisopropyl 4,5-dihydro-7*H*-furano[2,3-*c*]pyran **34** in good yield.[27]

A formal Pictet–Spengler condensation to give 2,3-dihydro-1*H*-2-benzazepine-3-carboxylic acid **36** was achieved in quantitative yield via a sigmatropic rearrangement of *cis*-2,3-methanophenylalanine **35** in the presence of paraformaldehyde and hydrochloric acid at room temperature.[29] It is interesting to note that homophenylalanine **38** did not cyclize to give **37**, even under vigorous reaction conditions.

9.14.5 Synthetic Utility
9.14.5.1 Stereochemical control
Several factors influence the diastereoselectivity of the Pictet–Spengler condensation to form 1,3-disubstituted and 1,2,3-trisubstituted tetrahydro-β-carbolines (**39** and **40**, respectively). The presence or absence of an alkyl substituent on the nitrogen of tryptophan has a large influence on the relative stereochemistry of the tetrahydro-β-carboline products formed from a condensation reaction with an aldehyde under various reaction conditions.

39 $R^2 = H$
40 $R^2 = $ alkyl

For the kinetically controlled formation of 1,3-disubstituted tetrahydro-β-carbolines, placing both substituents in equatorial positions to reduce 1,3-diaxial interactions resulted in the *cis*-selectivity usually observed in these reactions.[11] Condensation reactions carried out at or below room temperature in the presence of an acid catalyst gave the kinetic product distribution with the *cis*-diastereomer being the major product observed, as illustrated by the condensation of L-tryptophan methyl ester **41** with benzaldehyde. At higher reaction temperatures, the condensation reaction was reversible and a thermodynamic product distribution was observed. *Cis* and *trans* diastereomers were often obtained in nearly equal amounts suggesting that they have similar energies.[11]

Temp. (°C)	cis/trans
80	37:63
23	78:22
−70	83:17

Conversely, when *N*-alkyl tryptophan methyl esters were condensed with aldehydes, the *trans* diastereomers were observed as the major products.[11, 30] X-ray crystal structures of 1,2,3-trisubstituted tetrahydro-β-carbolines revealed that the C1 substituent preferentially adopted a pseudo-axial position, forcing the C3 substituent into a pseudo-equatorial orientation to give the kinetically and thermodynamically preferred *trans* isomer.[11] As the steric size of the C1 and N2 substituents increased, the selectivity for the *trans* isomer became greater. *N*-alkyl-L-tryptophan methyl ester **42** was condensed with various aliphatic aldehydes in the presence of trifluoroacetic acid to give predominantly the *trans* isomers.[30]

R^1	R^2	cis/trans
CH$_3$	Bn	12:88
n-Pr	Bn	11:89
c-hex	Bn	0:100
CH$_3$	CH(Ph)$_2$	0:100
n-Pr	CH(Ph)$_2$	0:100
c-hex	CH(Ph)$_2$	0:100

Stereoselectivity in the condensation reaction of 2-arylethylamines with carbonyl compounds to give 1,2,3,4-tetrahydroisoquinoline derivatives was somewhat dependent on whether acid catalysis or superacid catalysis was invoked. Particularly in the cases of 2-alkyl-N-benzylidene-2-phenethylamines, an enhanced stereoselectivity was observed with trifluorosulfonic acid (TFSA) as compared with the weaker acid, trifluoroacetic acid (TFA).[31] Compound **43** was cyclized in the presence of TFA to give modest to good *trans/cis* product ratios. The analogous compound **44** was cyclized in the presence of TFSA to give slightly improved *trans/cis* product ratios.

R	trans/cis
Me	80:20
Et	61:39
n-Bu	60:40

R	trans/cis
Me	88:12
Et	81:19
n-Bu	82:18

9.14.5.2 Asymmetric variations

One of the most common methods for introducing asymmetry into a Pictet–Spengler condensation reaction is to append a non-racemic auxiliary onto the indole amine or phenethylamine substrate. Various amino acids,[32–34] *N,N*-phthaloyl amino acids,[35] and α-arylethyl groups[36–38] have been utilized as asymmetric auxiliary substituents in the condensation reaction.   Carbamates derived from (–)-8-phenylmenthyl, such as compound **45**, reacted with several aliphatic aldehydes to give the condensation products in good yield and good diastereoselectivity.[39]

45

R = Me, Et, *i*-Pr, *n*-Bu, *c*-Hex
Yields = 72-97%
d.e.'s = 69-90%

N-sulfinyl chiral auxiliaries have been used to prepare enantiopure tetrahydro-β-carbolines[40] and tetrahydroisoquinolines[41] in good yields under mild reaction conditions. Both enantiomers of *N-p*-toluenesulfinyltryptamine **46** could be readily prepared from the commercially available Andersen reagents.[42] Compound **46** reacted with various aliphatic aldehydes in the presence of camphorsulfonic acid at –78 °C to give the *N*-sulfinyl tetrahydro-β-carbolines **47** in good yields.   The major diastereomers were obtained after a single crystallization.   Removal of the sulfinyl auxiliaries under mildly acidic conditions produced the tetrahydro-β-carbolines **48** as single enantiomers.

46   **47**   **48**

yields: 57-63%   R = alkyl
                                            e.e.'s > 98%

9.14.5.3   Natural product synthesis

The Pictet–Spengler condensation has been of vital importance in the synthesis of numerous β-carboline and isoquinoline compounds in addition to its use in the formation of alkaloid natural products of complex structure.   A tandem retro-aldol and Pictet–Spengler sequence was utilized in a concise and enantioselective synthesis of 18-pseudoyohimbone.[43] Amine **49** cyclized under acidic conditions to give the condensation product **50** in good yield.   Deprotection of the ketone produced the indole alkaloid **51**.

(−)-Eburnamonine was assembled utilizing a Pictet–Spengler cyclization of hydroxy-lactam **52** in the presence of trifluoroacetic acid at low temperature to give a mixture of diastereomers **53** in 95% yield.[44] These compounds were readily separated by chromatography and the α-epimer was further elaborated to give the natural product.

Model studies directed toward the synthesis of Ecteinascidin 743 employed an elegant Pictet–Spengler cyclization of phenethylamine **54** and the 1,2-dicarbonyl compound **55** to assemble the spiro tetrahydroisoquinoline **56** in a stereospecific fashion.[45,46] The silica-catalyzed condensation reaction provided **56** in excellent yield.

9.14.6   Experimental

57   **58**

(4S)-4-Isopropoxyphenyl-5-isopropoxy-6-methoxy-2-methyl-1,2,3,4-tetrahydroisoquinoline (58):[47]

A solution of amine **57** (300 mg, 0.84 mmol) and formaldehyde (37%, 2.17 mL) in EtOH (15 mL) was acidified with 0.7 mL of concentrated aqueous HCl. The mixture was heated to reflux for 6 h. The solvent and excess reagent were removed by rotary evaporation and the residue was dissolved in CHCl$_3$ (20 mL) and treated successively with a solution of aqueous ammonia 10% (20 mL), water (20 mL), and brine (20 mL). The dried solution (Na$_2$SO$_4$) was subjected to rotary evaporation and the residual solid was purified by flash column chromatography with CHCl$_3$/MeOH (93 : 7) as eluent to afford **58** as a pale yellow oil (232 mg, 75%). $[\alpha]_D^{20} = -3.4$ (c 0.76, CHCl$_3$). ^{1}H NMR (CDCl$_3$) δ 0.81 (d, $J = 6.1$ Hz, 3H), 1.09 (d, $J = 6.1$ Hz, 3H), 1.29 (d, $J = 6.0$ Hz, 6H), 2.32 (s, 3H), 2.60 – 2.71 (m, 2H), 3.35 (d, $J = 14.3$ Hz, 1H), 3.75 (s, 3H), 3.80 (d, $J = 14.3$ Hz, 1H), 4.49 (sept, $J = 6.0$ Hz, 1H), 4.60 (sept, $J = 6.1$ Hz, 1H), 6.74 (d, $J = 8.6$ Hz, 1H), 6.78 (s, 4H), 7.05 (d, $J = 8.6$ Hz, 1H).

9.14.7 References

1.   [R] Rozwadowski, M. D. *Heterocycles* **1994**, *39*, 903.
2.   [R] Hino, T.; Nakagawa, M. *Heterocycles* **1998**, *49*, 499.
3.   [R] Czerwinski, K. M.; Cook, J. M. *Adv. Heterocycl. Nat. Prod. Synth.* **1996**, *3*, 217.
4.   [R] Cox, E. D.; Cook, J. M. *Chem. Rev.* **1995**, *95*, 1797.
5.   [R] Ungemach, F.; Cook, J. M. *Heterocycles* **1978**, *9*, 1089.
6.   Pictet, A.; Spengler, T. *Ber.* **1911**, *44*, 2030.
7.   Tatsui, G. *J. Pharm. Soc. Jpn.* **1928**, *48*, 92.
8.   Yokoyama, A.; Ohwada, T.; Shudo, K. *J. Org. Chem.* **1999**, *64*, 611.
9.   Bailey, P. D.; *J. Chem. Res., Synop.* **1987**, 202.
10.   Bailey, P. D.; *Tetrahedron Lett.* **1987**, *28*, 5181.
11.   Bailey, P. D.; Hollinshead, S. P.; McLay, N. R.; Morgan, K.; Palmer, S. J.; Prince, S. N.; Reynolds, C. D.; Wood, S. D. *J. Chem. Soc., Perkin Trans. 1* **1993**, 431.
12.   Tóth, G. K.; Kele, Z.; Fülöp, F. *Tetrahedron Lett.* **2000**, *41*, 10095.
13.   Connors, R. V.; Zhang, A. J.; Shuttleworth, S. J. *Tetrahedron Lett.* **2002**, *43*, 6661.
14.   Wu, T. Y. H.; Schultz, P. G. *Org. Lett.* **2002**, *4*, 4033.
15.   Bonnet, D.; Ganesan, A. *J. Comb. Chem.* **2002**, *4*, 546.
16.   Orain, D.; Canova, R.; Dattilo, M.; Klöppner, E.; Denay, R.; Koch, G.; Giger, R. *Synlett* **2002**, 1443.
17.   Grimes, J. H. Jr.; Angell, Y. M.; Kohn, W. D. *Tetrahedron Lett.* **2003**, *44*, 3835.
18.   Wu, C.–Y.; Sun, C.–M. *Synlett* **2002**, 1709.
19.   Srinivasan, N.; Ganesan, A. *Chem. Commun.* **2003**, 916.
20.   Pal, B.; Jaisankar, P.; Giri, V. S. *Synth. Commun.* **2003**, *33*, 2339.
21.   Myers, A. G.; Lanman, B. A. *J. Am. Chem. Soc.* **2002**, *124*, 12969.
22.   Cutter, P. S.; Miller, R. B.; Schore, N. E. *Tetrahedron* **2002**, *58*, 1471.

23.   Miller, R. B.; Twang, T. *Tetrahedron Lett.* **1988**, *29*, 6715.
24.   Silveira, C. C.; Bernardi, C. R.; Braga, A. L.; Kaufman, T. S. *Tetrahedron Lett.* **2003**, *44*, 6137.
25.   Silveira, C. C.; Bernardi, C. R.; Braga, A. L.; Kaufman, T. S. *Tetrahedron Lett.* **2001**, *42*, 8947.
26.   Singh, K.; Deb, P. K.; Venugopalan, P. *Tetrahedron* **2001**, *57*, 7939.
27.   Miles, W. H.; Heinsohn, S. K.; Brennan, M. K.; Swarr, D. T.; Eidam, P. M.; Gelato, K. A. *Synthesis* **2002**, 1541.
28.   Guiso, M.; Marra, C.; Cavarischia, C. *Tetrahedron Lett.* **2001**, *42*, 6531.
29.   Martins, J. C.; Rompaey, K. V.; Wittmann, G.; Tömböly, C.; Tóth, G.; Kimpe, N. D.; Tourwé, D. *J. Org. Chem.* **2001**, *66*, 2884.
30.   Czerwinski, K. M.; Deng, L.; Cook, J. M. *Tetrahedron Lett.* **1992**, *33*, 4721.
31.   Nakamura, S.; Tanaka, M.; Taniguchi, T.; Uchiyama, M.; Ohwada, T. *Org. Lett.* **2003**, *5*, 2087.
32.   Schmidt, G.; Waldmann, H.; Henke, H.; Burkard, M. *Chem. Eur. J.* **1996**, *2*, 1566.
33.   Siwicka, A.; Wojtasiewicz, K.; Leniewski, A.; Maurin, J. K.; Maurin, J. K.; Czarnocki, Z. *Tetrahedron: Asymmetry* **2002**, *13*, 2295.
34.   Zawadzka, A.; Leniewski, A.; Maurin, J. K.; Wojtasiewicz, K.; Siwicka, A.; Blachut, D.; Czarnocki, Z. *Eur. J. Org. Chem.* **2003**, 2443.
35.   Waldmann, H.; Schmidt, G.; Henke, H.; Burkard, M. *Angew. Chem. Int. Ed. Engl.* **1995**, *34*, 2402.
36.   Soe, T.; Kawate, T.; Fukui, N.; Hino, T.; Nakagawa, M. *Heterocycles* **1996**, *42*, 347.
37.   Kawate, T.; Yamanaka, M.; Nakagawa, M. *Heterocycles* **1999**, *50*, 1033.
38.   Jiang, W.; Sui, Z.; Chen, X. *Tetrahedron Lett.* **2002**, *43*, 8941.
39.   Tsuji, R.; Nakagawa, M.; Nishida, A. *Tetrahedron: Asymmetry* **2003**, *14*, 177.
40.   Gremmen, C.; Willemse, B.; Wanner, M. J.; Koomen, G.-J. *Org. Lett.* **2000**, *2*, 1955.
41.   Gremmen, C.; Wanner, M. J.; Koomen, G.-J. *Tetrahedron Lett.* **2001**, *42*, 8885.
42.   Solladié, G.; Hutt, J.; Girardin, A. *Synthesis* **1987**, 173.
43.   Miyazawa, N.; Ogasawara, K. *Tetrahedron Lett.* **2002**, *43*, 4773.
44.   Wee, A. G. H.; Yu, Q. *J. Org. Chem.* **2001**, *66*, 8935.
45.   Zhou, B.; Guo, J.; Danishefsky, S. J. *Org. Lett.* **2002**, *4*, 43.
46.   Corey, E. J.; Gin, D. Y.; Kania, R. S. *J. Am. Chem. Soc.* **1996**, *118*, 9202.
47.   Couture, A.; Deniau, E.; Grandclaudon, P.; Lebrun, S. *Tetrahedron: Asymmetry* **2003**, *14*, 1309.

Jennifer M. Tinsley

9.15 Pomeranz–Fritsch Reaction

9.15.1 Description

The Pomeranz–Fritsch reaction involves the preparation of isoquinolines **4** *via* the acid-mediated cyclisation of the appropriate aminoacetal intermediate **3**.[1-2] The best yields are usually obtained when the benzaldehyde portion **1** has electron-donating substituents in the 3- or 3,4- positions relative to the aldehyde.

Of the well-known methods to prepare isoquinolines, including the Pictet–Spengler and Bischler–Napieralski cyclisation, the Pomeranz–Fritsch reaction is the only direct generally accepted method for the construction of the fully unsaturated isoquinoline ring system.

9.15.2 Historical perspective

Toward the end of the 19th century both Pomeranz and Fritsch independently reported the preparation of isoquinolines by the reaction of aminoacetaldehyde dimethyl acetal **2** (R = Me) with aromatic aldehydes **1** followed by cyclisation in acidic media.[3-4] Unfortunately yields were often poor and not always reproducible. This has prompted the search for various improvements and modifications on the original theme, including the use of reagents other than strong mineral acid which tends to destroy the intermediate imine.[5]

9.15.3 Mechanism

The most plausible mechanism involves condensation between aldehyde **1** and amine **5** to give the corresponding imine **6**. Cyclisation and subsequent elimination yields the fully unsaturated isoquinoline ring structure **4**.

9.15.4 Variations and Improvements
9.15.4.1 Bobbitt variation

The Bobbitt modification is the most widely used variation of the Pomeranz–Fritsch reaction. This modification involves cyclisation of benzylaminoacetal **10**, usually prepared from the classical Pomeranz–Fritsch imine **9**, to yield 4-hydroxy derivatives **11**. The success of this method can be attributed to avoiding treatment and thus (partial) destruction of imine **10** under strongly acidic conditions.

9.15.4.2 Jackson variation

The Jackson modification involves cyclisation of *N*-tosylated amine **12** and provides a complementary method to the classical Pomeranz–Fritsch reaction for entry into the fully unsaturated ring system **13**. Amine **12** can be prepared from either the Pomeranz–Fritsch–Bobbitt imine **10** or reaction of benzylhalide **14** and the corresponding sodium anion **15**.[6]

9.15.4.3 Schlittler–Muller variation

The Schlittler–Muller variation of the Pomeranz–Fritsch reaction involves reaction of diethoxyethanal **17** with benzylamine **16** to prepare the desired imine **18**. Intermediate **18** is subsequently cyclised to substituted isoquinoline **19**. The advantage here lies in the fact that the initial condensation can still take place between an aldehyde and an amine.

9.15.5 Synthetic utility
9.15.5.1 Classical Pomeranz–Fritsch

A number of Lewis acids have been utilized in the Pomeranz–Fritsch reaction, including polyphosphoric acid and boron trifluoride-trifluoroacetic anhydride.[5] Under the latter conditions yields were best when electron-donating groups were present in the 3- or 3, 4- position of imine **20**, whereas unactivated aldehydes failed to cyclise at all.[5]

R_1 = OMe	R_2 = OMe	60-82 %
R_1 = H	R_2 = OMe	73 %
R_1 = H	R_2 = H	0 %

In another example reaction of aldehyde 22 and amine 23 gave imine 24 which cyclised under strongly acidic conditions to yield the corresponding isoquinoline 25 in good yield.[7] It is interesting that the aldehyde portion 22 is not benzaldehyde derived.

9.15.5.2 Bobbitt variation

4-Hydroxyquinoline **28** was synthesized in excellent yield *via* cyclisation of the appropriate Pomeranz–Fritsch–Bobbitt imine **27**.[8] The desired amine **27** was prepared *via* hydride reduction of the classical Pomeranz–Fritsch imine.

The Pomeranz–Fritsch–Bobbitt reaction has been utilized for the preparation of 4-hydroxy tetrahydroisoquinoline **31** in excellent yield.[9] In this example 2,5-disubstituted benzaldehyde **29** has been successfully used as the reacting partner.

A concise total synthesis of (*R*)-Reticuline has been reported using the Pomeranz–Fritsch–Bobbitt reaction to synthesize tetrahydroisoquinoline portion **35**.[10-11] Pomeranz–Fritsch–Bobbitt imine **34** was prepared by reacting amine **33** with chiral epoxide **32**.

The Pomeranz–Fritsch–Bobbitt cyclisation of activated amino-acetal **38** yielded the desired 4-hydroxyquinoline **39** in acceptable yield. The non-obvious regioselectivity of the cyclisation can be attributed to the overriding *para*-directing effect of alkoxy groups.[12]

Condensation between aldehyde **40** and amine **29** followed by sodium borohydride reduction of the resultant imine and cyclisation yielded isoquinoline **41** in good yield.[13] Cyclisation occurred exclusively at the more electron-rich aromatic group.

Treatment of substituted pyrollidinone **42** with a Lewis acid, rather than simple protic acid, lead to a Pomeranz–Fritsch–Bobbitt type condensation to yield indolizinone **43** as a single diastereomer.[14]

Incorporation of a stereogenic center adjacent to the nitrogen of the isoquinoline ring has been achieved by the asymmetric addition of methyl lithium to prochiral iminoacetal **44** in the presence of chiral ligand **45**. Unfortunately only modest enantiomeric excesses were achieved with a high loading of the chiral ligand.[15]

9.15.5.3 Jackson variation

Tosylation of secondary amine **48** gave desired precursor **49** which was cyclised under prolonged acidic conditions to yield a mixture of linear and angular fully unsaturated isoquinolines **50** and **51**.[16]

9.15.5.4 Schlittler–Muller variation

2,7-Diazaphenanthrene **53** has been prepared *via* the Schlittler–Muller variation of the Pomeranz–Fritsch reaction in moderate yield.[17]

9.15.6 Experimental[7]

Aldehyde **22** and aminoacetaldehyde dimethyl acetal **23** (3eq.) were heated to reflux in toluene (Dean–Stark apparatus) until all of the starting material was consumed. The crystalline product was collected and washed with solvent to yield imine **24**, which was used without further purification.

Imine **24** was dissolved in sulfuric acid and the solution heated at 60 °C for 2h. The mixture was poured into cold water and the precipitate collected, washed with water and recrystallized from methanol to yield **25**.

9.15.7 References

1.   For a review see: [R] Gensler, W. J. *Organic Reactions* **1951**, *6*, 191.
2.   For a review on the synthesis of heterocycles using iminoacetals see: Bobbitt, J. M.; Bourque, A. J. *Heterocycles* **1987**, *25*, 601.
3.   Pomeranz, C. *Monatsch.* **1893**, *14*, 116.
4.   Fritsch, P. *Dtsch Chem. Ges Ber.* **1893**, *26*, 419.
5.   Bevis, M. J.; Forbes, E. J.; Naik, N. N.; Uff, B. C. *Tetrahedron* **1971**, *27*, 1253.
6.   Boger, D. L.; Brotherton, C. E.; Kelley, M. D. *Tetrahedron* **1981**, *37*, 3977.
7.   Chilin, A.; Manzini, P.; Confente, A.; Pastorini, G.; Guiotto, A. *Tetrahedron* **2002**, *58*, 9959.
8.   Kunitomo, J–I.; Miyata, Y.; Oshikata, M.; *Chem. Pharm. Bull.* **1985**, *33*, 5245.
9.   Mitscher, L. A.; Gill, H.; Filppi, J. A.; Wolgemuth, R. L. *J. Med. Chem.* **1986**, *29*, 1277.
10.   Hirsenkorn, R. *Tetrahedron Lett.* **1991**, *32*, 1775.
11.   Hirsenkorn, R. *Tetrahedron Lett.* **1990**, *31*, 7591.
12.   Schlosser, M.; Simig, G.; Geneste, H.; *Tetrahedron* **1998**, *54*, 9023.
13.   Capilla, A. S.; Romero, M.; Pujol, M. D.; Caignard, D. H.; Renard, P. *Tetrahedron* **2001**, *57*, 8297.

14.   Poli, G.; Baffoni, S. C.; Giambastiani, G.; Reginato, G. *Tetrahedron* **1998**, *54*,   10403.
15.   Gluszynska, A.; Rozwadowska, M. D. *Tetrahedron:Asymmetry* **2000**, *11*, 2359.
16.   Hall, R. J.; Dharmasena, P.; Marchant, J.; Oliveira–Campos, A–M. F.; Queiroz, M–J. R. P.; Raposa, M. M.; Shannon, P. V. R.; *J. Chem. Soc. Perkin Trans 1* **1993**,   1879.
17.   Gill, E. W.; Bracher, A. W. *J. Heterocyclic Chem.* **1983**, *20*, 1107.

Andrew Hudson

9.16 Riehm Quinoline Synthesis

9.16.1 Description

The Riehm synthesis involves the formation of quinolines *via* the reaction of an aniline hydrochloride salt and a ketone.[1–2]

9.16.2 Historical Perspective

In 1885 Riehm and Engler discovered that 2,4-dimethylquinoline could be synthesised by heating aniline hydrochloride with acetone over several days.[1]

9.16.3 Mechanism

The mechanism inevitably involves the loss of water and, remarkably, methane. The yields are usually low due to the high temperatures required for this transformation to occur. If the 1,2-dihydroquinoline is isolated the aromatization can be performed in a separate step under acidic or basic conditions.[3-4]

9.16.4 References

1.   Levin, J.; Riehm, P. *Ber.* **1886**, *19*, 1394.
2.   Elderfield, R. C.; McClenachan, E. C.; *J. Am. Chem. Soc.* **1960**, *82*, 1975 and references therein.
3.   Vaughan, W. E. In *Organic syntheses*; Wiley: New York, **1955**, Vol. III, pp 329.
4.   Craig, D.; *J. Am. Chem. Soc.* **1938**, *60*, 1458.

Andrew Hudson

9.17 Skraup/Doebner–von Miller Reaction

9.17.1 Description

The Skraup reaction involves the synthesis of quinoline **3** from the reaction of aniline **1** and glycerol **2** in the presence of a strong acid and an oxidant.

9.17.2 Historical Perspective

In 1880 Skraup discovered that quinoline could be synthesised by heating aniline, glycerine, sulfuric acid and an oxidizing reagent.[1-2] A year later Doebner and von Miller generalized Skraup's method for the synthesis of substituted quinolines by substituting 1,2-glycols or α,β-unsaturated aldehydes for glycerol.[3] As crotonaldehyde is probably an intermediate in the Skraup reaction, there is little difference between the two variants; as such they have been combined into one chapter.

9.17.3 Mechanism

Skraup proposed a simple mechanism involving imine formation followed by an acid-mediated cyclization.[1] Unfortunately the observed regioselectivity is not consistent with the proposed mechanism when, for example, electron-rich aniline **4** reacts with α,β-unsaturated aldehyde **5** to give quinoline **6**.[4]

Bischler has suggested that anilines undergo 1,4-addition followed by dehydration which would explain the inherent regioselectivity.[5] Mechanistic studies suggest that the reaction involves the reversible formation of diazetidinium ions **7** and there irreversible cyclization to quinolines.[6]

9.17.4 Variations and Improvements

Many people view the Skraup/Doebner–von Miller reaction as the worst 'witch's brew' of all the heterocyclic syntheses. The reaction can be violently exothermic. A variety of oxidizing reagents and additives have been added in an effort to improve yields, including iron (III) and tin (IV) salts, nitrobenzenes, iodine and various acids such as boric and arsenic. Cohn's conditions for the Skraup reaction using an iron salt and boric acid in concentrated sulfuric acid are frequently employed.[7]

9.17.5 Synthetic utility
9.17.5.1 Classical Skraup/Doebner–von Miller reaction — Quinolines

The preparation of 3-alkyl quinolines by traditional Skraup/Doebner–von Miller reaction typically results in very low yields.[8] When ethylacrolein (9) is condensed with *m*-toluidine (8) under typical Skraup/Doebner–von Miller conditions, the yield of 10 is only 25%, compared to 65% with di-acetyl acetal 11.

One of the drawbacks of the Skraup/Doebner–von Miller reaction is the isolation of the desired product from the starting aniline and co-formed alkyl anilines and 1,2,3,4-tetrahydroquinaldine.[9] Isolation can be simplified greatly by addition of one equivalent of zinc chloride at the end of the reaction; all of the basic products were precipitated. Washing the brown solids with 2-propanol removed all impurities and left the desired quinoline as a 2:1 complex with zinc chloride in yields of 42–55%.

Using Cohn's conditions fluorinated quinoline **16** was synthesized from the corresponding aniline **14**.[10] The yield was significantly improved when nitrobenzene was replaced by *m*-nitrobenzenesulfonic acid **15**.

Reaction of anilines **17** or **19** under similar reaction conditions gave the angular quinolines **18** and **20** respectively as the sole products.[11]

Acetylated aniline **21** was reacted under Skraup/Doebner–von Miller conditions in the presence of an arsenic salt to yield quinoline **22** with concurrent displacement of fluorine.[12]

A variety of oxidants have been used in the Skraup/Doebner–von Miller reaction between anilines and α,β-unsaturated aldehydes. For example, aniline **23** was reacted with 2-bromoacrolein in the presence of bromine to yield the corresponding quinoline **25** in high yield.[13]

The yield of substituted quinoline **27** was substantially improved when *p*-chloranil was used as the oxidant.[14]

3-Fluoroaniline **28** was reacted with crotonaldehyde **5** under similar Skraup/Doebner–von Miller conditions to yield quinoline **29** in good yield.[15]

The yields of Skraup/Doebner–von Miller reaction can be dramatically improved by running the reaction as a two-phase mixture.[16] Reaction of crotonaldehyde with **30** in acidic ethanol provides only 10% of quinoline **31**. However, when a toluene solution of crotonaldehyde is reacted with **30** (starting as the acetanilide) in 6M HCl at 100 °C for 2 h, quinoline **31** is isolated in 80% yield on 5kg scale.

In addition to two-phase conditions, phase transfer catalysts have been used to improve the Skraup/Doebner–von Miller reaction.[17] Condensation of **32** with **5** in a two-phase system of toluene/con HCl provides **33** in 47% yield. Addition of 5 mol% tetra-*n*-butyl ammonium chloride increased the yield to 57%.

32   **5**   **33**

Recently, solvent-free Skraup/Doebner–von Miller reactions have been developed under microwave radiation. For example, aniline **34** and enone **35** are reacted in the presence of silica gel impregnated with indium trichloride to give the corresponding quinoline **36** in good yield. It was subsequently shown that both electron-rich and electron-poor anilines undergo cyclization in a similar fashion.[18]

34   **36**   83 %

9.17.5.2   Modified Skraup/Doebner–von Miller reaction — Dihydroquinolines

A number of dihydroquinolines have been prepared by treating aniline derivatives with acetone or mesityl oxide in the presence of iodine. In these cases aromatization to the fully unsaturated quinoline would require the loss of methane, a process known as the Riehm quinoline synthesis. Such Skraup/Doebner–von Miller-type reactions are often low yielding due to large amounts of competing polymerization. For example, aniline **37** reacts with mesityl oxide to give dihydroquinolines **39**, albeit in low yield.[19]

37   **38**   **39**   12 %

In another example treating anilines **40** with acetone under similar conditions gave the desired quinoline **41** as a single regioisomer.[20–21] It has been reported that the addition of silylating reagents, in particular bis(trimethylsilyl)acetamide, may be beneficial for these types of substrates.[22]

40   **41**

Interestingly the Skraup/Doebner–von Miller reaction has been used to prepare a number of spiro-compounds. Aniline was reacted with enone **42** in the presence of iodine to yield dihydroquinoline **43** in acceptable yields.[23]

Skraup/Doebner–von Miller-type reactions with lanthanide catalysts under microwave radiation are efficient for a variety of different anilines.[24] For example, cyclisation of aniline **44** with acetone in the presence of scandium triflate gave the desired product **45** in excellent yield.

9.17.6 Experimental[16]

5,6,8-Trifluoro-2-methyl-quinoline (31)

After anilide **30** (1.12 g, 4.46 mmol) is hydrolized in 6 M HCl at 100 °C (by TLC analysis), toluene (5 mL) is added and then aldehyde **5** (0.74 mL, 8.92 mmol) is added dropwise at the same temperature. The reaction was stirred for 2 h and then cooled to room temperature. The aqueous layer is removed and neutralized with aqueous NaOH to afford **31** as a crystalline solid. The crude product is purified by silica gel chromatography (hexanes:ethyl acetate, 5:1) to give **31** (802 mg, 70%) as colorless crystals, mp 103 °C.

9.17.7 References

1. Skraup, Z. H. *Ber.* **1880**, *13*, 2086.
2. For a review see: [R] Manske, R. H. F.; Kulka, M. *Org.React.* **1953**, *7*, 80.
3. Doebner, O.; von Miller, W. *Ber. Dtsch Chem. Ges.* **1883**, *16*, 2464.
4. Choi, H–Y.; Lee, B. S.; Chi, D–Y.; Kim, D–J. *Heterocycles*, **1998**, *48*, 2647.
5. Bischler, A. *Ber. Dtsch Chem. Ges.* **1892**, *25*, 2864.
6. Eisch, J. J.; Dluzniewski, J. *J. Org. Chem.* **1989**, *54*, 1269.
7. Cohn, E. W. *J. Am. Chem. Soc.* **1930**, *52*, 3685.
8. Utermohlen, W. P. *J. Org. Chem.* **1943**, *8*, 544. Yields given include small amounts of the 3-ethyl-5-methyl substituted quinoline.
9. Leir, C. M. *J. Org. Chem.* **1977**, *42*, 911.
10. Oleynik, I. I.; Shteingarts, V. D. *J. Fluorine Chem.* **1998**, *91*, 25.
11. Fujiwara, H.; Kitagawa, K. *Heterocycles* **2000**, *53*, 409.
12. O' Neill, P. M.; Tingle, M. D.; Mahmud, R.; Storr, R. C.; Ward, S. A.; Park, B. K. *Bioorg. Med. Chem. Lett.* **1995**, *5*, 2309
13. Boger, D. L.; Boyce, C. W. *J. Org. Chem.* **2000**, *65*, 4088.
14. Song, Z.; Mertzman, M.; Hughes, D. L. *J. Heterocyclic Chem.* **1993**, *30*, 17.
15. Sprecher, A–v.; Gerspacher, M.; Beck, A.; Kimmel, S.; Wiestner, H.; Anderson, G. P.; Niederhauser, U.; Subramanian, N.; Bray, M. A. *Bioorg. Med. Chem. Lett.* **1998**, *8*, 965.

16.   Matsugi, M.; Tabusa, F.; Minamikawa, J. *Tetrahedron Lett.* **2000**, *41*, 8523.

17.   Li, X.–G.; Cheng, X.; Zhou, Q.–L. *Synth. Comm.* **2002**, *32*, 2477.

18.   Ranu, B. C.; Hajra, A.; Dey, S. S.; Jana, U.; *Tetrahedron* **2003**, *59*, 813.

19.   Johnson, J. V.; Rauckman, B. S.; Baccanari, D. P.; Roth, B. *J. Med. Chem.* **1989**, *32*, 1942.

20.   Ku, Y–Y.; Grieme, T.; Raje, P.; Sharma, P.; Morton, H. E.; Rozema, M.; King, S. A. *J. Org. Chem.* **2003**, *68*, 3238.

21.   Edwards, J. P.; West, S. J.; Marschke, K. B.; Mais, D. E.; Gottardis, M. M.; Jones, T. K. *J. Med. Chem.* **1998**, *41*, 303.

22.   Bender, R. H. W.; Edwards, P. J.; Jones, T. K. US Pat. 6,093,825; Jul 25, **2000**.

23.   Walter, v–H.; Sauter, H.; Winker, T. *Helvetica Chimica Acta*, **1992**, *75*, 1274.

24.   Theoclitou, M–E.; Robinson, L. A. *Tetrahedron Lett.* **2002**, *43*, 3907.

Andrew Hudson

Chapter 10 Other Six–Membered Heterocycles 495

10.1. Algar–Flynn–Oyamada Reaction

10.1.1 Description

The conversion of 2´-hydroxychalcones to 2-aryl-3-hydroxy-4H-1benzopyran-4-ones (flavonols) by alkaline hydrogen peroxide oxidation is known as the Algar–Flynn–Oyamada (AFO) reaction or AFO oxidation.[1-3]

10.1.2 Historical Perspective

In 1934 the transformation of 2´-hydroxychalcones to flavonols in the presence of hydrogen peroxide and sodium hydroxide was reported simultaneously by Algar and Flynn in Ireland[1] and Oyamada in Japan.[2] However, many reports following the original disclosures showed that the Algar–Flynn–Oyamada reaction could lead to several products including aurones **4**, dihydroflavonols **5**, 2-benzyl-2-hydroxydihydrobenzofuran-3-ones **6**, and 2-arylbenzofuran-3-carboxylic acids **7**.[3a,3c]

Despite the formation of several products, the AFO reaction has remained a popular method for the synthesis of flavonols.

10.1.3 Mechanism

The AFO mechanism has been studied for several decades.[3-7] It was originally postulated by Geismann and Fukushima that chalcone **8** is converted to epoxide **9**.[4] Intramolecular attack by the phenoxide anion can then proceed via two routes. Nucleophilic addition at the β-position of the keto epoxide delivers an intermediate dihydroflavonol **12** that is oxidatively converted to flavonol **13**. On the other hand α-attack of the phenoxide affords aurone **10**. More recently, Schlenoff and coworkers have also shown the epoxide **9** is a necessary intermediate for the traditional AFO, but also for the conversion of nitrogen analogues of 3-hydroxyflavones to quinolones.[5] However, over the past several decades this mechanism for flavonol formation has been challenged most notably by Dean[6] and Burke.[7] Under this mechanism, the dihydroflavonol **12** is delivered via two possible pathways. First, an intramolecular 1, 4 conjugate addition of **11** gives enolate **15** which subsequently reacts with hydrogen peroxide to yield **12**. In the second pathway, cyclization and oxidation occur concurrently (**11** to **12**). Despite disagreement concerning the generation of flavonols, the two schools of thought agree on the original pathway for aurone formation.

It is also hypothesized that formation of 2-benzyl-2-hydroxydihydrobenzofuran-3-ones **6** and 2-arylbenzofuran-3-carboxylic acids **7** are derived from an intramolecular attack of the phenoxide at the β-position.[3c] Despite the complex mechanism and multiple products, general trends have emerged through experimental results.[8] If the chalcone lacks a 6′-methoxy group but has a hydroxyl group at the C2 or C4 positions, flavonols are favored. However, if the 6′-methoxy group is present and no hydroxyl substituent is present at C2 or C4 aurones and flavonols are formed. Others have also shown that pH[10,11] and temperature[3c] influence the product distribution.

10.1.4 Variations and Improvements
The AFO reaction has seen very few variations since it was first reported in 1934. However, the most significant modification was reported in 1958 by Ozawa[12] and further elaborated by Smith[13a] and others.[13b–d] Prior to this modification the intermediate chalcones were purified and then subjected to hydrogen peroxide in a basic medium. With the modification, the chalcone was generated *in situ,* from an aldehyde and a hydroxyacetophenone, and then allowed to react with aqueous hydrogen peroxide in the presence of sodium hydroxide to deliver the flavonol. Smith and coworkers conducted a limited study to examine the scope and limitations of this modification.[13a] Flavonols were delivered in 51–67%; however, no flavonols were isolated with highly reactive aldehydes such as *p*-nitrobenzaldehyde and when 2-hydroxy-4-methoxyacetophenone was used.

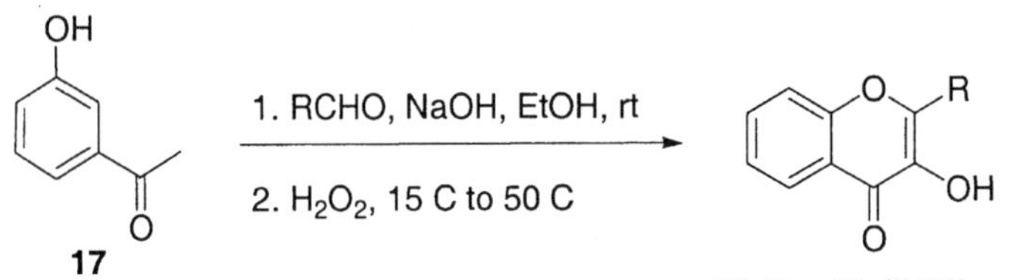

17

1. RCHO, NaOH, EtOH, rt

2. H_2O_2, 15 C to 50 C

18: R = Ph (54%)
19: R = 4-OMe-Ph (59%)
20: R = 3,4-diOMe-Ph (51%)
21: R = 4-Me-Ph (67%)
22: R = 4-Cl-Ph (52%)

As described earlier one of the possible products from the AFO reaction is dihydroxyflavonols. Simpson and coworkers took advantage of this outcome in their synthesis of the flavonol rhamnocitrin (**23**).[14] Chalcone **24** was subjected to the typical AFO conditions to deliver dihydroxyflavonol **25**. The isolated product was further subjected to hydrogen peroxide to afford flavonol **25a** in 30% yield. However, treatment of **25** with bismuth acetate, generated *in situ* from bismuth carbonate and acetic acid, gave **25a** in 77% yield for a respectable 52% overall yield over two steps. **25a** was then selectively demethylated with anilinium chloride to deliver rhamnocitrin (**23**).

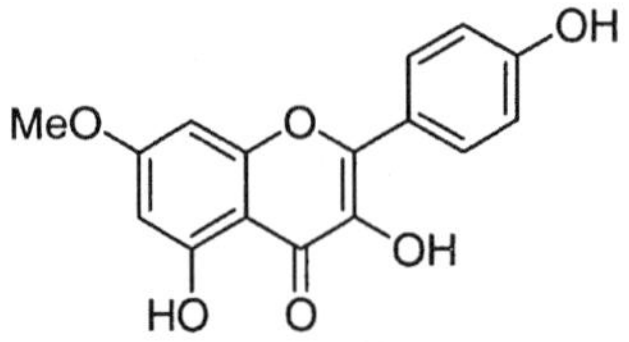

23 (rhamnocitrin)

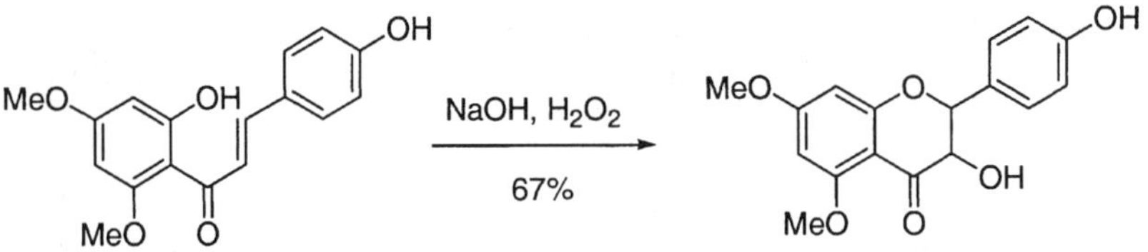

NaOH, H_2O_2

67%

24   **25**

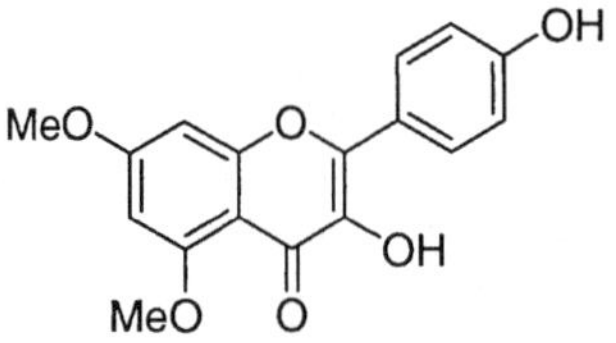

25a

Conditions	Yield of **25a**
H_2O_2, NaOH, 0 C	30%
BiCO₃, AcOH 2-ethoxyethanol, Δ	77%

A variant of the AFO reaction delivers nitrogen analogues of 3-hydroxyflavonol as described by Schlenoff and coworkers.[5] First methylamino acetophenone (**26**) is

condensed with benzaldehyde (**27**) to deliver the 2-methylamino chalcone **28** in 62%
yield. Oxidation of the aminochalcone with hydrogen peroxide in the presence of sodium
hydroxide afforded epoxide **29** in 50% yield. Upon heating **29** in refluxing ethanol,
intramolecular cyclization occurred at the β-position of epoxide **29** to give **30** which was
exposed to hydrogen peroxide and sodium hydroxide in methanol to deliver the 3-
hydroxydihydroquinolone **31** in 70% yield.

10.1.5 Synthetic Utility

Hundreds of flavonols have been isolated and characterized; many of them are
biologically active.[15] Hence a great synthetic interest has arisen. Some of the efforts have
concentrated on the synthesis of naturally occurring flavonols [15a, 16] while others have
focused on the synthesis of flavonol derivatives for structure activity relationships.[17]

Fukui and coworkers utilized an AFO reaction to synthesize and thus prove the
original structural assignment of the cytoxic agents eupatoretin (**32**) and eupatin (**33**).[16a]
Chalcone **34** was treated with 30% hydrogen peroxide in the presence of potassium
hydroxide to deliver flavonol **35** in 18% yield. Interestingly the reaction time was
extremely short- one minute. The authors never stated the reason for this abbreviated
reaction time, but this action is further evidence of the capricious nature of the AFO
reaction. Hydrogenolysis of **35** gave eupatoretin (**32**) which was elaborated to eupatin
(**33**) via selective demethylation of the C7 methoxy group.

32 (eupatoretin)   **33** (eupatin)

In the example below, Bhardwaj and coworkers synthesized tetramethoxyflavone **36**; this flavonol was believed to be the structure of a compound isolated from *Artemisia annua*.[16b] Methyl ketone **37** and aldehyde **38** were smoothly condensed to afford chalcone **39** in 73% yield. **39** was then converted to **40** under slightly modified AFO conditions in low yield. Selective demethylation of **40** gave **36**. However, spectral data and melting point data of **36** did not match up with the compound isolated from the plant. Hence, the original structure was misassigned and was not flavonol **36**.

Scriba and coworkers showed the utility of the AFO reaction by synthesizing a series of flavonols that exhibited anti-inflammatory activity.[17a] Two of the examples are depicted below. Aldehydes **41** were condensed with 2-hydroxyacetophenone (**42**) under anhydrous conditions to deliver intermediate chalcones which were immediately reacted with aqueous hydrogen peroxide to deliver **43** and **44** in 43% and 32% yields, respectively.

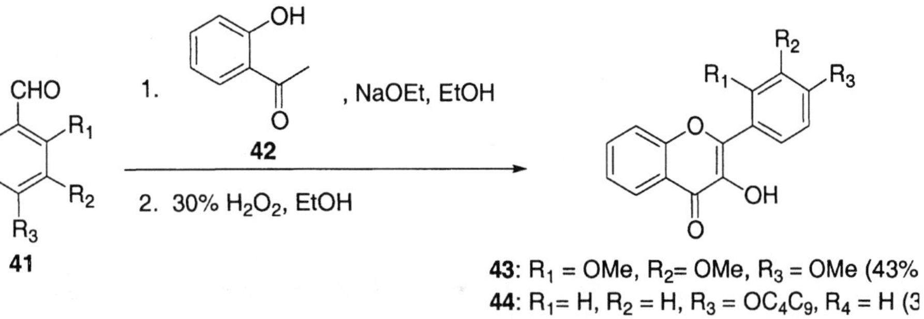

43: R_1 = OMe, R_2= OMe, R_3 = OMe (43%
44: R_1= H, R_2 = H, R_3 = OC_4C_9, R_4 = H (3

In addition, Pfister and coworkers investigated 3-hydroxyflavone-6-carboxylic acids as histamine induced gastric secretion inhibitors.[17b] After condensing 3-acetyl-4-hydroxybenzoic acid (**45**) with a variety of aldehydes **46** to deliver the chalcones **47**, these purified chalcones were then subjected to the standard AFO conditions to afford flavonols **48** in 51–80% yield. Subsequent alkylation of **48** with methyl iodide or isopropyl iodide followed by saponification of the corresponding esters gave the target compounds.

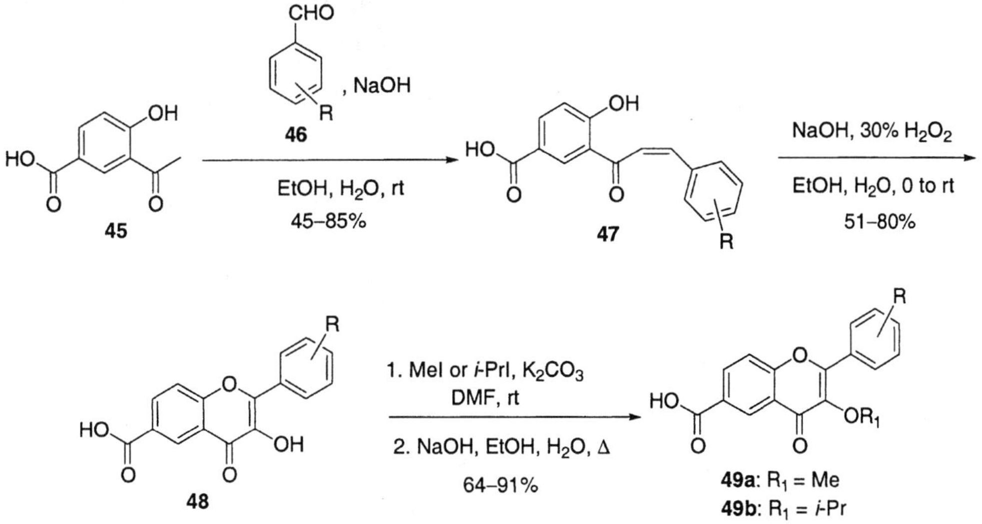

49a: R_1 = Me
49b: R_1 = i-Pr

10.1.6 Experimental

3-Hydroxy-2-(4-methoxyphenyl)-8, 8-dimethyl-8*H*-pyrano[2,3-*f*]chromen-4-one.[13c]
A vigorously stirred solution of 6-acetyl-5-hydroxy-2, 2-dimethylchromene (**50**, 0.800 g,
3.68 mmol) and *p*-anisaldehyde (**51**, 0.501 g, 3.68 mmol) in ethanol (5 mL) was treated
with aqueous sodium hydroxide (0.5 g) at room temperature to deliver a heavy precipitate.
The resulting mixture was then allowed to stand at room temperature overnight. Aqueous
sodium hydroxide (4.2 M, 7.5 mL) was then added to afford a solution. After cooling the
solution below 15°C, 30% H_2O_2 (0.42 mL, 3.68 mmol) was added quickly with stirring.
As the oxidation mixture stood, the temperature rose between 38–40°C. The mixture was
cooled after 30 min and solidified with dilute sulfuric acid. The solid was partially
dissolved by pouring the mixture into water and allowed to stand. The solid was filtered
and then recrystallized from methanol to deliver **52** (600 mg, 47%) as light yellow
needles: m.p. 137–138°C; [1] H NMR (CDCl₃) δ 1.36 (s, **6** H), 3.81 (s, 3 H), 5.51 (d, *J* = 9
Hz, 1 H), 6.31 (d, *J* = 9 Hz, 1 H), 6.70 (d, *J* =10.5 Hz, 1 H), 6.85 (d, *J* = 9 Hz, 2 H), 7.51
(d, *J* =9 Hz, 2 H), 7.62 (d, *J* = 10.5 Hz, 1 H), 13.86 (s, 1 H); Found: C = 71.7%, H = 5.4%;
calcd for $C_{21}H_{18}O_5$: C = 72.0%, H =5.1%.

10.1.7 References

1.   Algar, J.; Flynn, J. P. *Proc. Roy. Irish. Acad.* **1934**, *42B*, 1.
2.   (a) Oyamada, T. *J. Chem. Soc. Japan* **1934**, *55*, 1256. (b) Oyamada, T. *Bull. Chem. Soc. Jpn.* **1935**, *10*, 182.
3.   [R] (a) Wheeler, T.S. *Rec. of Chem. Prog.* **1957**, *18*, 133. (b) Donnelly, D. M. X.; Eades, J. F. K.; Philibin, E.
        M.; Wheeler, T. S. *Chem. Ind. (London)* **1961**, 1453. [R] (c) Cummins, B.; Donnelly, D. M. X; Eades, J. F.;
        Fletcher, H.; O'Cinnéide, F.; Philibin, E. M.; Swirski, J.; Wheeler, T.S.; Wilson, R. K. *Tetrahedron* **1963**, *19*,
        499.
4.   Geisman, T. A.; Fukushima, D. K. *J. Am. Chem. Soc.* **1948**, *70*, 1686.
5.   Gao, F.; Johnson, K. F.; Schlenoff, J . B. *J. Chem. Soc., Perkin Trans. 2* **1996**, 269.
6.   Dean F. M.; Podimuang, V. *J. Chem. Soc.* **1965**, 3978.
7.   (a) Brady, B. A., O'Sullivan, W. 1. *J. Chem. Soc., Chem. Comm.* **1970**, 1435. (b) Gormley, T. R.; O'Sullivan,
        W. I. *Tetrahedron* **1973**, *29*, 369. (c) Bennett, M.; Burke, A. J.; O'Sullivan, W. I. *Tetrahedron* **1996**, *52*, 7163.
8.   (a) Cummins, B.; Donnelly, E. M.; Philibin, E. M.; Swirski, J.; Wheeler, T. S.; Wilson, R. K. *Chem. Ind.
        (London)* **1960**, 393. (b) Donnelly, D. M. X.; Melody, D. P.; Philibin, E. M. *Tetrahedron Lett.* **1967**, 1023. (c)
        Cullen, W. P.; Donnelly, D. M. X.; Keenan, A. K.; Lavin, T. P.; Melody, D. P.; Philibin, E. M. *J. Chem. Soc.
        (C)* **1971**, 2848. (c) Cullen, W. P.; Donnelly, D. M. X.; Keenan, A. K.; Keenan, P. J. *J. Chem. Soc., Perkin
        Trans. 1* **1975**, 1671. (d) Hishmat, O. H.; El Ebrashi, N. M. A. *Indian J. Chem.* **1974**, *12*, 1052. (e) Gandhi, P.
        Indian J. Chem., Sec B **1976**, 14B, 1009. (f) related transformation, Rao, A. V. S.; Rao, N. V. S. *Curr. Sci.*
        1974, *43*, 477.
9.   Jain, A. C. Rohatgi, V. K.; Seshadri, T. R. *Curr. Sci.* **1966**, 35.
10.   Mulchandani, N . B.; Chadha, M .S. *Chem. Ind. (London)* **1964**, 1554.

11. Ferreira, D.; Brandt, E. V.; Volsteedt, F. du R.; Roux, D . G. *J. Chem. Soc., Perkin Trans. 1* **1975**, 1437.

12. Ozawa, H.; Okuda, T.; Matsumoto, S. *J. Pharm. Soc. Jpn.* **1958**, *71*, 1178.

13. (a) Smith, M. A.; Neumann, R. M.; Webb, R. A. *J. Heterocyl. Chem.* **1968**, *5*, 425. (b) Prasad, K. J. R.; Iyer, C. S. R.; Iyer, P. R. *Indian J. Chem., Sec. B* **1983**, *22B*, 693. (c) Jain, A. C.; Gupta, S. M.; Sharma, A. *Bull. Chem. Soc. Jpn.* **1983**, *56*, 1267. (d) Dharia, J, R.; Johnson, K. F.; Schlenoff, J. B. *Macromolecules* **1994**, *27*, 5167.

14. Guider, J. M.; Simpson, T. H.; Thomas, D. B. *J. Chem. Soc.* **1955**, 170.

15. (a) Wagner, H.; Farkas, L. In *The Flavonoids*; Harborne, J. B.; Mabry, T. J.; Mabry H., Eds.; Academic Press: New York, 1975; p 127. (b) Wollenweber, E. In *The Flavonoids: Advances in Research*; Harborne, J. B.; Mabry, T. J., Eds; Chapman and Hall: New York, 1982; p 189. (c) Wollenweber, E. In *The Flavonoids: Advances in Research since 1986*; Harborne, J. B., Ed.; Chapman and Hall: New York, 1994, p 259. (d) Bohm, B. A..; Stuessy, T. F. *Flavonoids of the Sunflower Family (Asteraceae)*; Springer-Verlag/Wiem: New York, 2001.

16. (a) Fukui, K.; Matsumoto, T.; Imai, S. *Bull. Chem. Soc. Jpn.* **1971**, *44*, 1698. (b) Bhardwaj, D. K.; Jain, S. C.; Sharma, G. C. *Indian J. Chem., Sec. B* **1977**, *15B*, 860. (c) Farkas, L.; Hörhammer, L.; Wagner, H. *Tetrahedron Lett.* **1963**, 727. (d) Kashikar, M . D.; Phatak, D. M.; Kulkarni, R. S.; Borkar, A. M.; Kulkarni, A. B. *Indian J. Chem.* **1964**, *2*, 485. (e) Farkas, L.; Hörhammer, L.; Wagner, H.; Rösler, H.; Gurniak, R. *Chem. Ber.* **1964**, *97*, 610. (f) Hörhammer, L.; Wagner, H.; Graf, E. *Tetrahedron Lett.* **1964**, 323. (g) Wagner, H.; Hörhammer, L.; Hitzler, G. *Tetrahedron Lett.* **1965**, 3849. (h) Murti, V. V. S.; Raman, P. V.; Seshadri, T. R. *Indian J. Chem.* **1966**, *4*, 396. (i) Raghunathan, K.; Rangaswami, S.; Seshadri, T. R. *Indian J. Chem.* **1974**, *12*, 1126. (j) Herz, W.; Anderson, G. D.; Wagner, Maurer, G.; Maurer, I.; Flores, G. *Tetrahedron* **1975**, *31*, 1577. (k) Wagner, H.; Maurer, I.; Farkas, L.; Strelisky, J. *Tetrahedron* **1977**, *33*, 1405.

17. (a) Sobottka, A. M.; Werner, W.; Blaschke, G.; Kiefer, W.; Nowe, U.; Dannhardt, G.; Schapoval, E. E. S.; Schenkel, E. P.; Scriba, G. K. E. *Arch. Pharm.* **2000**, *333*, 205. (b) Pfister, J. R.; Wymann, W. E.; Schuler, M . E.; Roszkowski, A. P. *J. Med. Chem.* **1980**, *23*, 335. (c) Raut, K.; Wender, S. H. *J. Org. Chem.* **1960**, *25*, 50. (d) Rao, a. V. S.; Rao, N. V. S. *Indian. J. Chem.* **1969**, *7*, 1091. (e) Rao, E. V. S. B; Rao, K. S. R. K. M.; Rao, N. V. S. *Curr. Sci.* **1973**, *42*, 498. (f) Wurm, G. *Arch. Pharm.* **1973**, *306*, 299. (g) Thakar, K. A.; Muley, P. R. *J. Indian Chem. Soc.* **1975**, *52*, 243. (h) Thakar, K. A.; Joshi, R. C. *J. Indian Chem. Soc.* **1980**, *57*, 1106. (i) Ankhiwala, M. D.; Naik, H. B. *J. Indian Chem. Soc.* **1989**, *66*, 482. (j) Bennett, C. J.; Caldwell, S. T.; McPhail, D. B.; Morrice, P. C.; Duthie, G. G.; Hartley, R. C. *Bioorg. Med. Chem.* **2004**, *12*, 2079.

Chris Limberakis

10.2   Beirut Reaction

10.2.1   Description

The Beirut reaction involves the condensation of benzofurazan oxide (BFO) **1** with an enamine **2** or an enolate anion **3** in an alcohol solvent to give the corresponding quinoxaline-1,4-dioxide **4**.[1–3]

10.2.2   Historical Perspective

In 1965, Haddadin and Issidorides, at the American University of Beirut, observed that combining **1** with morpholinocyclohexene **5** in methanol afforded quinoxaline-1,4-dioxide **6** in 48% yield.[4] Shortly thereafter, the same authors reported that **1** also reacts with 1,3-dicarbonyl compound **7** in the presence of triethylamine to give the quinoxaline-1,4-dioxide **8** in 38% yield.[5] This reaction has been referred to in the chemical literature as the Beirut reaction to acknowledge the city in which it was discovered.

10.2.3   Mechanism

Substituted benzofurazan oxides **9** and **11** have been studied by NMR at low temperature[6] and were observed as a mixture of tautomers, presumably interconverting via the *ortho*-dinitroso intermediate **10**. When R = Cl, MeO, or AcO, tautomer **9** is the more stable form and the stability is reversed when R = CO$_2$H or CO$_2$Et. Both tautomers are of equal stability when R is a methyl group.

         9   10   11

There is some debate in the literature as to the actual mechanism of the Beirut reaction. It is not clear which of the electrophilic nitrogens of BFO is the site of nucleophilic attack or if the reactive species is the dinitroso compound **10**. In the case of the unsubstituted benzofurazan oxide (R = H), the product is the same regardless of which nitrogen undergoes the initial condensation step. When R ≠ H, the nucleophilic addition step determines the structure of the product and, in fact, isomeric mixtures of quinoxaline-1,4-dioxides are often observed.[7,8] One report[9] suggests that N-3 of the more stable tautomer is the site of nucleophilic attack in accord with observed reaction products. However, a later study[10] concludes that the product distribution can be best rationalized by invoking the ortho-dinitrosobenzene form **10** as the reactive intermediate.

In the case of unsubstituted BFO **1** reacting with an enamine, the following mechanism is generally accepted in the literature. The first step is nucleophilic addition of an enamine **2** to electrophilic BFO **1** to form the intermediate **12**.[11] Ring closure occurs via condensation of the imino-oxide onto the iminium functionality to give **13**.[12] Finally, β-elimination of the dialkyl amine produces the quinoxaline-1,4-dioxide **4**.

         1   2   12

                  13   4

10.2.4 Variations and Improvements

Quinoxalinecarboxamide 1,4-dioxides were prepared in high yields from BFO and acetoacetamides by adding catalytic amounts of a calcium salt and ethanolamine to the condensation reaction.[13] Combining BFO **1** and acetoacetamide **14** in methanol in the presence of calcium chloride and ethanolamine afforded the quinoxalinecarboxamide **15** in excellent yield.

Quinoxaline 1,4-dioxides have also been prepared by condensation reactions carried out on the surface of solid catalysts such as silica gel,[14,15] molecular sieves,[15–17] or alumina.[14,17] As a representative example,[14] BFO **1** and the β-dicarbonyl compound **16** were combined with silica gel in methanol. The excess methanol was removed by evaporation and the silica gel with adsorbed reagents was allowed to stand for two weeks without drying. The quinoxaline 1,4-dioxide **17** was obtained in 90% yield after elution from a silica gel column.

10.2.5 Synthetic Utility

BFO reacted readily with 1,3-diketones to give 2,3-disubstituted quinoxaline 1,4-dioxides.[18] In the case of unsymmetrical 1,3-diketones, mixtures of isomeric quinoxaline dioxides were obtained, and the ratio of isomers was influenced by the steric bulk of the carbonyl substituent. When BFO **1** was combined with 1,3-diketone compounds **18** in the presence of triethylamine, the isomeric quinoxaline 1,4-dioxides **19** and **20** were obtained. When R = Me, **19** was the only product observed. As the steric bulk of R increased, increasing amounts of isomer **20** were observed. When R = *t*Bu, **20** was the only product detected in the reaction.

R	19 : 20
Me	100 : 0
Et	9 : 1
*i*Pr	1 : 2
*t*Bu	0 : 100

BFO **1** also reacted with 2-acetylbutyrolactone **21** to give the quinoxaline 1,4-dioxide **22** (n = 2) containing a primary hydroxyl group, which can be further functionalized.[19] The condensation of BFO and 4-hydroxybutanone gave the analogous (n = 1) quinoxaline 1,4-dioxide.[20]

BFO reacted with various phenolate anions to give phenazine 5,10-dioxides.[21,22] BFO **1** reacted with α-naphthol **23** in the presence of sodium methoxide to give a mixture of phenazine dioxides **24** and **25**, resulting from an initial *para* coupling and *ortho* coupling, respectively. When β-naphthol **26** was used as the condensation partner, phenazine dioxide **25** was the only product observed. The mechanism is thought to be analogous to that of the corresponding enamine condensation.

Heteroaromatic substituents can be incorporated onto the quinoxaline 1,4-dioxide ring system by condensing BFO with the appropriately substituted enamine,[23] cyanomethyl,[24] or 1,3-dicarbonyl[25] compound. 2-Cyanomethyl-1,3-benzothiazole[24] **27** reacted readily with BFO **1** in the presence of potassium carbonate to give the quinoxaline 1,4-dioxide **28** in good yield.

10.2.6 Experimental

2-Ethoxycarbonyl-3-methylquinoxaline 1,4-Dioxide (30):[26]

To a stirred mixture of BFO **1** (13.6 g, 0.1 mol) and ethyl acetoacetate **29** (13.0 g, 0.1 mol), cooled in an ice-water bath, morpholine (18.0 g, 0.2 mol) was added slowly dropwise. The ice-water bath was removed and the mixture was stirred for 10 hrs. The precipitate was collected by filtration, washed with cold ethanol, and recrystallized to give yellow crystals **30** (82%): mp 132–133°C (ethanol); IR (KBr, cm^{-1}) 1740 (C=O), 1600 (C=N), 1335 (N–O), 1290; H^1 NMR (DMSO-d$_6$) δ 1.40 (t, 3H), 2.50 (s, 3H), 4.55 (q, 2H), 7.80–8.10 (m, 2H), 8.20–8.50 (m, 2H).

10.2.7 References

1.   [R] Haddadin, M. J.; Issidorides, C. H. *Heterocycles* **1993**, *35*, 1503.
2.   [R] Haddadin, M. J.; Issidorides, C. H. *Heterocycles* **1976**, *4*, 767.
3.   [R] Ley, K.; Seng, F. *Synthesis* **1975**, 415.
4.   Haddadin, M. J.; Issidorides, C. H. *Tetrahedron Lett.* **1965**, 3253.
5.   Issidorides, C. H.; Haddadin, M. J. *J. Org. Chem.* **1966**, *31*, 4067.
6.   Boulton, A. J.; Katritzky, A. R.; Sewell, M. J.; Wallis, B. *J. Chem. Soc. B* **1967**, 914.
7.   Mufarrij, N. A.; Haddadin, M. H.; Issidorides, C. H. *J. Chem. Soc., Perkin Trans. 1* **1972**, 965.
8.   Haddadin, M. J.; Agopian, G.; Issidorides, C. H. *J. Org. Chem.* **1971**, *36*, 514.
9.   Mason, J. C.; Tennant, G. *Chem. Commun.* **1971**, 586.
10.   Abushanab, E.; Alteri, Jr., N. D. *J. Org. Chem.* **1975**, *40*, 157.
11.   Kluge, A. F.; Maddox, M. L.; Lewis, G. S. *J. Org. Chem.* **1980**, *45*, 1909.
12.   McFarland, J. W. *J. Org. Chem.* **1971**, *36*, 1842.
13.   Stumm, G.; Niclas, H. J. *J. Prakt. Chem.* **1989**, *331*, 736.
14.   Hasegawa, M.; Takabatake, T. *Synthesis* **1985**, 938.
15.   Takabatake, T.; Hasegawa, Y.; Hasegawa M. *J. Heterocycl. Chem.* **1993**, *30*, 1477.
16.   Takabatake, T.; Hasegawa, M. *J. Heterocycl. Chem.* **1987**, *24*, 529.
17.   Takabatake, T.; Miyazawa, T.; Kojo, M.; Hasegawa, M. *Heterocycles* **2000**, *53*, 2151.
18.   Haddadin, M. J.; Taha, M. U.; Jarrar, A. A.; Issidorides, C. H. *Tetrahedron* **1976**, *32*, 719.
19.   Usta, J. A.; Haddadin, M. J.; Issidorides, C. H.; Jarrar, A. A. *J. Heterocycl. Chem.* **1981**, *18*, 655.
20.   Edwards, M. L.; Bambury, R. E.; Ritter, H. W. *J. Med. Chem.* **1975**, *18*, 637.
21.   Ludwig, G. W.; Baumgärtel, H. *Chem. Ber.* **1982**, *115*, 2380.
22.   Abu El-Haj, M. J.; Dominy, B. W.; Johnston, J. D.; Haddadin, M. J.; Issidorides, C. H. *J. Org. Chem.* **1972**, *37*, 589.
23.   Tanaka, A.; Usui, T. *Chem. Pharm. Bull.* **1981**, *29*, 110.
24.   Borah, H. N.; Sandhu, J. S. *Heterocycles* **1986**, *24*, 979.
25.   Atfah, A.; Hill, J. *Tetrahedron* **1989**, *45*, 4557.
26.   Vega, A. M.; Gil, M. J.; Fernández-Alvarez, E. *J. Heterocycl. Chem.* **1984**, *21*, 1271.

Jennifer M. Tinsley

10.3 Biginelli Reaction

10.3.1 Description

The Biginelli reaction involves an one-pot reaction between aldehyde **1**, 1,3-dicarbonyl **2**, and urea **3a** or thiourea **3b** in the presence of an acidic catalyst to afford 3,4-dihydropyrimidin-2(1*H*)-one (DHPM) **4**.[1,2] This reaction is also referred to as the Biginelli condensation and Biginelli dihydropyrimidine synthesis. It belongs to a class of transformations called multi-component reactions (MCRs).

R = alkyl, aryl, het., R$_2$ = ester, amide, acyl, R$_3$ = alkyl, X = O, S

10.3.2 Historical Perspective

In 1893 Pietro Biginelli reported the first synthesis of 4-aryl-3,4-dihydropyrimidin-2(1*H*)-ones (DHPMs) via an one-pot process using three components.[1] Thus, DHPM **7** was synthesized by mixing benzaldehyde (**5**), ethyl acetoacetate (**6**), and urea (**3a**) in ethanol at reflux in the presence of a catalytic amount of HCl.

From the late 19th century through the mid-1970s few papers were published concerning this reaction. However, from the mid-1970s to the present, the utility of this reaction has grown rapidly, especially because some DHPMs possess significant therapeutic and pharmacological properties.[2a–c] Certain DHPMs have shown to act as calcium channel blockers, α_{1a}-adrenergic receptor antagonists, mitotic kinesin inhibitors, antihypertensive agents, neuropeptide Y (NPY) antagonists, antiviral, and antibacterial agents.[2c,3] In addition, several biologically active natural products contain these subunits, and thus a synthetic interest has grown.[2a–c] These synthetic interests have included both methodological improvements of the original Biginelli reaction conditions[2a,b] and total syntheses of natural products.[2b, 4]

10.3.3 Mechanism

Since the 1930s several mechanistic pathways have been proposed for the Biginelli reaction.[2a–b,5,6,7] In 1933, Folkers and Johnson reported that one of three intermediates **8–10,** was likely present in this reaction.[5] These included bisureide **8** which was formed by a condensation reaction between the aryl aldehyde and the urea followed by subsequent

attack of the resultant imine with another equivalent of urea. Also, 3-ureido ethyl acrylate
9 arose from a condensation reaction between the β-ketoester and urea. Finally, the
reaction of the β-ketoester and the aldehyde delivered the aldol adduct **10**.

Forty years after the initial proposal, Sweet and Fissekis proposed a more detailed
pathway involving a carbenium ion species.[6] According to these authors the first step
involved an aldol condensation between ethyl acetoacetate (**6**) and benzaldehyde (**5**) to
deliver the aldol adduct **11**. Subsequent dehydration of **11** furnished the key carbenium
ion **12** which was in equilibrium with enone **13**. Nucleophilic attack of **12** by urea then
delivered ureide **14**. Intramolecular cyclization produced a hemiaminal which underwent
dehydration to afford dihydropyrimidinone **15**. These authors demonstrated that the
carbenium species was viable through synthesis. After enone **13** was synthesized, it was
allowed to react with *N*-methyl urea to deliver the mono-*N*-methylated derivative of
DHPM **15**.

The mechanism was then reexamined 25 years later in 1997 by Kappe.[7] Kappe
used ¹H and ¹³C spectroscopy to support the argument that the key intermediate in the
Biginelli reaction was iminium species **16**. In the event, **5** reacted with **3a** to form an
intermediate "hemiaminal" **17** which subsequently dehydrated to deliver **16**. Iminium
cation **16** then reacted with **6** to give **14**, which underwent facile cyclodehydration to give
15. Kappe also noted that in the absence of **6**, bisureide **8** was afforded as a consequence
of nucleophilic attack of **16** by urea (**3a**). This discovery confirmed the conclusion of
Folkers and Johnson in 1933.[5] As far as the proposal from 25 years earlier by Sweet and
Fissekis,[6] Kappe saw no evidence by ¹H and ¹³C NMR spectroscopy that a carbenium ion
was a required species in the Biginelli reaction. When benzaldehyde (**5**) and ethyl

acetoacetate (**6**) were mixed under standard Biginelli conditions the requisite aldol product **11**, which was necessary for the formation of carbenium ion **12**, was not detected.

10.3.4 Variations and Improvements

Since the revised Biginelli mechanism was reported in 1997, numerous papers have appeared addressing improvements and variations of this reaction. The improvements include Lewis acid catalysis, protic acid catalysis, non-catalytic conditions, and heterogeneous catalysis. In addition, microwave irradiation (MWI) has been exploited to increase the reaction rates and yields.

The greatest number of reports has come from the Lewis acid field. The updated mechanism proposed by Kappe[7] suggests that the crucial intermediate in the mechanism is an acyliminium ion such as **16**. It is believed that a Lewis acid stabilizes this intermediate which results in higher yields of DHPMs.[7,8] These additives include $BF_3 \cdot OEt_2$,[9] $CeCl_3 \cdot 7H_2O$,[8] $FeCl_3 \cdot 6H_2O$,[10] $NiCl_2 \cdot 6H_2O$,[10] $InCl_3$,[11] $BiCl_3$,[12] $Bi(OTf)_3$,[13] $Yb(OTf)_3$,[14] PPE (polyphosphate ester),[2b,15] $SmCl_3 \cdot 6H_2O$,[16] $LiBr$,[17] $InBr_3$,[18] $LaCl_3 \cdot 7H_2O$,[19a,b] $CoCl_2 \cdot 6H_2O$,[19b] $La(OTf)_3$,[20] $Mn(OAc)_3 \cdot 2H_2O$,[21] $LiClO_4$,[22] and ionic liquids.[23] Typically Lewis acid catalyzed Biginelli reactions are run in ethanol, tetrahydrofuran, or acetonitrile at reflux to afford excellent yields of the DHPMs in typically 2–6 h. Traditional conditions require HCl in ethanol at reflux, and yields are moderate to poor.[24,25] A representative group of Lewis acids is listed below along with the traditional Biginelli conditions. The yields are greatly improved for aromatic aldehydes bearing an electron-donating or electron-withdrawing substituent. More impressive is the dramatic increase in yields for aliphatic aldehydes. In the case of $CeCl_3$-mediation, reactions may be run in ethanol or water without significant yield reduction when water is used.[8] Also, reactions may be run neat as the $Yb(OTf)_3$ examples indicate.[14]

R_1CHO + [diketoester R_2] + **3a** → [Lewis acid] → DHPM product (R_2O_2C, R_1, NH, O, methyl-substituted dihydropyrimidinone)

5: R_1 = Ph
18: R_1 = 4-OMe-Ph
19: R_1 = 2-NO$_2$-Ph
20: R_1 = 4-NO$_2$-Ph
21: R_1 = n-C$_3$H$_7$
22: R_1 = n-C$_4$H$_9$
23: R_1 = n-C$_6$H$_{13}$

6: R_2=Et
24: R_2= Me

15: R_1 = Ph, R_2 = Et
25: R_1 = Ph, R_2 = Me
26: R_1 = 4-OMe-Ph, R_2 = Me
27: R_1 = 4-OMe-Ph, R_2 = Et
28: R_1 = 2-NO$_2$-Ph, R_2 = Et
29: R_1 = 4-NO$_2$-Ph, R_2 = Et
30: R_1= n-C$_3$C$_7$, R_2 = Et
31: R_1 = n-C$_4$H$_9$, R_2 = Et
32: R_1 = n-C$_6$H$_{13}$, R_2 = Et

Products (% yield)

Conditions	15	25	26	27	28	29	30	31	32
CeCl$_3$·7H$_2$O, EtOH, Δ[8]	-	90	95	-	82	-	83	-	-
CeCl$_3$·7H$_2$O, H$_2$O, Δ[8]	-	88	90	-	78	-	76	-	-
FeCl$_3$·7H$_2$O, EtOH, Δ[10]	94	-	-	94	-	83	72	-	-
InCl$_3$, THF, Δ[11]	95	-	-	90	-	93	85	-	81
BiCl$_3$, CH$_3$CN, Δ[12]	95	-	-	90	-	90	72	-	50
Bi(OTf)$_3$, CH$_3$CN, rt[13]	90	-	-	95	-	85	-	-	58
Yb(OTf)$_3$, 100 °C[14]	98	-	-	96	-	94	-	87	-
LiBr, THF, Δ[17]	90	-	-	82	-	83	82	-	-
InBr$_3$, EtOH, Δ[18]	98	-	-	97	-	86	-	92	-
HCl, EtOH, Δ	78[24]	-	-	61[24]	-	58[24]	15[25]	-	8[24]

In the area of protic acids, several improvements have been reported with *p*-toluenesulfonic acid,[26] trifluoroacetic acid,[27] ammonium chloride,[28] and sulfamic acid.[29] In a related example using a Brönsted acid, Venkateswarlu and coworkers reported high yields of DHPMs from aromatic aldehydes by generating hydrogen bromide catalytically *in situ* from carbon tetrabromide in methanol at reflux.[30] The table below provides representative examples. In all cases, yields were significantly increased.

Products (% yield)

Conditions	15	27	29	30
p-TsOH, EtOH, Δ, 1–3 h[26]	91	90	86	85
NH$_4$Cl, 100 °C, 3 h[28]	90	84	83	78
NH$_2$SO$_3$H, EtOH, Δ, 1–3 h[29]	90	91	86	87
CBr$_4$, MeOH, Δ, 3 h[30]	90	92	80	-

Although acid catalysis is thought to be necessary for the Biginelli reaction, there has been a report disputing this requirement. Ranu and coworkers surveyed over 20 aldehydes and showed that excellent yields of DHPMs could be achieved at 100–105°C in 1 h in the absence of catalyst and solvent with no by-products formed.[31] In contrast Peng and Deng reported no significant formation of DHPM 15 when a mixture of benzaldehyde (5), ethyl acetoacetate (6), and urea (3a) was heated at 100°C for 30 min.[23]

6: R$_1$= Et   5: R$_2$=Ph   3a
24: R$_1$ = Me   18: R$_2$ = 4-OMe-Ph
                 20: R$_2$= 4-NO$_2$-Ph
                 21: R$_2$ = n-C$_3$C$_7$
                 23: R$_2$= n-C$_6$H$_{13}$

15: R$_1$=Et, R$_2$= Ph (81%)
27: R$_1$=Et, R$_2$ = 4-OMe-Ph (83%)
29: R$_1$= Et, R$_2$ = 4-NO$_2$-Ph (85%)
30: R$_1$= Et, R$_2$ = n-C$_3$C$_7$ (78%)
34: R$_1$= Me, R$_2$ = furyl (80%)

Some work has also been achieved with heterogeneous catalysis. These catalysts include Amberlyst-15,[23] Nafion-H,[23] montmorillonite KSF clay,[32] ferrihydrite silica gel aerogels containing 11–13% iron,[33] silica sulfuric acid,[34] and zeolites.[35]

Over the last several years research groups have also explored the use of microwaves to increase the reaction rate and efficiency of the Biginelli reaction.[2b] In one example, polyphosphate ester (PPE) was used as the promoter under microwave conditions to deliver a variety of DHPMs 38 in yields ranging from 65–95% yield with reaction times typically below 2 minutes.[36]

35   36   37   38

X = O,S

In addition to modification of the catalyst, several variants of the Biginelli reaction have emerged as viable alternatives; however, each method requires pre-formation of intermediates that are normally formed in the one-pot Biginelli reaction. First, Atwal and coworkers reported the reaction between aldol adducts **39** with urea **40a** or thiourea **40b** in the presence of sodium bicarbonate in dimethylformamide at 70°C to give 1,4-dihydropyrimidines **41**.[37] DHPM **42** was then produced by deprotection of **41**.

Furthermore, Shutalev and coworkers reported a two-step modification.[38] Urea **43a** or thiourea **43b** was condensed with **5** in the presence of *p*-toluenesulfonic acid to deliver α-tosylderivative **44**. The enolate of **6** was then allowed to react with **44** to give a substitution product which then cyclized to give the "hemiaminal" **45**. Dehydration of the hemiaminal with *p*-toluenesulfonic acid delivered **46**.

Moreover, Overman and Rabinowitz developed an intramolecular variant of the Biginelli reaction.[4a] This tethered Biginelli reaction has been important in the synthesis of a variety of natural products.[4a–e] This modification involved guanidine hemiaminal **47** reacting with 1,3-dicarbonyl **48** in the presence of a promoter such as morpholinium acetate or piperidinium acetate to deliver DHPM **49** with stereochemical control.[4a–e] For instance ureas (X = O) and *N*-arylsulfonylguanidines (X = NSO_2Ar) afforded the *cis*-stereochemistry, while the unprotected guanidine (X = NH_2^+) furnished the *trans*-geometry around the pyrrolidine ring.[4e]

X = O, NSO$_2$Ar, or NH$_2^+$X$^-$ **49**

Overman has extended his tethered Biginelli reaction to include alkenes and dienes instead of β-keto esters to deliver **51** diastereoselectively over **52** in the presence of Cu(OTf)$_2$.[4e]

10.3.5 Synthetic Utility

Since the early 1990s the Biginelli reaction has been utilized to deliver the DHPM core which was further elaborated to the target of interest.[2a–c] These reports are well documented in two reviews by Kappe in 2000.[2b,c] However, this section will address work primarily completed after these comprehensive reviews were published.

For example, Ghorab and coworkers exploited the classical Biginelli reaction to synthesize a variety of potentially active antifungal agents such as **56** from DHPM **55**.[39]

In addition, Namazi and coworkers expanded the DHPM core by constructing pyrrolo[3,4-*d*]pyrimidines via the classical approach.[40] First, DHPM **59** was delivered in 60% yield using the standard Biginelli conditions. **59** was then brominated in high yield to afford **60**. Substitution of bromide **60** with methylamine followed by cyclization of the intermediate amino ester furnished pyrrolo[3,4-*d*]pyrimidine **61** in 53% yield.

In recent years there has been some interest in monastrol (**62**), a potentially important chemotherapeutic for cancer which acts as an inhibitor of mitotic kinesin Eg5.[2c] Kappe and coworkers successfully synthesized racemic monastrol (**62**) using microwave-mediation in 60% yield from ethyl acetoacetate (**6**), 3-hydroxybenzaldehyde (**63**), and thiourea (**3b**) in the presence of PPE.[41]

However, Dondoni improved the synthesis by using Yb(OTf)$_3$ as the Lewis acid promoter in THF at reflux to deliver **62** in 95% yield.[42]

Dondoni has elaborated this methodology to include *C*-glycosylated dihydro-pyrimidines.[43] The sugar residue can be a subunit in the aldehyde, 1,3-dicarbonyl, or urea; consequently, substitution of the DHPM ring may occur in one of three places depending on which component originally contains the glycosidic residue. In the example below hydropyran carbaldehyde **64** was utilized to deliver **65** as the major product with moderate diastereoselection.

In the field of natural products, Overman and coworkers have published enantioselective syntheses of novel guanidine-containing alkaloids that are members of the crambescidin and batzelladine families.[4a–e] In each case a tethered Biginelli reaction, which was reported by Overman and Rabinowitz in 1993,[4a] was germane in the construction of the cores. For example, batzelladine F (**66**) presented an interesting challenge, for it contains two tricyclic guanidine subunits; hence, two stereoselective modified Biginelli reactions were required to complete the task.[4d] The first tethered reaction was between guanidine hemiaminal **67** and the 1,3-keto ester **68**. In the event, **67** was treated with **68** in the presence of morpholinium acetate to deliver **69** and its *anti*-diastereomer in 82% yield as a 5:1 diastereomeric mixture. The major diastereomer was then elaborated in several steps to 1,3-keto ester **70** which underwent a second modified Biginelli reaction with **71** to give the *anti*-diastereomer, pentacyclic bisguanidine **72**, as the major product in 59% yield. The *syn*-diastereomer was isolated albeit in less than or equal to 10% yield. Compound **72** was then transformed to **66** in several steps.

batzelladine F (**66**)

The Biginelli reaction has also been extended to solid phase and combinatorial synthesis.[2b,44]   In a recent combinatorial approach Kappe and coworkers used 4-chloroacetoacetate as a building block to create a library of diverse DHPMs under microwave conditions.[44e]   The DHPMs thus afforded were elaborated to three bicyclic systems: pyrimido[4,5,d]pyridazines 77, pyrrolo[3,4-d]pyrimidines 78, and furo[3,4-d]-pyrimidines 79.

10.3.6 Experimental

5-Ethoxycarbonyl-4-(4-methoxyphenyl)-6-methyl-3,4-dihydropyrimidin-2(1*H*)-one (26).[8]

A solution of methyl acetoacetate (24, 1.16 g, 10 mmol), 4-methoxybenzaldehyde (18, 1.36 g, 10 mmol), and urea (3a, 1.8 g, 30 mmol) in ethanol (5 mL) was heated under reflux in the presence of $CeCl_3 \cdot 7H_2O$ (931 mg, 25 mol %) for 2.5 h (monitored by TLC). The reaction mixture (after being cooled to room temperature) was poured onto crushed ice (30 g) and stirred for 5–10 min. The solid was filtered under suction (water aspirator), washed with ice-cold water (50 mL), and then recrystallized from hot ethanol to afford pure product 26 (2.62 g, 95%): m.p. 198–200°C; ^{1}H NMR ($CDCl_3$) δ 1.16 (t, J = 7.1 Hz, 3H), 2.24 (s, 3H), 3.75 (s, 3H), 3.98 (q, J = 7.1 Hz, 2H), 5.09 (d, J = 3.2 Hz, 1H), 6.78 (d, J = 8.7 Hz, 2H), 7.18 (d, J = 8.7 Hz, 2H), 7.24 (br s, 1H), 8.95 (br s, 1H).

10.3.7 References

1. Biginelli, P. *Gazz. Chim. Ital.* **1893**, *23*, 360.
2. [R] (a) Kappe, C. O. *Tetrahedron* **1993**, *49*, 6937. [R] (b) Kappe, C. O. *Acc. Chem. Res.* **2000**, *33*, 879. [R] (c) Kappe, C. O. *Eur. J. Med. Chem.* **2000**, *35*, 1043.
3. Yarim, M.; Saraç, S.; Kiliç, F. S.; Erol, K. *Farmaco* **2003**, *58*, 17.
4. (a) Overman, L. E.; Rabinowitz, M. H. *J. Org. Chem.* **1993**, *58*, 3235. (b) Coffey, D. S.; McDonald, A. I.; Overman, L. E.; Rabinowitz; Renhowe, P. A. *J. Am. Chem. Soc.* **2000**, *122*, 4893. (c) Coffey, D. S.; Overman, L. E.; Stappenbeck, F. *J. Am. Chem. Soc.* **2000**, *122*, 4904. (d) Cohen, F.; Overman, L. E. *J. Am. Chem. Soc.* **2001**, *123*, 10782. (e) Overman, L. E.; Wolfe, J. P. *J. Org. Chem.* **2001**, *66*, 3167. (f) Cohen, F.; Collins, S. K.; Overman, L. E. *Org. Lett.* **2003**, *5*, 4485. (g) Aron, Z. D.; Overman, L. E. *Chem. Commun.* **2004**, 253.
5. Folkers, K.; Johnson, T. B. *J. Am. Chem. Soc.* **1933**, *55*, 3784.
6. Sweet, F.; Fissekis, J. D. *J. Am. Chem. Soc.* **1973**, *95*, 8741.
7. Kappe, C. O. *J. Org. Chem.* **1997**, *62*, 7201.
8. Bose, D. S.; Fatima, L.; Mereyala, H. B. *J. Org. Chem.* **2003**, *68*, 587.
9. Hu, E. H.; Sidler, D. R.; Dolling, U.-H. *J. Org. Chem.* **1998**, *63*, 3454.
10. Lu, J.; Bai, Y. *Synthesis* **2002**, 466.
11. Ranu, B. C.; Hajra, A.; Jana, U. *J. Org. Chem.* **2000**, *65*, 6270.
12. Ramalinga, K.; Vijayalakshmi, P.; Kaimal, T. N. B. *Synlett* **2001**, 863.
13. Varala, R.; Alam, M. M.; Adapa, S. R. *Synlett* **2003**, 67.
14. Ma, Y.; Qian, C.; Wang, L.; Yang, M. *J. Org. Chem.* **2000**, *65*, 3864.
15. Kappe, C. O.; Falsone, S. F. *Synlett* **1998**, 718.
16. Fan, X.; Zhang, X.; Zhang, Y. *J. Chem. Res., Synop.* **2002**, 436.
17. (a) Baruah, P. P.; Gadhwal, S.; Prajapati, D.; Sandhu, J. S. *Chem. Lett.* **2002**, 1038. (b) Maiti, G.; Kundu, P.; Guin, C. *Tetrahedron Lett.* **2003**, *44*, 2757.
18. Fu, N.-Y.; Yuan, Y.-F. Cao, Z.; Wang, S.-W. Wang, J.-T.; Peppe, C. *Tetrahedron* **2002**, *58*, 4801.
19. (a) Lu, J.; Bai, Y.; Wang, Z.; Yang, B.; Ma, H. *Tetrahedron Lett.* **2000**, *41*, 9075. (b) Lu, J.; Bai, Y.-J.; Guo, Y.-H. Wang, Z.-J.; Ma, H.-R. *Chinese J. Chem.* **2002**, *20*, 681.

20.   Chen, R.-F.; Qiang, C.-T. *Chinese J. Chem.* **2002**, *20*, 427.
21.   Kumar, K. A.; Kasthuraiah, M.; Reddy, C. S.; Reddy, C. D. *Tetrahedron Lett.* **2001**, *42*, 7873.
22.   Yadav, J. S.; Reddy, B. V. S. Srinivas, R.; Venugopal, C.; Ramalingam, T. *Synthesis* **2001**, 1341.
23.   Peng, J.; Deng, Y. *Tetrahedron Lett.* **2001**, *42*, 5917.
24.   Folkers, K.; Harwood, H. J.; Johnson, T. B. *J. Am. Chem. Soc.* **1932**, *54*, 3751.
25.   Eynde, J. V.; Audiart, N.; Canonne, V.; Michel, S.; Haverbeke, Y. V.; Kappe, C. O. *Heterocycles* **1997**, *45*, 1967.
26.   Jin, T.; Zhang, S.; Li, T. *Synth. Comm.* **2002**, *32*, 1847.
27.   Bussolari, J. C.; McDonnell, P. A. *J. Org. Chem.* **2000**, *65*, 6777.
28.   Shaabani, A.; Bazgir, A.; Teimouri, F. *Tetrahedron Lett.* **2003**, *44*, 857.
29.   Jin, T.; Zhang, S.; Zhang, S.; Guo, J.; Li, T. *J. Chem. Res., Synop.* **2002**, 37.
30.   Reddy, A. V.; Reddy, V. L. N.; Ravinder, K.; Venkateswarlu, Y. *Heterocycl. Comm.* **2002**, *8*, 459.
31.   Ranu, B. C.; Hajra, A.; Dey, S. S. *Org. Process Res. Dev.* **2002**, *6*, 817.
32.   (a) Bigi, F.; Carloni, S.; Frullanti, B.; Maggi, R.; Sartori, G. *Tetrahedron Lett.* **1999**, *40*, 3465. (b) Lin, H.; Ding, J.; Chen, X.; Zhang, Z. *Molecules* **2000**, *5*, 1240.
33.   Martínez, S.; Meseguer, M.; Casas, L.; Rodríguez, E.; Molins, E.; Moreno-Mañas, M.; Roig, A.; Sebastián, R. M.; Vallribera, A. *Tetrahedron* **2003**, *59*, 1553.
34.   Salehi, P.; Dabiri, M.; Zolfigol, M. A.; Fard, M. A. B. *Tetrahedron Lett.* **2003**, *44*, 2889.
35.   Rani, V. R.; Srinivas, N.; Kishan, M. R.; Kulkarni, S. J.; Raghavan, K. V. *Green Chemistry* **2001**, *3*, 305.
36.   Kappe, C. O.; Kumar, D.; Varma, R. S. *Synthesis* **1999**, 1799.
37.   (a) O'Reilly, B. C.; Atwal, K. S. *Heterocycles* **1987**, *26*, 1185. (b) Atwal, K. S.; O'Reilly, B. C.; Gougoutass, J. Z.; Malley, M. F. *Heterocycles* **1987**, 1189. (c) Atwal, K. S.; Rovnyak, G. C.; O'Reilly, B. C.; Schwartz, J. *J. Org. Chem.* **1989**, *54*, 5898.
38.   Shutalev, A. D.; Kishko, E. A.; Sivova, N. V.; Kuznetsov, A. Y. *Molecules* **1998**, *3*, 100.
39.   Ghorab, M. M.; Abdel-Gawad, S. M.; El-Gaby, M. S. A. *Farmaco* **2000**, *55*, 249.
40.   Namazi, H.; Mirzaei, Y. R.; Azamat, H. *J. Heterocycl. Chem.* **2001**, *38*, 1051.
41.   Kappe, C. O.; Shishkin, O. V.; Uray, G.; Verdino, P. *Tetrahedron* **2000**, *56*, 1859.
42.   Dondoni, A.; Massi, A.; Sabbatini, S. *Tetrahedron Lett.* **2002**, *43*, 5913.
43.   (a) Dondoni, A.; Massi, A.; Sabbatini, S. *Tetrahedron Lett.* **2001**, *42*, 4495. (b) Dondoni, A.; Massi, A.; Sabbatini, S.; Bertolasi, V. *J. Org. Chem.* **2002**, *67*, 6979.
44.   (a) Kappe, C. O. *Bioorg. Med. Chem. Lett.* **2000**, *10*, 49. (b) Valverde, M. G.; Dallinger, D.; Kappe, C. O. *Synlett* **2001**, 741. (c) Dondoni, A.; Massi, A. *Tetrahedron Lett.* **2001**, *42*, 7975. (d) Stadler, A.; Kappe, C. O. *J. Comb. Chem.* **2001**, *3*, 624. (e) Pérez, R.; Beryozkina, T.; Zbruyev, O. I.; Hass, W.; Kappe, C. O. *J. Comb. Chem.* **2002**, *4*, 501. (f) Xia, M.; Wang, Y.-G. *Tetrahedron Lett.* **2002**, *43*, 7703. (g) Xia, M.; Wang, Y.-G. *Synthesis* **2003**, 262.

Chris Limberakis

10.4. Kostanecki–Robinson Reaction

10.4.1 Description

The conversion of *o*-hydroxyaryl ketones **1a** to chromones **2a** and/or coumarins **3a** with aliphatic acid anhydrides in the presence of the sodium or potassium salt of the corresponding acid[1] and the reaction between **1b** and aromatic acid anhydrides and the salt of the corresponding acid to form flavones **2b** (Allan–Robinson)[2] is called the Kostanecki–Robinson (K–R) reaction.[3]

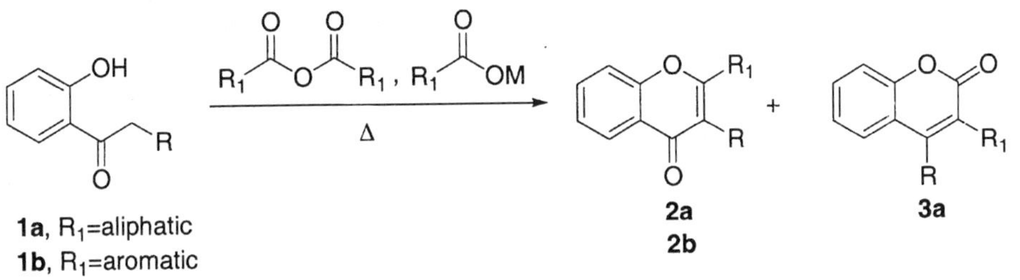

1a, R$_1$=aliphatic
1b, R$_1$=aromatic

10.4.2. Historical Perspective

In 1892 Nagai[4] and Tahara[5] independently reacted peonol with refluxing acetic anhydride in the presence of sodium acetate to form an unidentified heterocyclic compound. Nearly a decade later in 1901, Kostanecki and coworkers characterized this unidentified compound as a chromone when they successfully applied this methodology by converting resacetophenone and 4-ethoxy-2-hydroxyacetophenone to their corresponding chromones in refluxing acetic anhydride in the presence of sodium acetate (reactions 1 and 2).[6] This pioneering work was further elaborated in 1924 by Allan and Robinson.[7] They described the synthesis of a flavone by reacting ω-methoxyresacetophenone with benzoic anhydride in the presence of sodium benzoate (reaction 3). The latter transformation is also known as the Allan-Robinson reaction.[2]

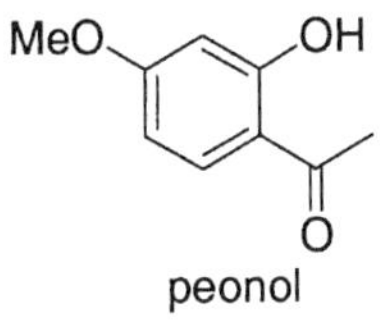

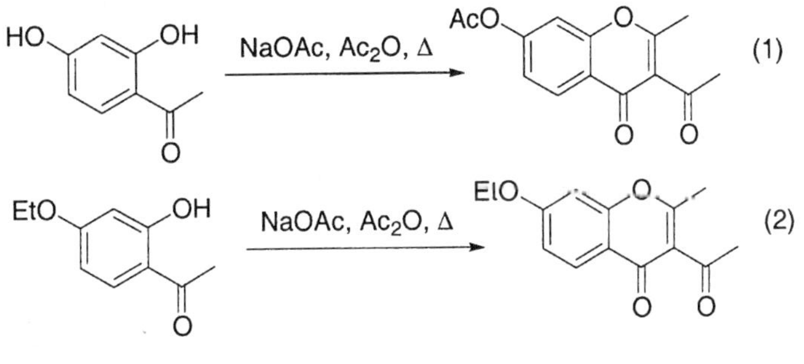

peonol

(1)

(2)

(3)

Although the literature refers to the formation of chromones/coumarins as the Kostanecki reaction (and often the Kostanecki–Robinson reaction) and the synthesis of flavones as the Allan-Robinson reaction, others have chosen to merge the two reactions and refer to both transformations as the Kostanecki–Robinson reaction.[3] This section will follow the latter school of thought, and use the Kostanecki–Robinson (K–R) nomenclature.

10.4.3 Mechanism
10.4.3.1. Chromones

The mechanism of the K–R reaction has been studied by several groups.[8-10] However, the mechanism proposed by Baker[9] and Széll[10] appears to be the most likely pathway. Széll and coworkers agreed with the original mechanism postulated by Baker based on product isolation, spectroscopy and kinetic studies. The o-hydroxyacetophenone **4** is first acylated to deliver ester **5**. Enolization of the phenone followed by acylation of the resulting enolate then affords enolacetate **6**.[11] **6** then undergoes ring closure followed by loss of water to deliver the γ-pyrone **8**. Széll suggested that the rate determining step of the K–R reaction is the ring closure and determined that sodium acetate was superior to triethylamine when the reaction is run at 180 °C.[10c] The converse is true when the reaction is run at temperatures below 100 °C.

10.4.3.2 Coumarins
Coumarin formation proceeds via an intramolecular attack by enol ester **9** on the ketone to give **10**.[3a] Dehydration of **10** then affords coumarin **11**. It has been observed that coumarins are favored when higher order homologs of acetic anhydride and their corresponding salts such as propionic anhydride/sodium propionate and butyric anhydride/sodium butyrate are used.

10.4.3.3 Flavones
Flavone formation is believed to proceed through a similar mechanism as the synthesis of chromones, albeit aromatic acid anhydrides and their corresponding salts are used.[10c] The first step is benzoylation of **12** to give the ester **14**. Enolization and *o*-alkylation then affords the enolbenzoate **15**. Enolbenzoate **15** then undergoes an acyl transfer to yield

dibenzoylmethane **16**. The formation of **16** is a likely intermediate, for others have isolated dibenzoylmethanes in K–R reactions.[10c,12] At this point the dibenzoylmethane may proceed through the same course as seen with chromones to give the flavone **18**. On the other hand, **16** may be converted to the 2-benzoyl 1,3-diketone **19**, and subsequent cyclization of **19** delivers the 3-benzoyl flavone **20**.

10.4.4. Variations and Modifications
10.4.4.1 Chromones

A modification of the K–R reaction was introduced by Mozingo.[13] This method involved reacting an *o*-hydroxyacetophenone with an ester in the presence of metallic sodium to form a 1,3-diketone. Treatment of the diketone with an acid then delivered the chromone via an intramolecular cyclization reaction. This method was applied to the preparation of 2-ethylchromone (**21**). *O*-hydroxyarylketone **22** was allowed to react with ethyl propionate (**23**) in the presence of sodium metal.[13a] The resulting sodium enolate was then quenched with acetic acid to deliver the 1,3-diketone **24**. Upon heating **24** in glacial acetic acid and hydrochloric acid, 2-ethylchromone (**21**) was delivered in 70–75% overall yield.

Typically, the K–R reaction is run at temperatures that often exceed 160 °C. However, a mild variation has been developed using acetic formic anhydride where the transformation occurs at ambient temperature.[1b,14] Okumara and coworkers smoothly converted hydroxyl ester **25** to chromone **26** in 76% yield with acetic formic anhydride and sodium formate at room temperature.[14a] Beckett and Ellis also used these conditions to synthesize two chromones in high yield when **27** and **28** were converted to **29** and **30**, respectively.[1b,14b]

27:R=NO$_2$
28:R=Ac

29: R=NO$_2$ (99%)
30: R=Ac (85%)

The use of acid chlorides instead of acid anhydrides has also been described. Wittig and coworkers converted propiophenone **31** to chromone **32** in 50% yield with chloroacetyl chloride in the presence of sodium chloroacetate at 190 °C.[1b,8] Despite the acid chloride's increased reactivity, a high temperature was still required.

31

50%

32

In addition to varying the electrophile, efforts have also focused on different bases.[10c,15] Yamaguchi and coworkers used a stoichiometric amount of DBU instead of sodium acetate when they synthesized a series of ethyl ω-(3-chromonyl)alkanoates **34** in 33–64% yield from **33**.[15c]

Ac$_2$O, DBU, Δ

33–64%

33
n=2,3,5,6

34
n=2,3,5,6

10.4.4.2 Flavones and Isoflavones

The synthesis of flavones has also seen modifications over the years. One of the primary modifications has been substituting the carboxylate salt for other bases.[16–19] Kohn and Löw showed that catalytic amounts of triethylamine allowed for the reaction to be run at 160 °C.[16] Looker and coworkers expanded on the Kohn and Löw modification by using amines as the solvent, and thus reduced the reaction temperatures.[17] They typically found that the reaction could be run at the refluxing temperatures of the amine. They showcased this modification by converting ω-methoxyphloroacetophenone (**35**) to the methyl ether of galangin (**36**) using a variety of amines in 60–75% yield with benzoic anhydride.

Base	b.p. of the amine	% yield
triethylamine	86	75
tripropylamine	156	60
N-ethylpiperidine	131	65

Seshradi and coworkers showed that potassium carbonate was an effective base in the K–R reaction for the conversion of resacetophenone (**37**) to 7-hydroxyflavone **39a** and 7-hydroxy-4'-methoxyflavone **39b** in 57% and 46% yield, respectively.[19] In addition, the reaction proceeded at a significantly lower temperature than the traditional K–R reaction by conducting the reaction in refluxing acetone.

39a: R = H (57%)
39b: R=OMe (46%)

Pivorarenko and Khilya investigated a series of organic bases including tribenzylamine, sodium methylate, sodium *tert*-butylate, *n*-methylmorpholine, trimethylamine, triethylamine, and tributylamine in the conversion of **40** to isoflavone **41** with acetic formic anhydride.[20] They found that the yields ranged from 88–95% when the latter four bases were used while the remaining bases gave disappointed yields ranging from 5–42%.

base	% yield
sodium methylate	<5
sodium *t*-butylate	31
tribenzylamine	42
N-methylmorpholine	88
trimethylamine	92
triethylamine	95
tributylamine	95

10.4.5 Synthetic Utility
10.4.5.1 Coumarins and chromones

The synthesis of chromones[1] and coumarins[21] has been reviewed over the years. Some trends have been established concerning the conditions that favor one class over the other. For instance, higher homologs of acetic anhydride such as propionic and butyric anhydrides tend to favor formation of coumarins while longer alkyl chains off of the aromatic ketone favor chromones such as o-hydroxypropiophenones over o-hydroxyaceto-phenones.[1,22] However, great care must be taken in characterizing the product, for some initial assignments have been incorrect.[1b] Hence, chemical derivatization and spectroscopic methods have emerged as prudent means for product identification.[23]

Over the years the literature is filled with examples where the initial characterization was incorrect.[1b] One example is illustrated below. In 1940, Sethna and Shah presumed that they synthesized coumarins **42** and **43** from a reaction between β-orcacetophenone (**44**) and its 4-O-methyl ether **45** under standard Kostanecki–Robinson conditions, respectively.[24] Three decades later Bose and Shah synthesized coumarin **43** via another route and concluded that the initial assignment made by Sethna and Shah was incorrect.[25] After the Bose and Shah findings were published, Ahluwalia and Kumar concluded that the Sethna and Shah products were actually chromones **46** and **47** based on proton NMR data and chemical derivatization.[26] Despite these shortcomings, the Kostanecki–Robinson reaction remains an effective method for formation of both coumarins [22, 27, 28] and chromones. [29–32]

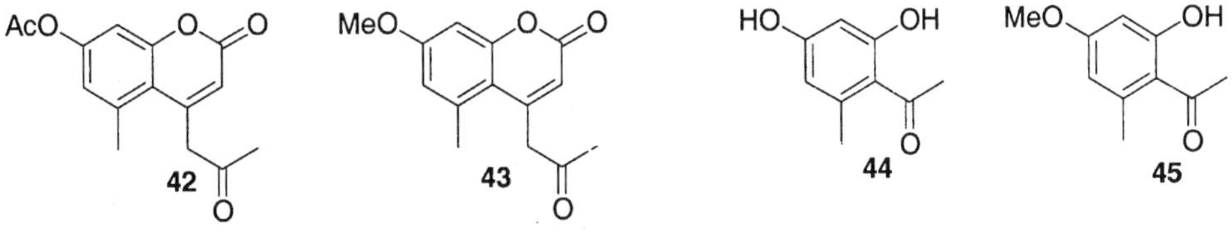

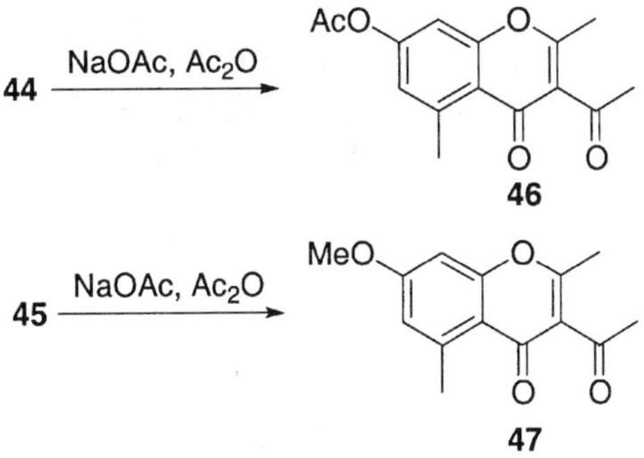

10.4.5.1.1 Coumarins

Sen and Kakaji synthesized a series of 4-butyrylnaphthocoumarins **48** from 1-butyryl-2-naphthols **49** using acetic anhydride and two homolog anhydrides in excellent yields.[27] They also showed that 1-propionyl-2-naphthols and 1-acctyl-2 naphthols could be converted to their corresponding coumarins using the same three anhydrides. However, 1-acetyl-2-naphthol in the presence of acetic anhydride and sodium acetate gave a chromone not a coumarin.

(RCO)$_2$O and RCOONa	% yield
R = CH$_3$	70
R = CH$_3$CH$_2$	88
R = CH$_3$CH$_2$CH$_2$	94

In addition to their work on naphthocoumarins, Sen and Kakaji showed that 4-*t*-butyl-2-hydroxyphenones **50** gave exclusively coumarins **51** when treated with various anhydrides in the presence of their corresponding sodium carboxylates.[22] They saw similar results with 4-*t*-amyl-2-hydroxyphenones.

R$_1$ = CH$_3$, CH$_2$CH$_3$, (CH$_2$)$_2$CH$_3$

R$_2$ = CH$_3$, CH$_2$CH$_3$, (CH$_2$)$_2$CH$_3$

10.4.5.1.2. Chromones

In the course of synthesizing DNA-gyrase inhibitors Högberg, Mitscher, and coworkers determined that an effective means of constructing the core of their inhibitors was via a K–R reaction.[29] Under mild conditions, keto ethylester **52** was acylated using acetic formic anhydride in the presence of sodium formate to deliver chromone **53** in 75% yield.

Gadaginamath and Kavali synthesized a series of novel 4H-pyrano[2,3-*f*]indole derivatives that exhibited varying degrees of antibacterial and antifungal activity.[30] 6-benzoylacetyl-5-hydroxyindoles **54** were treated with acetic anhydride in the presence of sodium acetate to deliver chromones **55** in modest yields.

54a: R$_1$ = H, R$_2$ = Me
54b: R$_1$ = Cl, R$_2$ = Me
54c: R$_1$ = NO$_2$, R$_2$ = Me
54d: R$_1$ = H, R$_2$ = Cl
54e: R$_1$= Cl, R$_2$ = Cl
54f: R$_1$ = NO$_2$, R$_2$ = Cl

In the realm of natural product synthesis, Kepler and Rehder utilized the K–R reaction to synthesize (±)-calanolide A (**56**), a potent non-nucleosidal human immunodeficiency virus (HIV-1) specific reverse transcriptase inhibitor.[31] Propiophenone **57** was allowed to react with acetic anhydride in the presence of sodium acetate to afford benzopyranone **58** in 56% yield; subsequent deacetylation of **58** gave **59**. Flavone **59** was then transformed to (±) calanolide A (**56**) over several steps.

(±) calanolide A (**56**)

10.4.5.2 Flavones and Isoflavones
10.4.5.2.1 Flavones

Over the years the venerable K–R reaction has been instrumental in constructing flavones for SAR programs[33,34] and natural product synthesis.[2, 18, 35–37]

In the area of medicinal chemistry, Haemers and coworkers synthesized a series of 4´-hydroxy-3-methoxyflavones that exhibited antiviral activity against poliomyelitis and rhinoviruses.[33]   A representative number of compounds is shown below. First, *O*-hydroxyacetophenones **61** were converted to the corresponding flavones **64** using standard conditions in yields of 74–92%. Cleavage of the benzyloxy groups of **64** was then achieved under acidic conditions to deliver the requisite flavones **65**.

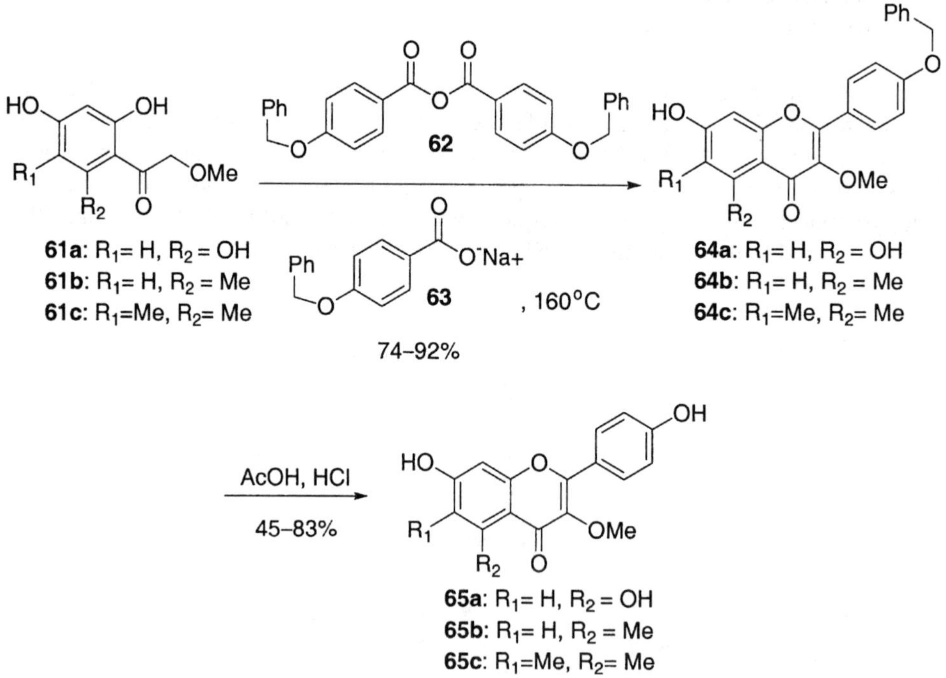

61a: R₁= H, R₂ = OH
61b: R₁= H, R₂ = Me
61c: R₁=Me, R₂= Me

62

63 , 160°C

74–92%

64a: R₁= H, R₂ = OH
64b: R₁= H, R₂ = Me
64c: R₁=Me, R₂= Me

AcOH, HCl
45–83%

65a: R₁= H, R₂ = OH
65b: R₁= H, R₂ = Me
65c: R₁=Me, R₂= Me

Horie and coworkers synthesized a series of flavones that showed promising inhibitory activity against archidonate 5-lipooxygenase.[34] This enzyme is responsible for the initiation of bioactive leukotrienes that are chemical mediators of anaphylaxis and inflammation. Under standard K–R conditions *o*-hydroxyarylketone **66** and anhydride **67** in presence of the corresponding anhydride **68** delivered flavones **69** in yields of 42–65%. Subsequent hydrogenation of **69** afforded the flavone inhibitors **70**.

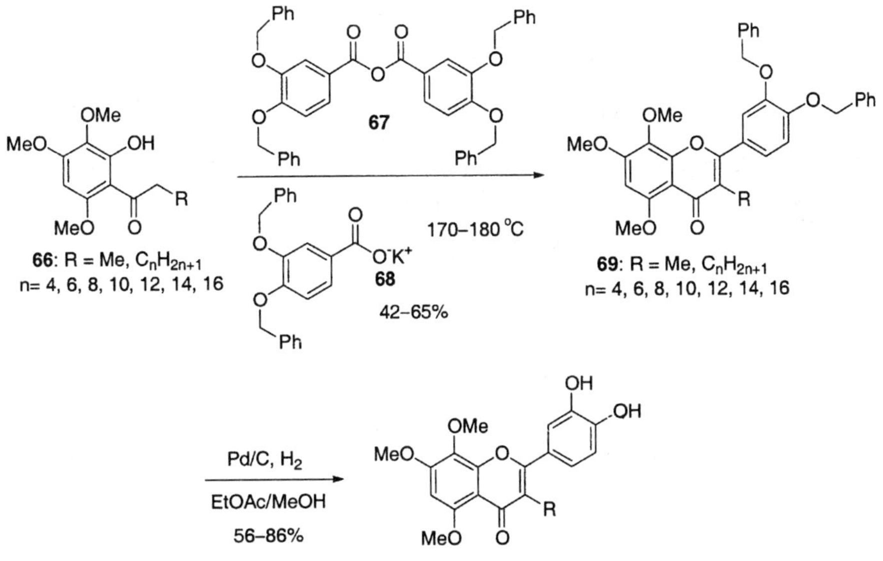

66: R = Me, C_nH_{2n+1}
n= 4, 6, 8, 10, 12, 14, 16

67

68

42–65%

170–180 °C

69: R = Me, C_nH_{2n+1}
n= 4, 6, 8, 10, 12, 14, 16

Pd/C, H_2

EtOAc/MeOH

56–86%

70: R = Me, C_nH_{2n+1}
n= 4, 6, 8, 10, 12, 14, 16

The K–R reaction has also been useful for structural confirmation of natural products such as tambulin (**71**), a flavonoid isolated from the seeds of *Xanthoxylum acanthopodium*.[35] In the critical reaction ω-ethoxyphloroacetophenone (**72**) was allowed to react with anisic anhydride (**38b**) in the presence of sodium anisate (**73**) at 170 °C to deliver flavone **74** in 65% yield. Flavone **74** was then converted after multiple steps to diethyl ether **75** which corresponded to the diethyl ether of tambulin (**71**).

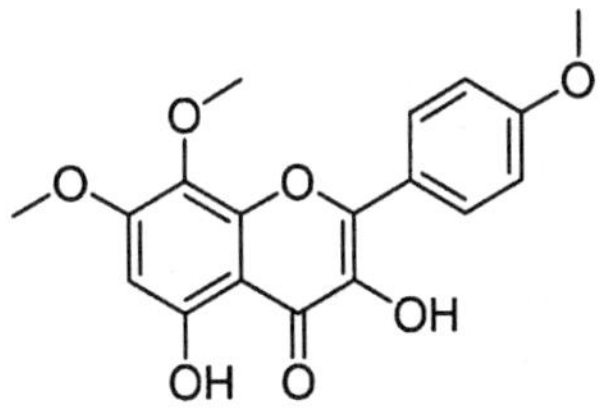

tambulin (**71**)

Lemière and coworkers synthesized the antipicornavirus agent 3-*O*-Methylquercetin (**76**).[36] A key transformation was the conversion of acetophenone **61a** to 3-methoxyflavone **79**. In the event, **61a** and 3,4-dibenzyloxybenzoic anhydride (**77**) were allowed to react at 160 °C in the presence of sodium carboxylate **78** to deliver the penultimate intermediate in 78% yield. Debenzylation of **79** in the presence of Pearlman's catalyst delivered the natural product in 99% yield.

3-*O*-Methylquercetin (**76**)

10.4.5.2.2 Isoflavones

Over the years there have many reports of isoflavone syntheses utilizing the K–R reaction.[38,39] One example was reported by Liu and Cheng where a crucial isoflavone intermediate was required to synthesize a series of antineoplastic agents.[38] In the event acetophenone

80a was allowed to react with acetic formic anhydride to deliver isoflavone **81a** in 70% yield. They achieved similar results with the conversion of **80b** to **81b**.

80a: R = OH
80b: R = H

81a: R = OH (70%)
81b: R = H (76%)

10.4.6 Experimental

NaOAc, Ac$_2$O, reflux

62 %

82   **83**

6,6,10,11-Tetramethyl-4-propyl-2*H*,6*H*,12*H*-benzo[1,2-*b*:3,4-*b*':5,6-*b*'']tripyran-2,12-dione (83).[40]

A mixture of the chromene **82** (1.76 g, 5.11 mmol) and sodium acetate (419 mg, 5.11 mmol) in acetic anhydride (12 mL) was refluxed for 4 h, and then the solvent was removed *in vacuo*. The residue was purified by chromatography on a silica gel column eluting first with 25% ethyl acetate/hexane followed by 50% ethyl acetate/hexane to provide chromone **83** (1.16 g, 62%) as a yellow solid: m.p. 209–209.5 °C (recrystallized from ethyl acetate); ^{1}H NMR (DMSO-d$_6$) d 1.00 (t, *J* = 7.2 Hz, 3 H), 1.51 (s, 6 H), 1.60 (apparent sextet, *J* = 7.6 Hz, 2 H), 1.88 (s, 3 H), 2.39 (s, 3 H), 2.90 (t, *J* = 7.7 Hz, 2 H), 5.89 (d, *J* = 10.0 Hz, 1 H), 6.26 (s, 1 h), 6.75 (d, *J* =10.2 Hz, 1 H); Found C = 72.14 %, H = 6.15%; calcd for:C$_{22}$H$_{22}$O$_5$: C = 72.12%, H = 6.05%.

10.4.7 References

1. [R] (a) Hauser, C. R.; Swamer, F. W.; Adams, J. T. *Org. React.* **1954**, *8*, 59. [R] (b) Ellis, G. P., *Chromenes, Chromanones, and Chromones from The Chemistry of Hetereocylic Compounds*, Weissberger, A. and Taylor, E. C., eds John Wiley & Sons, 1977, vol. 31, New York, p.495. Note: The author in the former reference refers to the formation of chromones, coumarins, and flavones as the Kostanecki acylation while the latter author calls the formation of chromones and coumarins the Kostanecki-Robinson reaction.

2. [R] (a) Wagner, H.; Farkas, L. In *The Flavonoids*; Harborne, J. B.; Mabry, T. J.; Mabry H., Eds.; Academic Press: New York, 1975; p 127. [R] (b) Staunton, J. In *Comprehensive Organic Chemistry The Synthesis and Reactions of Organic Compounds*; Sammes, P. G.; Pergamon Press: New York, 1979, vol.4; p. 659.

3. (a) Coffey, S., ed In *Rodd's Chemistry of Carbon Compounds*; Elsevier Scientific Publishing Company: New York, 1977, vol. IV (E); p.140. [R] (b) Hepworth, J. D.; Gabbutt, C. D.; Heron, B. M. In *Comprehensive Heterocyclic Chemistry A Review of the Literature 1982-1995*; McKillop, A. ed.; Elsevier Science Ltd.. Oxford, 1996, vol. 5; p. 351.

4. Nagai, W. N. *Ber.* **1892**, *25*, 284.

5. (a) Tahara, Y. *Ber.* **1892**, *25*, 1292. (b) *Ibid*, 1306.

6.   Kostanecki, S.v.; Różycki, A. *Chem. Ber.* **1901**, *34*, 102.

7.   Allan, J.; Robinson, R. *J. Chem. Soc.* **1924**, *125*, 2192.

8.   Wittig, G.; Bangert, F.; Richter, H. E. *Liebigs. Ann.* **1925**, *446*, 155.

9.   Baker, W. *J. Chem. Soc.* **1933**, 1381.

10.   (a) Széll, T. *J. Chem. Soc. C: Organic* **1967**, 2041. (b) Széll, T.; Kovács, K.; Zarándy, M. S.; Erdőhelyi, Á. *Helv. Chim. Acta* **1969**, *52*, 2636. (c) Széll, T.; Dózsai, L.; Zarándy, M.; Menyhárth *Tetrahedron* **1969**, *25*, 715.

11.   (a) Buggy, T.; Ellis, G. P. *J. Chem. Res. Synop.* **1980**, 317. (b) Ibid, *J. Chem. Res. Miniprint* **1980**, 3875.

12.   (a) Doporto, M. L.; Gallagher, K. M.; Gowan, J. E.; Hughes, A. C.; Philibin, E. M.; Swain, T.; Wheeler, T. S. *J. Chem. Soc.* **1955**, 4249. (b) Jermanowska, Z. I.; Michalska, M. *J. Chem. Ind.* **1958**, 132. (c) Da Re, P.; Cimatoribus, L. *Experientia* **1962**, *18*, 67.

13.   (a) Mozingo, R. *Org. Synth., Coll. Vol. III* **1955**, 387. (b) Banerji, K. D.; Poddar, D. *J. Indian Chem. Soc.* **1976**, *53*, 1119.

14.   (a) Okumura, K.; Kondo, K.; Oine, T.; Inoue, I. *Chem. Pharm. Bull.* **1974**, *22*, 331. (b) Becket, G. J. P.; Ellis, G. P. *Tetrahedron Lett.* **1976**, 719.

15.   (a) Huebner, C. F.; Link, K. P. *J. Am. Chem. Soc.* **1945**, *67*, 99. (b) Huffman, K. R.; Loy, M.; Ullman, E. F. *J. Am. Chem. Soc.* **1965**, *87*, 5417. (c) Yamaguchi, S.; Mutoh, M.; Shimakura, M.; Tsuzuki, K.; Kawase, Y. *J. Heterocycl. Chem.* **1991**, *28*, 119.

16.   Kuhn, R.; Löw, I. *Chem. Ber.* **1944**, *77*, 202.

17.   (a) Looker, J. H.; McMechan, J. H.; Mader, J. W. *J. Org. Chem.* **1978**, *43*, 2344.

18.   (a) Fukui, K.; Matsumoto, T.; Nakamura, S.; Nakayama, M.; Horie, T. *Bull. Chem. Soc. Jpn.* **1968**, *41*, 1413. (b) Wagner, h.; Maurer, G.; Farkas, l. Hänsel, R.; Ohlendorf, D. *Chem. Ber.* **1971**, *104*, 2381.

19.   Grover, S. K.; Jain, A. C.; Mathur, S. K.; Seshardi, T. R. *Indian J. Chem.* **1963**, *1*, 382.

20.   Pivovarenko, V. G.; Khilya, V. P. *Chem. Heterocycl. Compd.* **1992**, *28*, 497.

21.   [R] Sethna, S. M.; Shah, N. *Chem. Rev.* **1945**, 1.

22.   Sen, A. B.; Kakaji, T. N. *J. Indian Chem. Soc.* **1952**, *29*, 127.

23.   [R] Ellis, G. P., *Chromenes, Chromanones, and Chromones from The Chemistry of Hetereocyclic Compounds*, Weissberger, A. and Taylor, E. C., eds John Wiley & Sons, 1977, vol. 31, New York, p.481.

24.   (a) Sethna, S. M.; Shah, R. C. *J. Indian Chem. Soc.* **1940**, *17*, 239. (b) *Ibid.*, 487.

25.   Bose, J. L.; Shah. R. C. *Indian J. Chem.* **1973**, *11*, 729.

26.   (a) Ahluwalia, V. K.; Kumar, D. *Indian J. Chem.* **1975**, *13*, 981. (b) Ibid. **1976**, *14B*, 326.

27.   Sen, A. B.; Kakaji, T. N. *J. Indian Chem. Soc.* **1952**, *29*, 950.

28.   (a) Shah, D. N.; Shah, N. M. *J. Am. Chem. Soc.* **1955**, *77*, 1699. (b) Thanawalla, C. B.; Trivedi, P. L. *J. Indian. Chem. Soc.* **1959**, *36*, 49. (c) Pardanani, N. H.; Trivedi, K. N. *J. Indian Chem. Soc.* **1972**, *49*, 599. (d) Kuriakose, A. P.; Sethna, S. *J. Indian Chem. Soc.* **1972**, *49*, 1155. (e) Pardanani, J. H.; Sethna, S. *J. Indian Chem. Soc.* **1978**, *55*, 806.

29.   Högberg, T.; Vora, M.; Drake, S. D.; Mitscher, L. A.; Chu, D. T. W. *Acta Chem. Scan. B* **1984**, *38*, 359.

30.   Gadaginamath, G. S.; Kavali, R. R. *Indian J. Chem.* **1999**, *38B*, 178.

31.   Rehder, K. S.; Kepler, J. A. *Synth. Comm.* **1996**, *26*, 4005.

32.   (a) Rao, C. B.; Subramanyam, G.; Venkateswarlu, V. *J. Org. Chem.* **1959**, *24*, 683. (b) Mustafa, A.; Hsihmat, O. H.; Zayed, S. M. A. D.; Nawar, A. A. *Tetrahedron* **1963**, *19*, 1831. (c) Save, S. R.; Trivedi, P. L. *J. Indian Chem. Soc.* **1971**, *48*, 675. (d) Chakravarti, D.; Saha, M.; Das, R. *J. Indian Chem. Soc.* **1971**, *48*, 765. (e) Save, S. R.; Trivedi, P. L. *J. Indian Chem. Soc.* **1972**, *49*, 25. (f) Shaikh, Y. A.; Trivedi, K. N. *J. Indian Chem. Soc.* **1972**, *49*, 715. (g) Thomas, T. C.; Sethna, S. *J. Indian Chem. Soc.* **1973**, *50*, 326. (h) Lakshmi, M. V.; Rao, N. V. S. *Curr. Sci.* **1973**, *42*, 19. (i) Caputo, O.; Cattel, L.; Viola, F.; Biglino, G. *Gazz. Chim. Ital.* **1979**, *109*, 339. (j) Iyer, P. R.; Yer, C. S. R.; Prasad, K. J. R. *Indian. J. Chem.* **1983**, *22B*, 1055. (k) Zammatio, F.; Brion, J. D.; Belachmi, L.; Le Baut, G. *J. Heterocycl. Chem.* **1991**, *28*, 2013. (l) Saraf, B. D.; Wadodkar, K. N. *J. Indian Chem. Soc.* **1993**, *70*, 643. (m) Rao, Y. J.; Krupadanam, G. L. D. *Bull. Chem. Soc. Jpn.* **1994**, *67*, 1972. (n) Lacova, M.; El-Shaaer, H. M.; Loos, D.; Mutulova, M.; Chovancova, J.; Furdik, M. *Molecules* **1998**, *3*, 120. (o) Rossollin, V.; Lokshin, V.; Samat, A.; Gugliemetti, R. *Tetrahedron* **2003**, *59*, 7725.

33.   DeMeyer, N.; Haemers, A.; Mishra, L.; Pandey, H-K.; Pieters, L. A. C.; Vanden Berghe, D. A.; Vlietinick, A. J. *J. Med. Chem.* **1991**, *34*, 736.

34.   Horie, T.; Tominaga, H.; Kawamura, Y.; Hada, T.; Ueda, N.; Amano, Y,; Yamamoto, S. *Ibid.*, 2169.

35.   Ahuja, M.; Bandopadhyay, M.; Seshadri, T. R. *Indian J. Chem.* **1975**, *13*, 1134.

36.   Boers, F.; Deng, B-L; Lemière, G.; Lepoivre, J.; De Groot, A.; Dommisse, R.; Vlietnick, A. J. *Arch. Pharm. Med. Chem.* **1997**, *330*, 313.

37.   (a) Baker, W.; Nodzu, R.; Robinson, R. *J. Chem. Soc.* **1929**, 74. (b) Baker, W.; Robinson, R. *Ibid.*, 152. (c) Shah, R. C.; Mehta, C. R.; Wheeler, T. S. *J. Chem. Soc.* **1938**, 1555. (d) Fukui, K.; Matsumoto, T.; Matsuzaki, S. *Bull. Chem. Soc. Jpn.* **1964**, *37*, 265. (e) Krishnamurti, M.; Seshadri, T. R.; Shankaran, P. R. *Curr. Sci.* **1965**, *34*, 559. (f) Krishnamurti, M.; Seshadri, T. R.; Shankaran, P. R. *Indian. J. Chem.* **1967**, *5*, 137. (g) Farkas, L.; Nógrádi, M.; Strelisky, J. *Tetrahedron Lett.* **1965**, 4563. (h) Farkas, L.; Nogradi, M.;

Strelisky, J. *Chem. Ber.* **1966**, *99*, 3218. (i) Bahl, C. P.; Parthasarathy, M. R.; Seshardi, T. R. *Curr. Sci.* **1966**, *35*, 281. (j) Ahluwalia, V. K.; Sachdev, G. P.; Seshadri, T. R. *Indian J. Chem.* **1966**, *4*, 456. (k) Sim, K. Y. *J. Chem. Soc. (C)* **1967**, 976. (l) Berti, G. Livi, O.; Segnini, D.; Cavero, I. *Tetrahedron* **1967**, *23*, 2295. (m) Hänsel, R.; Rimpler, H.; Schwarz, R. *Tetrahedron Lett.* **1967**, 735. (n) Fukui, K.; Matsumoto, T.; Nakamura, S.; Nakayama, M. *Bull. Chem. Soc. Jpn.* **1968**, *41*, 1413. (o) Fukui, K.; Matsumoto, T.; Nakayama, M.; Horie, T. *Experientia* **1969**, *25*, 349. (p) Fukui, K.; Nakayama, M. *Bull. Chem. Soc. Jpn.* **1969**, *42*, 1649. (q) Wagner, H.; Maurer, G.; Farkas, L.; Hänsel, R.; Ohlendorf, D. *Chem. Ber.* **1971**, *104*, 2381. (r) Krishnamurti, M.; Seshaddri, T. R.; Sharma. N. D. *Indian J. Chem.* **1973**, *11*, 201. (s) Kasim, S. M.; Neelakantan, S.; Raman, P. V. *Curr. Sci.* **1974**, *43*, 476. (t) Malik, S. B.; Sharma, P.; Seshaadri, T. R. *Indian J. Chem.* **1977**, *15B*, 536. (u) Patwardhan, S. A.; Gupta, A.S. *J. Chem. Res. (M)* **1984**, 3791.

38. Liu, D. F.; Cheng, C. C. *J. Heterocycl. Chem.* **1991**, *28*, 1641.

39. (a) Baker, W.; Robinson, R. *J. Chem. Soc.* **1929**, 152. (b) Baker, W.; Robinson, R. *Ibid.*, 2713. (c) Shriner, R. L.; Stephenson, R. W. *J. Am. Chem. Soc.* **1942**, *64*, 2737. (d) Mehta, A. C.; Seshardi, T. R. *J. Chem. Soc.* **1954**, 3823. (e) Kawase, Y.; Fujino, Y.; Ichioka, Y.; Fukui, K. *Bull. Chem. Soc. Jpn.* **1957**, *30*, 689.

40. Flavin, M. T.; Rizzo, J. D.; Khilevich, A.; Kucherenko, A.; Sheinkman, A. K.; Vilaychack, V.; Lin, L.; Chen, W.; Greenwood Mata, e.; Pengsuparp, T.; Pezzuto, J. M.; Hughes, S. H.; Flavin, T. M.; Cibulski, M; Boulanger, W. A.; Shone, R. L.; Xu, Z-Q. *J. Med. Chem.* **1996**, *39*, 1303.

Chris Limberakis

10.5 Pinner Pyrimidine Synthesis

10.5.1 Description

The condensation of 1,3-dicarbonyl compounds **1** with amidines **2** catalyzed by acids or bases to give pyrimidine derivatives **3** is regarded as the Pinner pyrimidine synthesis.[1,2]

10.5.2 Historical Perspective

In the 1880s, Pinner found that the amidine derivative **2** reacted with acetoacetic ester (**4**) to furnish 2-substituted-6-hydroxy-4-methylpyrimidine **5**. The condensation of amidine derivative **2** with other β-keto esters, malonic esters, and β-diketones proceeded similarly (see the following pages for examples).[3-5]

10.5.3 Mechanism

Although the Pinner pyrimidine synthesis was discovered a century ago only a few reports on the reaction mechanism have appeared.[6,7] The condensation of acetylacetone, methyl acetoacetate, or dimethyl malonate with acetamidine (**6**) has been studied by Katritzky *et al.* and the reaction mechanisms for these processes have been proposed by these authors.[7] Outlined below is the proposed mechanism of the condensation of methyl acetoacetate (**4**) with acetamidine (**6**).[7]

8   **9**

10.5.4 Synthetic Utility

Many pyrimidine derivatives have been prepared *via* the Pinner procedure. Amidines react with 1,3-dicarbonyl compounds to form 2, 4, 6-trisubstituted pyrimidines.[8]

10   **6**   **11**

Amidines react with β-keto esters to provide hydroxypyrimidines. The synthesis of the 2,6-dimethyl-4-hydroxylpyrimidine (**9**) has been improved dramatically[9] by combining Pinner's procedure with that of Donleavy *et al.* for the synthesis of 6-methyluracil.[10]

4   **6**   (placed in a desiccator over H_2SO_4)   **9**

The 1,3-dicarbonyl components can be replaced by an enol ether, which can be prepared by Claisen condensation from an *ortho* ester and a reactive methylene compound.[11]

12   **6**   **13**

Amidines can also react with malonic ester derivatives to provide pyrimidines.[12] A side reaction occurs sometimes to give by-products which have not cyclized.[12] Basic conditions favor the formation of pyrimidine derivative. The nature and amount of base employed could affect the course of the reaction as expected.[13]

Ghosh *et al.* reported a modified procedure for the preparation of highly substituted pyrimidines by condensation of a 1,3-dicarbonyl compound **22** with *tri*-(trimethylsilyl)amidine (**23**) in good yield as compared to the lesser yield obtained from employing a classical Pinner procedure.[6]

	R₁	R₂	X	Yield
a	Me	Me	H	89
b	Ph	Ph	H	83
c	Me	Me	Et	78

10.5.5 Experimental

Preparation of 2-(4-chlorophenyl)-4-hydroxy-6-phenylpyrimidine[14]

25 **26** **27**

4-Chlorobenzamidine (**26**) (15 g, 78 mmol), ethyl benzoylacetate (**25**) (20 g, 129 mmol), sodium carbonate (15 g, 141 mmol) and water (30 mL) were mixed and this mixture was brought to homogeneity by adding ethanol. The solution which resulted was stirred at rt for 16 h. The thick mixture was diluted with water (50 mL) and the solid was collected and washed with ethanol. The crude product was crystallized from acetic acid which furnished 9g of the title compound **27** (30%). m.p. 305–307°C.[14]

10.5.6 References

1. [R] Kenner, G. W.; Todd, A. Pyrimidine and Its Derivatives. *Heterocyclic Compounds*; John Wiley & Sons: New York, 1957.
2. Brown, D. J. *The Pyrimidines*; John Wiley & Sons: New York, 1994.
3. Pinner, A. *Ber.* **1893**, *26*, 2122.
4. Pinner, A. *Ber.* **1884**, *17*.
5. Pinner, A. *Ber.* **1908**, *41*, 3517.
6. Ghosh, U.; Katzenellenbogen, J. A. *J. Heterocycl. Chem.* **2002**, *39*, 1101.
7. Katritzky, A. R.; Yousaf, T. I. *Can. J. Chem.* **1986**, *64*, 2087.
8. Bowman, A. *J. Chem. Soc.* **1937**, 494.
9. Snyder, H. R.; Foster, H. M. *J. Am. Chem. Soc.* **1954**, *76*, 118.
10. Donleavy, J. J.; Kise, M. A. *Org. Synth.* **1943**, *Coll. Vol. II*, John Wiley & Sons, 422.
11. Todd, A. R.; Bergel, F. *J. Chem. Soc.* **1937**, 364.
12. Kenner, G. W.; Lythcoe, A. R.; Topham, A. *J. Chem. Soc.* **1943**, 388.
13. Maggiolo, A.; Phillips, A. P.; Hitchings, G. *J. Am. Chem. Soc.* **1951**, *73*, 106.
14. Gillespie, S. P.; Acharya, S. P.; Davis, R. E. *J. Heterocycl. Chem.* **1972**, *9*, 931.

Jin Li and James M. Cook

10.6 von Richter Cinnoline Synthesis

10.6.1 Description

The thermal cyclization of 2-alkynylbenzenediazonium salts represented by **2** to provide the 4-hydroxycinnoline derivatives **3** respectively is regarded as the von Richter cinnoline synthesis.[1-4]

10.6.2 Historical Perspective

The first synthesis of cinnoline was reported by von Richter in 1883.[5] The diazonium chloride **5** which was obtained from o-aminophenylpropiolic acid (**4**), was heated in water at 70°C to provide the 4-hydroxycinnoline-3-carboxylic acid (**6**). When this acid **6** was heated above its melting point, carbon dioxide was liberated and 4-hydroxycinnoline (**7**) was obtained. Distillation of 4-hydroxycinnoline (**7**) with zinc dust furnished a small amount of oil, which was assumed to be cinnoline (**8**). The preparation of 4-hydroxycinnoline (**7**) was repeated by Busch and Klett,[6] although in lower yield when compared to the original report. Busch and Rast later converted the 4-hydroxycinnoline (**7**) successfully to cinnoline (**8**) via the 4-chlorocinnoline (**9**).[7]

10.6.3 Mechanism

The mechanism of the von Richter cinnoline synthesis has been discussed in several reports. Earlier papers[8-10] suggest that the von Richter process proceeds through pathway

B (11→ 12→14). Examination of a more recent study suggests the von Richter reaction probably involves simultaneous attack of the triple bond on the diazo group accompanied by attack by a halide ion but not by a water molecule.[11] The halocinnoline 13 is then hydrolyzed under the von Richter reaction conditions to furnish the 4-hydroxycinnoline derivative 14 (11 →12→ 13 → 14, pathway A).

10.6.4 Synthetic Utility

Only a limited number of cinnoline derivatives have been prepared *via* the von Richter cinnoline synthesis.[8,9,12]

Vasilevsky *et al.* reported that diazotization of phenylethynylaminopyrazole 19 in hydrochloric or hydrobromic acid furnished the 4-chloro- or 4-bromo cinnoline derivative

20, respectively, as the major product, which is in support of pathway A for the mechanism of the von Richter reaction.[11]

X = Cl, 65%
X = Br, 77%

19   **20**

The von Richter cinnoline process was further extended to solid-phase synthesis.[13] The route began from benzylaminomethyl polystyrene and the required diverse o-haloaryl resins represented by **21** were prepared from substituted o-haloanilines. A Pd-mediated cross-coupling reaction with **21** and the alkynes provided the alkynylaryl derivatives represented by alkyne **22**. The von Richter cyclization reaction with hydrobromic or hydrochloric acid in acetone/H$_2$O and cleavage from the resin occurred in the same step to furnish the cinnoline derivatives **23** in 47–95% yield and 60–90% purity (no yield reported for each entry).

21   **22**   **23**

	R	R$_1$	R$_2$
23a	H	H	OH or Cl
23b	Ph	H	OH or Cl, Br
23c	SiMe$_3$	6-CHCHSiMe$_3$	Br
23d	C$_5$H$_{11}$	6,8-difluoro	Br or Cl

10.6.5 Experimental

Preparation of 4-hydroxycinnoline-3-carboxylic acid 6[8]

4   **5**   **6**

The *o*-aminophenylpropiolic acid **4** (20 g) in water (60 mL) and aqueous ammonia (9 mL, d = 0.88) was added with shaking during 15 minutes to a mixture prepared from ferrous sulfate (220 g), water (440 mL), and aqueous ammonia (110 mL, d = 0.88). After 45 minutes, with occasional shaking but no external cooling, the suspension was filtered. The residue was washed with water, and the combined filtrates were treated with ammonium acetate (60 g) and made weakly acidic with acetic acid. The solution was then cooled to 0°C by addition of crushed ice, and then made acidic to Congo-red with concentrated hydrochloric acid (70–80 mL). Additional hydrochloric acid (20 mL, 2 N) was immediately added, and the turbid solution which resulted was diazotized with 20% aqueous sodium nitrite, after which the mixture was kept at 70°C. The cinnoline acid **6** was separated over 45 minutes as a dark brown, granular solid (12.5 g), m.p. 260–265°C.[8]

10.6.6 References

1.   [R] Simpson, J. C. E. Cinnolines. In *The Chemistry of Heterocyclic Compounds. Condensed Pyridazine and Pyrazine rings (Cinnolines, Phthalazines, and Quinoxalines)*; Weissberger, A. Ed.; Interscience Publishers: New York-London, 1953; p3.
2.   [R] Jacobs, T. L. Cinnolines and Related Compounds. In *Heterocyclic Compounds*; Elderfield, R. C. Ed.; Interscience Publishers: New York, 1957; p136.
3.   [R] Singerman, G. M. Cinnolines. In *Heterocyclic Compounds. Condensed Pyridazines Including Cinnolines and Phthalazines*; Weissberger, A., Taylor, E. C. Eds.; John Wiley & Sons: New York-London, 1973; p1.
4.   [R] Leonard, N. J. *Chem. Rev.* **1945**, *37*, 269.
5.   von Richter, V. *Ber.* **1883**, *16*, 677.
6.   Busch, M.; Klett, M. *Ber.* **1892**, *25*, 2847.
7.   Busch, M.; Rast, A. *Ber.* **1897**, *30*, 521.
8.   Schofield, K.; Simpsin, J. C. E. *J. Chem. Soc.* **1945**, 512.
9.   Schofield, K.; Swain, T. *J. Chem. Soc.* **1949**, 2393.
10.   Osborn, A. R.; Schofield, K. *J. Chem. Soc.* **1956**, 4207.
11.   Vasilevsky, S. F.; Tretyakov, E. V. *Liebigs Ann.* **1995**, *5*, 775.
12.   Leonard, N. J.; Boyd, S. N. *J. Org. Chem.* **1946**, *11*, 419.
13.   Bras, S.; Dahmen, S.; Heuts, J. *Tetrahedron Lett.* **1999**, *40*, 6201.

Jin Li and James M. Cook